Erik Stålberg
Hari Shanker Sharma
Yngve Olsson (eds.)

Spinal Cord Monitoring

Basic Principles, Regeneration, Pathophysiology, and Clinical Aspects

SpringerWienNewYork

Erik Stålberg, M.D.
Department of Clinical Neurophysiology, University Hospital, Uppsala University, Uppsala, Sweden

Hari Shanker Sharma, Ph.D.
Laboratory of Neuroanatomy, Department of Medical Cell Biology, Biomedical Centre, Uppsala University, Uppsala, Sweden

Yngve Olsson, M.D.
Laboratory of Neuropathology, Department of Pathology, University Hospital, Uppsala University, Uppsala, Sweden

This publication is made possible by partial financial assistance from
Medtronic Dantec Medical A/S, Denmark.

Typesetting: Thomson Press (India) Ltd., New Delhi
Printing: Adolf Holzhausens Nfg., A-1070 Wien
Binding: Fa. Papyrus, A-1100 Wien
Cover design: Bernhard Kollmann

Printed on acid-free and chlorine-free bleached paper
SPIN: 10558160

With 152 partly coloured Figures

Library of Congress Cataloging-in-Publication Data
Spinal cord monitoring : basic principles, regeneration,
pathophysiology, and clinical aspects / edited by Erik Stålberg,
Hari Shanker Sharma, Yngve Olsson.
p. cm.
Includes bibliographical references and index.
ISBN 3-211-82923-7 (alk. paper)
1. Spinal cord–Pathophysiology. 2. Spinal cord–Diseases-
-Diagnosis 3. Spinal cord–Regeneration. 4. Patient monitoring.
I. Stålberg, Erik. II. Sharma, H. S. (Hari Shanker) III. Olsson,
Yngve.
[DNLM: 1. Spinal Cord–physiopathology. 2. Nerve Regeneration-
-physiology. 3. Monitoring, Intraoperative. 4. Evoked Potentials-
-physiology. WL 400S757785 1998]
RC400.S66
616.8–dc21
DNLM/DLC 98-25818
for Library of Congress CIP

ISBN 3-211-82923-7 Springer-Verlag Wien New York

Preface

Spinal cord monitoring. Basic principles, regeneration, pathophysiology and clinical aspects

Research on spinal cord began about three centuries ago, i.e. before the advent of microscopical methods[1]. At that time, some scientists thought that the spinal cord is simply an elongation of the brain[2]. Others proposed the opposite view, i.e. that the brain is an outgrowth of the spinal cord[3]. With the introduction of light microscopical methods it was shown that the spinal cord contains nerve cells and processes forming a highly differentiated structure. Electron microscopical methods considerably increased our knowledge about the fine structure of the cord demonstrating for instance different types of synapses[4,5]. Various neuroanatomical and neurophysiological methods characterised sensory and motor pathways in the cord connecting the brain with the peripheral and autonomic nervous systems[3–5].

During the last decades, research on spinal cord has attracted a great deal of attention because of problems such as sensory-motor and autonomic dysfunctions associated with traumatic and other injuries. According to the vital statistics, trauma to the brain and spinal cord is rated as the second largest killer in the World after cardiovascular diseases[6]. It is an important problem causing many individuals life-long disability, and is a heavy burden for the afflicted individual, for the families and for the entire society. It is known that trauma to the spinal cord induces injuries by direct mechanical lesions followed by a number of secondary injury factors[7]. The secondary lesions occur with some delay and may cause cell damage even in parts of the cord which are not directly influenced by the mechanical trauma[8]. The pathophysiology of secondary injuries is an important field of study since it can be anticipated that secondary injury mechanisms may be influenced by various forms of treatment and in this way minimising the final disability after trauma.

Electrophysiologists are engaged in identifying ways to understand the spinal cord function employing several non-invasive techniques[9]. Recording of spinal cord evoked potentials is one most promising approach in this direction[8–10]. The technique is widely used to study spinal cord dysfunction[11]. However, the details of the evoked potentials and their significance in various pathological conditions are not yet fully characterised. Our knowledge about the nature of spinal cord bioelectrical activity and its disturbances in various disease conditions is therefore still incomplete.

Only a few volumes in the past have described this subject up to some extent and provided data on a probable relationship between electrical activity and neurological recovery in experimental or in clinical conditions[8–11]. On the basis of these findings, the usefulness of monitoring spinal cord electrical activity in spinal cord trauma remained controversial. On the other hand, some observations in the past relate the electrical activity to local blood flow disturbances and indicate good correlation between them[7,12]. This information provides some bases for the understanding of the relationship between electrophysiology and function of the spinal cord.

One principle reason of incongruence between neurophysiological findings and neurological outcome is their occurrence at different time intervals following trauma in either clinical or in experimental conditions[12]. In general, the electrical disturbances occur immediately after trauma, whereas the neurological dysfunction can be seen at a time when most of the electrical activity is already in a recovery phase. These observations give an impression that the early electrical disturbances and later worsening of neurological symptoms may be unrelated. However, it seems quite likely that early electrical disturbances in the spinal cord following trauma reflect an altered neurochemical and or perturbed neuronal microenvironment[8,9]. Thus, the magnitude and severity of primary insult will set the stage for a series of biological events which can be seen immediately by recording of early bioelectrical activity[10,11]. These early disturbances in the cord microenvironment will initiate secondary injury mechanisms which, if allowed to continue for longer time, will result in development of cell injury leading to pathology of the spinal cord[12,13].

Experiments carried out in our laboratory in the past focusing on early changes in the spinal cord evoked potentials following trauma provide strong indication to this effect[14]. Thus it is possible to predict the later outcome of edema formation on the basis of these early disturbances of the spinal cord bioelectrical activity[15]. These findings open a new vista of research in the field and stimulate further research in this area. However, the main challenge in this subject is to find a therapeutic time window when the post-trauma intervention with pharmacological agents can still improve the bioelectrical disturbances and attenuate the later pathological outcome in terms of cell injury, a feature which requires additional investigation.

Research on the spinal cord is thus rapidly progressing. An update of knowledge focusing on neurophysiological alterations will help us to understand the fundamental principles underlying spinal cord dysfunction in various clinical situations. The present book summarises new knowledge in the field of electrophysiology and relates this knowledge to pathology and regeneration research. To our knowledge, this book is the first of its kind, which deals with various ways to monitor spinal cord function in experimental and clinical situations in great details. It also provides an up-to-date knowledge regarding spinal cord bioelectrical activity and its modification with pharmacological agents. The book also covers new aspects of regeneration studies, particularly the role of myelin associated inhibitory molecules[12]. Finally, a section of the book represents clinical studies dealing with electrical activity, pathology and current therapeutic measures.

The book is divided in 6 sections comprising an outstanding refereed collection of 22 chapters written by eminent scientists. Section I is focused on new aspects of spinal cord blood flow, neurochemistry and sensory function of the cord and is divided in 5 chapters. Spinal cord blood flow (SCBF) is reduced dramatically following trauma to the cord and correlates well with the spinal cord metabolism of oxygen and glucose consumption. SCBF is measured using either radiolabelled microspheres or by autoradiographic techniques[12,13]. In some studies hydrogen clearance was applied to measure SCBF. All these methods have some limitations and thus the field of SCBF is still a challenging one and new insight into the regulation of SCBF is highly warranted. Chapter 1 (Marsala) provides current state of the art of SCBF measurement in the cord in relation to anatomical and physiological investigations. In addition, new results on SCBF and metabolism in ischemic injuries of the cord are presented.

Trauma to the spinal cord is associated with various events leading to disturbances of the fluid microenvironment of the cord. Thus, disruption of the blood-spinal cord barrier (BSCB) and leakage of plasma proteins are followed by increased intracellular calcium, vasogenic edema and cell injury[12–14]. Recently, several proteins in the CNS have been identified as calcium-binding proteins. These proteins are involved in important molecular events during cell injury. Chapter 2 (Ren and Ruda) describes physiological significance of calcium binding proteins in the CNS, which seems to be very important for our basic understanding of spinal cord function.

Spinal cord is rich in several neurochemical transmitters including neuropeptides[16]. Most of these neurochemicals are often co-localised with several neuropeptides in the spinal cord[17]. Thus, serotonin, the amine which is found in the highest concentrations in the cord is often co-localised with substance P, a neuropeptide involved in pain and sensory mechanism[16,17]. There are evidences that substance P is also altered following traumatic injuries to the spinal cord[18]. However, its function as a neurotransmitter or neuromodulator in the spinal cord is still not well characterised. Chapter 3 (Henry) reviews this subject. The emerging concept regarding this neuropeptide in sensory mechanisms of the spinal cord is presented.

Involvement of excitatory amino acid neurotransmitters and their receptors were long identified to play important roles in many physiological or pathological processes in the spinal cord. Recently, the role of inhibitory neurotransmitters like GABA and glycine in modulation of spinal cord function is also emerging. Thus, glycine has been suggested to play important role in the basic mechanisms of spinal shock following trauma. In Chapter 4 (Simpson Jr) new role of glycine in pain and spasticity is reviewed and new evidences supporting the involvement of this neurotransmitter in spinal cord function is presented.

Apart from biogenic amines, neuropeptides and amino acid neurotransmitters, the spinal cord is rich in prostaglandins[16]. The enzymes responsible for prostaglandin synthesis and degradation are present in the spinal cord and recently several prostaglandin receptors have been identified in the cord. Such observations support the concept that prostaglandins are important for the function of the spinal cord. Previously, there are some observations that prostaglandins are involved in traumatic injuries to the spinal cord[19] influencing spinal cord conduction. However, the detail mechanisms of prostaglandin system and its receptors are not well known. Chapter 5 (Matsumura *et al.*) reviews this subject and presents some possible functions of prostaglandins and their receptors.

Regeneration studies in the CNS have recently accumulated a wealth of new information, which seems to be relevant in our understanding of spinal cord repair mechanisms following trauma. Section II is devoted to this subject and is covered by 3 chapters. Chapter 6 (Hagg and Oudega) reviews the state of knowledge regarding neurotrophic factors and their receptors in CNS regeneration and suggestions for further research. In Chapter 7 (Palladini and Caronti) new experimental evidences regarding the influence of cyclosporine A and neonatal immune desensitisation on axonal regeneration are presented. Data accumulated in the past suggest that neurotrophic factors can provide neuroprotection in ischemia and brain injury. However, the mechanisms by which neurotrophins induce neuroprotection are not well known. Recently, a few studies indicate an involvement of neurotrophins in influencing nitric oxide (NO) function *in vitro*. Since NO is a free radical gas and is involved in the pathophysiology of spinal cord injury, cell changes and signal transduction mechanisms[20], it appears that neurotrophins can influence NO metabolism *in vivo* situations as well. Chapter 8 (Sharma *et al.*) provides new experimental evidences which suggest that the neuroprotective effects of brain derived neurotrophic factors (BDNF) and insulin-like growth factor-1 (IGF-1) following spinal cord trauma are mediated by their inhibitory influence on nitric oxide synthase (NOS), an enzyme responsible for NO production.

Development of new methods and various techniques to record spinal cord bioelectrical disturbances is dealt with in Section III that is divided into 4 chapters. In Chapter 9 (Hurlbert) recording of cerebellar evoked potentials is described as a new tool to assess spinal cord function and new ideas to monitor spinal cord in disease conditions are discussed. Recently, the use of magnetic evoked potentials in monitoring brain function has generated new information, which can be utilised to understand spinal cord function as well. Chapter 10 (Kitagawa *et al.*) describes the fundamentals of magnetic evoked potentials, its source and generation in the spinal cord in the light of recent developments in the field. In Chapter 11 (Tsuji) magnetic stimulation of spinal nerve root and recording of somatosensory cortical evoked potentials is presented as a new tool to understand the underlying CNS function. In Chapter 12 (Liu) a combined technique of microdialysis, electrophysiology and histological methods to explore secondary injury mechanisms is presented.

The possible functional significance of bioelectrical potentials in the spinal cord in relation to the pathophysiology of spinal cord is presented in Section IV that is divided in two chapters. In Chapter 13 (Winkler *et al.*) a new method to record spinal cord evoked potentials from the epidural space is described and new data are presented which demonstrate a good correlation between early disturbances in spinal cord evoked potentials and the development of edema and cell changes following trauma. In Chapter 14 (Li and Tator) a close correlation between SCBF changes and evoked potentials following trauma to the spinal cord is presented. Such observations strongly support the concept that evoked potentials are very useful in understanding the underlying spinal cord dysfunction.

Obviously, recording of evoked potential can be used in the clinical situation. Section V is organised to assimilate new clinical observations on spinal cord pathology and bioelectrical activity and is covered by 3 chapters. Chapter 15 (Kakulas *et al.*) describes white matter changes in human spinal cord injury based on a large sample of human spinal cord material obtained from various kinds of trauma. Chapter 16 (Dimitrijevic) describes neurophysiological techniques to assess human spinal cord injury in relation to pathophysiological aspects of spinal cord dysfunction. Chapter 17 (Deletis and Kothbauer) provides another promising approach to monitor spinal cord function using intraoperative neurophysiological techniques. In Chapter 18 (Burke and Hicks) a technique using simultaneous recording of descending corticospinal volleys evoked by transcranial electrical or magnetic stimulation and ascending somatosensory volleys from the spinal cord is described which suggest a new promising path for future research.

Based on basic studies, several attempts have been made to develop suitable therapeutic measures to treat various kinds of human spinal cord injuries. Section VI deals with new aspects of therapeutic measures currently being examined at several clinical centres and is represented by 4 chapters. In Chapter 19 (Young *et al.*) some novel approaches to treat post-compressive spinal cord injury in human spinal cord trauma is presented with a stimulating discussion. Chapter 20 (Nuwer) surveys the multicentre outcome of spinal cord monitoring in relation to neurological scores and suggests that monitoring of spinal cord functions helps to control neurological deficits. In Chapter 21 (Andoh and Okumura) some new insights on the influence of anaesthetics on somatosensory and motor evoked potentials in humans is summarised. This aspect is quite important in the clinical situation for the understanding of evoked potentials while assessing spinal cord function. Finally, Chapter 22 (Tamaki) gives an overall view on spinal cord monitoring in clinical settings.

We hope that this volume will be useful to various basic and clinical neuroscientists engaged in the field of neurology, neurophysiology, neurochemistry, neurosurgery, neuropathology, neurorehabilitation and related disciplines in order to understand basic functions of the spinal cord and will stimulate further research in this rapidly advancing field.

Uppsala
October 1998

Erik Stålberg
Hari Shanker Sharma
Yngve Olsson

References

1. Willis T (1664) Cerebri Anatome. Martyn and Allestry, London
2. Monro A (secundus) (1783) Observations on the Structure and Functions of the Nervous System. Creech, Edinburgh
3. Grainger R D (1837) Observations on the Structure and Functions of the Spinal Cord. Highly, London
4. Raymón y Cajal S (1928) Degenration and Regeneration of the Nervous System, vols. 1 and 2 (translated by May RM). Oxford University Press, London
5. Réxed BA (1954) Cytoarchitectonic atlas of the spinal cord of the cat. J Comp Neurol 100: 297–379
6. Stover SL, Fine PR (1986) Spinal cord injury. The facts and figures. University of Albama, Birmingham, pp 8–11
7. Tator CH, Fehlings MG (1991) Review of the secondary injury theory of acute spinal cord trauma with emphasis on vascular mechanisms. J Neurosurg 75: 15–26
8. Willis WD Jr (1980) Spinal cord potentials, In: Windle WF (ed) Spinal Cord and its Reaction to Traumatic Injury. Marcel Dekker, New York, pp 159–187
9. Homma S, Tamaki T (1984) Fundamentals and Clinical Applications of Spinal Cord Monitoring. Saikon Tokyo, pp 1–396
10. Schramm J, Jones S (1985) Spinal Cord Monitoring. Springer, Berlin Heidelberg New York Tokyo, pp 1–350
11. Ducker TB, Brown RH (1989) Neurophysiology and standards of spinal cord monitoring. Elsevier, Amsterdam
12. Schwab ME, Bartholdi D (1996) Degeneration and regeneration of axons in the lesioned spinal cord. Physiol Rev 76: 319–370
13. Faden AI (1993) Experimental neurobiology of central nervous system trauma. Crit Rev Neurobiol 7: 175–186
14. Sharma HS, Winkler T, Stålberg E, Olsson Y, Dey PK (1991) Evaluation of traumatic spinal cord edema using evoked potentials recorded from the spinal epidural space. An experimental study in the rat. J Neurol Sci 102: 150–162
15. Winkler T (1994) Evaluation of spinal cord injuries using spinal cord evoked potentials. An experimental study in the rat. Acta Univ Ups 467: 1–45
16. Nyberg F, Sharma HS, Wisenfeld-Hallin Z (1995) Neuropeptides in the Spinal Cord, Prog Brain Res. Amsterdam, Elsevier, 104: 1–416
17. Hökfelt T, Elde R, Johansson O, Ljungdahl Å, Schultzberg M, Fuxe K, Goldstein M, Nilsson G, Pernow B, Terenius L, Ganten D, Jeffocote SL, Rehfeld J, Said S (1978) Distribution of Peptide Containing Neurons. In: Lipton MA, DiMascio A, Killam KF (eds) Psychopharmacology: A Generation of Progress. Raven Press, New York, pp 39–66
18. Sharma H S, Nyberg F, Olsson Y, Dey PK (1990) Alteration in substance P in brain and spinal cord following spinal cord injury. An experimental study in the rat. Neuroscience 38: 205–212
19. Hsu CY, Hogan EL, Gadsen RH, Spicer KM, Shi MP, Cox RD (1985) Vascular permeability in experimental spinal cord injury. J Neurol Sci 70: 275–282
20. Sharma H S, Westman J, Olsson Y, Alm P (1996) Involvement of nitric oxide in acute spinal cord injury: An immunohistochemical study using light and electron microscopy in the rat. Neurosci Res 24: 373–384

Contents

List of contributors

Alm, Per, Department of Pathology, University Hospital, Lund University, Lund, Sweden
Andoh, Tomio, M.D., Department of Anesthesiology, Yokohama City University School of Medicine, 3–9 Fukuura, Kanazawa-ku, Yokohama 236, Japan. Fax.: 45-787-2916 or 45-785-8353
Burke, David, Professor, Prince of Wales Medical Research Institute, High Street, Randwick, N.S.W. 2031, Australia. Tel.: Int +61 2 382 2671, Facsimile: Int +61 2 382 2724, e-mail: D.Burke@unsw.edu.au
Cao, Chunyu, Dept. of Neuroscience, Osaka Biosciences Institute, Suita, Osaka 565, Japan
Caronti, Brunella, Dipartimento di Scienze Neurologiche, Università 'La Sapienza', Viale dell'Università, 30-00185 Roma, Italy
Deletis, Vedran, M.D., Ph.D., Institute of Neurology and Neurosurgery, Beth Israel Medical Center North Division, 170 East End Avenue, New York, NY 10128, USA. Tel.: 212-8709684, Fax.: 212-8709690
Dimitrijevic, Milan R., M.D., D.Sc., Baylor College of Medicine, Division of Restorative Neurology and Human Neurobiology, One Baylor Plaza (Suite 800), Houston 77030, Texas, USA. Fax.: 713-7983683
Dominique, Devanand M.D., Department of Neurosurgery, Temple University Hospital, 3401, N. Broad St., Philadelphia, PA 19140. Tel.: 215-707-6843, Fax.: 215-707-3831
Goodman, J. Clay Department of Pathology, Baylor College of Medicine, Houston, Texas, USA
Gordh, Torsten Department of Anaesthesiology, University Hospital, Uppsala University, S-751 23 Uppsala, Sweden
Gubbay, Adam D., Department of Neuropathology, Royal Perth Hospital and the Australian Neuromuscular Research Institute and the University of Western Australia, Perth, W. Australia
Hagg, Theo, Department of Anatomy and Neurobiology, Faculty of Medicine, Dalhousie University, Halifax, Nova Scotia, B3H 4H7, Canada
Henry, James L., Department of Physiology, Department of Psychiatry, McGill University, 3655 Drummond St. Montreal (Quebec), Canada H3G 1Y6. Tel.: 514-398-6003, Fax.: 514-398-4370, e-mail: jhenry@physio.mcgill.ca
Hicks, Richard, Department of Neurology, Prince of Wales Hospital, High Street, Randwick, N.S.W. 2031, Australia
Hurlbert, R. John, M.D., Ph.D., FRCSC, From the Division of Neurosurgery and Department of Clinical Neurosciences, Foothills Hospital, 1403 29th St. N.W., Calgary, Alberta, Canada T2N 2T9. Tel.: 403-670-4495, Fax.: 403-283-5559
Kakulas, Byron A., Department of Neuropathology, Royal Perth Hospital and the Australian Neuromuscular Research Institute and the University of Western Australia, Perth, W. Australia. Fax.: 619 224 2556
Kawaguchi, Yoshiharu, M.D., Department of Orthopaedic Surgery, Faculty of Medicine, Toyama Medical and Pharmaceutical University, 2630 Sugitani, Toyama City, 930-01 Japan
Kitagawa, Hideki, M.D., Department of Orthopaedic Surgery, Faculty of Medicine, Toyama Medical and Pharmaceutical University, 2630 Sugitani, Toyama City, 930-01

Japan. Tel.: 011-81-764-34-2281 ext. 3036, Fax.: 011-81-764-34-503520, e-mail: hk@ms.toyama-mpu.ac.jp

Kothbauer, Karl, M.D., Institute of Neurology and Neurosurgery, Beth Israel Medical Center North Division, 170 East End Avenue, New York, NY 10128, USA. Tel.: 212-8709684, Fax.: 212-8709690

Li, S., Division of Neurosurgery and Canadian Paraplegic Association Spinal Cord research laboratory, Toronto Hospital and University of Toronto, Western Division, 399 Bathtrust Street, Toronto, Ontario, M5T 2S8, Canada

Liu, Danxia, Hall Magnet 1143, Marine Biomedical Institute, University of Texas Medical Branch, Galveston, TX 77555-1143. Tel.: 409-772-6399, Fax.: 409-747-2182

Lorimer, Rachel L., Department of Neuropathology, Royal Perth Hospital and the Australian Neuromuscular Research Institute and the University of Western of Western Australia, Perth, W. Australia

Marsala, Martin, M.D., Anaesthesiology Research Laboratory, 0818, University of California, 9500 Gilman Drive, San Diego, CA 92093. Tel.: 619-543-5446, Fax.: 619-543-5446, e-mail: mmarsala@ucsd.edu

Matsumura, Kiyoshi, Subfemtomol Biorecognition Project, Japan Science and Technology Corporation; Dept. of Neuroscience, Osaka Bioscience Institute, 6-2-4 Furuedai, Suita 565, Osaka 565, Japan. Fax.: 81-6-872-0240

Nakamura, Hiroshi, M.D., Department of Orthopaedic Surgery, Faculty of Medicine, Toyama Medical and Pharmaceutical University, 2630 Sugitani, Toyama City, 930-01 Japan

Nakatoh, Shinichi, M.D., Department of Orthopaedic Surgery, Faculty of Medicine, Toyama Medical and Pharmaceutical University, 2630 Sugitani, Toyama City, 930-01 Japan

Nuwer, Marc R., M.D., Ph.D., Professor, Head, Department of Clinical Neurophysiology, Department of Neurology, UCLA School of Medicine, UCLA Medical Center, Reed Neurological Research Center 710 Westwood Plaza, Los Angeles, California 90024, USA. Tel.: 1-310-206-3093, Fax.: 1-310-825-3167

Nyberg, Fred, Professor, Department of Pharmaceutical Biosciences, Biomedical Centre, Uppsala University, Uppsala, Sweden

Okumura, Fukuichiro, M.D., Department of Anaesthesiology, Yokohama City University School of Medicine, 3-9 Fukuura, Kanazawa-ku, Yokohama 236, Japan, Fax.: 45-787-2916 or 45-785-8353

Oudega, Martin, The Miami Project to Cure Paralysis, University of Miami School of Medicine, Miami, Florida 33136, USA

Ozaki, Masashi, Dept. of Neuroscience, Osaka Bioscience Institute, Suita, Osaka 565, Japan

Palladini, Guido, Professor, Dipartimento di Scienze Neurologiche, Università degli Studi 'La Sapienza', Viale dell'Università, 30-00185 Roma, Italy. Tel.: +39-6-49914500, Fax.: +39-6-4440790

Ren, K., Department of Oral and Craniofacial Biological Sciences, University of Maryland Dental School, Baltimore, MD 21201, USA

Robertson, Claudia S., Department of Neurosurgery, Baylor College of Medicine, One Baylor Plaza, Houston, Texas 77030. Tel.: 713-798-4695, Fax.: 713-798-3739

Ruda, M.A., Neurobiology and Anäesthesiology Branch, National Institute of Dental Research, National Institutes of Health Building 49, Room 1A11, 49 Convent Drive, MSC 4410, Bethesda, MD 20892-4410, USA. Tel.: 301-402-4980, Fax.: 301-402-0067, e-mail: ruda@yoda.nidr.nih.gov

Sharma, Hari Shanker, Ph.D., Laboratory of Neuroanatomy, Department of Medical Cell Biology, Biomedical Centre, Post Box 571, Uppsala University, S-751 23 Uppsala, Sweden. Tel.: +46-18-17 44 33, Fax.: +46-18-24 38 99, e-mail: Sharma@anatomi.uu.se

Simpson, Jr., Richard K., Department of Neurosurgery, Baylor College of Medicine, One Baylor Plaza, Houston, Texas 77030. Tel.: 713-798-4695, Fax.: 713-798-3739

Stålberg, Erik, Professor and Head, Department of Clinical Neurophysiology, University Hospital, Uppsala University, S-751 85 Uppsala, Sweden

Takeuch, Kazuhisa, The 2nd Dept. of Internal Medicine, Tohoku University, School of Medicine, Sendai 980, Japan

Tamaki, Tetsuya, Professor, Chairman, Department of Orthopaedic Surgery, Wakayama Medical College, Japan. Tel.: +81 784 26 8765, Fax.: +81 734 33 2996

Tator, C.H., Professor and Head, Division of Neurosurgery and Canadian Parapiegic Association Spinal Cord Research Laboratory, Toronto Hospital and University of Toronto, Western Division, 399 Bathtrust Street, Toronto, Ontario, M5T 2S8, Canada

Tsuji, Haruo, M.D., Department of Orthopaedic Surgery, Faculty of Medicine, Toyama Medical and Pharmaceutical University, 2630 Sugitani, Toyama City, 930-01 Japan

Tsuji, Sadatoshi, M.D., Professor, Department of Neurology, University of Occupational and Environmental Health, School of Medicine, Yahatanishi-ku, Kitakyushu, Fukuoka 807, Japan. Tel.: 81-093-691-7438, Fax.: 81-93-693-9842, e-mail: tsujisa@med.uoeh-u.ac.jp

Tuma, Ronald F., Ph.D., Department of Neurosurgery, Temple University Hospital, 3401, N. Broad St., Philadelphia, PA 19140. Tel.: 215-707-6843, Fax.: 215-707-3831

Vasthare, Usha S., Ph.D., Department of Neurosurgery, Temple University Hospital, 3401, N. Broad St., Philadelphia, PA 19140. Tel.: 215-707-6843, Fax.: 215-707-3831

Watanabe, Yasuyoshi, Subfemtomol Biorecognition Project, Japan Science and Technology Corporation; Dept. of Neuroscience, Osaka Bioscience Institute, Suita, Osaka 565, Japan

Watanbe, Yumiko, Subfemtomol Biorecognition Project, Japan Science and Technology Corporation; Dept. of Neuroscience, Osaka Bioscience Institute, Suita, Osaka 565, Japan

Westman, Jan, Laboratory of Neuroanatomy, Department of Medical Cell Biology, Box 571, Biomedical Center, University Hospital, Uppsala University, S-751 23 Uppsala, Sweden

Winkler, Tomas, Department of Clinical Neurophysiology, University Hospital, S-751 85 Uppsala, Sweden. Tel.: +46-18-66 34 36, Fax.: +46-18-55 61 06, e-mail: Tomas.Winkler@neurofys.uu.se

Yamagata, Kanato, Dept. of Molecular Neurobiology, Tokyo Metropolitan Institute of Neurosciences, Fuchu-shi, Tokyo 183, Japan

Young, William F.M.D., Department of Neurosurgery, Temple University Hospital, 3401, N. Broad St., Philadelphia, PA 19140. Tel.: 215-707-6843, Fax.: 215-707-3831

Section I

Spinal cord blood flow, neurochemistry and sensory function

Spinal Cord Monitoring
Eds: E. Stålberg, H. S. Sharma, Y. Olsson

Chapter 1

Spinal cord blood flow and metabolism in transient spinal ischemia

M. Marsala

Anesthesiology Research Laboratory, University of California, San Diego, USA

Introduction

Spinal cord ischemia and the resulting irreversible loss of neurological function (i.e. the development of spastic or flaccid paraplegia) represent serious complications associated with transient aortic cross-clamp (as for repair of aortic aneurysm). While the precise mechanism leading to the irreversible spinal neuronal degeneration after spinal ischemia is not well understood, current experimental studies suggest that the progressive evolution of spinal injury after aortic cross clamp may be considered in at least two categories: i) spinal cord blood flow changes, and, ii) metabolic changes resulting from critical reduction of substrate (glucose/oxygen) delivery and corresponding release of active factors (excitatory amino acids, prostaglandins, oxygen free radicals etc.) into the spinal extracellular space.

Immediately after aortic occlusion, there is a rapid but incomplete reduction in distal aortic pressure (i.e. below the level of aortic cross-clamp) and a corresponding decrease in spinal cord blood flow (Kaplan *et al.*, 1987; Marini *et al.*, 1989; Bower *et al.*, 1989; Marsala *et al.*, 1994). In both clinical and animal studies, incomplete reduction in SCBF is believed to reflect the presence of significant intrathoracic and paraspinal arterial collateral systems which may provide sufficient nutritive flow for a limited period after aortic occlusion (Gray, 1993; Schievink *et al.*, 1988; Evans, 1993). Accordingly, corresponding with the conditions of incomplete ischemia, the critical ischemic time for spinal cord (i.e. the interval of aortic occlusion which leads to the irreversible loss of neurological function) is significantly longer then in global cerebral ischemia models (Zivin *et al.*, 1980; Marsala *et al.*, 1991; Homma *et al.*, 1979; Vanicky *et al.*, 1992; Smith *et al.*, 1984; Pulsinelli *et al.*, 1982).

With increasing intervals of ischemia, a progressive decline in high energy phosphates in both spinal gray and white matter and a corresponding loss of neuronal excitability are observed (Danielisova *et al.*, 1987; Marsala *et al.*, 1989; Cheng *et al.*, 1984). After ischemia, depending on the duration of the ischemic insult, hyperemia and a loss of the CO_2 response of spinal vasculature are noted (Bower *et al.*, 1989; Marsala *et al.*, 1994). The magnitude of the post-reflow hyperemia, suggesting loss of pressure autoregulation, appears to correlate with the degree of neurological deficit (Bower *et al.*, 1989; Svensson *et al.*, 1989). After the hyperemic period, as in supraspinal structures, hypoperfusion is seen where SCBF is reduced to a fraction of pre-ischemic values (Bower *et al.*, 1989, Aadahl *et al.*, 1990; Follis *et al.*, 1995).

More detailed understanding of the hemodynamic changes and the assessment of the corresponding metabolic failure during and after the period of transient aortic cross-clamp will likely provide important information on the role of the above variables in the development of the postischemic picture. In the following paragraphs, a brief overview of spinal cord vascular anatomy, SCBF and metabolism (high energy phosphate, glucose) under control and ischemic-reperfusion period are provided. In addition, recent experimental data describing possible treatment approaches targeted to improve intraischemic SCBF or modulate spinal metabolic demands and their effect on neurological outcome after spinal ischemia are briefly described.

Anatomy of spinal vasculature

In human, as in several animal species (monkey, dog, sheep, pig, and rat) a typical heterosegmental organization of the spinal cord blood supply can be found (Gray, 1993; Schievink *et al.*, 1988; Evans, 1993; Tveten, 1976), meaning that two or more spinal cord segments receive their principal blood flow from one radicular artery. Accordingly, these radicular arteries may branch at different segmental level from the thoracoabdominal aorta. In contrast, it is believed that the rabbit possesses a homosegmental aorta in which individual spinal segments are supplied by individual radicular arteries branching from the thoracic and abdominal aorta at each segmental level, with minimal to no intraspinal collateral system (Zivin *et al.*, 1980). As will be outlined in the following sections, the above observations on the anatomical differences between several species has no overall significant effect on the baseline spinal cord blood flow (see section SCBF) and functionality, however, it can potentially be important in determining the outcome after transient aortic occlusion.

In general, in humans (Gray, 1993) and other animal species with the heterosegmental configuration (Schievink *et al.*, 1988; Evans, 1993) the principal blood supply of the ventral cervical spinal cord is derived from the anterior spinal artery, which is formed rostrally by the union of two branches of the intracranial portions of the vertebral arteries. The anterior spinal artery usually extends along the entire length of the spinal cord, feeding the anterior two thirds of the spinal cord and receiving blood from radicular arteries. In the cervical area, the radicular arteries are derived from branches of the vertebral arteries as well as from ascending cervical arteries. In the upper part of thoracic cord, the segmental arteries are derived from deep cervical and ascending cervical arteries. The middle and lower thoracic spinal cord is generally termed a "watershed" area because of the inconsistent presence of the anterior spinal artery and the wide spacing of the radicular feeders. The lower thoracic and lumbar part of the anterior spinal artery is supplied by the great anterior radicular artery (artery of Adamkiewicz) which is believed (based on the level of its origin) to provide a major portion of nutritive flow to the lower thoracic and lumbar spinal cord segments. This artery is usually accompanied by one to four other principal arteries within two or more spinal cord levels. While several reports on the origin of the artery of Adamkiewicz show certain degree of variability, more precise angiographic studies localize the origin of this artery to levels T9 to T12 in the majority (75%) of cases (Szilagyi *et al.*, 1978; Doppman *et al.*, 1968; Faure *et al.*, 1966).

The two posterior spinal arteries, supplying the posterior third of the spinal cord, originate rostrally from the vertebral or posterior inferior cerebellar arteries and are supplemented along their descent by posterior radicular arteries. The anterior and two posterior spinal arteries make up the terminal division from which white and gray matter are supplied. The superficial rim of the cord is supplied by dorsal and ventral vessels and by a peripheral anastomotic ring. The central part of the spinal cord (gray matter) is penetrated by vessels branching from the sulcal arteries. No anastomoses are found within the cord and the penetrating arteries represent end arterioles (Woolen and Millen, 1955; Rubinstein and Arbit, 1990).

Anatomy of spinal vasculature and spinal ischemic models

The presence of either heterosegmental or a homosegmental aorta represents an important variable which defines the surgical approach necessary to induce spinal ischemia in various animal species. Thus, in heterosegmental species (dog, cat, baboon and rat), it is necessary to occlude the descending thoracic aorta in order to achieve significant spinal cord ischemia in lower thoracic and LS segments. However, because of the presence of significant collateral system in the heterosegmental species the critical ischemic time may range between 20–60 min (Bower *at al.*, 1989; Homma *et al.*, 1979; Marini *et al.*, 1993; Svensson *et al.*, 1986; Marsala *et al.*, 1994).

In contrast, in the rabbit which possesses a homosegmental aorta, it is sufficient to occlude only the abdominal aorta to achieve comparable spinal cord ischemia in lumbosacral segments. In this model, consistent paraplegia is achieved after ischemic intervals as short as 10–15 min (Zivin and DeGirolami, 1980; Cheng *et al.*, 1984; Vacanti *et al.*, 1984; Marsala *et al.*, 1992; Jacobs *et al.*, 1992). While it is believed that the rabbit has a minimally developed intraspinal intersegmental collateral system, the most recent experimental data show that permanent occlusion of even 3–4 lumbar radicular arteries has no significant effect on neurological function (Kwun and Vacanti, 1995). We have obtained identical data (Dinda and Marsala-unpublished observation). These observations would indicate the presence of intersegmental collaterals which may extend to at least 3–4 spinal segments.

Spinal cord blood flow (SCBF)

Spinal cord blood flow under control condition

a) Techniques for measurement of SCBF. During the last 2–3 decades several techniques for measurement of spinal cord blood flow have been developed. In general, these techniques could mechanistically be considered as those which will permit on-line (laser Doppler) and repetitive, region specific measurements (hydrogen clearance) or those which will permit the concurrent assessment of SCBF in anatomic segments throughout the whole spinal cord (radioactive microspheres, iodo-[14C]-antipyrine). The use of these several techniques, however, has provided relatively homogenous data on SCBF and its hemodynamic characteristic in a variety of experimental species under control conditions (see Table 1).

b) Anatomic variability in SCBF. Rostrocaudal distribution. Corresponding with the functional and anatomical organization of the spinal cord and consistent with the irregular distribution of radicular arteries described above a significant degree of variability in SCBF as measured at different segmental levels has been reported. Nystrom and Norlen (1983), for example, obtained SCBF values in the rat in cervical, upper thoracic, lower thoracic and lumbar gray matter with values 48, 42, 74 and 98 ml/100g/min respectively, using microspheres. Using the hydrogen clearance method the thoracic gray SCBF was detected to be 41 ml/100g/min and lumbar SCBF was 52 ml/100g/min (Rubinstein and Arbit, 1990). Comparable differences in segmental SCBF in several other experimental species (dog, baboon) have been described (Bower *et al.*, 1989, Svensson *et al.*, 1989). Lower blood flow in the thoracic region and higher blood flow in the cervical and lumbar spinal segments appears to be proportional to the density of spinal vasculature in these regions (i.e. the presence of a "watershed" area in the mid-thoracic region and the presence of the major radicular blood supply in the cervical and lumbar areas respectively).

Dorsoventral distribution: Detailed SCBF measurements have revealed the highest SCBF to be in the intermediate gray matter (lamina VII), (66–79 ml/100g/min) as compared to 53–54 ml/100g/min found in the dorsal horn and 32–56 ml/100g/min measured in the ventral quadrant (Torre, 1984; Crosby *et al.*, 1990; Rubinstein and Arbit, 1990). White matter flow in the dorsal, lateral or ventral funiculus has been reported to be

Table 1. Summary of SCBF (ml/100g/min) data under control condition in several animal species

Reference	Measur. meth.	Cervical	Thoracic	Lumbar	Species	Anesth.
Nystrom and Norlen, 1983	radioactive microspheres	48 (*g*)	42–74 (*g*)	98 (*g*)	SD rat	thiobarbital
Crosby *et al*., 1990	iodo-[14C]-antipyrine	⟶	⟶	62–129/ 23–36 (*)	SD rats	awake
Rubinstein and Arbit, 1990	hydrogen clearance	54/27 (*g*/*w*)	41/22 (*g*/*w*)	52 (*g*)	SD rats	haloth./chloral.
Lenzi *et al*., 1986	radioactive microspheres	22 (total)	12 (total)	19 (total)	rabbit	awake
Kaplan *et al*., 1987	radioactive microspheres	116 (*g*)	71 (*g*)	109 (*g*)	sheep	isoflur.
Gordh *et al*., 1986	radioactive microspheres	18 (total)	20 (total)	24 (total)	pig	nit.oxi. hypnodil

g Gray matter; *w* white matter; (*) average data obtained from 5 different spinal levels.

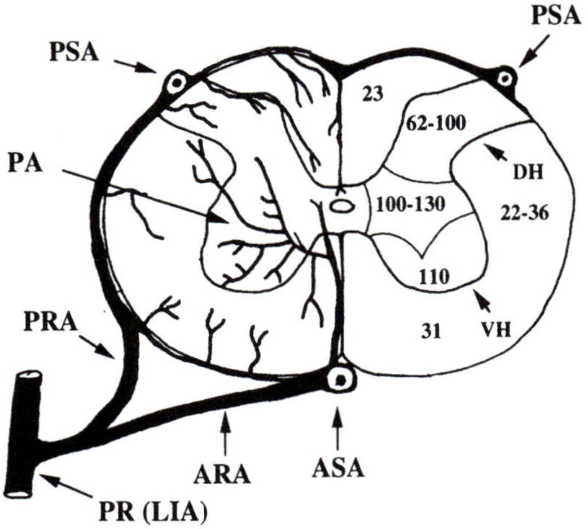

Fig. 1. Schematic drawing of the spinal cord vasculature and laminar distribution of SCBF (ml/100g/min) in lumbar spinal cord (*PSA* posterior spinal artery; *PA* penetrating artery; *ASA* anterior spinal artery; *ARA* anterior radicular artery; *PRA* posterior radicular artery; *PR (LIA)* posterior ramus of lumbar intercostal artery); *VH* ventral horn; *DH* dorsal horn

relatively homogenous, ranging from 23 (dorsal funiculus) to 31–36 (lateral and ventral funiculus)ml/100g/min (Torre, 1984; Crosby *et al.*, 1990; Rubinstein and Arbit, 1990).

The rostro-caudal gradient of SCBF as well as the differences in dorsal versus ventral perfusion likely reflect different metabolic demands of certain neuronal pools. Such differences can be of importance particularly under conditions of partial ischemia, as during thoracic aorta occlusion. It is likely that the spinal regions which display the highest baseline blood flow (central gray matter) will be the most sensitive to transient ischemic insult. Such a selective neuronal vulnerability then can be expected to correspond with the clinical deficit resulting from the specific loss of function associated with these classes of neurons. For example, selective degeneration or loss of inhibitory interneurons in the central gray matter (laminae V–VII) of LS segments will be clinically expressed as spastic paraplegia and hyperreflexia, while after the loss of A-motoneurons the presence of flaccid paraplegia (i.e. lower motor neuron lesion) will be noted.

c) Spinal cord blood flow autoregulation and CO_2 responsivity. Pressure autoregulation: As in supraspinal structures, the spinal cord vasculature possesses the ability to autoregulate, thus maintaining constant SCBF despite significant changes in perfusion pressure. In several animal species, using the hydrogen clearance or radiolabeled microspheres method, it has been firmly established that spinal cord maintains its normal flow at arterial pressures varying between 50–140 mmHg (Hickey *et al.*, 1986; Rubinstein and Arbit, 1990; Senter *et al.*, 1979; Kobrine *et al.*, 1975). Such effective

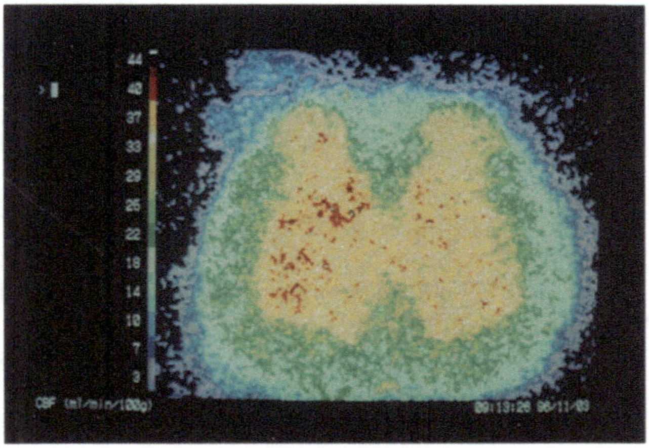

Fig. 2. Spinal cord blood flow (iodo-[14C] antipyrine method) in control animal (L4 spinal segment), (Ishikawa and Marsala in preparation)

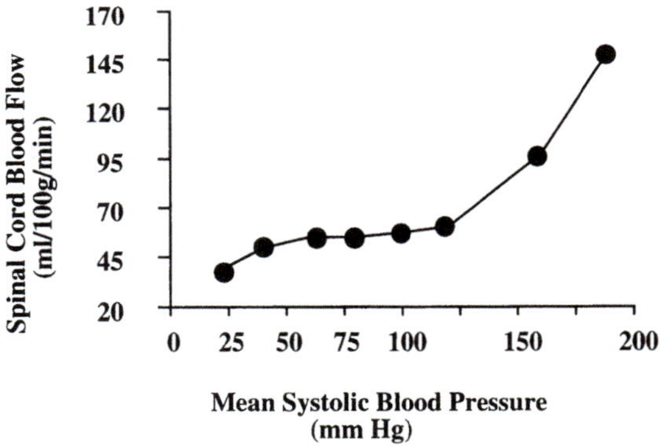

Fig. 3. Spinal cord blood flow autoregulation in the rat spinal cord. As indicated, SCBF is maintained constant in the range of 50–120 mm Hg of mean arterial blood pressure. (Adopted from Hicket *et al.*, 1986; Rubinstein and Arbit, 1990)

autoregulation by the spinal vasculature represents an important hemodynamic variable which can potentially modulate the degree of SCBF reduction under conditions of incomplete spinal ischemia (e.g. aortic cross-clamp). It is likely that in species with heterosegmental aortas and well developed paraspinal collateral systems, maintenance of normal proximal arterial pressure (i.e. systemic arterial pressure proximal to the level of the aortic occlusion) can be effective in providing a certain level of nutritive flow to the collateral-dependent areas. This interaction (i.e. collateral distribution and perfusion head) can influence the duration of the critical ischemic time (i.e. duration of the ischemic episode associated with significant loss of neurological function after reperfusion). Accordingly, we

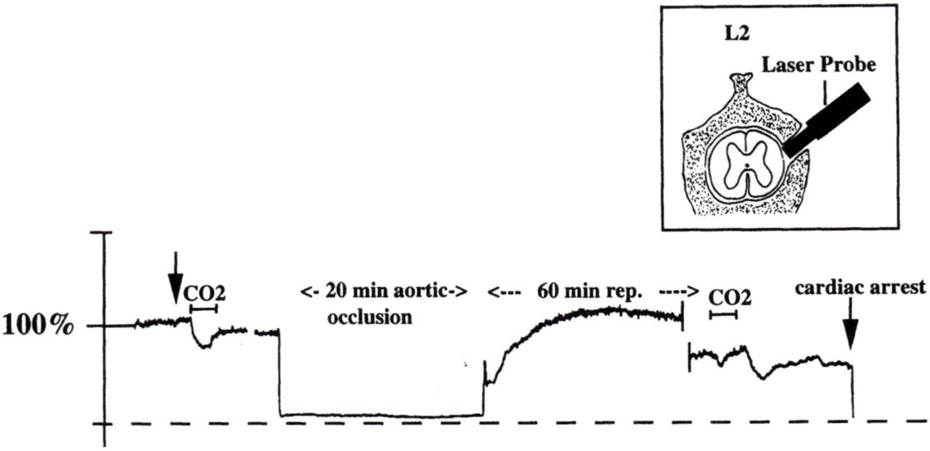

Fig. 4. Polygraph tracing of the output of the laser Doppler probe (see inset) under normal conditions ($PaCO_2 = 38$ mmHg) and during decreases in $PaCO_2$ ($PaCO_2 = 26$ mmHg) induced by hyperventilation before and after the initiation of 20 minutes of aortic occlusion in the rat. Note a significant decrease in SCBF and loss of CO_2 responsivity at 60 min of reperfusion (Marsala *et al.*, 1994)

have recently found that systemic hypotension (40 mmHg) during the period of aortic occlusion significantly decreases residual SCBF and subsequently shortens critical ischemic time in rat spinal cord ischemic model (Taira and Marsala, 1996).

Chemosensitivity. The SCBF shows clear responsivity to $PaCO_2$ changes. Hypercarbia ($PaCO_2 = 70$ mmHg) evokes an increase in gray matter blood flow of about 75%, while hypocarbia ($PaCO_2 = 25$ mmHg) is associated with a decrease in local SCBF of approximately 25 % (Griffiths *et al.*, 1975). Using the laser Doppler technique, we have observed comparable SCBF reduction in lumbar spinal segment in the rat after induction of hypocarbia ($PaCO_2 = 25$–30 mmHg) achieved by hyperventilating the animals for 2 min (Marsala *et al.*, 1994), (Fig. 4).

Effect of aortic occlusion and reperfusion on SCBF and CO_2 responsivity

Because of the collateral perfusion system described above, aortic occlusion typically leads to incomplete spinal ischemia. However, there can be considerable differences in the magnitude of the reduction in spinal cord blood flow at different segmental levels. As above, such differences may depend on the level of the aortic occlusion, and, more importantly, the localization of major radicular arteries supplying lower thoracic and lumbar spinal segments. Thus, the data from several experimental studies show that immediately after aortic cross-clamp, there is a significant decrease in SCBF with the highest degree of reduction measured in the most caudal (typically lumbar) spinal segments (Kaplan *et al.*, 1987; Bower *et al.*, 1989; Svensson *et al.*, 1986). Such a decrease may reach 1–20% of the baseline flow. After cross-clamp, in some cases, stable reduction is observed during the whole period of occlusion while in others, after initial reduction, gradual increase during the period of occlusion is observed and the levels may increase about 10–20% of initial post-clamp value (Kaplan *et al.*, 1987). Again, such variability likely reflects individual differences in collateral systems, the localization of major radicular arteries, and, the ability of collaterals to open during the period of cross clamp. However, because of technical limitations of SCBF measurements, particularly when low flow values are to be measured, it is very difficult to draw the line which will precisely predict the critical reduction in SCBF and the duration of such a critical reduction which will eventually be associated with loss of neurological function after reperfusion.

In general, experimental data indicate that reduction in SCBF to levels around 10% of control is associated with a rapid alteration in spinal evoked neuronal activity. When maintained for an additional 20–30 min a significant incidence of postischemic neurological deficit is noted. In contrast, reduction to about 20% of baseline has minimal or no effect on SCEP and can be tolerated for 40–60 min without significant neurological deficit after reperfusion (Bower *et al.*, 1989; Kaplan *et al.*, 1987). However, experimental and clinical data have indicated that absolute "safe" time limit of spinal ischemia may be as short as 18 min while ischemic periods in excess of 60 min are associated with near 100% incidence of postischemic deficit (Marini and Cunningham, 1993). In accordance with the above observation, we have reported similar results using a rat spinal cord ischemic model and laser Doppler blood flow measurements. In this study we have shown that aortic occlusion in rat induced by the inflation of an aortic balloon catheter in the descending thoracic aorta will cause an immediate SCBF decrease to about 6–7% of baseline flow (Marsala *et al.*, 1994). Such a reduction when maintained for period in excess of 20 min, is associated with complete loss of spinal CO_2 reactivity at 60 min of reperfusion, and, as defined in behavioral part of the study, with a high incidence of neurological deficit.

After ischemia, a significant increase in SCBF (reactive hyperemia) can be observed (Table 2). The magnitude of postischemic hyperemia is inversely related to the magnitude and the duration of SCBF reduction during the period of occlusion (i.e. the deeper and longer the reduction in SCBF during ischemia the higher the hyperemic response during reperfusion). Importantly, several independent experimental studies, using radiolabeled microsphere techniques which permit SCBF measurement and the assessment of neurological status in the same animal, show clear correlations between the magnitude of postischemic hyperemia and the incidence of neurological deficit after reperfusion (Bower *et al.*, 1989; Svensson *et al.*, 1989; Marini *et al.*, 1989). While, at present, it is still not

Table 2. Summary of SCBF (ml/100*g*/min) during periischemic period

Reference	Measur. meth.	Baseline Th./L.	Cross-clamp Th./L. (min)	Reperfusion Th./L. (min/hr)	Species	Anesth.
Bower *et al.*, 1988	radioactive microspheres	23/27 (*g*)	4/2.5 (60 min), (*g*)	83/137 (5 min), (*g*)	dog	haloth.
Marini *et al.*, 1989	radioactive microspheres	12/19 (total)	6/6 (10 min), (total)	16/65 (5 min), (total)	dog	pentob.
Kaplan *et al.*, 1987	radioactive microspheres	71/109 (*g*)	7/3 (30 min), (*g*)	85/146 (30 min), (*g*)	sheep	isoflur.
Svensson *et al.*, 1986	radioactive microspheres	19/26 (total)	6/2 (60 min), (total)	18/39 (30 min), (total)	baboon	thiop.
Jacobs *et al.*, 1992	iodo-[14C]-antipyrine	32/60 (*g*)	NR	60/146 (12 hr), (*g*)	rabbit	pentob.
Marsala *et al.*, 1994	laser Doppler	—	6% (20 min) (% baseline), (*w*)	124% (120 min) (% baseline), (*w*)	rat	haloth.

g Gray matter; *w* white matter; *NR* no reported; *Th* thoracic segments; *L* lumbar segments; *total* (gray + white matter).

clear whether or not such a postischemic hyperemia has direct potentiating effect on spinal neuronal degeneration, it has been demonstrated that during the period of reactive hyperemia, transient opening of the blood brain barrier with protein extravasation occurs which may contribute to the development of spinal edema and resulting secondary hypoperfusion (Jacobs *et al.*,1987; Orendacova *et al.*, 1989).

Of particular importance appears to be the observation that the neurodegenerative changes observed in spinal gray matter during reperfusion period show similarity of spatial distribution not only to the segmental but also to the laminar differences in SCBF under control and ischemic condition. Thus, in animals with fully developed spastic paraplegia, interneuronal pools localized in lumbosacral segments between laminae IV–VII (i.e. the areas with the highest blood flow values under control conditions and with the biggest reduction in blood flow during aortic occlusion) are typically affected. These neurodegenerative changes, in comparison with central structures have very rapid onset. In several animal species we have observed, as have others, that clear signs of irreversible neuronal degeneration can be seen as soon as 30–120 min after reperfusion. These changes are expressed as the presence of heavy somatodentritic neuronal argyrophilia, (Nauta silver impregnation technique) which corresponds with the electron-dense form of neuronal degeneration as analyzed by electron microscopy (Marsala *et al.*, 1989; Marsala *et al.*, 1991; Marsala *et al.*, 1994, Jacobs *et al.*, 1987).

Spinal cord metabolism

Glucose metabolism under control conditions

Under control conditions, with fully preserved substrate (glucose, oxygen) delivery, the segmental and laminar variation in spinal metabolic rate (as measured by 2-deoxyglucose reuptake) shows remarkable similarities to those of SCBF previously described (Table 3).

a) Rostrocaudal distribution of 2-DG uptake. It has been reported that glucose consumption is the highest in cervical and lumbar segments (39 vs 40 μmol. 100 g^{-1}. min^{-1}) while slightly lower levels are found in thoracic segments (25–33 μmol. 100 g^{-1}. min^{-1}), (Cole *et al.*, 1990).

b) Dorsoventral distribution of 2-DG uptake. As for SCBF, significantly higher levels of 2-DG activity were detected in spinal gray matter than in spinal white matter. Thus, the 2-DG reuptake in lumbar dorsal horn gray matter has been reported to range between 30–52 (μmol. 100 g^{-1}. min^{-1}) while in ventral horn, levels have been reported ranging between 40–58 (μmol.100 g^{-1}. min^{-1}). Measured in dorsal, lateral or ventral white matter 2-DG uptake ranged between 16–32 (μmol.100 g^{-1}. min^{-1}).

It is important to note that manipulations which are associated with increased spinal receptor activity (peripheral nociceptive stimulation, intrathecal glutamate receptor agonist administration or postischemic

Table 3. Summary of glucose metabolic activity in spinal lumbar gray and white matter (μmol. 100 g^{-1}. min^{-1})

Reference	Dorsal gray (laminae I–VI)	Central gray (lamina VII)	Ventral gray (lam. VIII/IX)	White (dors/lat/ant)	Species	Anesth.
Crosby *et al.*, 1990	35–48	57	52–58	16/26/32	SD rat	awake
Porro *et al.*, 1991*	35–52	50	45–46	21/26–28	SD rats	awake
Cole *et al.*, 1990	30–40	NR	40	13/17/19	SD rats	awake
Cavazzuti *et al.*, 1991	42	47	41	16/NR/NR	SD rats	awake
Abram and Kostreva, 1986	19	—	21	9.4 (whole white)	cat	pentobarb.
Kadekaro *et al.*, 1985	36	—	—	—	SD rats	barbital

* Data expressed as metabolic activity index (for details see Porro *et al.*, 1991); *NR* not reported; *Dors* dorsal; *Lat* lateral; *Ant* anterior.

reperfusion) typically evoke a marked increase in glucose uptake in the areas of spinal gray matter corresponding with the localization of activated neurons. Thus, it has been shown that formalin injection or radiant heat stimulation (a powerful nociceptive stimulus) of the paw evokes a significant and predominantly ipsilateral increase in spinal gray matter 2-DG uptake in lumbosacral segments (Porro *et al.*, 1991; Abram and Kostreva, 1986). Comparably, we have observed a significant increase in 2-DG uptake in lumbar spinal segments after intrathecal kainic acid (1 μg) injection (Fig. 5). In contrast, drug treatment or manipulation serving to suppress or block spinal neuronal activity has been reported to significantly decrease glucose metabolism. For example it has been shown that propofol anesthesia (8 mg. kg^{-1}) reduces glucose metabolism in both spinal white and gray matter more than 50% in comparison to awake animals (Cavazzuti *et al.*, 1991). Similar suppression was reported in the rat after intrathecal injection of tetracaine (local anesthetic) or clonidine (alpha 1 and alpha 2 adrenergic receptor agonist), (Crosby *et al.*, 1990; Cole *et al.*, 1990). We have also observed that selective spinal cord hypothermia (28 °C), a manipulation known to provide a powerful suppression of metabolic rate in supraspinal structures (Steen *et al.*,1983), significantly decreases 2-DG uptake in lumbosacral spinal segments (Fig. 6).

The above characteristics of metabolic changes after specific and non-specific spinal neuronal activation or receptor blockade suggest the presence of conventional receptor-effect coupling. In addition, as observed for SCBF, it is likely that the spinal gray matter layers with the highest metabolic rate (i.e. central gray matter) will display the highest degree of

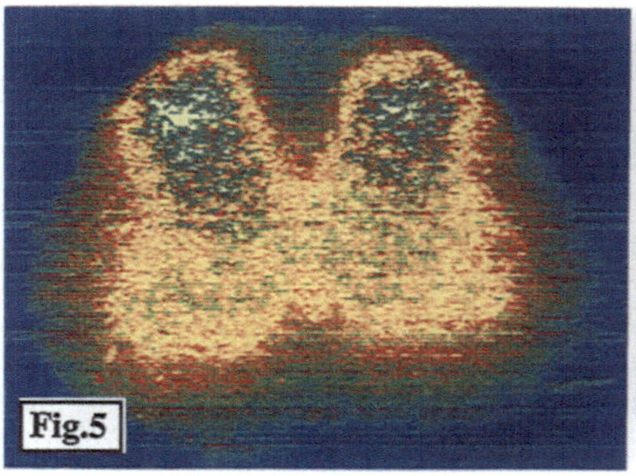

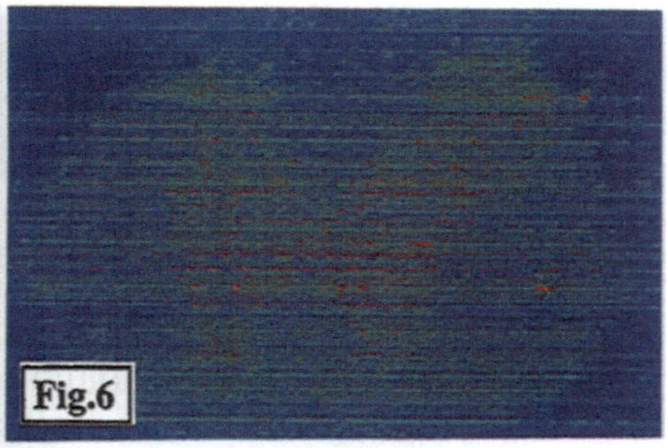

Fig. 5. Computerized pseudocolor aotoradiogram of 2-deoxyglucose uptake in rat transverse lumbar spinal cord section after intrathecal kainic acid (1 μg) injection and 45 min survival. Note the highest 2-dg uptake in both dorsal horns (Marsala and Ishikawa, 1996; in preparation)

Fig. 6. Computerized pseudocolor aotoradiogram of 2-deoxyglucose uptake in rat transverse lumbar spinal cord section after 45 min of spinal cord hypothermia (27°C) in halothane anesthetized rat. In comparison with animal injected intrathecally with kainic acid a profound decrease in 2-dg uptake in both gray and white matter can be seen (Marsala and Ishikawa, 1996; in preparation)

sensitivity to the events associated with rapid depletion of energy substrates such as transient spinal ischemia. As it will be outlined in the following sections, such a shortage will then be expressed as a significant drop in spinal glucose concentration and, correspondingly an acceleration of anaerobic metabolism and significant increase in the concentration of anaerobic metabolic product (lactate).

High energy phosphates (HEP)

Consistent with the ongoing oxygen and glucose delivery under control-non-ischemic conditions, the spinal cord maintains relatively stable levels of high energy phosphates (ATP-adenosine triphosphate; ADP-adenosine diphosphate; AMP-adenosine monophosphate). Data from several studies show that the concentration of ATP ranges between 1.6–2.2 nmol/g of wet tissue in lumbar spinal cord segments (Anderson *et al.*, 1980; Danielisova *et al.*, 1987; Allen *et al.*, 1994) . Due to a rapid rephosphorylation of ADP and AMP, these levels are kept constant even under conditions of transient metabolic stress associated with increased neuronal activity.

Effect of transient spinal ischemia and reperfusion on spinal glucose and HEP metabolism

a) Events associated with the ischemic period. Depending on the completeness of the ischemic insult, spinal cord ischemia causes a rapid decline in spinal glucose and ATP concentration. Thus, glucose concentrations are decreased to about 10% of baseline after 20–30 min of spinal ischemia in rabbit spinal cord ischemic model. In parallel with the gradual loss of glucose, a gradual decline in ATP concentrations is noted: ATP is decreased to about 40% after 10 min of spinal ischemia and is about 5% of control values at 40 min of aortic occlusion (Danielisova *et al.*, 1987; Marsala *et al.*, 1989). While not identical, comparable decreases in other studies have been reported (Anderson *et al.*, 1983; Allen *et al.*, 1994).

Importantly, by using bioluminescence methods, which permit the assessment of the dorsoventral distribution of high energy phosphates significant differences in ATP depletion after induction of spinal ischemia

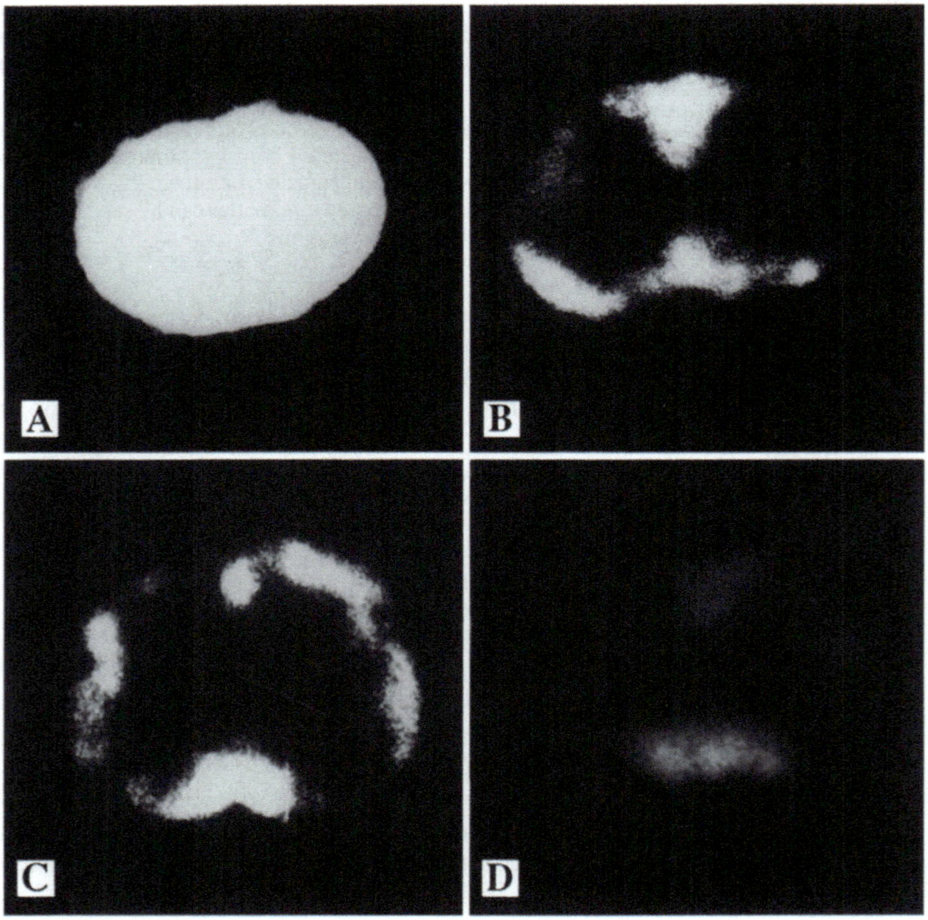

Fig. 7. ATP induced bioluminescence in lumbar transverse spinal cord sections after increasing intervals (control (**A**); 10 min isch. (**B**); 20 min isch. (**C**); and 40 min isch. (**D**)) of aortic occlusion in rabbit. Note the presence of measurable ATP activity in dorsal and ventral white matter even after 20 min of spinal ischemia and complete loss of ATP activity at 40 min after aortic occlusion (adopted from Danielisova *et al.*, 1987)

have been measured (Danielisova *et al.*, 1987). In this study, it was shown that after 10–20 min after aortic occlusion in rabbit, there is a near complete loss of ATP-induced bioluminescence in both dorsal and ventral gray matter, while white matter still retains measurable ATP. After 40 min of ischemia complete loss of ATP-induced bioluminescence in both white and gray matter was seen (Fig. 7).

With progressive intervals of spinal ischemia, and corresponding in time with significant loss of ATP, a pronounced increase in the concentration of metabolites associated with anaerobic metabolism is observed. Thus a significant (2.5–5.5 fold) increase in lactate concentration and increase in lactate: pyruvate ratio (to estimate tissue redox state) can be observed between 10–40 min of aortic occlusion in both rabbit and cat spinal cord ischemia models (Danielisova *et al.*, 1987; Marsala *et al.*, 1989; Anderson *et al.*, 1983; Allen *et al.*, 1994).

b) Effect of postischemic reperfusion. Despite relatively prolonged periods of spinal ischemia (10–40 min), rapid normalization of HEP, decrease in lactate concentration and normalization of glucose metabolism is observed. It has been reported that despite variability in the duration of the ischemic insult (i.e. 10, 20 or 30 min of ischemia) ATP levels recovered to 86% of baseline values after 90 min of reperfusion in cat. Comparably, total adenylates recovered to 84% (Anderson *et al.*, 1983). Similar recovery in the rabbit ischemia model were reported during initial 24 hr of the reperfusion (Allan *et al.*, 1994). These data indicate that even periods of prolonged spinal ischemia have no significant effect on mitochondrial function during the initial hours (1.5–24 hrs) of reperfusion. However, a significantly different picture emerges after longer (4 days) intervals of reperfusion. As we have described in previous studies, rapid spinal neuronal degeneration, typically affecting small and medium-sized interneurons in central gray matter (lamina VII) is observed (Jacobs *et al.*, 1987; Marsala *et al.*, 1991; Marsala *et al.*, 1992). These initial changes are then followed by the complete loss of affected neurons and the development of gray matter necrosis localized at the same laminar level (Zivin and DeGirolami, 1980; Marsala and Yaksh, 1994). Such advanced neurodegenerative changes (central gray matter necrosis) are fully developed between 2–4 days after reperfusion and correspond clinically with fully developed spastic or flaccid paraplegia (Fig. 8C).

Corresponding with the distribution of gray matter necrosis a loss of HEP and glucose uptake is observed at the same laminar and segment levels. As it is shown in Fig. 8A, 40 min of spinal ischemia followed by 4 days of reperfusion evokes near complete loss of ATP-induced bioluminescence in lumbar spinal gray matter, while white matter ATP shows near complete recovery. Similarly, 2-DG uptake is specifically affected in the necrotic areas (Fig. 8B), delineating the borders between necrotic and near normal tissue. However, it is likely that after longer survival periods when the necrotic areas are replaced with gliosis, near normal ATP concentration and normal glucose metabolism would be found. Such a "normalization",

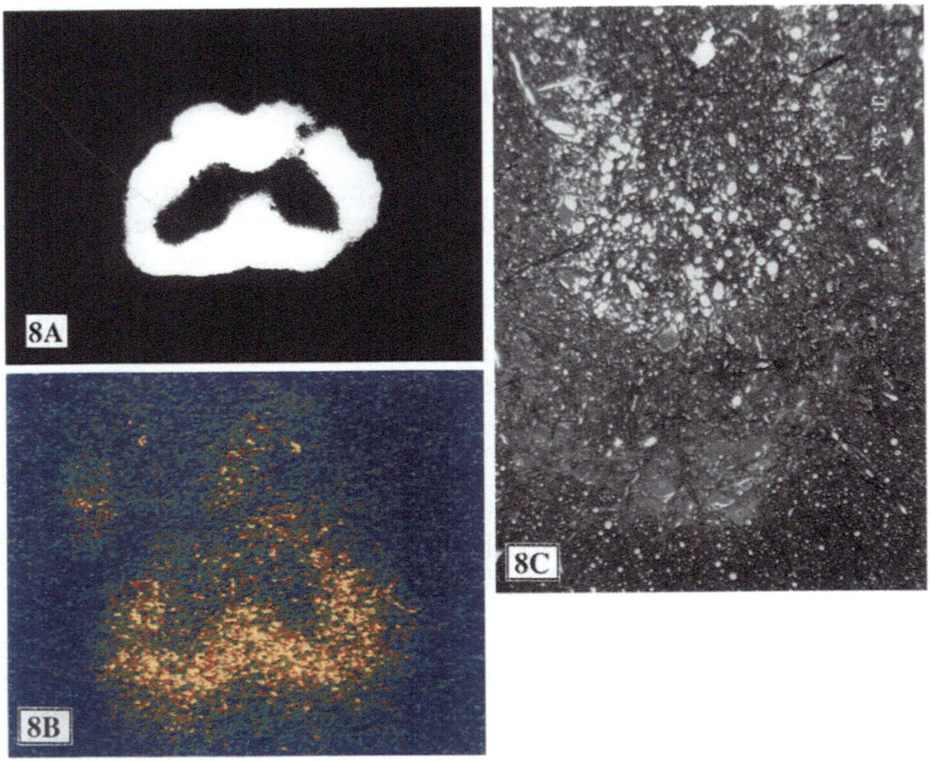

Fig. 8. (**A**) ATP-induced bioluminescence in lumbar transverse spinal cord section after 40 min of aortic occlusion and 4 days survival in an animal with flaccid paraplegia. Note a complete loss of ATP activity in central and ventral gray matter (according Danielisova *et al.*, 1987). (**B**) Computerized pseudocolor aotoradiogram of 2-deoxyglucose uptake in rat transverse lumbar spinal cord section after 12 min of aortic occlusion and 48 hr survival with fully developed spastic paraplegia. Comparably as for ATP, significant loss of glucose uptake in the areas of central gray matter necrosis can be seen (arrows), (Ishikawa and Marsala, 1996; in preparation). (**C**) Light microphotograph of lumbar transverse spinal cord section after 10 min of aortic occlusion and 2 days survival in rat. Consistent with the presence of fully developed spastic paraplegia near complete loss of spinal interneurons in central gray matter can be seen (arrows), (Nissl staining, original magnification 2x)

will simply reflect the presence of cellular elements possessing the ability to synthesize ATP, however, which are without any functional association with the neuronal pools previously localized in the same areas.

c) Functional consequences and the relevance of critical energy depletion. As described above, continuous and stable substrate delivery and corresponding ATP synthesis are the basic conditions for the maintenance of neuronal ion homeostasis and for providing adequate energy support associated with the functional properties of spinal neurons (i.e. depolarization and repolarization). The high requirements of this process and the ability to activate specific spinal neuronal pools (spinal cord evoked potentials–SCEP) after peripheral nerve activation thus can provide a simple method for an indirect assessment of the energy state of these neurons.

Under control conditions, repetitive peripheral nerve stimulation will evoke depolarization of spinal neurons in the areas corresponding with the termination of activated afferents. Because of ongoing availability of energy supplies such repetitive depolarization can be evoked for relatively prolonged periods, without any detectable change in the latency or amplitude of the response. However, as mentioned above, because of a high energy dependency of this process the ability of neurons to respond to repetitive stimulation can be expected to be significantly altered under conditions of energy depletion (such as spinal cord ischemia). By using spinal cord evoked potentials, it has been shown that after induction of ischemia, there is a rapid decrease in the amplitude and increase in the latency of spinal cord evoked potentials. It has also been shown that the postsynaptic waves of the SCEP complex (i.e. waves N3, N4, P), believed

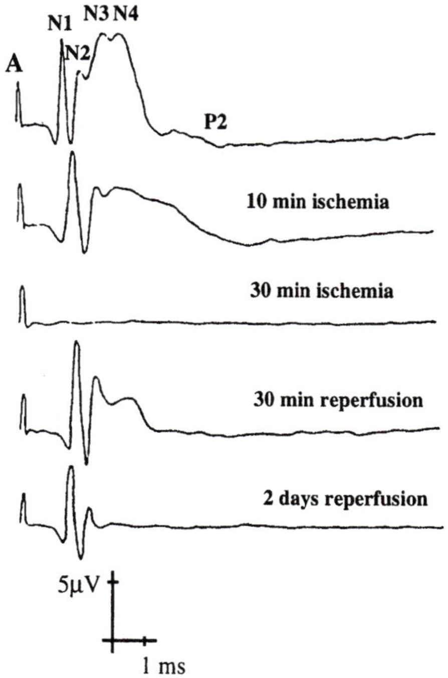

Fig. 9. Spinal cord evoked potentials before and after 30 min of spinal cord ischemia in rabbit. Evoked potentials were evoked by repetitive sciatic nerve stimulation (0.1 ms; 3V; 4.1Hz) and recorded by silver-wire recording electrodes placed into small holes drilled into he L-6 and L-5 vertebrae. Note a significant decrease in postsynaptic components of SCEP complex (N3 and N4 wave) at 10 min after ischemia. Incomplete recovery of postsynaptic complex at 30 min of reperfusion followed by complete loss of N3, N4 waves at 48 hrs of reperfusion can also be seen (adopted from Malatova *et al.*, 1995)

to reflect the functional integrity of spinal interneurons and A-motoneurons, are the most sensitive to ischemia. In contrast, presynaptic components (N1, N2) corresponding to the activity of primary afferents, show a relatively high resistance to ischemia and can still be evoked even after 15–20 min of aortic occlusion (Cheng *et al.*, 1984; Malatova *et al.*, 1995), (Fig. 9). It is interesting to note that such apparent differences in the functional sensitivity of gray vs. white matter appears to correlate with the time course of dorsoventral ATP depletion after induction of ischemia (i.e. more rapid ATP depletion in gray vs. white matter after ischemia (see Fig. 7). Corresponding with a gradual loss of neuronal excitability and the onset of neuronal depolarization, increased concentrations of extracellular potassium and, as measured by spinal microdialysis, significant increases in excitatory amino acids (glutamate) are seen (Simpson *et al.*, 1989; Marsala *et al.*, 1994; Rokkas *et al.*, 1995).

As described above, during the initial hours of reperfusion, there is a rapid recovery of the ATP concentration in both gray and white matter, followed later, by near complete ATP depletion in the areas of gray matter necrosis. Accordingly, the timeline of the recovery of spinal evoked potentials shows a very similar pattern. Thus, there is rapid but incomplete recovery of pre- and postsynaptic components of SCEP during the initial 30–120 min of reperfusion, however, over the period of 24–48 hrs, a selective loss of postsynaptic waves (N3, N4, P) is observed, corresponding in time with the development of gray matter necrosis (Fig. 9).

Significance and therapeutic implications

The above data demonstrate that transient aortic occlusion is associated with rapid but incomplete reduction in SCBF. With increasing intervals of ischemia, a progressive loss of high energy phosphates and an increase in metabolites of anaerobic metabolism (lactate) in both white and gray matter are observed. The loss of energy substrates corresponds in time with the loss of neuronal excitability, followed later by anoxic depolarization of neurons and excessive release of excitatory amino acids into the extracellular space. The time spent under conditions of anoxic depolarization correlates significantly with loss of neurological function after reperfusion. The above characteristics of intra-ischemic SCBF and metabolic changes suggest that therapeutic manipulations which serve to improve SCBF or decrease metabolic demands during the period of incomplete ischemia should predictably improve outcome.

Modulation of SCBF during spinal cord ischemia

Therapeutic approaches to improve SCBF during the period of aortic occlusion can be considered in two categories: I) Bypass the level of aortic occlusion. Several surgical techniques (shunts, left atriofemoral bypass, partial cardiopulmonary bypass) are aimed at providing adequate distal

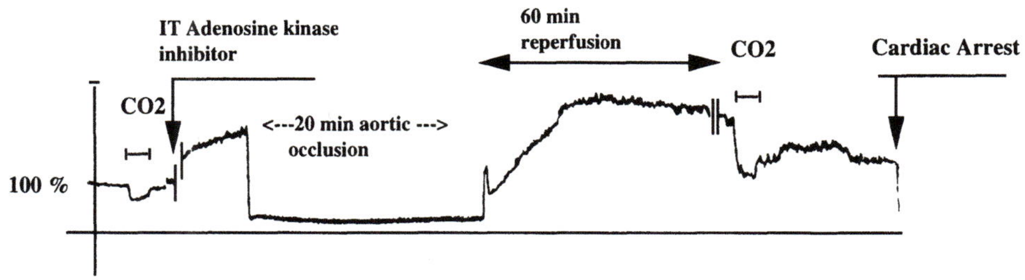

Fig. 10. Polygraph tracing of the output of the laser Doppler signal after intrathecal injection of adenosine kinase inhibitor and 20 min of aortic occlusion. Note a significant increase in SCBF 10 min after intrathecal drug administration. Prolonged postischemic hyperemia in GP-B pretreated animals but with preservation of CO_2 responsivity can also be seen (*IT* intrathecal; Wo and Marsala, in preparation)

perfusion and correspondingly sufficient spinal cord blood flow during aortic cross-clamp (Marini and Cunningham, 1993). However, a number of clinical studies have shown that the incidence of paraplegia is not significantly improved by these techniques. The efficacy of such treatments appears to be determined by individual differences in spinal vascular anatomy (i.e. the presence of major radicular arteries) and by the segmental level of the aortic cross-clamp. II) Systemic or intrathecal treatment with vasodilatatory drugs. Intrathecal papaverine, magnesium or local anesthetics have been shown to provide significant protection in several experimental studies (Svensson *et al.*, 1986; Simpson *et al.*, 1994; Breckwoldt *et al.*, 1991). We have recently found that intrathecal treatment with adenosine kinase inhibitor, which serves to increase endogenous adenosine levels, provides a powerful spinal vasodilatatory effect 10–15 min after IT injection. More importantly, significant increases in SCBF during the period of aortic occlusion were observed when SCBF reached 24% of baseline values which was significantly higher then in saline injected animals (6–7% of baseline; Fig. 10). However, it has to be stressed that the therapeutic efficacy of treatments targeted to increase the level of spinal perfusion during ischemia will be determined by the individual differences in the residual flow after aortic cross clamp. It is likely that in the individuals with higher residual flow after aortic cross-clamp, better vasodilatatory responses can be expected, while in those with poorly developed collaterals and resulting low residual flow during ischemia, minimal or no effect will be observed. Such a therapeutic dependency of the above treatments may thus result in a high variability in the outcome between individual studies and in individual subjects.

Modulation of metabolic activity during spinal cord ischemia

Because the loss of high energy phosphates and corresponding loss of spinal excitability are functions of altered SCBF, manipulation targeted to decrease metabolic demands will represent second intention therapy. However, as it will be outlined below such therapy can be very effective even under conditions of near complete ischemia. In general, decrease in spinal metabolic activity can be achieved i) by means of drugs (barbiturates, local anesthetics), (Cavazzuti *et al.*, 1991; Crosby *et al.*, 1990) or ii) spinal cord hypothermia (see Fig. 6). Based on the experimental and clinical data so far published, it appears to be generally accepted that spinal hypothermia provides the most effective protective tool against transient spinal ischemia. We have recently reported, as have others, that selective spinal cord hypothermia provided complete protection against prolonged intervals of aortic occlusion as evidenced by the recovery of neurological function, spinal evoked potentials, and suppression of histopathological changes (Marsala *et al.*, 1994; Berguer *et al.*, 1992; Ueno *et al.*, 1994; Wisselink *et al.*, 1994). A comparable protective effect of epidural cooling after 60 min of aortic occlusion in the rabbit spinal cord ischemia model was also seen (Vanicky *et al.*, 1994). In contrast to the global cerebral ischemia model, in which hypothermia-mediated protection was not associated with significant differences in HEP, in the rabbit spinal cord, deep hypothermia (23°C) as achieved by intraaortic perfusion has been shown to slow HEP depletion during 20 min of aortic occlusion (Allen *et al.*, 1994). This would indicate that spinal cord hypothermia may be particularly effective under conditions of incomplete spinal ischemia such as after aortic cross-clamping. However, in accordance with the previous comments it must be noted that while deep spinal hypothermia can be expected to provide powerful protection of spinal cord tissue during ischemic period, the overall long-term neurological outcome will depend on the adequacy of spinal cord blood flow (i.e. reimplantation of critical intercostal arteries) during the period of rewarming/reperfusion.

Summary

The above data indicate that aortic occlusion and the corresponding loss of distal perfusion lead to a significant but incomplete loss of spinal cord blood flow and resulting spinal ischemia. The completeness of ischemia depends on individual differences in spinal vascular anatomy and the presence of paraspinal collateral systems. With a progressive loss of spinal perfusion, gradual decreases in high energy phosphates and glucose reuptake are observed with the most rapid loss observed in the gray matter in the lumbosacral segments. During the same period as a result of accelerated anaerobic metabolism a massive increase in gray matter lactate concentration is measured. Corresponding with high energy phosphate depletion and resulting membrane failure loss of neuronal excitability is seen followed by excessive release of several active factors including excitatory amino acids from spinal parenchyma. Importantly, the time spent under conditions of neuronal depolarization shows clear correlations with the degree of neurological deficit after reperfusion. Therapeutic manipulations which serve to improve intraischemic SCBF or suppress metabolic demands can provide a certain degree of protection against a limited periods of spinal ischemia. The efficacy of such protection will depend on the ability to provide adequate blood flow into collateral-dependent areas during the period of ischemia as well as adequate long term reperfusion of the previously ischemic spinal cord segments.

References

Aadahl P, Saether OD, Stenseth R, Myhre HO (1990) Microcirculation of the spinal cord during proximal aortic cross-clamping. Europ J Vasc Surg 4: 5–10

Abram SE, Kostreva DR (1986) Spinal cord metabolic response to noxious radiant heat stimulation of the cat hind footpad. Brain Res 385: 143–147

Allen BT, Davis CG, Osborne D, Karl I (1994) Spinal cord ischemia and reperfusion metabolism: The effect of hypothermia. J Vasc Surg 19: 332–339

Anderson DK, Behbehani MM, Means ED, Waters TR, Green ES (1983) Susceptibility of feline spinal cord energy metabolism to severe incomplete ischemia. Neurology 33: 722–731

Berguer R, Porto J, Fedoronko B, Dragovic L (1992) Selective deep hypothermia of the spinal cord prevents paraplegia after aortic cross-clamping in the dog model. J Vasc Surg 15: 62–71

Bower TC, Murray MJ, Gloviczki P, Yaksh TL, Hollier LH, Pairolero PC (1989) Effects of thoracic aortic occlusion and cerebrospinal fluid drainage on regional spinal cord blood flow in dogs: Correlation with neurologic outcome. J Vasc Surg 9: 135–144

Breckwoldt WL, Genco CM, Connolly RJ, Cleveland RJ, Diehl JT (1991) Spinal cord protection during aortic occlusion: Efficacy of intrathecal tetracaine. Ann Thorac Surg 51: 959–963

Cavazzuti M, Porro CA, Barbieri A, Galetti A (1991) Brain and spinal cord metabolic activity during propofol anaesthesia. Br J Anaesth 66: 490–495

Cheng MK, Robertson C, Grossman RG, Foltz R, Williams V (1984) Neurological outcome correlated with spinal evoked potentials in spinal cord ischemia model. J Neurosurg 60: 786–795

Cole DJ, Lin DM, Drummond JC, Shapiro HM (1990) Spinal tetracaine decreases central nervous system metabolism during somatosensory stimulation in the rat. Can J Anaesth 37: 231–237

Crosby G, Russo MA, Szabo MD, Davies KR (1990) Subarachnoid Clonidine reduces spinal cord blood flow and glucose utilization in conscious rats. Anesthesiology 73: 1179–1185

Danielisova V, Chavko M, Kehr J (1987) Adenine nucleotide levels and regional distribution of ATP in rabbit spinal cord after ischemia and recirculation. Neurochem Res 12: 241–245

Doppman JL, DiChiro G, Morton DL (1968) Arteriographic identification of spinal cord blood supply prior to aortic surgery. JAMA 204: 174–175

Evans HE (1993) Miller's Anatomy of the Dog, 3rd edn. Saunders, Philadelphia, pp 648–649

Faure C, Djindjian R, Lefebvre J (1966) [Apropos of spinal cord hazards in abdominal aortography]. Ann Radiol (Paris) 9: 523–530

Follis F, Miller K, Scremin OU, Pett S, Kessler R, Temes T, Wernly JA (1995) Experimental delayed postischemic spinal cord hypoperfusion after aortic cross-clamping. Can J Neurol Sci 22: 202–207

Gordh T Jr, Feuk U, Norlen K (1986) Effect of epidural clonidine on spinal cord blood flow and regional and central hemodynamics in pigs. Anesth Analg 65: 1312–1318

Gray H. (1973) Edited by Charles Mayo Goss. Anatomy of the Human Body, 29th edn. Lea & Febiger, pp 964–971

Griffiths IR, Rowan JO, Crawford RA (1975) Spinal cord blood flow measured by a hydrogen clearance technique. J Neurol Sci 26: 529–544

Hickey R, Albin MS, Bunegin L, Gelineau J (1986) Autoregulation of spinal cord blood flow: Is the cord a microcosm of the brain. Stroke 17: 1183–1189
Homma S, Suzuki T, Murayama S, Otsuka M (1979) Amino acid and substance P contents in spinal cord of cats with experimental hind-limb rigidity produced by occlusion of spinal cord blood supply. J Neurochem 32: 601–698
Jacobs TP, Kempski O, McKinley D, Dutka AJ, Hallenbeck JM, Feuerstein G (1992) Blood flow and vascular permeability during motor dysfunction in a rabbit model of spinal cord ischemia. Stroke 23: 367–373
Jacobs TP, Shoami E, Baze W, Burgard E, Gunderson C, Hallenbeck JM, Fuerstein G (1987) Deteriorating stroke model: histopathology, edema, and eicosanoid changes following spinal cord ischemia in rabbit. Stroke 18: 741–750
Kadekaro M, Crane AM, Sokoloff L (1985) Differential effects of electrical stimulation of sciatic nerve on metabolic activity in spinal cord and dorsal root ganglion in the rat. Proc Nat Acad Sci USA 82: 6010–6030
Kaplan BJ, Friedman WA, Gravenstein N, Richards R, Davis RF (1987) Effects of aortic occlusion on regional spinal cord blood flow and somatosensory evoked potentials in sheep. Neurosurgery 21: 668–675
Kobrine AI, Doyle TF, Martins AN (1975) Autoregulation of spinal cord blood flow. Clin Neurosurg 22: 573–581
Kwun BD, Vacanti F (1995) Mild hypothermia protects against irreversible damage during prolonged spinal cord ischemia. J Surg Res 59: 780–782
Lenzi P, Cianci T, Guidalotti PL, Franzini C (1986) Regional spinal cord blood flow during sleep-waking cycle in rabbit. Am J Physiol 251: H957–960
Malatova Z, Vanicky I, Galik J, Marsala M (1995) Epidural perfusion cooling protects against spinal cord ischemia in rabbits. An evaluation of cholinergic function. Mol Chem Neuropath 25: 81–96
Marini CP, Cunningham JN (1993) Issues surrounding spinal cord protection. In: Advances in Cardiac Surgery, vol 4. Mosby, St. Louis, pp 89–107
Marini CP, Grubbs PE, Toporoff B, Woloszyn TT, Coons MS, Acinapura AJ, Cunningham JN (1989) Effect of sodium nitroprusside on spinal cord perfusion and paraplegia during aortic cross-clamping. Ann Thorac Surg 47: 379–783
Marsala M, Danielisova V, Chavko M, Hornakova A, Marsala J (1989) Improvement of energy state and basic modifications of neuropathological damage in rabbits as a result of graded postischemic spinal cord reoxygenation. Exper Neurol 105: 93–103
Marsala J, Sulla I, Santa M, Marsala M, Zacharias L, Radonak J (1991) Mapping of the canine lumbosacral spinal cord neurons by nauta method at the end of the early phase of paraplegia induced by ischemia and reperfusion. Neuroscience 45: 479–494
Marsala M, Vanicky I, Radonak J, Kliesenbauerova E, Marsala J (1992) Postischemic hyperoxia enhances vulnerability in the rabbit spinal cord ischemia model. Rest Neurol Neurosci 6: 283–291
Marsala M, Vanicky I, Galik J, Radonak J, Kundrat I, Marsala J (1993) Panmyelic epidural cooling protects against ischemic spinal cord damage. J Surg Res 55: 21–31.
Marsala M, Vanicky I, Yaksh TL (1994) Effect of graded hypothermia (27–34°C) on behavioral function, histopathology and spinal blood flow following spinal ischemia in rat. Stroke 25: 2038–2046
Nystrom B, Norlen K (1983) Regional spinal cord and brain blood flows in the rat. Neurol Res 5: 91–101
Orendacova J, Marsala M, Marsala J (1991) The blood-brain barrier permeability in graded postischemic spinal cord reoxygenation in rabbits. Neurosci Lett 128: 143–146
Porro CA, Cavazzuti M, Galetti A, Sassatelli L, Barbieri GC (1991) Functional activity mapping of the rat spinal cord during formalin-induced noxious stimulation. Neuroscience 41: 655–665
Pulsinelli WA, Brierley JB, Plum F. (1982) Temporal profile of neuronal damage in a model of transient forebrain ischemia. Ann Neurol 11: 491–498

Rokkas CK, Cronin CS, Nitta T, Helfrich LR Jr, Lobner DC, Choi DW, Kouchoukos NT (1995) Profound systemic hypothermia inhibits the release of neurotransmitter amino acids in spinal cord ischemia. J Thorac Cardiovasc Surg 110: 27–35
Rubinstein A, Arbit E (1990) Spinal cord blood flow in the rat under normal physiological conditions. Neurosurgery 27: 882–886
Schievink WI, Luyendijk W, Los JA (1988) Does the artery of Adamkiewicz exist in the albino rat? J Anat 161: 95–101
Senter HJ, Venes JL (1979) Loss of autoregulation and post-tarumatic ischemia following experimental spinal cord trauma. J Neurosurg 50: 198–206
Simpson JI, Eide TR, Schiff GA, Clagnaz JF, Hossain I, Tverskoy A, Koski G (1994) Intrathecal magnesium sulfate protects the spinal cord from ischemic injury during thoracic aortic cross-clamping. Anesthesiology 81: 1493–1499
Simpson RK Jr, Robertson CS, Goodman JC (1990) Spinal cord ischemia-induced elevation of amino acids: extracellular measurement with microdialysis. Neurochem Res 15: 635–639
Smith ML, Auer RN, Siesjo BK (1984) The density and distribution of ischemic brain injury in the rat following 2–10 min of forebrain ischemia. Acta Neuropathol (Berl) 64: 319–332
Steen PA, Newberg L, Milde JH, Michenfelder JD (1983) Hypothermia and barbiturates: individual and combined effects on canine cerebral oxygen consumption. Anesthesiology 58: 527–532
Svensson LG, Von Ritter CM, Groeneveld HT, Rickards ES, Hunter SJ, Robinson MF, Hinder RA (1986) Cross-clamping of the thoracic aorta. Influence of aortic shunts, laminectomy, papaverine, calcium channel blocker, allopurinol, and superoxide dismutase on spinal cord blood flow and paraplegia in baboons. Ann Surg 204: 38–47
Szilagyi DE, Hageman JH, Smith RF, Elliott JP (1978) Spinal cord damage in surgery of the abdominal aorta. Surgery 83: 38–45
Taira Y, Marsala M (1996) Effect of proximal arterial perfusion pressure on function, spinal cord blood flow, and histopathological changes after increasing intervals of aortic occlusion in the rat. Stroke 27: 1850–1858
Torre de la J C (1984) Spinal cord injury models. Prog Neurobiol 22: 289–344
Tveten L (1976) Spinal cord vascularity: the spinal cord arteries in the rat. Acta Radiol Diagn 17: 385–398
Ueno T, Itoh T, Hirahara K, Sakai M, Naitoh K (1994) Protection against spinal cord ischaemia: one-shot infusion of hypothermic solution. Cardiovasc Surg 2: 374–378
Vacanti FX; Ames A 3d (1984) Mild hypothermia and Mg^{++} protect against irreversible damage during CNS ischemia. Stroke 15: 695–698
Vanicky I, Marsala M, Galik J, Marsala J (1993) Epidural perfusion cooling protection against protracted spinal cord ischemia in rabbits. J Neurosurg 79: 736–741
Vanicky I, Marsala M, Orendacova J, Marsala J (1992) Silver impregnability of ischemia-sensitive neocortical neurons after 15 minutes of cardiac arrest in the dog. Anat Embryol 186: 167–173
Wisselink W, Becker MO, Nguyen JH, Money SR, Hollier LH (1994) Protecting the ischemic spinal cord during aortic clamping: the influence of selective hypothermia and spinal cord perfusion pressure. J Vasc Surg 19: 788–795
Woolen D H M, Millen J W (1955) The arterial supply of the spinal cord and its significance. J Neurol Neurosurg Psychiatry 18: 97–102
Zivin JA, DeGirolami U (1980) Spinal cord infarction: a highly reproducible stroke model. Stroke 11: 200–202

Correspondence: Dr. M. Marsala, Anesthesiology Research Laboratory, 0818, University of California, 9500 Gilman Drive, San Diego, CA 92093, USA.

Spinal Cord Monitoring
Eds: E. Stålberg, H. S. Sharma, Y. Olsson

Chapter 2

Calcium-binding proteins in the spinal cord: physiological significance

K. Ren[1] and **M. A. Ruda**[2]

[1]Department of Oral and Craniofacial Biological Sciences, University of Maryland Dental School, Baltimore, MD, USA
[2]Neurobiology and Anesthesiology Branch, National Institute of Dental Research, National Institutes of Health, Bethesda, MD, USA

Introduction

There has been increased interest in the presence of calcium-binding proteins (CBPs) in the spinal cord over the past few years (Yamamoto *et al.*, 1989; Antal *et al.*, 1990; Celio, 1990; Yoshida *et al.*, 1990; Zhang *et al.*, 1990; Arvidsson *et al.*, 1992; Menétrey *et al.*, 1992b; Heppelmann and Emson 1993; Ren *et al.*, 1993; Ren and Ruda, 94; also see Andressen *et al.*, 1993, for a review). This interest is partly due to the unique advantage of some CBPs as neuronal markers in the spinal cord. As in other parts of the nervous system, several CBPs exhibit a distinct distribution pattern, suggesting that they may be involved in a variety of physiological activities, including initial sensory processing and final motor control. CBPs not only couple the elevation of intracellular Ca^{2+} in response to an extracellular signal with the subsequent physiological response, they also modulate the intracellular Ca^{2+} level to help maintain Ca^{2+} homeostasis. The latter action of the CBPs is especially important as the perturbation of the intracellular Ca^{2+} homeostasis may precede the development of cytotoxicity and irreversible injury (Nicotera *et al.*, 1992).

The prototype CBP is calmodulin with a molecular weight of approximately 17 kDa. Calmodulin was identified in 1967 and has since been intensively studied (Cheung, 1980; Means *et al.*, 1982; Gnegy, 1993, 1995, reviews). Calmodulin acts as a primary intracellular Ca^{2+} receptor and is involved in multiple cellular signal transduction pathways. The activity of more than twenty enzymes are regulated by Ca^{2+}-calmodulin complexes (Means *et al.*, 1991). Calmodulin immunoreactivity is widely distributed in the brain (Lin *et al.*, 1980; Wood *et al.*, 1980; Seto-Ohshima *et al.*, 1983, 1985; Tabuchi *et al.*, 1984), and it has been localized subcellularilly to the cell nucleus, postsynaptic densities, axonal and dendritic microtubules, endoplasmic reticulum, mitochondria and the plasma membrane (Lin *et al.*, 1980; Wood *et al.*, 1980; Caceres *et al.*, 1983). In the spinal cord, calmodulin immunoreactivity in ventral horn motoneurons has attracted the most attention (Wood *et al.*, 1980; Seto-Oshima *et al.*, 1985; Ren and Ruda, 1994).

The most studied CBPs in the spinal cord, namely calbindin-D28k (CB-D28k), parvalbumin, calretinin, calmodulin, and calpain, belong to a superfamily of the structurally related proteins, the EF-hand family of CBPs (Persechini *et al.*, 1989). The "EF-hand" is a consensus amino acid sequence involved in calcium-binding activity. The typical domain of the EF-hand consists of 29 residues in a helix (E)-loop-helix (F) configuration (Persechini *et al.*, 1989). The binding site for Ca^{2+} is in the loop and the EF-helices coordinate calcium binding with dissociation constants of about 1 μM (see Persechini *et al.*, 1989, for a review). A CBP can possess multiple EF-hand domains. However, not all EF-hands are active in binding Ca^{2+}. For example, both calmodulin and CB-D28k bind four Ca^{2+}, but they have 4 and 6 EF-hand domains, respectively. It should be mentioned that some CBPs also present in the nervous system are not members of the EF-family, such as annexins and protein kinase C.

A number of CBPs, including CB-D28k, parvalbumin, and calretinin, have recently been found to be widely distributed in the central nervous system (CNS) (Braun, 1990; Baimbridge *et al.*, 1992, review). CB-D28k was first found in chick intestine as a vitamin D-dependent CBP (Wasserman *et al.*, 1966) and was later localized to mammalian brain (Jande *et al.*, 1981; Séquier *et al.*, 1990). Parvalbumin was first found in muscular tissue (Lehky *et al.*, 1974) and then also in the brain (Celio and Heizmann, 1981). The EF-hand structure was first identified in parvalbumin (Persechini *et al.*, 1989, review). Calretinin was recently discovered (Rogers, 1987, 1991; Winsky *et al.*, 1989) and is closely related to CB-D28k but appears to be a neuronal specific CBP (Parmentier, 1989; Rogers, 1987). Using primarily the immunohistochemical technique, interesting distribution patterns of these CBPs have been described in the CNS. CB-D28K and parvalbumin are present for the most part in different groups of neurons in the cerebral cortex (Hendry *et al.*, 1989; Van Brederode *et al.*, 1991), hippocampus (Slovitar, 1989), superior colliculus (Mize *et al.*, 1992), thalamic nuclei (Jones and Hendry, 1989; Pritz and Stritzel, 1991; Rausel and Jones, 1991), medullary dorsal horn (Bennett-Clarke *et al.*, 1992), and spinal cord (Yamamoto *et al.*, 1989; Yoshida *et al.*, 1990; Antal *et al.*, 1990; Ren and Ruda, 1994). Calretinin, when compared to CB-D28k and parvalbumin, is also present in different sets of neurons, although the cell groups of the three CBPs do overlap to some extent (Carr *et al.*, 1989a,b; Résibois *et al.*, 1989; Rogers and Résibois, 1992). The differential distribution of these CBPs in certain regions of the CNS suggests that each CBP may perform distinct functions. For instance, CB-D28k and parvalbumin are located in distinct nuclear groups in the thalamus where they may be involved in relaying sensory information of different modalities (Rausell and Jones, 1991; Rausell *et al.*, 1992).

Calpain, a calcium-activated neutral protease, is also a member of the EF-hand family of CBPs with two pairs of EF-hand domains in each of its large and small subunits (Wang and Yuen, 1994). Ca^{2+} binding induces dissociation of the calpain dimer into its catalytic 80 kDa and regulatory 30 kDa subunits (Yoshizawa *et al.*, 1995). The activation of calpain, or the release of the active 80 kDa subunit, leads to irreversible proteolysis of substrate proteins (Suzuki *et al.*, 1995). Calpain was originally isolated from the brain (Guroff, 1964), and recent studies have found that calpain represents a family of proteins that can be further divided into two groups: ubiquitous (calpain I and II) and tissue-specific (n-calpain) (Sorimachi *et al.*, 1995; Suzuki *et al.*, 1995, review). The tissue specific calpain is expressed in skeletal or smooth muscles (Suzuki *et al.*, 1995). The term calpain will refer to the ubiquitous group of calpains in the following discussion. Calpain-like immunoreactivity (LI) is found in glia and virtually all neurons (Siman *et al.*, 1985; Perlmutter *et al.*, 1990). In the spinal cord, ventral horn motoneurons exhibit the highest level of immunoreactive labeling (Siman *et al.*, 1985; Perlmutter *et al.*, 1990).

The spinal cord is an important site for processing nociceptive input and integrating sensorimotor functions. The immunohistochemical distribution of CB-D28k, calretinin, parvalbumin and calmodulin in the spinal cord has been studied in detail (Yamamoto *et al.*, 1989; Antal *et al.*, 1990; Yoshida *et al.*, 1990; Zhang *et al.*, 1990; Arvidsson *et al.*, 1992; Résibois and Rogers, 1992; Ren *et al.*, 1993; Ren and Ruda, 1994; also see Andressen *et al.*, 1993, for a review). The CBPs clearly exhibit differential and complementary distribution patterns in the spinal cord. The precise function of these CBPs in the spinal cord, however, is still largely unknown. Resembling neurodegenerative disorders, calcium-mediated excitotoxicity in the spinal cord may also be related to the response to spinal cord injury and development of abnormal pain sensations such as allodynia (Dubner, 1991). Thus, studying spinal CBPs may provide insight into the mechanisms of spinal injury and persistent pain and hyperalgesia. Based primarily on immunocytochemical studies, we will summarize our current knowledge about the localization of several CBPs, CB-D28k, calretinin, parvalbumin, calmodulin and calpain, in the spinal cord. The physiological significance of their distribution will also be discussed.

Comparison of calcium-binding proteins in the spinal cord

CB-D28k, calretinin, parvalbumin, calmodulin and calpain are abundant in the spinal cord where a distinct distribution pattern has been identified, especially in the superficial (laminae I and II) and ventromedial (medial portion of laminae V and VI) dorsal horn, the ventral horn (especially lamina IX), and the dorsal gray commissure (DGC) (see Andressen *et al.*, 1993; Ren and Ruda, 1994, for reviews). Extensive immunostaining is also

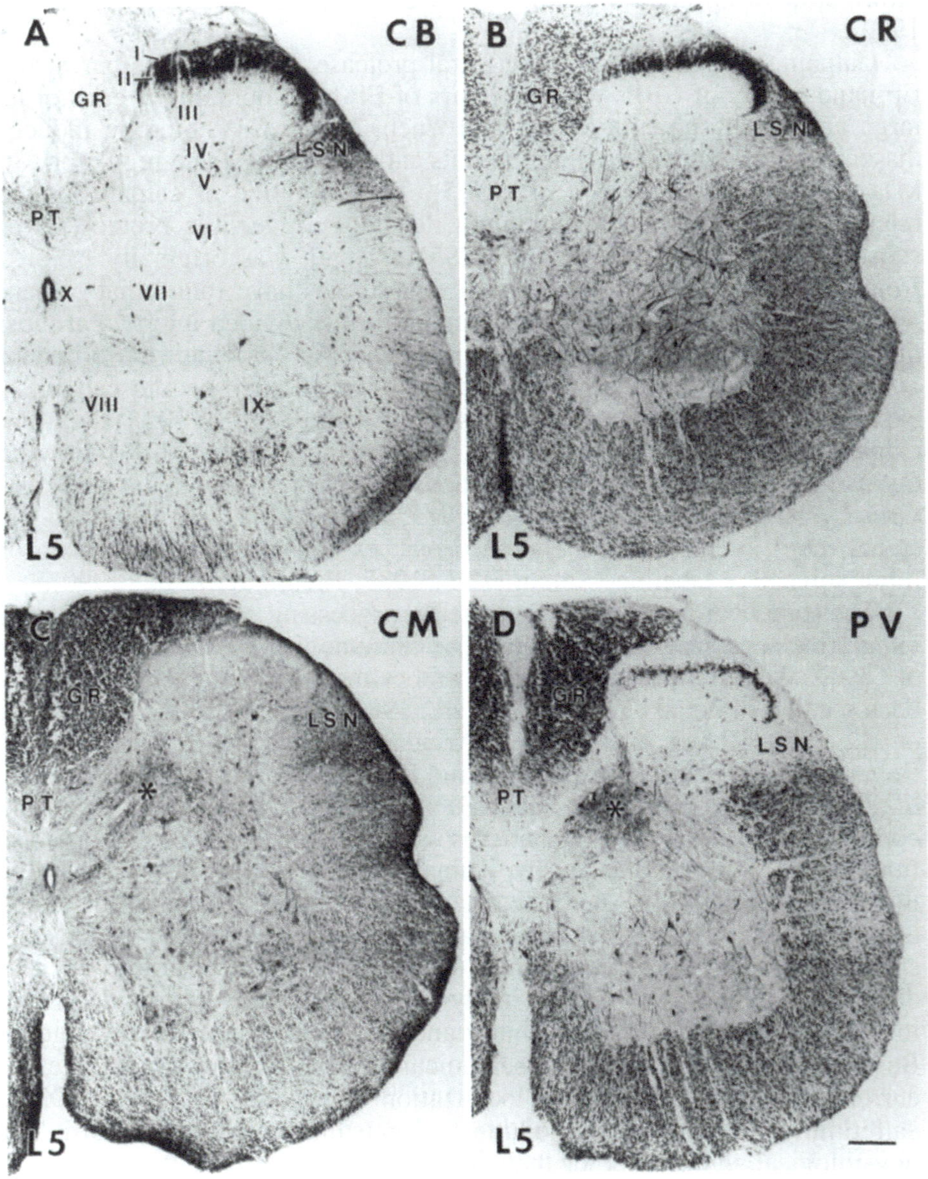

Fig. 1. The distribution of calcium-binding protein-like immunoreactivity in the L5 spinal cord. **(A)** Calbindin-D28K (*CB*), mouse anti-chicken gut CB-D28K (monoclonal IgG1, Clone No. CL-300; Lot 060H4816, Sigma). **(B)** Calretinin (*CR*), rabbit anti-guinea pig CR, Winsky *et al.*, 1989. **(C)** Calmodulin (*CM*), mouse anti-bovine CM (monoclonal IgG1, Lot 10905, Upstate Biotechnology, Inc.). **(D)** Parvalbumin (*PV*), mouse anti-carp muscle parvalbumin (monoclonal IgG1, Clone No. PA-235; Lot 49F4824, Sigma). Note the differential staining patterns in the superficial laminae of the dorsal horn and the lateral spinal nucleus (*LSN*). Also, note the dense calmodulin- and parvalbumin-like immunoreactivity in the ventromedial dorsal horn (medial laminae V/VI, asterisks). *GR* gracile fasciculus; *PT* pyramidal tract; Roman numerals I-X indicate spinal cord laminae. Diaminobenzidine dihydrochloride (DAB, Sigma) reaction. Scale bar (same for A-D) = 200 μm. (From Ren and Ruda, 1994)

seen in the spinal white matter throughout all segments. A brief summary of the CBPs in the spinal cord will be given in the following section with emphasis on the differential and complementary nature of their laminar location. Detailed morphological analysis can be found elsewhere (see Andressen *et al.*, 1993, for a list of references).

Distribution

Calbindin-D28k: CB-D28k-LI is most concentrated in Rexed laminae I and II, or superficial layers of the dorsal horn (Fig. 1A) where densely packed CB-D28k-immunoreactive (IR) cell bodies and fibers form a distinct immunoreactive band (Yamamoto *et al.*, 1989; Antal *et al.*, 1990; Celio, 1990; Yoshida *et al.*, 1990; Ren and Ruda, 1994). CB-D28K is evenly distributed dorsoventrally in lamina II such that there is no distinct border between the inner and outer parts of the lamina (Fig. 1A). A similar superficial laminar distribution of CB-D28k-LI is found throughout all segments of the spinal cord (Yamamoto *et al.*, 1989; Yoshida *et al.*, 1990). The lateral spinal nucleus (LSN) (Fig. 1A) and lateral cervical nucleus (LCN) are also intensely labeled by CB-D28k antibodies. Only a few scattered CB-D28k-IR neurons are found in laminae III and IV (Fig. 2A). Many neurons in lamina V are CB-D28k-IR (Fig. 3A). Very few CB-D28k-IR cell bodies and fibers were found in lamina VI (Fig. 3A), while a moderate number of cells in the intermediolateral cell column at the thoracic and sacral levels exhibit CB-D28k-LI.

In the ventral horn, numerous CB-D28k-IR cell bodies are found in laminae VII and VIII, but large motoneurons in lamina IX essentially lack CB-D28k-LI (Fig. 1A) (Lunam, 1989; Ince *et al.*, 1993). Although few neurons in lamina IX show light punctate staining for CB-D28k (see panel C of Fig. 6), there appears to be a developmental down-regulation of CB-D28k in spinal motoneurons. In embryonic rat and chick spinal cord, motoneurons demonstrate strong CB-D28k-LI (Zhang *et al.*, 1990; Antal and Polgar, 1993). Dénizot *et al.* (1988) have shown CB-D28k-LI in spinal motoneurons of teleost fish, thus suggesting a species variability in CB-D28k-LI. Many CB-D28k-IR cell bodies are found in lamina X. The ependymal cells surrounding the central canal are darkly labeled for CB-D28k. At the electron microscopic level, CB-D28k-IR dendrites and axons are found in synaptic contacts with unlabeled profiles (Antal *et al.*, 1990). In the monkey spinal ventral horn, CB-D28k-IR terminals make contact with dendrites and cell bodies of motoneurons in the lateral motor nucleus, suggesting that CB-D28k-IR neurons may function as Renshaw cells that are involved in recurrent inhibition (Arvidsson *et al.*, 1992).

Calretinin: Compared to CB-D28k, fewer neurons are labeled by the calretinin antibody in lamina I. Calretinin-IR large Waldeyer type neurons are frequently encountered in lamina I (Fig. 2B, also see Ren *et a*l., 1993). Similar to CB-D28k, a dense IR band of calretinin-IR cell bodies and neuropil are distributed throughout lamina II (Fig. 1B) (Résibois and

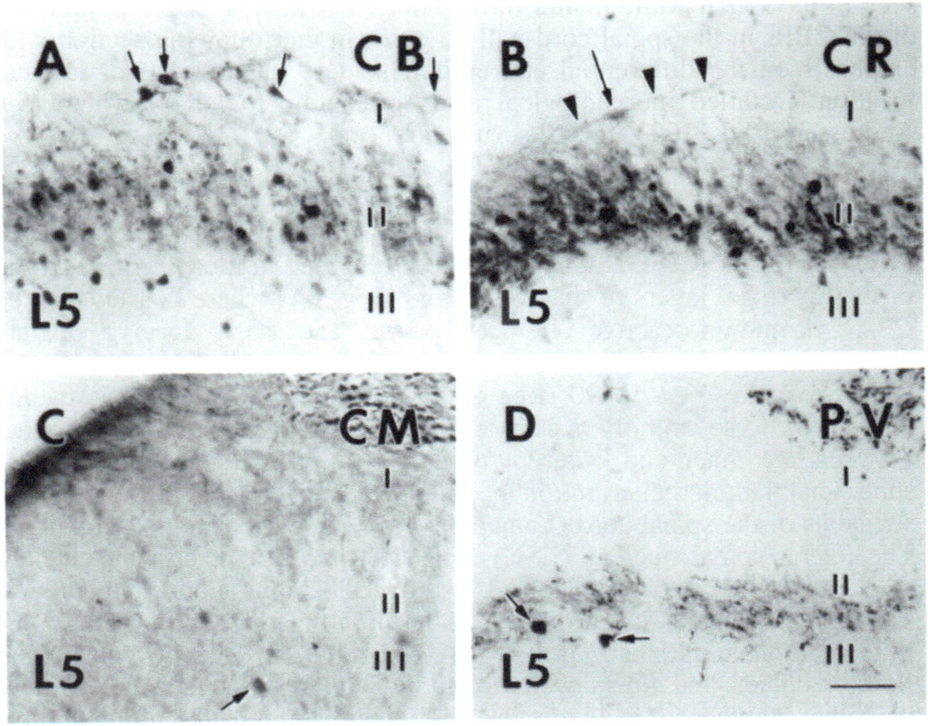

Fig. 2. Comparison of calcium-binding protein-like immunoreactivity in the superficial dorsal horn of the L5 spinal cord. **(A)** Calbindin-D28K (*CB*). **(B)** Calretinin (*CR*). **(C)** Calmodulin (*CM*). **(D)** Parvalbumin (*PV*). Some lamina I calbindin-D28K- and calretinin-immunoreactive neurons are marked (arrows in **A** and **B**). Arrowheads in **B** indicate the dendrites of a Waldeyer type lamina I neuron. Note the absence of CM-IR cell bodies and fibers in laminae I and II, while a few varicose fibers and labeled cells (arrow) can be identified in lamina III (**C**). Parvalbumin primarily labeled fiber networks in inner lamina II (**D**). Two parvalbumin-IR cell bodies are indicated (arrows in **D**). DAB reaction. Scale bar (same for **A–D**) = 50 μm. (From Ren and Ruda, 1994)

Rogers, 1992). The morphology of some lamina II CR-IR neurons is reminiscent of previously characterized stalked cells and islet cells (Fig. 4) (Gobel 1978; Cervero and Iggo, 1980). Compared to CB-D28k, parvalbumin, calmodulin, and calpain, there are relatively more calretinin-IR cell bodies and fibers in laminae V and VI, primarily in the middle third of laminae V and VI (Fig. 1B) (Ren *et al.*, 1993). Many medium to large sized neurons are calretinin-IR in laminae VII and VIII. Lamina IX motoneurons exhibit neither calretinin-LI (Ren *et al.*, 1993; also see panel D of Fig. 6) nor calretinin mRNA (Heppelmann and Emson, 1993).

In lamina X, at the level of the sacral spinal cord, calretinin-IR cell bodies are mainly found in the area dorsal to the central canal, the dorsal gray commissure (DGC) (Fig. 1B). In the DGC, an area dorsal to the central canal that extends laterally beyond the borders of lamina X, a large

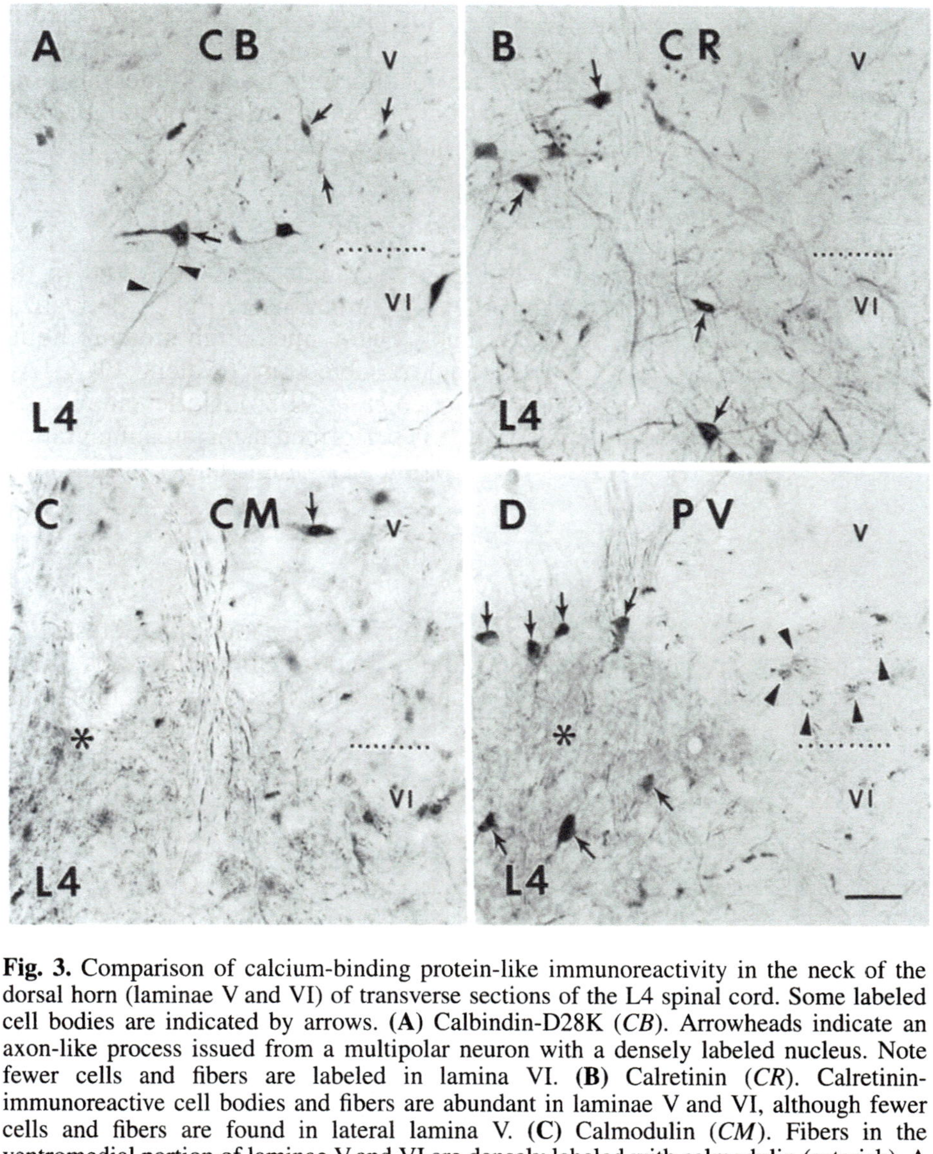

Fig. 3. Comparison of calcium-binding protein-like immunoreactivity in the neck of the dorsal horn (laminae V and VI) of transverse sections of the L4 spinal cord. Some labeled cell bodies are indicated by arrows. **(A)** Calbindin-D28K (*CB*). Arrowheads indicate an axon-like process issued from a multipolar neuron with a densely labeled nucleus. Note fewer cells and fibers are labeled in lamina VI. **(B)** Calretinin (*CR*). Calretinin-immunoreactive cell bodies and fibers are abundant in laminae V and VI, although fewer cells and fibers are found in lateral lamina V. **(C)** Calmodulin (*CM*). Fibers in the ventromedial portion of laminae V and VI are densely labeled with calmodulin (asterisk). A bipolar neuron with a densely stained nucleus in lamina V is indicated (arrow). **(D)** Parvalbumin (*PV*). Fibers and cell bodies exhibit strong immunoreactivity to parvalbumin in the ventromedial portion of laminae V and VI (asterisk). Many rostrocaudal running small fiber bundles in lateral lamina V show parvalbumin-like immunoreactivity (arrowheads). Approximate border between laminae V and VI is indicated by a short dotted line. Tissue orientation: *top* dorsal; *right* lateral. DAB reaction. Scale bar (same for **A–D**) = 50 μm. (From Ren and Ruda, 1994)

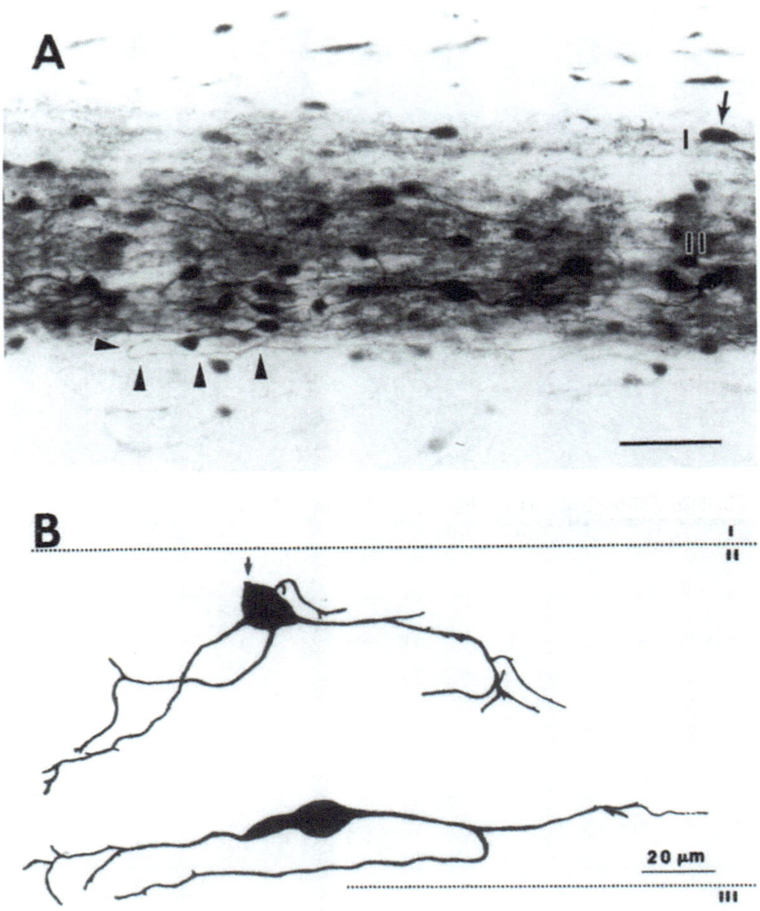

Fig. 4. (**A**). Calretinin-like immunoreactivity in laminae I and II of the spinal cord viewed in the sagittal plane. Many labeled fibers rostrocaudally. Arrow indicates the large cell body of a lamina I neuron. A recurrent process (arrowheads) was issued by a labeled neuron located at the bottom of lamina II. DAB reaction. Scale bar = 50 mm, ×200. (**B**) Camera lucida drawings from two different sagittal sections showing lamina II neurons stained for calretinin. The top neuron resembles a lamina II stalked cell and the bottom neuron resembles an islet cell. Arrow indicates what may be an axon initial segment. The dotted lines indicate the borders of lamina II. *Top* dorsal. (From Ren *et al.*, 1993)

cluster of medium to large sized calretinin-IR cell bodies and fibers form a dense calretinin-IR network. From this calretinin-labeled cell group, many fibers run dorsally to reach the edge of the gray matter; other fibers travel laterally and horizontally to interconnect the intermediolateral cell columns (Fig. 5A). The CB-D28K-LI in the DGC is similar to calretinin, except that the cell bodies and fibers were less densely packed.

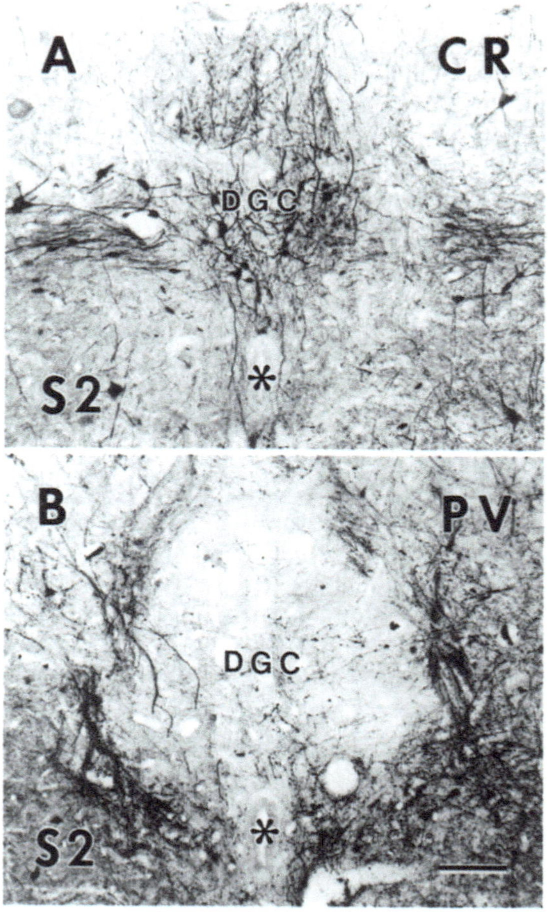

Fig. 5. Distinct complementarity of calretinin (CR) (**A**) and parvalbumin (*PV*) (**B**) like immunoreactivity in the dorsal gray commissure (*DGC*) of the S2 spinal cord. The central canal is indicated by asterisks. *Top* dorsal. Nickel enhanced DAB reaction. Scale bar (same for **A** and **B**) = 100 μm. (From Ren and Ruda, 1994)

Parvalbumin: The most striking pattern of parvalbumin-LI in the spinal cord is the appearance of a dense narrow parvalbumin-IR band that consists of parvalbumin-IR cell bodies and terminals in the superficial dorsal horn (Fig. 1D). The parvalbumin-IR band is about 25–60 μm thick and straddles the border between laminae II and III (Yamamoto *et al.*, 1989; Celio, 1990; Yoshida *et al.*, 1990). The morphology of parvalbumin-IR cells in this region resembles that of islet cells of lamina II (Celio, 1990; Yoshida *et al.*, 1990). Parvalbumin-LI on the border of lamina II/III is not uniformly distributed mediolaterally at the cervical and lumbar enlargement. The most medial portion of the parvalbumin-IR band is usually less intensely stained than the lateral part (Fig. 1D). In sharp contrast to CB-D28k, lamina I, the outer portion of lamina II, and the lateral spinal nucleus do not exhibit parvalbumin-LI (Fig. 1D). In the ventral horn, large motor neurons in lamina IX generally lack parvalbumin-LI (Fig. 1D) (Ince *et al.*, 1993) and parvalbumin mRNA expression (Elliott and Snider, 1995).

In the medial half of laminae V and VI (ventromedial dorsal horn) at the level of lumbar enlargement, strong parvalbumin-LI is noteworthy (Figs. 1D and 3D). Parvalbumin-LI is present as both labeled cell bodies and fibers (Figs. 1D and 3D). Parvalbumin-IR fibers are seen to enter from the dorsal columns and radiate towards the ventral gray matter (Figs. 1D and 3D). In the lateral reticular portion of lamina V, small rostrocaudal running bundles of parvalbumin-IR fibers are frequently found (Fig. 3D).

In the DGC, parvalbumin-IR fiber networks encircle the region and appear to define the borders of the DGC (Fig. 5B). Only a few varicose parvalbumin-IR fibers are seen in the center of the DGC. The pattern of parvalbumin-LI in the DGC is clearly complementary to that of calretinin (Fig. 5).

At the rostral cervical spinal cord, intensely labeled parvalbumin-IR cell bodies and fibers are densely packed in the internal basilar nucleus (IBN) and central cervical nucleus (CCN), making the borders of the two nuclei clearly distinguishable. The IBN and CCN also are lightly labeled by calmodulin and calretinin antibodies, but these two nuclei generally lack CB-D28k-LI. Similar relative levels of CBP immunoreactivity are exhibited in Clarke's column.

Calmodulin: Calmodulin-IR cell bodies are not found in lamina I (Figs. 1C, 2C). In contrast to CB-D28k and calretinin, lamina II is not immunoreactive for calmodulin, leaving a blank band in the superficial dorsal horn (Fig. 1C). In laminae III and IV, calmodulin-LI is found in some neurons (Fig. 2C). A few varicose calmodulin-IR fibers are also seen. There is strong calmodulin-LI in the ventromedial dorsal horn, particularly at the level of lumbar enlargement, (Figs. 1C and 3C). Calmodulin-LI in this region consists primarily of a labeled fiber network and punctate staining. Although all four CBPs demonstrate some immunoreactive staining in lamina IX, large motoneuronal cell bodies are stained primarily, if not exclusively, by the calmodulin antisera (Fig. 4) (Wood *et al.*, 1980; Seto-Ohshima *et al.*, 1985). Calmodulin-IR cell bodies are rarely seen in lamina X. Scattered small calmodulin labeled cell bodies are distributed in the sacral dorsal gray commisure (Ren and Ruda, 1994).

In the dorsal columns, calmodulin-LI is uniformly present in both fasciculi gracilis and cuneatus, from sacral through cervical spinal cords (Ren and Ruda, 1994). There appears to be weak immunostaining for calmodulin in the pyramidal tract of the rat spinal cord (Ren and Ruda, 1994).

Calpain: Compared to the above described CBPs, less information on the distribution of calpain in the spinal cord is available, although in humans, the highest calpain activity is found in the spinal cord (Bannay-Schwartz *et al.*, 1994). Due to its ubiquitous nature, calpain-IR staining is found in both neuronal perikarya and glial cells throughout the spinal gray and white matter (Siman *et al.*, 1985; Perlmutter *et al.*, 1990). Calpain-LI is localized throughout the somatic cytoplasm and extends into dendritic processes. Axons are also stained for calpain. The specialized glia of the spinal cord pial surface is highly immunoreactive to calpain antisera. Ultrastructural analysis indicates that microglial cells, but not myelin, are labeled with calpain antibodies (Perlmutter *et al.*, 1990). In contrast to CB-D28k, calretinin and parvalbumin, but similar to calmodulin, large motoneurons in lamina IX exhibit dense calpain-LI. Calpain does not appear to have a distinct localization in the different laminae of the spinal cord.

Colocalization of CBP's in spinal cord neurons

The CBPs Calretinin and CB-D28k have a 50–60% cDNA and amino acid sequence homology (Rogers, 1987; Parmentier, 1989; Winsky *et al.*, 1989; Parmentier and Lefort, 1991). It is likely some CB-D28k antibodies detected calretinin in early immunocytochemical studies (Pochet *et al.*, 1985, also see Parmentier, 1989). Fortunately calretinin antibodies do not appear to crossreact with CB-D28k (Rogers, 1991; Arai *et al.*, 1991), especially a CB-D28k monoclonal antibody (Celio *et al.*, 1990). It has been confirmed that for the most part calretinin and CB-D28k are localized to different sets of neurons in the spinal cord (Ren and Ruda, 1994), although they are colocalized to a subpopulation of neurons (Fig. 6). The rat appears to have more spinal neurons double-labeled by calretinin and CB-D28k antisera when compared with chick spinal cord, indicating a phylogenic difference (Rogers *et al.*, 1989).

A small number of parvalbumin-IR spinal neurons are reported to contain CB-D28k (Antal *et al.*, 1990, 1991). However, Yoshida *et al.* (1990) have observed no colocalization of CB-D28k and parvalbumin in neurons of the superficial dorsal horn.

CBPs have been colocalized with neuropeptides and other transmitters. Although neurotensin has a similar immunocytochemical staining pattern as parvalbumin in the superficial dorsal horn (Gibson *et al.*, 1981; also see Ruda *et al.*, 1986), they do not appear to colocalize in the spinal dorsal horn (Yoshida *et al.*, 1990). On the other hand, 13% of CB-D28k neurons in the superficial dorsal horn also stain for neurotensin (Yoshida *et al.*, 1990). Parvalbumin is colocalized with GABA in laminae II and III, but not with aspartate (Antal *et al.*, 1991). In contrast, CB-D28k is colocalized with aspartate, but not GABA (Antal *et al.*, 1991). Parvalbumin and GABA double-labeled spinal neurons also show glycine-LI but not neuropeptide Y-LI or reduced nicotinamide adenine dinucleotide phosphate (NADPH)-diaphorase activity (Laing *et al.*, 1994). About 10% of CB-D28k-IR neurons in laminae I–III colocalize with either enkephalin, substance P or somatostatin, but no parvalbumin-IR neurons in this region contained these peptides (Yoshida *et al.*, 1990).

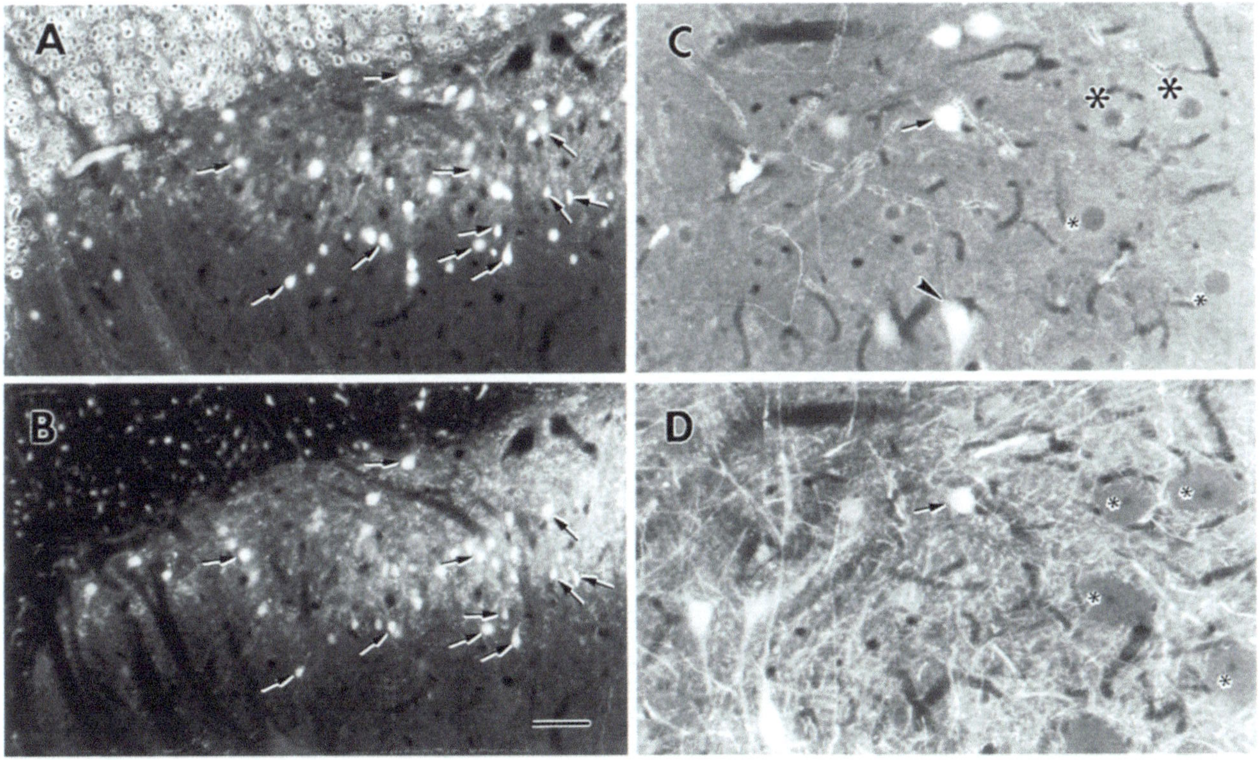

Fig. 6. Colocalization of calbindin-D28k (**A** and **C**) with calretinin (**B** and **D**) in the spinal cord of the rat demonstrated by double immunofluorescence labeling of the same tissue section. A and B: Medial superficial dorsal horn (laminae I and II). Dorsal is top and medial to the left. **C** and **D**: Lamina IX of the ventral horn. Arrows indicate neurons labeled for both calbindin-D28k- and calretinin-like immunoreactivity. Arrowheads in **A** and **B** indicate singly labeled neurons. An arrowhead in **C** indicates neurons with a densely labeled nucleus and unlabeled nucleolus. Small asterisks in **C** and **D** indicate unlabeled large neurons in lamina IX. Large asterisks in C overlie two large ventral horn neurons with light punctate staining in the cytoplasm for CB-D28k. Scale bar (same for **A–D**) = 40 μm

Projection neurons

After injection of Fast Blue dye into the rostral cervical spinal cord, approximately 32% of lamina I CB-D28k-IR neurons at the lumbar level, including large Waldeyer-type cells, are retrogradely labeled, suggesting these neurons project supraspinally (Yoshida *et al.*, 1990). The target areas of spinal CB-D28k projection neurons include the lateral reticular nucleus/A1 catecholaminergic group, parabrachial area, nucleus of the solitary tract and periaqueductal gray (Menétrey *et al.*, 1992b). Most CB-D28k projection neurons are located in the superficial laminae, especially lamina I, and the lateral spinal nucleus. A small number of CB-D28k projection neurons are found in the neck of the dorsal horn, the area around the central canal, and the sacral parasympathetic nucleus (Menétrey *et al.*, 1992b). In the trigeminal subnucleus caudalis some CB-D28k-IR neurons send their axons to the thalamus (Bennett-Clarke *et al.*, 1992), parabrachial area or the nucleus of the solitary tract (Menétrey *et al.*, 1992a). In lower vertebrates, such as the teleost fish, a spinocerebellar projection originates from CB-D28k-IR neurons in the spinal ventral horn (Szabo *et al.*, 1990). The location of calretinin-IR neurons in the spinal gray matter matches in part that of the cells of origin of several ascending nociceptive tracts, i.e., spinothalamic, spinoreticular and spinomesencephalic tracts (Willis, 1985). It remains to be determined whether some projection neurons in the spinal cord contain calretinin. Parvalbumin-IR neurons in the superficial dorsal horn do not appear to project to supraspinal sites (Yoshida *et al.*, 1990).

Effects of primary afferent deafferentation

Studies have been carried out to identify the primary afferent origin of CBP-LI in the spinal cord (Yamamoto *et al.*, 1989; Yoshida *et al.*, 1990; Ren *et al.*, 1993; Ren and Ruda, 1994). Following unilateral multiple ganglionectomies in rats, the common results for CB-D28k, calretinin, parvalbumin and calmodulin staining is a substantial loss of immunoreactivity in the dorsal column gracile fasciculus at related segments ipsilateral to the injury (Figs. 7 and 8). Calmodulin-LI is almost depleted in the ipsilateral gracile fasciculus (Fig. 7A) and only scattered punctate staining remains in the dorsal column for parvalbumin (Fig. 7B). In spinal gray matter, a great reduction of calmodulin- and parvalbumin-LI is observed at the medial portion of laminae V and VI (ventromedial dorsal horn) ipsilateral to the ganglionectomy (Fig. 7). Ipsilateral to the surgery, the parvalbumin-IR band at the inner part of lamina II, especially the medial part, is less intensely stained when compared to the contralateral side (Fig. 7B). Calmodulin-LI is also reduced in ipsilateral laminae III, IV and IX. Comparatively, changes in CB-D28K- and calretinin-LI in spinal gray matter are less apparent after ganglionectomy. Calretinin-LI is reduced slightly in Clarke's column rostral to the surgery levels and increased slightly in the medial third of ipsilateral lamina II at the surgical

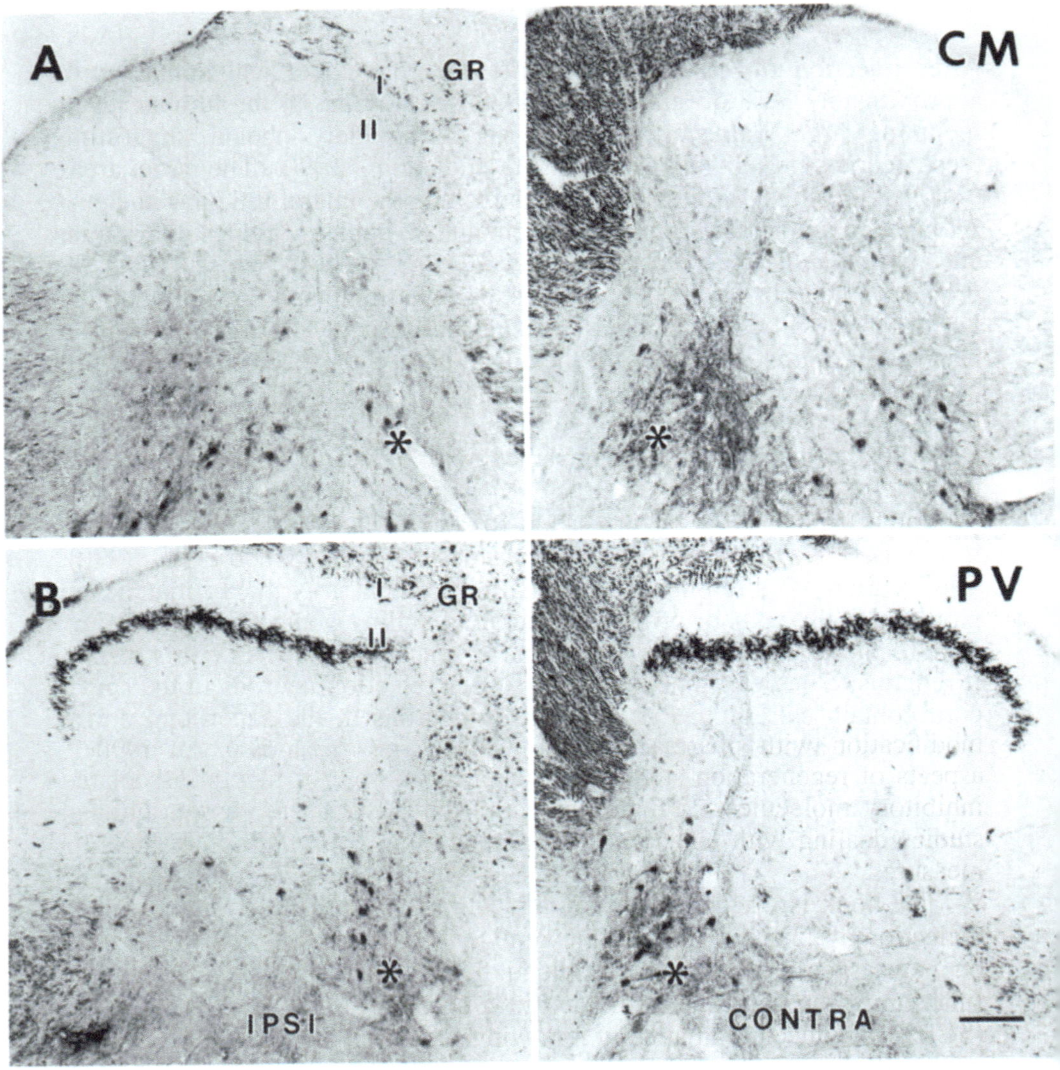

Fig. 7. The effects of multiple unilateral dorsal root ganglionectomies at the lumbar level (L2-5) on calmodulin (CM) (**A**) and parvalbumin (*PV*) (**B**) like immunoreactivity in the L5 segment of the spinal cord. CONTRA. contralateral; *GR* gracile fasciculus; *IPSI* ipsilateral. *I–II* spinal cord laminae I and II. Note large reduction of immunoreactivity in the GR and ventromedial (asterisks) dorsal horn IPSI to the ganglionectomy. The intensity of parvalbumin-IR band in the superficial dorsal horn was slightly reduced. Nickel enhanced DAB reaction. Scale bar (same for **A** and **B**) = 200 μm. (From Ren and Ruda, 1994)

level ipsilateral to the ganglionectomy (Ren *et al.*, 1993). The staining intensity and packing density of calretinin-IR cell bodies was slightly increased, especially in the medial end of lamina II (Fig. 8). Dorsal root rhizotomy also produces a significant reduction of CB-D28k- and

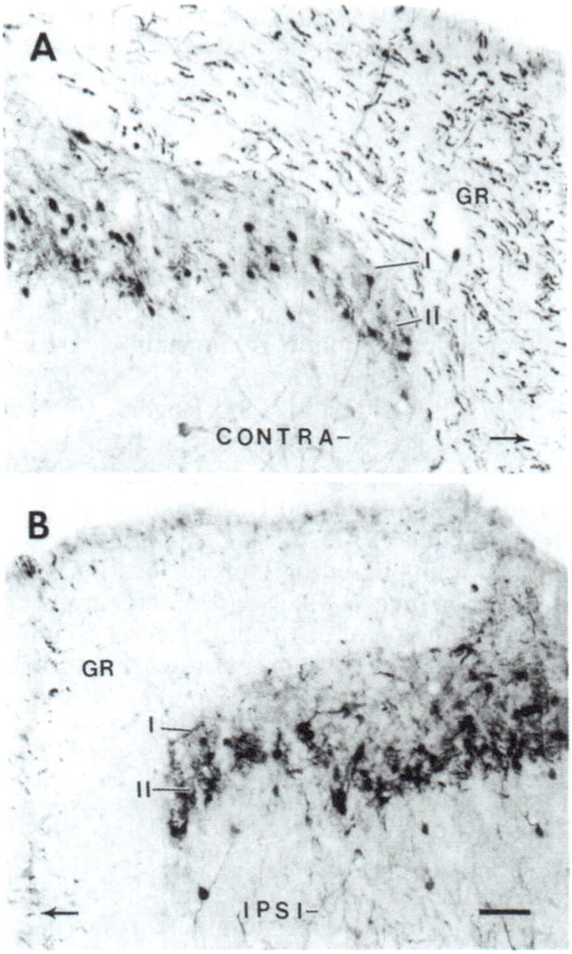

Fig. 8. The effects of multiple unilateral dorsal root ganglionectomies at the lumbar level (L2-5) on calretinin-like immunoreactivity (*CR-LI*) in the L5 segment of the spinal cord. **(A)** and **(B)** are views taken from the same section. Comparing ipsilateral (*IPSI-*) and contralateral (*CONTRA-*) sides, CR-IR fibers were largely reduced in the fasciculus gracilis (*GR*) on the ipsilateral side after ganglionectomy. However, CR-LI appeared to increase in lamina II ipsilateral to ganglionectomy. Arrows in **A** and **B** point to the medial direction. DAB reaction. Scale bar (same for **A** and **B**) = 50 μm. (From Ren *et al.*, 1993)

parvalbumin-LI in the dorsal columns ipsilateral to the injury, but has little or no effect on CB-D28k- and parvalbumin-LI in spinal gray matter (Yamamoto *et al.*, 1989; Yoshida *et al.*, 1990). Selective destruction of small primary afferent C-fibers by neonatal capsaicin treatment produces no detectable changes in staining intensity for CB-D28k and parvalbumin at any level of the rat spinal cord (Yamamoto *et al.*, 1989). These findings indicate that primary afferent terminals in the spinal superficial dorsal horn do not contain CBPs CB-D28k, calretinin, and calmodulin. Most CBPs in the dorsal column pathway are of primary afferent origin, while the superficial dorsal horn exhibits intrinsic CBP-LI. It needs to be determined whether descending afferents contribute to spinal CBP-LI.

Functional significance of spinal calcium-binding proteins

Superficial dorsal horn

Immunocytochemical localization studies have consistently found intense CB-D28k-, calretinin- and parvalbumin-LI in the superficial laminae of the dorsal horn, although parvalbumin-LI is mainly restricted to the inner portion of lamina II (Yamamoto *et al.*, 1989; Antal *et al.*, 1990; Celio, 1990; Yoshida *et al.*, 1990; Résibois and Rogers, 1992; Ren *et al.*, 1993; Ren and Ruda, 1994). The superficial dorsal horn is an important relay station for nociceptive input since most Aδ- and C- nociceptive primary afferents terminate in laminae I and II. In comparing the laminar location of CBPs, calretinin and CB-D28k are most likely to play a role in nociceptive neural circuits based on laminar location. Calretinin- and CB-D28k-IR neurons in lamina II likely are interneurons that may play an important role in modulation of nociceptive neural circuits. Outer lamina II stalked cells and islet cells have been shown to respond to nociceptive input, while inner lamina II islet cells respond exclusively to low threshold mechanoreceptive input (Bennett *et al.*, 1980). Laminae II/III parvalbumin-IR neurons, including those apparent islet cells, may use GABA and glycine as their transmitters (Antal *et al.*, 1991; Laing *et al.*, 1994). Some lamina I CB-D28k-IR neurons, especially the large cells, have been shown to project supraspinally (Yoshida *et al.*, 1990; Menétrey *et al.*, 1992b). CB-D28k-IR neurons in the superficial dorsal horn have been shown to receive synaptic contacts from axons that originate from the locus coeruleus, subcoeruleus and A5 cell group (Antal *et al.*, 1994). These brainstem noradrenergic cell groups are sources of descending control of spinal nociceptive transmission. Thus, CB-D28k-IR neurons in the superficial dorsal horn are one component of an ascending-descending loop that may be involved in nociceptive processing. The CB-D28k-, calretinin- and most parvalbumin-LI in the superficial dorsal horn do not appear to originate from primary afferents since following multiple ganglionectomies, or dorsal root rhizotomy, the dense IR fiber plexi remain essentially unchanged (Yamamoto *et al.*, 1989; Yoshida *et al.*, 1990; Ren *et al.*, 1993; Ren and Ruda, 1994).

Calmodulin-LI appears to be absent in the superficial laminae of the dorsal horn. This is unexpected in that a brain nitric oxide synthase (NOS), which has been found to be calmodulin-dependent (Bredt and Snyder, 1990), has been localized to that region of the dorsal horn (Bredt *et al.*, 1991). NADPH-diaphorase has been found largely co-localized with brain NOS (Bredt *et al.*, 1991). The NOS in neurons may be responsible for NADPH-diaphorase staining (Dawson *et al.*, 1991; Hope *et al.*, 1991).

Dense NADPH-diaphorase staining is found in the superficial dorsal horn (Valtschanoff *et al.*, 1992), resembling the staining pattern achieved with antisera specific for NOS (Bredt *et al.*, 1991). The apparent absence of calmodulin-LI in lamina II is puzzling as far as the calmodulin-dependency of NOS is concerned period. It is unlikely that an isoform of calmodulin may exist which is not detected by the antibody used. Also, although there is a family of calmodulin genes, they all encode a calmodulin protein with identical amino acid sequence in rats (Nojima *et al.*, 1987; SenGupta *et al.*, 1987). Winsky and Kuznicki (1996) report that antibody recognition of CBPs CB-D28k, calretinin and parvalbumin depends on their calcium-binding status, with better antibody-binding in a higher calcium environment. The antibody used for the calmodulin study (Ren and Ruda, 1994) identifies both Ca^{2+}-bound and Ca^{2+}-free rat calmodulin, although the antibody exhibits higher affinity for Ca^{2+}-free calmodulin (Sacks *et al.*, 1991). Thus, the chance of a false negative staining due to Ca^{2+}-binding status seems unlikely. One unknown factor is that calmodulin may have its antigenic sites blocked by interactions with other proteins and thus cannot be detected (see discussion in Caceres *et al.*, 1983). This possibility needs to be further examined. Interestingly CB-D28k, calretinin and parvalbumin are specifically distributed in the superficial dorsal horn, but it is not clear at this time if CBP's other than calmodulin could regulate NOS activity in the spinal cord. However, the apparent absence of calmodulin and presence of CB-D28k, calretinin and parvalbumin in the NOS containing superficial laminae suggest that they may play an alternative role in NOS dependency. Further studies are necessary to elucidate the role of CBPs in NOS activity in the superficial dorsal horn.

Dorsal gray commissure

The distribution pattern of calretinin- and CB-D28k-LI in the sacral DGC is quite similar to that of NADPH-diaphorase, and some peptides, especially dynorphin, enkephalin and somatostatin. NADPH-diaphorase positive cell bodies and fiber networks are densely stained in the DGC (Valtschanoff *et al.*, 1992). Dynorphin, enkephalin, somatostatin, cholecystokinin, avian pancreatic polypeptide, vasoactive intestinal polypeptide, substance P, and neurotensin, have been localized immunohistochemically to cell bodies and fibers of the sacral DGC (Kivipelto and Panula, 1991; Sasek *et al.*, 1984). Neurons in the sacral DGC may be important for processing of visceral afferent information, since these neurons receive input from visceral primary afferents (Ciriello and Calaresu, 1983) and send axons to reach the intermediolateral cell columns (Ren and Ruda, 1994). Neurons surrounding the central canal are also nociceptive and project supraspinally (Nahin *et al.*, 1983). Based on their location, calretinin and CB-D28k are likely involved in visceral regulation and/or nociceptive processing in the DGC. Interestingly, most parvalbumin-LI avoid the area of calretinin staining and exhibit an unique distribution compared to other known substances in the DGC (Gibson *et al.*, 1981; Sasek *et al.*, 1984). The unique labeling pattern of calretinin and parvalbumin suggests that they are differentially involved in functional activities of DGC neurons. Further studies are necessary to identify the significance of the complementary distribution of calretinin and parvallbumin in the DGC.

Ventral horn

Large motoneurons in the lateral and ventromedial spinal motor nuclei (lamina IX) exhibit immunoreactivity to calmodulin (Wood *et al.*, 1980; Seto-Ohsima *et al.*, 1985) and calpain (Siman *et al.*, 1985; Perlmutter *et al.*, 1990). The distinct immunostaining of spinal motoneurons by calmodulin antisera forms a striking contrast to the apparent lack of calmodulin-LI in the superficial dorsal horn and little immunoreactivity in laminae III and IV (Ren and Ruda, 1994). Comparatively, these studies suggest a prominent role for calmodulin and calpain in spinal motoneuronal activity. CB-D28k, calretinin and parvalbumin, on the other hand, may not be directly involved in spinal motoneuronal activity. The abundant CB-D28k-, calretinin- and parvalbumin-LI in laminae VII and VIII suggests they may play a role in coordinating ventral horn motor-related neural circuits, such as the Renshaw cells that mediate recurrent inhibition of motoneurons (Celio, 1990; Arvidsson *et al.*, 1992).

Dorsolateral funiculus and related nuclei

Two functionally and anatomically distinct nuclei, the LSN and LCN, are found in the dorsolateral funiculus (DLF) of the rat (Giesler and Elde, 1985), as well as ascending (spinocervical, spinocerebellar and spinomesencephalic) and descending (from raphe and dorsal column nuclei) axons (Tracey, 1985). In general, CB-D28k/calretinin and CM/parvalbumin exhibit two different patterns of labeling in the DLF (Ren and Ruda, 1994) which is most conspicuous at the rostral cervical level. At the C1 spinal cord, the dorsal portion of the DLF is predominantly stained by CB-D28k and calretinin antibodies. Calmodulin-LI is most prominent in the dorsolateral part of the DLF, and parvalbumin mainly labeled an area ventral to the LSN and LCN.

Each pathway in the DLF may use different CBPs for its distinct functional activity. For example, descending serotoninergic fibers travel within the dorsal portion of the DLF (Bullitt and Light, 1989). Comparing the DLF distribution of the CBPs, fibers labeled by calretinin, CB-D28k and probably calmodulin, but not parvalbumin, are candidates for descending modulation of pain sensation mediated by serotoninergic pathways. Parvalbumin may be closely related to the dorsal spinocerebellar tract which ascends in the DLF and originates primarily from CCN and Clarke's column (Matsushita and Hosoya, 1979; Tracey, 1985). Dense staining of cell bodies and fibers by parvalbumin in the CCN and Clarke's column has been described (Ren and Ruda, 1994).

Dorsal columns

CB-D28k, calretinin, parvalbumin and calmodulin IR labeling in the dorsal columns of the rat have been systematically compared (Antal *et al.*, 1990; Ren *et al.*, 1993; Ren and Ruda, 1994). Calretinin and parvalbumin antibodies similarly labeled the gracile fasciculus at the lumbar level of the spinal cord, and labeled the cuneate, but not the gracile fasciculus at the cervical level of the spinal cord. CB-D28k-LI is marginally detectable in fasciculus gracilis. The fasciculus cuneatus is devoid of CB-D28k-LI. Calmodulin labeled the gracile and cuneate fasciculi throughout all spinal cord segments.

The differential distribution of CBPs in the dorsal column may be related to the anatomical and functional diversity of dorsal column pathways. The dorsal column fasciculi consist of several fiber tracts of different origin (see Tracey, 1985, for a review). These include the ascending collaterals of primary afferents (direct dorsal column pathway), the axons of post synaptic dorsal column neurons (second order dorsal column pathway), and the descending axons from the dorsal column nuclei. Many of the fibers that enter the dorsal columns do not reach the dorsal column nuclei. A portion of ascending fibers in the fasciculus gracilis has been shown to terminate on neurons in Clarke's column (Ganchrow and Bernstein, 1981). The difference in calretinin- and parvalbumin-LI in the fasciculus gracilis between the rostral and caudal spinal cord segments suggests that most calretinin and parvalbumin-IR fibers in the fasciculus gracilis at the lumbosacral levels terminate below the level of the nucleus gracilis. Dense calretinin-fiber networks are localized to Clarke's column, and a reduction in calretinin-LI is found in Clarke's column rostral to the level of a ganglionectomy (Ren *et al.*,1993). Fibers in the gracile fasciculus show light CB-D28k-IR labeling throughout the rostrocaudal extent of the spinal cord (Ren and Ruda, 1994). CB-D28k also showed the least immunoreactivity in Clarke's column when compared to calmodulin, calretinin and parvalbumin. This comparison suggests that a subpopulation of DRG neurons that contain CB-D28k may project directly to the dorsal column nuclei. Thus, the distribution of CBPs will provide additional markers for the study of the dorsal column pathways.

The ascending projections in the dorsal columns to the dorsal column gracile and cuneate nuclei include primarily axons of large DRG neurons (Giuffrida and Rustioni, 1992). The selective labeling of dorsal column pathways suggests that CBPs are involved in proprioception. Most CBP-LI in the dorsal column gracile fasciculus is of primary afferent origin, since the CBP-LI in the gracile fasciculus is greatly reduced following primary afferent deafferentation (Yamamoto *et al.*, 1989; Ren *et al.*, 1993; Ren and Ruda, 1994). The dorsal column calretinin- and parvalbumin-IR fibers appear to be from medium to large dorsal root ganglion neurons (Carr *et al.*, 1989a; Ren *et al.*, 1993; Ichikawa *et al.*, 1994), as the calretinin- and parvalbumin-LI in the dorsal columns are not sensitive to neonatal capsaicin treatment (Yamamoto *et al.*, 1989). The involvement of CBPs in proprioception is further supported by the fact that the majority of CB-D28k-, calretinini- and parvalbumin-IR DRG neurons are also positive for carbonic anhydrase (Carr *et al.*, 1989b; Ichikawa *et al.*, 1993, 1994), an enzyme that has been used as a specific marker for DRG proprioceptive neurons (Szabolcs *et al.*, 1989; Ichikawa *et al.*, 1994).

Subcellular staining

Neuronal cell nuclei are immunoreactive to many CBPs in the spinal cord. The nuclear staining is typically more intense than that of the cytoplasm (Fig. 6C). The intranuclear immunohistochemical localization of CBPs appears to be exclusive of the nucleolus (Fig. 6C). Abundant CBPs in the cell nucleus suggest that Ca^{2+} concentrations in the cytoplasm and nucleus may be differentially or independently regulated. In fact, the intranuclear Ca^{2+} concentration has been shown to rise faster than that of the cytoplasm (Przywara *et al.*, 1991). CBPs may participate in many important functional activities in the nucleus. For example, calmodulin has been shown to be involved in nuclear functions such as DNA replication, gene transcription and DNA repair (Bachs *et al.*, 1992, review).

Biochemical considerations

Calcium-binding proteins may be divided functionally into buffer proteins that modulate intracellular Ca^{2+} levels and trigger proteins that induce or regulate a physiological response (Dalgarno *et al.*, 1984). CB-D28k, parvalbumin and calretinin are considered to be calcium-buffer proteins while calmodulin and calpain belong to the class of trigger CBPs. The biochemical activity of trigger CBPs such as calmodulin, calpain, troponin C, and myosin light chains, has been studied in detail. Whether calcium buffer proteins such as CB-D28k, calretinin and parvalbumin interact with other proteins upon binding Ca^{2+} is not known (Rogers, 1991). Calretinin appears to inhibit a calcium-stimulated protein phosphorylation (Yamaguchi *et al.*, 1991).

Pathological implications

The role of Ca^{2+} in cytotoxicity has become a major focus of study in recent years. After exposure to a neurotoxin, or excessive and continuous excitation, extracellular Ca^{2+} can enter the cell via receptor-operated or voltage-operated Ca^{2+} channels. Intracellular Ca^{2+} can also be mobilized through second messenger pathways such as inositol 1,4,5-trisphosphate and diacylglycerol. An elevation of cytosolic Ca^{2+} can eventually lead to irreversible cell injury and result in a variety of neurological disorders (see Nicotera *et al.*, 1992, for a review). Some CBPs such as calmodulin and calpain are targets of increased cytosolic Ca^{2+}. Other CBPs such as CB-D28k and parvalbumin may buffer Ca^{2+} to maintain a physiological concentration.

Although possibly not always related, alterations in CBPs have been found in neurodegenerative disorders such as Alzheimer's disease, Parkinson's disease and epilepsy (Heizmann and Braun, 1992; Andressen *et al.*, 1993; reviews). The presence of CBPs may protect neurons from chemical- or ischemia-induced neurotoxicity (Nitsch *et al.*, 1989; Mattson *et al.*, 1991; Iacopino *et al.*, 1992; Pike and Cotman, 1995). In the rat hippocampus, neurons containing either CB-D28K or parvalbumin are insensitive to seizure activity (Sloviter, 1989), whereas in genetically epilepsy-prone rats lower levels of CB-D28k and calretinin mRNA expression have been found in the brain (Montpied *et al.*, 1995). Calretinin-IR cortical neurons are resistant to neurodegenerative processes (Hof *et al.*, 1993). CBP activity may also be related to the development of morphine tolerance since chronic morphine treatment results in an increase in CB-D28k-LI in rat striatum (Garcia and Harlan, 1993).

Calcium-binding proteins and chronic pain

Peripheral nerve injury may result in persistent pain in humans as predicted in experimental animals (Wall *et al.*, 1979; Wiesenfeld and Lindblom, 1980; Levitt, 1985, review). Rausell *et al.* (1992) report that in normal monkeys parvalbumin selectively labels the non-nociceptive somatosensory pathways, while CB-D28k labels pathways related to nociceptive transmission. After long-term (11 years) peripheral deafferentation in monkeys, there is an increase in CB-D28k-LI and a decrease in parvalbumin-LI in the deafferented zone of the thalamus (Rausell *et al.*, 1992). Persistent pain is also apparent in these monkeys. It appears that an imbalance in somatosensory pathways is produced by the deafferentation and some of these pathways can be selectively marked by CB-D28k and parvalbumin. The alterations in sensory inputs may lead to release of inhibition of nociceptive neural circuits and development of persistent pain. In an animal model of neuropathic pain (Bennett and Xie, 1988), there was a significant reduction of parvalbumin-LI in the superficial dorsal horn, resulting in the disappearance of the most medial portion of the characteristic parvalbumin-IR band (Ren and Ruda, unpublished observations).

Calpain and spinal injury

Calpain activity appears to play a role in a variety of pathological situations including stroke, brain trauma, Alzheimer's disease, and spinal cord injury (Wang and Yuen, 1994). One underlying mechanism of calpain action may involve glutamate-induced excitotoxicity. Due to neuronal activity related to cerebral ischaemia or enhanced primary afferent barrage, excessive glutamate is released and results in excess activation of postsynaptic excitatory amino acid receptors and Ca^{2+} channel opening. A surge of Ca^{2+} influx will raise intracellular Ca^{2+} levels leading to persistent activity of several Ca^{2+}-dependent enzymes including calpain, calmodulin, protein kinase C and phospholipase A_2. Although the normal selective and limited proteolytic activity functions to regulate certain enzymes, to induce cytoskeletal rearrangements, and to involve intracellular signaling and structural stabilization (Nixon *et al.*, 1994, review), the abnormally high levels of calpain-related proteolytic activity may induce irreversible cell damage.

Following spinal cord injury calpain has been associated with the resultant neuronal pathology. In the spinal cord, calpain mediates degradation of myelin proteins (Banik *et al.*, 1985; Berlet, 1987). In a model of mechanical spinal cord injury, there is a progressive increase of neutral proteinase activity as indicated by an increase in degradation of myelin basic protein, a substrate of calpain (Banik *et al.*, 1986). The application of a protease inhibitor significantly decreased axonal degeneration due to spinal cord injury (Iizuka *et al.*, 1986), prevented axotomy-induced axonal degeneration (George *et al.*, 1995), and reduced capsaicin-induced cell death (Chard *et al.*, 1995).

Parvalbumin and motoneuron disease

In human motoneuron diseases such as amyotrophic lateral sclerosis (ALS), there is a selective degeneration of spinal motoneurons. Recent studies find that the ALS-sensitive motoneurons (e.g. lumbar ventral horn) are usually devoid of parvalbumin-LI (Ince *et al.*, 1993) and mRNA expression (Elliott and Snider, 1995). In contrast, the ALS-resistant motor pools (e.g. bulbar oculomotor neurons) exhibit high levels of parvalbumin mRNA and immunoreactivity (Ince *et al.*, 1993; Elliott and Snider, 1995). In ALS disease, the parvalbumin-containing spinothalamic tract axons are relatively well preserved while the adjacent anterolateral funiculus shows myelin pallor (Okamoto *et al.*, 1995). These results suggest that the lack of Ca^{2+} buffering may contribute to selective vulnerability of motoneurons in ALS.

References

Antal M, Freund TF, Polgár E (1990) Calcium-binding proteins, parvalbumin- and calbindin-D 28K-immunoreactive neurons in the rat spinal cord and dorsal root ganglia: a light and electron microscopic study. J Comp Neurol 295: 467–484

Antal M, Polgár E (1993) Development of calbindin-D28k immunoreactive neurons in the embryonic chick lumbosacral spinal cord. Eur J Neurosci 5: 782–794

Antal M, Polgár E, Berki Á Cs (1994) The innervation of the dorsal and ventral horns of the rat spinal cord by axons descending from the locus coeruleus/subcoeruleus complex and A5 cell group. Soc Neurosci Abstr 20: 1584

Antal M, Polgár E, Chalmers J, Minson JB, Llewellyn-Smith I, Heizmann CW, Somogyi P (1991) Different populations of parvalbumin- and calbindin-D28k-immunoreactive neurons contain GABA and accumulate ^{3}H-D-aspartate in the dorsal horn of the rat spinal cord. J Comp Neurol 314: 114–124
Andressen C, Blümcke I, Celio MR (1993) Calcium-binding proteins: selective markers of nerve cells. Cell Tiss Res 271: 181–208
Arai R, Winsky L, Arai M, Jacobowitz DM (1991) Immunohistochemical localization of calretinin in the rat hindbrain. J Comp Neurol 310: 21–44
Arvidsson U, Ulfhake B, Cullheim S, Ramírez V, Shupliakov O, Hökfelt T (1992) Distribution of calbindin D28k-like immunoreactivity (LI) in the monkey ventral horn: do Renshaw cells contain calbindin D28k-LI? J Neurosci 12: 718–728
Bachs O, Agell N, Carafoli E (1992) Calcium and calmodulin function in the cell nucleus. Biochim Biophys Acta 1113: 259–270
Baimbridge KG, Celio MR, Rogers JH (1992) Calcium-binding proteins in the nervous system. Trends Neurosci 15: 303–308
Banay-Schwartz M, DeGuzman T, Palkovits M, Lajtha A (1994) Calpain activity in adult and aged human brain regions. Neurochem Res 19: 563–567
Banik NL, Hogan EL, Powers JM, Smith KP (1986) Proteolytic enzymes in experimental spinal cord injury. J Neurol Sci 73: 245–256
Banik NL, McAlhaney WW, Hogan EL (1985) Calcium-stimulated proteolysis in myelin: evidence for a Ca^{2+}-activated neutral proteinase associated with purified myelin of rat CNS. J Neurochem 45: 581–588
Bennett GJ, Abdelmoumene M, Hayashi H, Dubner R (1980) Physiology and morphology of substantia gelatinosa neurons intracellularly stained with horseradish peroxidase. J Comp Neurol 194: 809–827
Bennett GJ, Xie YK (1988) A peripheral mononeuropathy in rat that produces disorders of pain sensation like those seen in man. Pain 33: 87–107
Bennett-Clarke CA, Chiaia NL, Jacquin MF, Rhoades RW (1992) Parvalbumin and calbindin immunocytochemistry reveal functionally distinct cell groups and vibrissa-related patterns in the trigeminal brainstem complex of the adult rat. J Comp Neurol 320: 323–338
Berlet HH (1987) Calcium-dependent neutral protease activity of myelin from bovine spinal cord: evidence for soluble cleavage products of myelin proteins. Neurosci Lett 73: 266–270
Braun K (1990) Calcium-binding proteins in avian and mammalian central nervous system: localization, development and possible functions. Prog Histochem Cytochem 21: 1–64
Bredt DS, Glatt CE, Hwang PM, Fotuhi M, Dawson TM, Snyder SH (1991) Nitric oxide synthase protein and mRNA are discretely localized in neuronal populations of the mammalian CNS together with NADPH diaphorase. Neuron 7: 615–624
Bredt DS, Snyder SH (1990) Isolation of nitric oxide synthetase, a calmodulin-requiring enzyme. Proc Natl Acad Sci USA 87: 682–685
Bullitt E, Light AR (1989) Intraspinal course of descending serotoninergic pathways innervating the rodent dorsal horn and lamina X. J Comp Neurol 286: 231–242
Caceres A, Bender P, Snavely L, Rebhun LI, Steward O (1983) Distribution and subcellular localization of calmodulin in adult and developing brain tissue. Neuroscience 10: 449–461
Carr PA, Yamamoto T, Karmy G, Baimbridge KG, Nagy JI (1989a) Parvalbumin is highly colocalized with calbindin D28k and rarely with calcitonin gene-related peptide in dorsal root ganglia neurons of rat. Brain Res 497: 163–170
Carr PA, Yamamoto T, Karmy G, Baimbridge KG, Nagy JI (1989b) Analysis of parvalbumin and calbindin D28k-immunoreactive neurons in dorsal root ganglia of rat in relation to their cytochrome oxidase and carbonic anhydrase content. Neuroscience 33: 363–371
Celio MR (1990) Calbindin D-28k and parvalbumin in the rat nervous system. Neuroscience 35: 375–475
Celio MR, Heizmann CW (1981) Calcium-binding protein parvalbumin as a neuronal marker. Nature 293: 300–302

Celio MR, Baier W, Schärer L, Gregersen HJ, de Viragh PA, Norman AW (1990) Monoclonal antibodies directed against the calcium binding protein calbindin D-28k. Cell Calcium 11: 599–602

Cervero F, Iggo A (1980) The substantia gelatinosa of the spinal cord: A critical review. Brain 103: 717–772

Chard PS, Bleakman D, Savidge JR, Miller RJ (1995) Capsaicin-induced neurotoxicity in cultured dorsal root ganglion neurons: involvement of calcium-activated proteases. Neuroscience 65: 1099–1108

Cheung WY (1980) Calmodulin plays a pivotal role in cellular regulation. Science 207: 19–27

Ciriello J, Calaresu FR (1983) Central projections of afferent renal fibers in the rat: An anterograde transport study of horseradish peroxidase. J Auton Nerv Syst 8: 273–285

Conti F, De Biasi S, Giuffrida R, Rustioni A (1990) Substance P-containing projections in the dorsal columns of rats and cats. Neuroscience 34: 607–621

Dalgarno D, Klevit RE, Levine BA, Willians RJP (1984) The calcium receptor and trigger. Trends Neurosci 4: 266–271

Dawson TM, Bredt DS, Fotuhi M, Hwang PM, Snyder SH (1991) Nitric oxide synthase and neuronal NADPH diaphorase are identical in brain and peripheral tissues. Proc Natl Acad Sci USA 88: 7797–7801

Dénizot JP, Brattoy BO, Brehier A, Thomasset M (1988) Immunohistochemical demonstration of Calbindin-D-28k (CaBP28k) in the spinal cord motoneurons of teleost fishes. Cell Tiss Res 254: 629–634

Dubner R (1991) Neuronal plasticity and pain following peripheral tissue inflammation or nerve injury. In: Bond MR, Charlton JE, Woolf CJ (eds) Proceedings of the VIth World Congress on pain. Elsevier, Amsterdam, pp 263–276

Elliott JL, Snider WD (1995) Parvalbumin is a marker of ALS-resistant motor neurons. Neuroreport 6: 449–452

Ganchrow D, Bernstein JJ (1981) Projections of caudal fasciculus gracilis to nucleus gracilis and other medullary structures, and Clarke's nucleus in the rat. Brain Res 205: 383–390.

Garcia MM, Harlan RE (1993) Chronic morphine increases calbindin D_{28k} in rat striatum: possible NMDA receptor involvement. Neuroreport 5: 65–68

George EB, Glass JD, Griffin JW (1995) Axotomy-induced axonal degeneration is mediated by calcium influx through ion-specific channels. J Neurosci 15: 6445–6552

Gibson SJ, Polak JM, Bloom SR, Wall PD (1981) The distribution of nine peptides in rat spinal cord with special emphasis on the substantia gelatinosa and on the area around the central canal (Lamina X). J Comp Neurol 210: 65–79

Giesler Jr GJ, Elde RP (1985) Immunocytochemical studies of the peptidergic content of fibers and terminals within the lateral spinal and lateral cervical nuclei. J Neurosci 5: 1833–1841

Giuffrida R, Rustioni A (1992) Dorsal root ganglion neurons projecting to the dorsal column nuclei of rats. J Comp Neurol 316: 206–220

Gnegy ME (1993) Calmodulin in neurotransmitter and hormone action. Annu Rev Pharmacol Toxicol 32: 45–70

Gnegy ME (1995) Calmodulin: Effects of cell stimuli and drugs on cellular activation. Prog Drug Res 45: 33–65

Gobel S (1978) Golgi studies of the neurons in layer I of the dorsal horn of the medulla (trigeminal nucleus caudalis). J Comp Neurol 180: 395–414

Guroff G (1964) A neutral calcium-activated protease from the soluble fraction of rat brain. J Biol Chem 239: 149–155

Heizmann CW, Braun K (1992) Changes in Ca^{2+}-binding proteins in human neurodegenerative disorders. Trends Neurosci 15: 259–264

Hendry SHC, Jones EG, Emson PC, Lawson DEM, Heizmann CW, Streit P (1989) Two classes of cortical GABA neurons defined by differential calcium binding protein immunoreactivities. Exp Brain Res 76: 467–472

Heppelmann B, Emson PC (1993) Distribution of calretinin mRNA in rat spinal cord and dorsal root ganglia: a study using non-radioactiv in situ hybridization histochemistry. Brain Res 624: 312–316

Hof PR, Nimchinsky EA, Celio MR, Bouras C, Morrison JH (1993) Calretinin-immunoreactive neocortical interneurons are unaffected in Alzheimer's disease. Neurosc Lett 152: 145–149

Hope BT, Michael GJ, Knigge KM, Vincent SR (1991) Neuronal NADPH diaphorase is a nitric oxide synthase. Proc Natl Acad Sci USA 88: 2811–2814

Ince P, Stout N, Shaw P, Slade J, Hunziker W, Heizmann CW, Baimbridge KG (1993) Parvalbumin and calbindin D-28k in the human motor system and in motor neuron disease. Neuropathol Appl Neurobiol 19: 291–299

Iacopino A, Christakos S, German D, Sonsalla PK, Altar CA (1992) Calbindin-D_{28K}-containing neurons in animal models of neurodegeneration: possible protection from excitotoxicity. Molec Brain Res 13: 251–264

Ichikawa H, Jacobowitz DM, Sugimoto T (1993) Calretinin-immunoreactive neurons in the trigeminal and dorsal root ganglia of the rat. Brain Res 617: 96–102

Ichikawa H, Deguchi T, Nakago T, Jacobowitz DM, Sugimoto T (1994) Parvalbumin, calretinin and carbonic anhydrase in the trigeminal and spinal primary neurons of the rat. Brain Res 655: 241–245

Iizuka H, Iwasaki Y, Yamamoto T, Kadoya S (1986) Morphometric assessment of drug effects in experimental spinal cord injury. J Neurosurg 65: 92–98

Jande SS, Maler L, Lawson DEM (1981) Immunohistochemical mapping of vitamin D-dependent calcium-binding protein in brain. Nature 294: 765–767

Jones EG, Hendry SHC (1989) Differential calcium binding protein immunoreactivity distinguishes classes of relay neurons in monkey thalamic nuclei. Eur J Neurosci 1: 222–246

Kivipelto L, Panula P (1991) Origin and distribution of neuropeptide-FF-like immunoreactivity in the spinal cord of rats. J Comp Neurol 307: 107–119

Laing I, Todd AJ, Heizmann CW, Schmidt HH (1994) Subpopulations of GABAergic neurons in laminae I-III of rat spinal dorsal horn defined by coexistence with classical transmitters, peptides, nitric oxide synthase or parvalbumin. Neuroscience 61: 123–132

Lehky P, Blum HE, Stein EA, Fischer EH (1974) Isolation and characterization of parvalbumin from skeletal muscle of higher vertebrates. J Biol Chem 249: 4332–4334

Levitt M (1985) Dysesthesias and self-mutilation in humans and subhumans: a review of clinical and experimental studies. Brain Res Rev 10: 247–290

Lin CT, Dedman JR, Brinkley BR, Means AR (1980) Localization of calmodulin in rat cerebellum by immunoelectron microscopy. J Cell Biol 85: 473–480

Lunam CA (1989) Calbindin immunoreactivity in the neurons of the spinal cord and dorsal root ganglion of the domestic fowl. Cell Tissue Res 257: 149–153

Matsushita M, Hosoya Y (1979) Cells of origin of the spinocerebellar tract in the rat, studied with the method of retrograde transport of horseradish peroxidase. Brain Res 173: 185–200

Mattson MP, Rychlik B, Chu C, Christakos S (1991) Evidence for calcium-reducing and excitoprotective roles for the calcium-binding protein calbindin-D_{28K} in cultured hippocampal neurons. Neuron 6: 41–51

Means AR, VanBerkum MFA, Bagchi I, Lu KP, Rasmussen CD (1991) Regulatory functions of calmodulin. Pharmac Ther 50: 255–270

Means AR, Tash JS, Chafouleas JG (1982) Physiological implications of the presence, distribution, and regulation of calmodulin in eukaryotic cells. Physiol Rev 62: 1–39

Menétrey D, De Pommery J, Baimbridge KG, Thomasset M (1992a) Calbindin-D28K (CaBP28k)-like immunoreactivity in ascending projections. I. Trigeminal nucleus caudalis and dorsal vagal complex projections. Eur J Neurosci 4: 61–69

Menétrey D, De Pommery J, Thomasset M, Baimbridge KG (1992b) Calbindin-D28K (CaBP28k)-like immunoreactivity in ascending projections II. spinal projections to brain stem and mesencephalic areas. Eur J Neurosci 4: 70–76

Mize RR, Luo Q, Butler G, Jeon C-J, Nabors B (1992) The calcium binding proteins parvalbumin and calbindin-D 28K form complementary patterns in the cat superior colliculus. J Comp Neurol 320: 243–256

Montpied P, Winsky L, Dailey JW, Jobe PC, Jacobowitz DM (1995) Alteration in levels of expression of brain calbindin D-28k and calretinin mRNA in genetically epilepsy-prone rats. Epilepsia 36: 911–921

Nahin RL, Madsen AM, Giesler Jr GJ (1983) Anatomical and physiological studies of the gray matter surrounding the spinal cord central canal. J Comp Neurol 220: 321–335

Nicotera P, Bellomo G, Orrenius S (1992) Calcium-mediated mechanisms in chemically induced cell death. Annu Rev Pharmacol Toxicol 32: 449–470

Nitsch C, Scotti A, Sommacal A, Kalt G, (1989) GABAergic hippocampal neurons resistant to ischemia-induced neuronal death contain the Ca^{2+}-binding protein parvalbumin. Neurosci Lett 105: 263–268

Nixon RA, Saito K-I, Grynspan F, Griffin WR, Katayama S, Honda T, Mohan PS, Shea TB, Beermann M (1994) Calcium-activated neutral proteinase (calpain) system in aging and Alzheimer's disease. Ann NY Acad Sci 747: 77–91

Nojima H, Kishi K, Sokabe H (1987) Multiple calmodulin mRNA species are derived from two distinct genes. Molec Cell Biol 7: 1873–1880

Okamoto K, Hirai S, Nakagawa T, Sakurai A, Morita M, Yanagisawa T (1995) Parvalbumin-positive small myelinated fiber bundles in anterior marginal areas in the human cervical cord. Neurosci Lett 185: 79–82

Parmentier M (1989) Structure of the human cDNAs and genes coding for calbindin D28K and calretinin. In: Pochet R, Lawson DEM, Heizmann CW (eds) Calcium Binding Proteins in Normal and Transformed Cells. Plenum Press, New York, pp 27–34

Parmentier M, Lefort A (1991) Structure of the human brain calcium-binding protein calretinin and its expression in bacteria. Eur J Biochem 196: 79–85

Patterson JT, Chung K, Coggeshall RE (1992) Further evidence for the existence of long ascending unmyelinated primary afferent fibers within the dorsal funiculus: effects of capsaicin. Pain 49: 117–120

Perlmutter LS, Gall C, Baudry M, Lynch G (1990) Distribution of calcium-activated protease calpain in the rat brain, J Comp Neurol 296: 269–276

Persechini A, Moncrief ND, Kretsinger RH (1989) The EF-hand family of calcium-modulated proteins. Trends Neurosci 12: 462–467

Pike CJ, Cotman CW (1995) Calretinin-immunoreactive neurons are resistant to β-amyloid toxicity in vitro. Brain Res 671: 293–298

Pochet R, Parmentier M, Lawson DEM, Pasteels JL (1985) Rat brain synthesizes two "vitamin D-dependent" calcium-binding proteins. Brain Res 345: 251–256

Pritz MB, Stritzel ME (1991) Calcium binding protein immunoreactivity in a reptilian thalamic reticular nucleus. Brain Res 554: 325–328

Przywara DA, Bhave SV, Bhave A, Wakade TD, Wakede AR (1991) Stimulated rise in neuronal calcium is faster and greater in the nucleus than the cytosol. FASEB J 5: 217–222

Rausell E, Cusick CG, Taub E, Jones EG (1992) Chronic deafferentation in monkeys differentially affects nociceptive and nonnociceptive pathways distinguished by specific calcium-binding proteins and down-regulates gamma-aminobutyric acid type. A receptors at thalamic levels. Proc Natl Acad Sci USA 89: 2571–2575

Rausell E, Jones EG (1991) Chemically distinct compartments of the thalamic VPM nucleus in monkeys relay principal and spinal trigeminal pathways to different layers of the somatosensory cortex. J Neurosci 11: 226–237

Ren K, Ruda MA (1994) A comparative study of the calcium-binding proteins calbindin-D28K, calretinin, calmodulin and parvalbumin in the rat spinal cord. Brain Res Rev 19: 163–179

Ren K, Ruda MA, Jacobowitz DM (1993) Immunohistochemical localization of calretinin in the dorsal root ganglion and spinal cord of the rat. Brain Res Bull 31: 13–22

Résibois A, Blachier F, Rogers JH, Lawson DEM, Pochet R (1989) Comparison between rat brain calbindin- and calretinin-immuno-reactivities. In: Pochet R, Lawson DEM, Heiz-

mann CW (eds) Calcium Binding Proteins in Normal and Transformed Cells. Plenum Press, New York, pp 211–214
Résibois A, Rogers JH (1992) Calretinin in rat brain: an immunohistochemical study. Neuroscience 46: 101–134
Rogers JH (1987) Calretinin: a gene for a novel calcium-binding protein expressed principally in neurons. J Cell Biol 105: 1343–1353
Rogers JH (1991) Calretinin. In: Heizmann CW (ed) Novel Calcium-Binding Proteins. Springer, Berlin Heidelberg New York Tokyo, pp 251–276
Rogers JH, Khan M, Ellis J (1989) Calretinin and other CaBPs in the nervous system. In: Pochet R, Lawson DEM, Heizmann CW (eds) Calcium Binding Proteins in Normal and Transformed Cells. Plenum Press, New York, pp 195–203
Rogers JH, Résibois A (1992) Calretinin and calbindin-D_{28k} in rat brain: patterns of partial co-localization. Neuroscience 51: 843–865
Ruda MA, Bennett GJ, Dubner R (1986) Neurochemistry and neural circuitry in the dorsal horn. Prog Brain Res 66: 219–268
Sacks DB, Porter SE, Ladenson JH, McDonald JM (1991) Monoclonal antibody to calmodulin: Development, characterization, and comparison with polyclonal anti-calmodulin antibodies. Anal Biochem 194: 369–377
Sasek CA, Seybold VS, Elde RP (1984) The immunohistochemical localization of nine peptides in the sacral parasympathetic nucleus and the dorsal gray commissure in rat spinal cord. Neuroscience 12: 855–873
SenGupta B, Friedberg F, Detera-Wadleigh SD (1987) Molecular analysis of human and rat calmodulin complementary DNA clones. J Biol Chem 262: 16663–16670
Séquier JM, Hunziker W, Andressen C, Celio MR (1990) Calbindin D-28k protein and mRNA localization in the rat brain. Eur J Neurosci 2: 1118–1126
Serratosa J, Pujol MJ, Bachs O, Carafoli E (1988) Rearrangement of nuclear calmodulin during proliferative liver cell activation. Biochem Biophys Res Commun 150: 1162–1169
Seto-Ohshima A, Kitajima S, Sano M, Kato K, Mizutani A (1983) Immunohistochemical localization of calmodulin in mouse brain. Histochemistry 79: 251–257
Seto-Ohshima A, Sano M, Mizutani A (1985) Characteristic localization of calmodulin in human tissues: Immunohistochemical study in the paraffin sections. Acta Histochem Cytochem 18: 275–282
Siman R, Gall C, Perlmutter LS, Christian C, Baudry M, Lynch G (1985) Distribution of calpain I, an enzyme associated with degenerative activity, in rat brain. Brain Res 347: 399–403
Sloviter RS (1989) Calcium-binding protein (calbindin-D_{28K}) and parvalbumin immunocytochemistry: localization in the rat hippocampus with specific reference to the selective vulnerability of hippocampal neurons to seizure activity. J Comp Neurol 280: 183–196
Sorimachi H, Tsukahara T, Okada-Ban M, Sugita H, Ishiura S, Suzuki K (1995) Identification of a third ubiquitous calpain species-chicken muscle expresses four distinct calpains. Biochim Biophys Acta 1261: 381–393
Suzuki K, Sorimachi H, Yoshizawa T, Kinbara K, Ishiura S (1995) Calpain: Novel family members, activation, and physiological function. Biol Chem Hoppe Seyler 376: 523–529
Szabo T, Libouban S, Denizot JP (1990) A well defined spinocerebellar system in the weakly electric teleost fish Gnathonemus petersii. A tracing and immuno-hitochemical study. Arch Ital Biol 128: 229–47
Szabolcs MJ, Kopp M, Schaden GE (1989) Carbonic anhydrase activity in the peripheral nervous system of rat: the enzyme as a marker for muscle afferents. Brain Res 492: 129–138
Tabuchi K, Ohnishi R, Nishimoto A, Isobe T, Okuyama T (1984) Reverse cellular distribution of calmodulin to S-100 protein in primate brain. Brain Res 298: 353–357
Tracey DJ (1985) Ascending and descending pathways in the spinal cord. In: Paxinos G (ed) The Rat Nervous System, vol 2. Hindbrain and Spinal Cord. Academic Press, Sydney, pp 311–324

Valtschanoff JG, Weinberg RJ, Rustioni A (1992) NADPH diaphorase in the spinal cord of rats. J Comp Neurol 321: 209–222

Van Brederode JFM, Helliesen MK, Hendrickson AE (1991) Distribution of the calcium-binding proteins parvalbumin and calbindin-D28k in the sensorimotor cortex of the rat. Neuroscience 44: 157–171

Wall PD, Devor M, Inbal R, Scadding JW, Schonfeld D, Seltzer Z, Tomkiewicz MM (1979) Autotomy following peripheral nerve lesions: experimental anaesthesia dolorosa. Pain 7: 103–113

Wang KKW, Yuen P-W (1994) Calpain inhibition: an overview of its therapeutic potential. Trends Pharmacol Sci 15: 412–419

Wasserman RH, Taylor AN (1966) Vitamin D_3-induced calcium-binding protein in chick intestinal mucosa. Science 152: 791–793

Wiesenfeld Z, Lindblom U (1980) Behavioral and electrophysiological effects of various types of peripheral nerve lesions in the rat: A comparison of possible models for chronic pain. Pain 8: 285–298

Willis WD (1985) The pain system. Karger, Basel

Winsky L, Kuznicki J (1996) Antibody recognition of calcium-binding proteins depends on their calcium-binding status. J Neurochem 66: 764–771

Winsky L, Nakata H, Martin BM, Jacobowitz DM (1989) Isolation, partial amino acid sequence, and immunohistochemical localization of a brain-specific calcium-binding protein. Proc Natl Acad Sci USA 86: 10139–10143

Wood JG, Wallace RW, Whitaker JN, Cheung WY (1980) Immunocytochemical localization of calmodulin in regions of rodent brain. Ann NY Acad Sci 356: 75–82

Yamaguchi T, Winsky L, Jacobowitz DM (1991) Calretinin, a neuronal calcium binding protein, inhibits phosphorylation of a 39 kDa synaptic membrane protein from rat brain cerebral cortex. Neurosci Lett 131: 79–82

Yamamoto T, Carr PA, Baimbridge KG, Nagy JI (1989) Parvalbumin- and calbindin D28K-immunoreactive neurons in the superficial layers of the spinal cord dorsal horn of rat. Brain Res Bull 23: 493–508

Yoshida S, Senba E, Kubota Y, Hagihira S, Yoshiya I, Emson PC, Tohyama M (1990) Calcium-binding proteins calbindin and parvalbumin in the superficial dorsal horn of the rat spinal cord. Neuroscience 37: 839–848

Yoshizawa T, Sorimachi H, Tomioka S, Ishiura S, Suzuki K (1995) Calpain dissociates into subunits in the presence of calcium ions. Biochem Biophys Res Commun 208: 376–383

Zhang J-H, Morita Y, Hironaka T, Emson PC, Tohyama M (1990) Ontological study of calbindin-D_{28K}-like and parvalbumin-like immunoreactivities in rat spinal cord and dorsal root ganglia. J Comp Neurol 302: 715–728

Correspondence: Dr. M. A. Ruda, NAB/NIDR/NIH, Bldg. 49, Rm. 1A11, 49 Convent Dr., MSC 4410, Bethesda, MD 20892-4410, U.S.A.

Spinal Cord Monitoring
Eds: E. Stålberg, H. S. Sharma, Y. Olsson

Chapter 3

Substance P transmitter or regulator in spinal sensory mechanisms?

J. L. Henry

Department of Physiology and Psychiatry, McGill University, Montreal, Canada

"The suggested solutions for the difficulties disclosed by observation form ideas. Data (facts) and ideas (suggestions, possible solutions) thus form the two indispensable and correlative factors of all reflective activity. . . . Mere facts or data are dead, as far as mind is concerned, unless they are used to suggest and test some idea, some way out of a difficulty. Ideas, on the other hand, are mere ideas, idle speculations, fantasies or dreams, unless they are used to guide new observations of and reflections upon actual situations, past, present or future."

from: *Of How We Think*, 1933, John Dewey

Introduction

Evidence is reviewed which suggests multifunctional roles of substance P in spinal nociceptive pathways. There remains little challenge today to the notion that substance P is involved in excitatory regulation of nociception in the region of the first sensory synapse at spinal and brain stem levels. Criteria seem to be satisfied for substance P to qualify as a neurotransmitter. Actions of substance P on dorsal horn nociceptive neurones do seem to include relatively fast changes which can be consistent with those induced by a neurotransmitter (Otsuka and Yoshioka, 1993; Henry, 1993).

However, there are also actions linked to intracellular transduction mechanisms (Tanabe *et al.*, 1996; Koyano *et al.*, 1993) which suggest that substance P may play an additional role in regulating the efficacy of synaptic transmission. In fact, substance P has been shown to increase the responses to at least some other transmitter candidates (Dougherty *et al.*, 1995; De Koninck *et al.*, 1994), including both excitatory and inhibitory transmitters, and this may be via its effects on these intracellular mechanisms.

What is less settled, though, is the precise mechanism by which regulation is expressed. This is due to the fact that such a plethora of possible mechanisms have been implicated, from simply depolarization of the cell membrane allowing removal of the Mg^{2+} block of the NMDA channel, to complementary activation of excitatory intracellular transduction mechanisms, to potentiation of all synaptic inputs to a dorsal horn neurone. What are needed, then, are further studies on the particular mechanisms regulating substance P function at this synapse, perhaps under different conditions such as those in different animal models of pain.

Recent evidence linking substance P to longer-term changes is also important for us to understand fully its role in nociception. Substance P downregulation after peripheral axotomy (Hökfelt *et al.*, 1994; Rydh-Rinder *et al.*, 1996; Maeda *et al.*, 1996), after spinal trauma (Sharma *et al.*, 1990) and in diabetes (Brewster *et al.*, 1994) suggests that it may play a role in maintenance of normal cell function or in adaptive changes in cell function. Yet substance P has also been shown to contribute to post-trauma cell damage in peripheral tissues (Chiao and Caldwell, 1996; Poncet *et al.*, 1996) and it may be implicated in at least some central neurodegenerative disorders (Khawaja and Rogers, 1996; Kowall *et al.*, 1991). In fact, withdrawal of this type of role, thus serving a permissive type of function, may be part of the adaptive responses to neuronal stress following nerve injury (Hökfelt *et al.*, 1994), during ageing (Bergman *et al.*, 1996), and possibly during other disturbances. Furthermore, the link between growth factors and substance P function (Verge *et al.*, 1995; Zhang *et al.*, 1995; McLeod *et al.*, 1997) provides additional support for a possible role in long-term governance of neuronal function.

Background

Substance P is perhaps the one peptide for which there exists the most evidence pertaining to its role in sensory physiology. Yet a clear consensus regarding this role remains to be found. This may be partly due to the fact that substance P effects on central neurones are so varied in nature. Another may be that its actions diverge from those of "classical" neurotransmitters, such as the excitatory and inhibitory amino acids, which have relatively simple and predictable postsynaptic effects. There may also be an element due to our lack of full understanding of the role of the type of sensory fibre with which substance P is mainly associated, the unmyelinated C-fibre. An attempt to survey all the aspects of substance P involvement in sensory function would yield a document of prohibitive length. Therefore, the focus of this survey will be restricted to the role of substance P in synaptic function at the central terminals of primary afferent fibres, particularly those projecting to the spinal cord. Thus, out of necessity, this paper must bypass the interesting and important role of substance P in nociceptive mechanisms in peripheral tissues. We must also bypass reference to the fascinating molecular biology of the NK-1 receptor and its functional parts, as well as the interesting number of new NK-1 receptor antagonists.

What we shall dwell on, then, are physiological data directing us to the particular role of substance P in the sensory spinal cord, the actions of substance P on dorsal horn neurones, the functional significance of the colocalization of substance P with other chemical mediators of synaptic transmission and the intracellular signal transduction or effector mechanisms which may make substance P unique in its actions. Despite limiting the survey in such a way, to avoid going into too much detail, invariably and necessarily some important contributions to these topics are excluded.

In the following section, a case will be presented supporting a role of substance P as a neurotransmitter. Yet, despite such evidence, a number of actions of substance P cannot be explained fully by such a role only, and therefore, further information is reviewed indicating a broader role in neurotransmission. Such evidence suggests that substance P may function to regulate the level of excitability of neurones in nociceptive pathways. An awareness of the precise conditions under which substance P expresses these effects is now only dawning, but this allows an opportunity for us to speculate more broadly on the role of substance P in spinal nociceptive mechanisms. We are also drawn to evidence from the clinical domain to help us in our speculations. What the final understanding will be is too far from view at the present time, but the journey will be adventuresome!

Criteria for neurotransmission

To begin with, it is important to go quickly through some of the criteria which have been fulfilled by substance P, or any chemical for that matter, in order for us to consider it as a chemical mediator of synaptic transmission. A growing body of evidence implicates substance P in processing of sensory information, in particular, nociceptive information. High levels of substance P are observed in the CSF of chronic pain patients (Almay, 1988), while in patients with familial dysautonomia, characterized by loss of pain sensitivity, substance P is depleted from the substantia gelatinosa (Pearson *et al.*, 1982). *Immunoreactive* substance P is found in either unmyelinated or thinly myelinated sensory fibres that terminate mainly in the dorsal part of the spinal cord (Hökfelt *et al.*, 1975; Cuello and Kanazawa, 1978) and lamina V (Ruda *et al.*, 1986). High levels of substance P are found in the superficial regions of the dorsal horn (Takahashi and Otsuka, 1975), where nociceptive information is integrated and where small diameter nociceptive afferents terminate (Light and Perl, 1979; Kumazawa and Perl, 1977; Sugiura *et al.*, 1986). Dorsal rhizotomy decreases dorsal horn levels of substance P (Takahashi and Otsuka, 1975), indicating that at least some origin is in primary afferents. Substance P-containing terminals preferentially appose nociceptive neurones while non-nociceptive neurones receive few if any such terminals (De Koninck *et al.*, 1992). Substance P is *released* in the spinal cord *in vivo* upon activation of

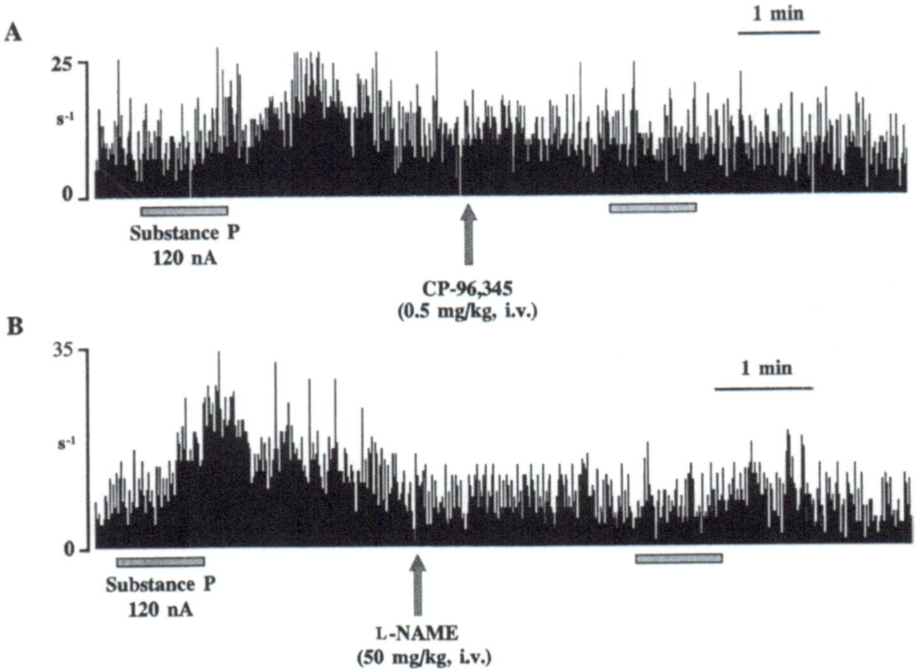

Fig. 1. Effects of substance P on two wide dynamic range neurones in the cat spinal dorsal horn. (**A**) The response to substance P is blocked by systemic administration of the NK-1 receptor antagonist, CP-96, 345. (**B**) In another neurone, systemic administration of the nitric oxide synthase inhibitor, L-NAME, blocks the response to substance P. The periods of substance P application are indicated by the horizontal bars below the records. Ejection currents are in nA

nociceptive fibres (Theriault *et al.*, 1979; Brodin *et al.*, 1987; Go and Yaksh, 1987; Duggan *et al.*, 1988) and physiological studies have demonstrated cellular and reflex responses which are blocked by NK-1 (substance P) receptor antagonists (Yashpal and Henry, 1984; Cridland and Henry, 1988a; Radhakrishnan and Henry, 1991; De Koninck and Henry, 1991). NK-1 *receptors* are concentrated where immunoreactive material is highest (Yashpal *et al.*, 1990) and dorsal rhizotomy induces upregulation of these receptors (Yashpal *et al.*, 1991a), indicating a predominant input from primary afferents. As shown in Figure 1, *iontophoretic* application of substance P induces a slow, prolonged excitatory effect on dorsal horn neurones in the cat spinal cord in vivo (Henry *et al.*, 1975). This effect is observed preferentially on nociceptive neurones, as substance P lacks effects on non-nociceptive neurones (Henry, 1976; Henry *et al.*, 1977; Salter and Henry, 1991). In fact, neurokinin A, which derives from β and γ preprotachykinin, two of the three species of mRNA which encode substance P (Krause *et al.*, 1987; Nawa *et al.*, 1983), lacks the selectivity of substance P, inducing excitation in both nociceptive and non-nociceptive dorsal horn neurones (Salter and Henry, 1991). Fig. 1A provides an example of the block of the substance P-induced excitation by the non-peptide NK-1 receptor antagonist, CP-96, 345, and this is seen also with other NK-1 receptor antagonists (Radhakrishnan *et al.*, 1997; Radhakrishnan and Henry, 1995). Substance P also facilitates responses of nociceptive neurones to noxious cutaneous stimuli (Henry and Ben Ari, 1976). Importantly, the late, slow synaptically-elicited excitatory responses of dorsal horn nociceptive neurones to noxious thermal (Fig. 2A) and noxious mechanical (Fig. 3A) are also depressed by CP-96, 345, indicating the particular involvement of NK-1 receptors in this type of input (De Koninck and Henry, 1991; Radhakrishnan and Henry, 1995). *Physiological* studies on spinal reflexes have also indicated a role of substance P in nociceptive pathways. *Intrathecal* administration of substance P to the lumbo-sacral cord in the rat produces a transient facilitation of the tail withdrawal from a noxious radiant heat stimulus (Cridland and Henry, 1986; Cridland and Henry, 1988b; Yashpal and Henry, 1983; Yashpal and Henry, 1984; Yashpal *et al.*, 1982). Interestingly, NKA has a much less potent effect in this paradigm than does substance P (Cridland and Henry, 1986; Yashpal *et al.*, 1996). This supports the idea that substance P facilitates transmission of nociceptive information at this level, as suggested originally (Henry, 1976). Noxious cutaneous stimulation of the tail produces a response similar to substance P and this facilitation is blocked by a substance P analogue antagonist (Cridland and Henry, 1988a) as well as by a non-peptide antagonist (Yashpal *et al.*, 1993). Transgenic mice overexpressing substance P in the spinal cord exhibit both allodynia and hyperalgesia in response to noxious mechanical stimulation (McLeod *et al.*, 1997). Systemic administration of CP-96, 345 also selectively depresses the second phase in the formalin test (Yashpal *et al.*, 1993). Thus, block of NK-1 receptors prevents nociceptive responses to noxious thermal, noxious mechanical and noxious chemical stimuli.

It is reasonable in view of such evidence to consider that substance P does fulfil the criteria as a neurotransmitter. But does it do more? How is it that in the presence of an NK-1 receptor antagonist, excitatory synaptic responses can be elicited in dorsal horn neurones (De Koninck and Henry, 1991) and spinal reflexes, such as the tail withdrawal reflex, are unaltered (Yashpal *et al.*, 1993)?

Actions of substance P at the first sensory synapse

At the cellular level, substance P may be operating via more than a single mechanism to account for the slow, prolonged effects seen (Otsuka and Yoshioka, 1993). Some of the earliest studies on actions of substance P in the central nervous system raised the possibility that the excitatory effects of substance P on neurones was due to depression of a K^+ current (Krnjevic, 1977) and this suggestion is consistent with the changes we have reported in the membrane properties and lack of a reversal potential of dorsal horn neurones in response to high intensity electrical stimulation of C-fibre inputs in vivo (De Koninck and Henry, 1991). In whole-cell patch-clamp studies substance P inhibits an inwardly rectifying K^+ current in cultured forebrain and locus coeruleus neurones (Yamaguchi *et al.*, 1990) and this inhibition is mediated through a pertussis toxin-insensitive G-protein mechanism (Nakajima *et al.*, 1988; Nakajima *et al.*, 1991; Macdonald *et al.*, 1996), although an additional G-protein mechanism has been suggested (Koyano *et al.*, 1993). A substance P-induced inhibition of a voltage-dependent background K^+ current has also been reported in bullfrog dorsal root ganglion cells (Ishimatsu, 1994). Substance P inhibits the M-current in mammalian spinal neurones (Murase *et al.*, 1986; Nowak and MacDonald, 1982) and in bullfrog dorsal root ganglion (Ishimatsu, 1994) and sympathetic neurones (Adams *et al.*, 1983). Substance P also inhibits a calcium-activated potassium current in guinea-pig submucosal neurones (Shen and Surprenant, 1993).

In rat dorsal horn neurones *in vitro* substance P has been reported to enhance a slow, persistent, calcium-sensitive current (Murase *et al.*, 1986) and low- and high-voltage activated Ca^{2+} currents (Ryu and Randic, 1990). In mammalian and non-mammalian sympathetic neurones, a suppression of voltage-gated calcium currents by substance P has also been reported (Bley and Tsien, 1990; Shapiro and Hille, 1993). Womack *et al.* (Womack *et al.*, 1988; Heath *et al.*, 1994) reported an entry of Ca^{2+} by substance P in single isolated postnatal rat dorsal horn neurones. In Chinese hamster ovary cells substance P has been found to induce a rapid release of Ca^{2+} from intracellular stores followed by an influx of Ca^{2+} (Mochizuki-Oda *et al.*, 1994; Garland *et al.*, 1996); similar results have been found in human neutrophils (Tanabe *et al.*, 1996). Recently, we have found that substance P-induced excitation of dorsal horn neurones and nociceptive responses of these neurones in the cat are blocked by administration of the L-type Ca^{2+} channel blockers verapamil and diltiazem (Radhakrishnan and Henry, 1994). Furthermore, we have found that substance P-induced hyperalgesia in the rat is blocked by verapamil and diltiazem, but it is also blocked by adminstration of thapsigargin which inhibits the Ca^{2+} pump on the endoplasmic reticulum (Henry *et al.*, 1996). Therefore, our physiological data are consistent with reports that substance P expresses at least some of its effects via both intracellular Ca^{2+} release and Ca^{2+} influx.

As an overview, then, substance P fulfils at least the major criteria as a transmitter but it seems to be coupled with other mechanisms which may lead to longer-term changes in neuronal excitability. Thus, it is important to focus on its precise role in synaptic transmission because as outlined above, a substance P antagonist does not block excitatory synaptic inputs or spinal reflexes. Therefore, it is necessary to take a more incisive view of the evidence available and go beyond its role as a neurotransmitter to examine its role in neurotransmission. With the introduction of evidence of increased intracellular levels of Ca^{2+}, this introduces a number of intracellular transduction mechanisms. There are a number of pathways

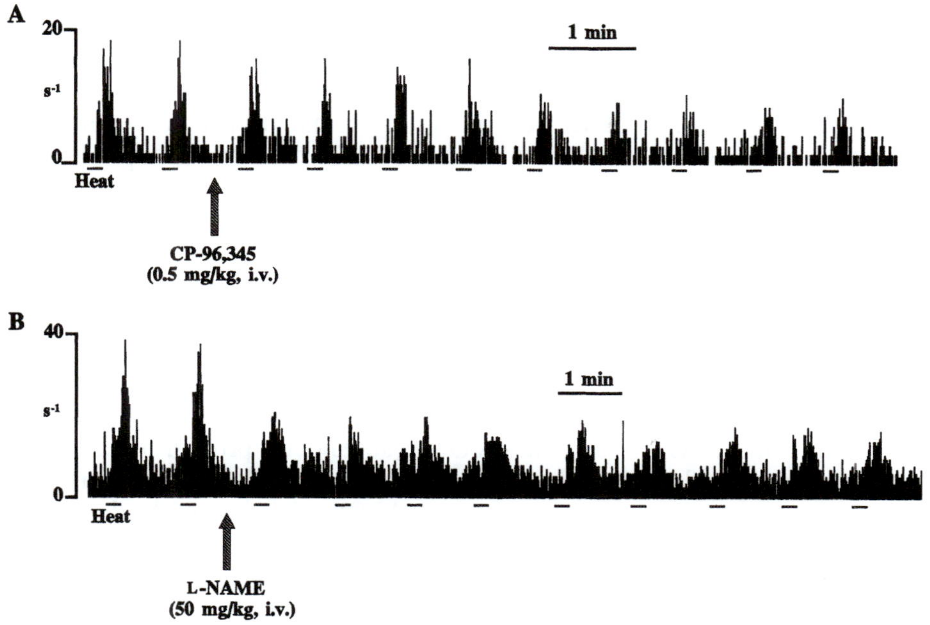

Fig. 2. Nociceptive responses of two wide dynamic range neurones to noxious thermal stimulation of the cutaneous receptive field are depressed by systemic administration of CP-96,345 (**A**) and of L-NAME (**B**). The short horizontal bars below the records indicate the periods of application of the noxious stimulus

downstream from Ca^{2+} which have been implicated in substance P action. The major enzymes implicated are NOS, PLA_2 and PKC. Thus, the role of substance P as a neuroregulator becomes a more realistic possibility.

Intracellular signal transduction mechanisms

In fact, there is evidence that NK-1 receptors are coupled via a G-protein (Luber-Narod *et al.*, 1990) to a number of intracellular signal transduction pathways including a phospholipase C, 1,4,5-inositol trisphosphate, diacyl glycerol (DAG), a protein kinase C (PKC), nitric oxide synthase (NOS), soluble guanylyl cyclase and cyclic guanosine monophosphate (cGMP), cyclic adenosine monophosphate (cAMP), and phospholipase A_2 and the arachidonic acid pathway.

Phospholipase C is coupled in a number of tissues to NK-1 receptors by the α-subunit of a pertussis-insensitive G-protein (Taylor *et al.*, 1986), specifically G_q and G_{11} (Khawaja and Rogers, 1996; Macdonald *et al.*, 1996). Phospholipase C hydrolyses inositol-containing phospholipids, such as phosphatidylinositol-4,5-bisphosphate (PIP_2) to inositol-1,4,5-trisphosphate (IP_3) and diacyl glycerol. NK-1 receptor activation has been shown to increase intracellular levels of IP_3 in a number of tissues (Tanabe *et al.*,

1996; Sagan *et al.*, 1996; Raddatz *et al.*, 1995; Khawaja and Rogers, 1996). IP_3 in turn acts on its receptor to inhibit a Ca^{2+}-ATPase in the endoplasmic reticulum and thus release Ca^{2+} into the cytosol. A substance P-induced release of Ca^{2+} from intracellular stores (Womack *et al.*, 1988) has been reported in a number of tissues including spinal dorsal horn neurones, although this was contradicted by a later study by the same group (Heath *et al.*, 1994). We have recently found that substance P-induced hyperalgesia in the rat is blocked by spinal intrathecal administration of thapsigargin (Henry *et al.*, 1996), which has been reported to increase intracellular levels of Ca^{2+}, possibly by inhibiting uptake of leaked Ca^{2+} into the endoplasmic reticulum, and to decrease substance P-induced intracellular release of Ca^{2+} in an astrocyte cell line (Bordey *et al.*, 1994).

NOS stimulates the generation of nitric oxide (NO) from L-arginine when Ca^{2+} binds to a site on NOS. NO activates soluble guanylyl cyclase to yield cGMP, which activates cGMP-dependent protein kinases and protein phosphorylation, and generates other intracellular mediators such as cyclic adenosine-monophosphate (cAMP). Nitric oxide has been implicated in a number of central nervous system functions but although it has been linked to nociceptive mechanisms at the spinal level its role in mediating substance P-induced effects in the central nervous system has been limited; a substance P-NO link has been established, though, in a number of other tissues (Enokibori *et al.*, 1994). NOS has been found in neurones in the superficial dorsal horn (Dun *et al.*, 1993; Liuzzi *et al.*, 1993; Terenghi *et al.*, 1993) and it is increased in the spinal dorsal horn after noxious peripheral stimulation (Herdegen *et al.*, 1994). Figure 1B shows nociceptive neurone responses to iontophoretic application of substance P are reduced by administration of the NOS inhibitor, L-NAME, and Figures 2B and 3B show that synaptically-elicited responses to noxious stimulation of the cutaneous receptive fields are also depressed by L-NAME (Radhakrishnan and Henry, 1993). In the formalin test, substance P-induced enhancement of the nociceptive response is reversed by L-NAME, although the inhibitor had an effect by itself on the nociceptive scores and it was not established whether the effects were simply additive (Coderre and Yashpal, 1994). Substance P-induced thermal hyperalgesia in the tail-flick test is depressed by L-NAME (Radhakrishnan *et al.*, 1995) and the hyperalgesia attributed to noxious stimulation-induced release of endogenous substance P (Yashpal *et al.*, 1993; Yashpal and Henry, 1984) is also depressed by L-NAME (Radhakrishnan *et al.*, 1995). Interestingly, a non-nociceptive response to intrathecal administration of substance P, a scratching, biting and licking response, is not affected by L-NAME (Sakurada *et al.*, 1996).

Arachidonic acid derives from two sources, mostly by the action of PLA_2 to convert phosphatidyl choline but some also from diacyl glycerol. Arachidonic acid is relatively rapidly broken down into prostaglandins, leukotrienes and thromboxane. While the latter two have not been tightly linked to nociceptive mechanisms, prostaglandins have been implicated, in

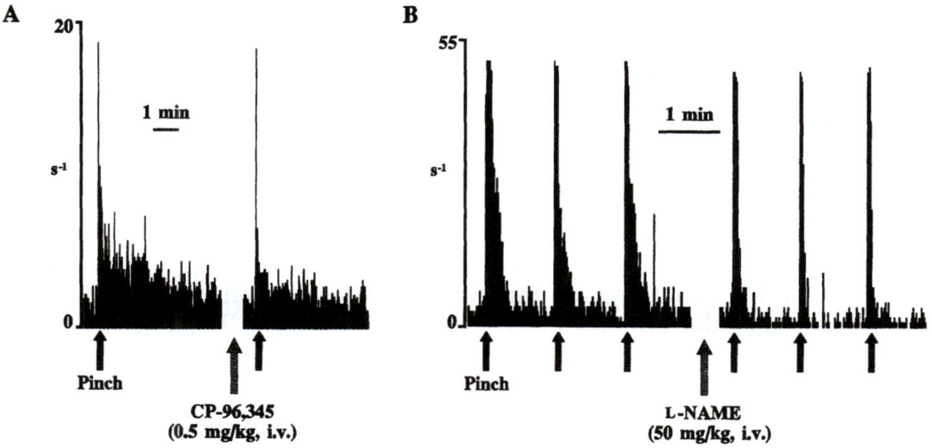

Fig. 3. Nociceptive responses of two wide dynamic range neurones to noxious mechanical stimulation (pinch) of the cutaneous receptive field are depressed by systemic administration of CP-96,345 (**A**) and of L-NAME (**B**). The filled arrows below the records indicate the periods of application of the noxious stimulus

particular because non-steroidal antiinflammatory drugs, or NSAIDs, which inhibit the cyclooxygenase enzyme (COX) which converts arachidonic acid to prostaglandins, are widely used as analgesics. Substance P action has been reported to liberate arachidonic acid from mast cells via a PLA_2-dependent mechanism (Mousli *et al.*, 1992) and from Chinese hamster ovary cells (Garcia *et al.*, 1994). Little evidence links arachidonic acid or prostaglandins to substance P actions in the spinal cord, although we have recently found that the COX inhibitor, indomethacin, depresses responses of dorsal horn neurones to iontophoretic application of substance P and to noxious stimulation of the cutaneous receptive field (Pitcher and Henry, 1996) and iontophoretic application of arachidonic acid increases dorsal horn neurone responses to application of substance P (Pitcher and Henry, 1997).

PKC translocation from a soluble cytosolic fraction to a plasma membrane-bound fraction is a Ca^{2+}-dependent process and is generally considered to be an indication of activation of the enzyme. PKC has been implicated in a number of intracellular functions including, via protein phosphorylation/dephosphorylation, modulation of receptor sensitivity and modulation of release mechanisms. Substance P has been reported to stimulate the translocation of the alpha and beta isozymes of PKC in lactotroph-enriched anterior pituitary cell cultures (Mochizuki-Oda *et al.*, 1994), to promote tyrosine phosphorylation of PKCδ in parotid acinar cells of rat salivary gland (Soltoff and Toker, 1995). Accumulation of inositol phosphates in human astrocytoma cells by substance P is mimicked by PKC activation (Lee *et al.*, 1992), although in rat KNRK cells (Mitsuhashi *et al.*, 1992), substance P-induced mobilization of intracellular Ca^{2+} is unaffected by phorbol esters, which activate PKC. In rat submucosal neurones PCK inhibition reduces substance P-induced depolarization (Barajas-López, 1994). Nociceptive responses to intraplantar injection of formalin, which have been attributed at least partially to activation of NK-1 receptors (Yashpal *et al.*, 1993), are reduced by PKC inhibition (Yashpal *et al.*, 1995). Formalin injection also produces an increase in dorsal horn binding of phorbol 12, 13-dibutyrate (PDBu), which reflects the translocation of PKC (Yashpal *et al.*, 1995).

Therefore, it appears that substance P-induced responses may be mediated via a number of different effector mechanisms. Studies focused on the intracellular effector mechanisms of substance P action at the first sensory synapse are still relatively few. It is far from clear which pathways are activated by substance P in these neurones. It is also far from clear the intracellular regulatory mechanisms which activate one vs. another of these intracellular mechanisms when the conditions shift in the primary afferents (Brewster *et al.*, 1994; Hökfelt *et al.*, 1994) or in the spinal cord (Sharma *et al.*, 1990; Bergman *et al.*, 1996).

Interactions of substance P with other chemical mediators of synaptic transmission

Opioids

Interaction has been found between substance P and chemicals released by *intrinsic spinal neurones*. Substance P administration (Yashpal and Henry, 1983) and noxious cutaneous stimulation (Cridland and Henry, 1988a), both produce hyperalgesia, but this is followed by a long-lasting depressant effect on the nociceptive reflex which is blocked by opiate antagonists (Yashpal and Henry, 1983). Nerve terminals containing substance P and other nerve terminals containing enkephalin are found apposed onto the same spinal nociceptive neurones (De Koninck *et al.*, 1993) and some terminals contain both substance P and enkephalin (Ma *et al.*, 1997), indicating a postsynaptic interaction of the two peptides. Substance P containing terminals also appose spinal nociceptive neurones containing enkephalin (Ribeiro-da-Silva *et al.*, 1992), indicating a hitherto unanticipated relationship between the functions of these two peptides. Transgenic mice which overexpress substance P in the spinal cord and which exhibit exaggerated substance P-mediated responses (McLeod *et al.*, 1997) also exhibit exaggerated opioid responses (Ribeiro-da-Silva *et al.*, 1997).

Purines

In studies on purinergic inputs to the spinal cord, activation of Pacinian afferents by cutaneously applied vibration was found to provoke an inhibition of nociceptive neurones (Salter and Henry, 1990). This inhibition is blocked by adenosine receptor antagonists but not by antagonists at other classical receptors (Salter and Henry, 1987a). Substance P was found to potentiate adenosine-induced depression of nociceptive neurones (Salter and Henry, 1987b), it potentiates the vibration-induced depression of these neurones (Salter and Henry, 1988) and a noxious stimulus which activates a substance P-mediated excitation of dorsal horn neurones (Radhakrishnan and Henry, 1991), also potentiates vibration induced depression (De Koninck *et al.*, 1994). These studies clearly indicate interaction of endogenously-released substance P with endogenously-released purines in the region of the *primary afferent* synapse.

Excitatory amino acids

Substance P has been reported to induce the release of glutamate and aspartate in spinal cord in vivo (Skilling *et al.*, 1992; Sluka and Westlund, 1993) and in vitro (Kangrga and Randic, 1990; Kawagoe *et al.*, 1986). Substance P coexists with glutamate in primary afferent terminals and in dorsal root ganglion cells (De Biasi and Rustioni, 1988; Merighi *et al.*, 1991).

There is an interaction of substance P with glutamate receptor-mediated responses in the dorsal horn: substance P prolongs excitatory effects of *N*-methyl-D-aspartate (NMDA) on spinothalamic dorsal horn neurones (Dougherty and Willis, 1991; Dougherty *et al.*, 1993), both substance P and glutamate elevate intracellular Ca^{2+} in the same dorsal horn neurones and both effects are blocked by the glutamate receptor antagonist, kynurenate (Womack *et al.*, 1988; Heath *et al.*, 1994). Substance P-induced facilitation of the tail withdrawal reflex is blocked by intrathecal administration of the specific NMDA-receptor antagonist 2-amino-5-phosphonovaleric acid (APV) (Yashpal *et al.*, 1991b). Substance P enhances ionic currents induced in vitro by the activation of NMDA, and α-amino-3-hydroxy-5-methyl-4-isoxazolepropionic acid (AMPA) but not of kainate receptors (Randic *et al.*, 1990; Rusin *et al.*, 1993b; Rusin *et al.*, 1993a; Rusin *et al.*, 1992). Finally, facilitatory effects of both exogenously administered substance P and endogenously-released substance P by noxious cutaneous stimulation are blocked by APV, suggesting that among the intrinsic neurones activated by substance P is a population of interneurones in spinal reflex pathways which mediate their effects via NMDA receptor activation (Yashpal *et al.*, 1991b).

Intrathecal administration of substance P decreases the latency of the nociceptive tail withdrawal reflex (Yashpal *et al.*, 1982) and this was suggested to be a facilitation of an excitatory process, rather than simply an effect by itself. In fact, the original suggestion from electrophysiological results was that substance P may be a regulator of excitability at the primary afferent synapse in nociceptive pathways (Henry, 1976). These suggestions have been borne out by more recent experiments where intrathecal administration of a non-peptide NK-1 receptor antagonist, CP-96, 345, blocked effects of substance P on a nociceptive withdrawal reflex but had no effect on the baseline withdrawal latency (Yashpal *et al.*, 1993). This indicates that substance P plays no role *per se* in the nociceptive response to a brief noxious stimulus. It does, however, appear to play a role in promoting mechanisms eliciting the response to a brief noxious stimulus; in other words, substance P leads to hyperalgesia and allodynia without participating itself in the fast response to a brief noxious stimulus.

Our initial paper on this same antagonist (De Koninck and Henry, 1991) indicated similarly that when a single, high intensity electrical stimulus was applied to a sensory nerve, an EPSP occurred at a latency consistent with conduction in C fibres. The same stimulus applied repetitively for a few seconds elicited additionally a slow, prolonged depolarization and an associated afterdischarge. Administration of the antagonist blocked the effects which outlasted the stimulus but had no effect on the EPSP in response to a single stimulus. These observations were interpreted as indicating that the EPSP was mediated via a fast-acting chemical, as the reversal potential was similar to conventional EPSPs. The reversal potential of the prolonged excitation was consistent with depression of a K^+ current, as seen with substance P in vitro (Nowak and MacDonald, 1982; Murase and Randic, 1984). Thus, it appears that substance P is released only upon repetitive activation of the C fibres and that a single action potential predominantly releases a fast-acting transmitter, presumably glutamate.

When our three studies (Yashpal *et al.*, 1991b; Yashpal *et al.*, 1993; De Koninck and Henry, 1991) are taken together, it appears that substance P does not normally take part in nociceptive responses to brief noxious stimuli. However, under conditions of a more prolonged noxious stimulus, substance P appears to be released and it acts to potentiate the excitatory effects of the fast-acting transmitter, which is probably glutamate. Similar conclusions regarding substance P interaction with glutamate-elicited effects have been reached by other laboratories. In the isolated spinal cord of the young rat the evoked C fibre response in ventral roots is potentiated in an animal model of peripheral injury and this potentiation is blocked by APV and by the NK-1 receptor antagonists, CP-96, 345 and RP 67580, while the response to single pulse afferent stimulation is antagonized only by APV (Thompson *et al.*, 1994). Substance P also potentiates the ventral root response to application of NMDA or of quisqualate (Urban *et al.*, 1994). Substance P potentiates the excitatory effects of NMDA receptor activation of STT cells in the spinal dorsal horn (Dougherty and Willis, 1991; Dougherty *et al.*, 1993); upon scrutiny of the figures presented in these papers, though, it appears that the effects are not clearly more than just additive. In vitro experiments on the other hand do provide more convincing support for a positive interaction of substance P on NMDA receptor mediated responses. In freshly isolated dorsal horn neurones of the rat, when NMDA was applied up to 20 min after substance P the amplitude of the early transient component of the NMDA current more than doubled and there was also a small increase in the steady-state component of this current (Rusin *et al.*, 1992). Subsequently it was reported that the potentiating effect of substance P was mimicked by intracellular injection of cAMP and by a catalytic subunit of cAMP-dependent protein kinase A. The conclusion was drawn that the potentiation was mediated via the adenylate cyclase pathway (Cerne *et al.*, 1993).

Conclusions

Exciting new ideas are emerging regarding the role of substance P in spinal nociceptive pathways; these ideas may contribute to our understanding of the etiology of at least some types of chronic pain. Among these ideas some may prove to be valid. Some may fall along the wayside. However, it is important that our concepts remain dynamic and up-to-date with all the evidence. It is also important to remain abreast of developments of the other neuroactive chemicals participating in neurotransmission in spinal nociceptive pathways, as substance P remains only one of the players and its role must be assessed within the context of the whole synaptic soup.

References

Adams PR, Brown DA, Jones SW (1983) Substance P inhibits the M-current in bullfrog sympathetic neurones. Br J Pharmacol 79: 330–333

Almay BGL (1988) CSF levels of substance P in chronic pain patients. Pain 33: 3–9

Barajas-Lopez C (1994) Interactions between inhibitory and excitatory modulatory signals in single submucosal neurons. Am J Physiol Cell Physiol 267: C1359–C1365

Bergman E, Johnson H, Zhang X, Hökfelt T, Ulfhake B (1996) Neuropeptides and neurotrophin receptor mRNAs in primary sensory neurons of aged rats. J Comp Neurol 375: 303–320

Bley KR, Tsien RW (1990) Inhibition of Ca^{2+} and K^{+} channels in sympathetic neurons by neuropeptides and other ganglionic transmitters. Neuron 2: 379–391

Bordey A, Feltz P, Trouslard J (1994) Mobilization of intracellular calcium by substance P in a human astrocytoma cell line (U-373 MG). GLIA 11: 277–283

Brewster WJ, Diemel LT, Leach RM, Tomlinson DR (1994) Reduced sciatic nerve substance P and calcitonin gene-related peptide in rats with short-term diabetes or central hypoxaemia co-exist with normal messenger RNA levels in the lumbar dorsal root ganglia. Neuroscience 58: 323–330

Brodin E, Linderoth B, Cazelius B, Ungerstedt V (1987) *In vivo* release of substance P in cat dorsal horn studied with microdialysis. Neurosci Lett 76: 357–362

Cerne R, Rusin KI, Randic M (1993) Enhancement of the *N*-methyl-D-aspartate response in spinal dorsal horn neurons by cAMP-dependent protein kinase. Neurosci Lett 161: 124–128

Chiao H, Caldwell RW (1996) The role of substance P in myocardial dysfunction during ischemia and reperfusion. Naunyn Schmiedebergs Arch Pharmacol 353: 400–407

Coderre TJ, Yashpal K (1994) Intracellular messengers contributing to persistent nociception and hyperalgesia induced by L-glutamate and substance P in the rat formalin pain model. Eur J Neurosci 6: 1328–1334

Cridland RA, Henry JL (1986) Comparison of the effects of substance P, neurokinin A, physalaemin and eledoisin in facilitating a nociceptive reflex in the rat. Brain Res 381: 93–99

Cridland RA, Henry JL (1988a) Facilitation of the tail-flick reflex by noxious cutaneous stimulation in the rat: antagonism by a substance P analogue. Brain Res 462: 15–21

Cridland RA, Henry JL (1988b) Intrathecal administration of substance P in the rat: spinal transection or morphine blocks the behavioural responses but not the facilitation of the tail flick reflex. Neurosci Lett 84: 203–208

Cuello AC, Kanazawa I (1978) The distribution of substance P immunoreactive fibers in the rat central nervous system. J Comp Neurol 178: 129–156

De Biasi S, Rustioni A (1988) Glutamate and substance P coexist in primary afferent terminals in the superficial laminae of spinal cord. Proc Natl Acad Sci USA 85: 7820–7824

De Koninck Y, Ribeiro-da-Silva A, Henry JL, Cuello AC (1992) Spinal neurons exhibiting a specific nociceptive response receive abundant substance P-containing synaptic contacts. Proc Natl Acad Sci USA 89: 5073–5077

De Koninck Y, Ribeiro-da-Silva A, Henry JL, Cuello AC (1993) Ultrastructural immunocytochemistry combined with intracellular marking of physiologically identified neurons. In: Cuello AC (ed) Immunocytochemistry. Wiley, Chichester, pp 369–393

De Koninck Y, Salter MW, Henry JL (1994) Substance P released endogenously by high-intensity sensory stimulation potentiates purinergic inhibition of nociceptive dorsal horn neurons induced by peripheral vibration. Neurosci Lett 176: 128–132

De Koninck Y, Henry JL (1991) Substance P-mediated slow excitatory postsynaptic potential elicited in dorsal horn neurons *in vivo* by noxious stimulation. Proc Natl Acad Sci USA 88: 11344–11348

Dougherty PM, Palecek J, Zorn S, Willis WD (1993) Combined application of excitatory amino acids and substance P produces long-lasting changes in responses of primate spinothalamic tract neurons. Brain Res Rev 18: 227–246

Dougherty PM, Palecek J, Paleckova V, Willis WD (1995) Infusion of substance P or neurokinin A by microdialysis alters responses of primate spinothalamic tract neurons to cutaneous stimuli and to iontophoretically released excitatory amino acids. Pain 61: 411–425

Dougherty PM, Willis WD (1991) Enhancement of spinothalamic neuron responses to chemical and mechanical stimuli following combined micro-iontophoretic application of *N*-methyl-D-aspartic acid and substance P. Pain 47: 85–93

Duggan AW, Hendry IA, Morton CR, Hutchison WD, Zhao ZQ (1988) Cutaneous stimuli releasing immunoreactive substance P in the dorsal horn of the cat. Brain Res 451: 261

Dun NJ, Dun SL, Wu SY, Förstermann U, Schmidt HHHW, Tseng LF (1993) Nitric oxide synthase immunoreactivity in the rat, mouse, cat and squirrel monkey spinal cord. Neuroscience 54: 845–857

Enokibori M, Okamura T, Toda N (1994) Mechanism underlying substance P-induced relaxation in dog isolated superficial temporal arteries. Br J Pharmacol 111: 77–82

Garcia M, Sakamoto K, Shigekawa M, Nakanishi S, Ito S (1994) Multiple mechanisms of arachidonic acid release in Chinese hamster ovary cells transfected with cDNA of substance P receptor. Biochem Pharmacol 48: 1735–1741

Garland AM, Grady EF, Lovett M, Vigna SR, Frucht MM, Krause JE, Bunnett NW (1996) Mechanisms of desensitization and resensitization of G protein-coupled neurokinin$_1$ and neurokinin$_2$ receptors. Mol Pharmacol 49: 438–446

Go VLW, Yaksh TL (1987) Release of substance P from the cat spinal cord. J Physiol (Lond) 391: 141–167

Heath MJS, Womack MD, Macdermott AB (1994) Substance P elevates intracellular calcium in both neurons and glial cells from the dorsal horn of the spinal cord. J Neurophysiol 72: 1192–1198

Henry JL, Krnjevic K, Morris ME (1975) Substance P and spinal neurones. Can J Physiol Pharmacol 53: 423–432

Henry JL (1976) Effects of substance P on functionally identified units in cat spinal cord. Brain Res 114: 439–451

Henry JL, Hu JW, Lucier GE, Sessle BJ (1977) Responses of units in the trigeminal sensory nuclei to oral-facial stimuli and to substance P. In: Anderson DJ, Mathews B (eds) Pain in the Trigeminal Region. Elsevier, Amsterdam, pp 295–306

Henry JL (1993) Participation of substance P in spinal physiological responses to peripheral aversive stimulation. Regul Pept 46: 138–143
Henry JL, Yashpal K, Reid J, Samulack DD (1996) Calcium mediation of hyperalgesic effects of substance P in rat spinal cord. Soc Neurosci Abstr 22: 875
Henry JL, Ben Ari Y (1976) Actions of the *p*-chlorophenyl derivative of GABA, Lioresal, on nociceptive and non-nociceptive units in the spinal cord of the cat. Brain Res 117: 540–544
Herdegen T, Rüdiger S, Mayer B, Bravo R, Zimmermann M (1994) Expression of nitric oxide synthase and colocalisation with *Jun*, *Fos* and *Krox* transcription factors in spinal cord neurons following noxious stimulation of the rat hindpaw. Mol Brain Res 22: 245–258
Hökfelt T, Kellerth J-O, Nilsson G, Pernow B (1975) Substance P: localization in the central nervous system and in some primary sensory neurons. Science 190: 889–891
Hökfelt T, Zhang X, Wiesenfeld-Hallin Z (1994) Messenger plasticity in primary sensory neurons following axotomy and its functional implications. Trends Neurosci 17: 22–30
Ishimatsu M (1994) Substance P produces an inward current by suppressing voltage-dependent and -independent K^+ currents in bullfrog primary afferent neurons. Neurosci Res 19: 9–20
Kangrga I, Randic M (1990) Tachykinins and calcitonin gene-related peptide enhance release of endogenous glutamate and aspartate from the rat spinal dorsal horn slice. J Neurosci 10: 2026–2038
Kawagoe R, Onodera K, Takeuchi A (1986) The release of endogenous glutamate from the newborn rat spinal cord induced by dorsal root stimulation and substance P. Biomed Res 7: 253–259
Khawaja AM, Rogers DF (1996) Tachykinins: Receptor to effector. Int J Biochem Cell Biol 28: 721–738
Kowall NW, Beal MF, Busciglio J, Duffy LK, Yankner BA (1991) An in vivo model for the neurodegenerative effects of β amyloid and protection by substance P. Proc Natl Acad Sci USA 88: 7247–7251
Koyano K, Velimirovic BM, Grigg JJ, Nakajima S, Nakajima Y (1993) Two signal transduction mechanisms of substance P-induced depolarization in locus coeruleus neurons. Eur J Neurosci 5: 1189–1197
Krause JE, Chirgwin JM, Carter MS, Xu ZS, Hershey AD (1987) Three rat preprotachykinin mRNAs encode the neuropeptides substance P and neurokinin A. Proc Natl Acad Sci USA 84: 881–885
Krnjevic K (1977) Effects of substance P on central neurons in cats. In: von Euler US, Pernow B (eds) Substance P. Raven Press, New York, pp 217-230
Kumazawa T, Perl ER (1977) Primate cutaneous receptors with unmyelinated (C) fibres and their projection to the substantia gelatinosa. J Physiol (Paris) 73: 287–304
Lee CM, Tung WL, Young JD (1992) Tachykinin-stimulated inositol phospholipid hydrolysis and taurine release from human astrocytoma cells. J Neurochem 59: 406–414
Light AR, Perl ER (1979) Spinal termination of functionally identified primary afferent neurons with slowly conducting myelinated fibers. J Comp Neurol 186: 133–150
Liuzzi FJ, Wu W, Scoville SA, Schinco FP (1993) Development of nitric oxide synthase expression in the superficial dorsal horn of the rat spinal cord. Exp Neurol 121: 275–278
Luber-Narod J, Boyd ND, Leeman SE (1990) Guanine nucleotides decrease the affinity of substance P binding to its receptor. Eur J Pharmacol Mol Pharmacol 188: 185–191
Ma W, Ribeiro-da-Silva A, De Koninck Y, Radhakrishnan V, Cuello AC, Henry JL (1997) Substance P and enkephalin immunoreactivities in axonal boutons presynaptic to physiologically identified dorsal horn neurons. Neuroscience 77: 793–811
Macdonald SG, Dumas JJ, Boyd ND (1996) Chemical cross-linking of the substance P (NK-1) receptor to the α subunits of the G proteins G_q and G_{11}. Biochemistry 35: 2909–2916
Maeda K, Fernyhough P, Tomlinson DR (1996) Regenerating sensory neurones of diabetic rats express reduced levels of mRNA for GAP-43, gamma-preprotachykinin and the nerve growth factor receptors, trkA and $p75^{NGFR}$. Mol Brain Res 37: 166–174

Mcleod AL, Ritchie J, Cuello AC, Julien J-P, Ribeiro-da-Silva A, Henry JL (1997) Transgenic mice over-expressing nerve growth factor exhibit mechanical allodynia and hyperalgesia which are reversed by substance P and NMDA receptor antagonists. Soc Neurosci Abstr 23: 1806

Merighi A, Polak JM, Theodosis DT (1991) Ultrastructural visualization of glutamate and aspartate immunoreactivities in the rat dorsal horn, with special reference to the co-localization of glutamate, substance P and calcitonin-gene related peptide. Neuroscience 40: 67–80

Mitsuhashi M, Ohashi Y, Shichijo S, Christian C, Sudduth-Klinger J, Harrowe G, Payan DG (1992) Multiple intracellular signaling pathways of the neuropeptide substance P receptor. J Neurosci Res 32: 437–443

Mochizuki-Oda N, Nakajima Y, Nakanishi S, Ito S (1994) Characterization of the substance P receptor-mediated calcium influx in cDNA transfected chinese hamster ovary cells. A possible role of inositol 1,4,5-trisphosphate in calcium influx. J Biol Chem 269: 9651–9658

Mousli M, Fischer T, Landry Y (1992) Role of phospholipase A_2 and pertussis toxin-sensitive G proteins in histamine secretion induced by substance P. Agents Actions 36 [Suppl] C: C305–C307

Murase K, Ryu PD, Randic M (1986) Substance P augments a persistent slow inward calcium-sensitive current in voltage-clamped spinal dorsal horn neurons of the rat. Brain Res 365: 369–376

Murase K, Randic M (1984) Actions of substance P on rat spinal dorsal horn neurones. J Physiol (Lond) 346: 203–217

Nakajima Y, Nakajima S, Inoue M (1988) Pertussis toxin-insensitive G protein mediates substance P-induced inhibition of potassium channels in brain neurons. Proc Natl Acad Sci USA 85: 3643–3647

Nakajima Y, Nakajima S, Inoue M (1991) Substance P induced inhibition of potassium channels via a pertussis toxin-insensitive G protein. Ann NY Acad Sci 632: 103–111

Nawa H, Hirose T, Takashima H, Imayama S, Nakanishi S (1983) Nucleotide sequences of cloned cDNAs for two types of bovine brain substance P precursors. Nature 306: 32–36

Nowak LM, Macdonald RL (1982) Substance P ionic basis for depolarizing responses of mouse spinal cord neurons in cell culture. J Neurosci 2: 1119–1128

Otsuka M, Yoshioka K (1993) Neurotransmitter functions of mammalian tachykinins. Physiol Rev 73: 229–308

Pearson J, Brandeis L, Cuello AC (1982) Depletion of substance P-containing axons in substantia gelatinosa of patients with diminished pain sensitivity. Nature 295: 61–63

Pitcher GM, Henry JL (1996) Cyclooxygenase involvement in excitatory responses to synaptic inputs, excitatory amino acids and substance P in rat spinal dorsal horn neurones in vivo. Soc Neurosci Abstr 22: 1369

Pitcher GM, Henry JL (1997) Arachidonic acid increases rat dorsal horn neuron responses to noxious and innocuous peripheral stimuli and to application of substance P, NMDA and quisqualate. Soc Neurosci Abstr 23: 1535

Poncet L, Denoroy L, Dalmaz Y, Pequignot JM, Jouvet M (1996) Alteration in central and peripheral substance P- and neuropeptide Y-like immunoreactivity after chronic hypoxia in the rat. Brain Res 733: 64–72

Raddatz R, Crankshaw CL, Snider RM, Krause JE (1995) Similar rates of phosphatidylinositol hydrolysis following activation of wild-type and truncated rat neurokinin-1 receptors. J Neurochem 64: 1183–1191

Radhakrishnan V, Yashpal K, Hui-Chan CWY, Henry JL (1995) Implication of a nitric oxide synthase mechanism in the action of substance P: L-NAME blocks thermal hyperalgesia induced by endogenous and exogenous substance P in the rat. Eur J Neurosci 7: 1920–1925

Radhakrishnan V, Iyengar S, Henry JL (1998) The non-peptide antagonists LY303870 and LY306740 block responses of spinal dorsal horn neurons to substance P and to peripheral noxious stimuli. Neuroscience 83: 1251–1260

Radhakrishnan V, Henry JL (1991) Novel substance P antagonist, CP-96, 345, blocks responses of spinal dorsal horn neurons to noxious cutaneous stimulation and to substance P. Neurosci Lett 132: 39–43

Radhakrishnan V, Henry JL (1993) L-NAME blocks responses to NMDA, substance P and noxious cutaneous stimuli in cat dorsal horn. Neuroreport 4: 323–326

Radhakrishnan V, Henry JL (1994) Effect of L-type calcium channel antagonists on the responses of dorsal horn neurons to peripheral cutaneous stimuli and to substance P in cats. Soc Neurosci Abstr 20: 964

Radhakrishnan V, Henry JL (1995) Antagonism of nociceptive responses of cat spinal dorsal horn neurons in vivo by the NK-1 receptor antagonists CP-96, 345 and CP-99, 994, but not by CP-96, 344. Neuroscience 64: 943–958

Randic M, Hecimovic H, Ryu PD (1990) Substance P modulates glutamate-induced currents in acutely isolated rat spinal dorsal horn neurones. Neurosci Lett 117: 74–80

Ribeiro-da-Silva A, De Koninck Y, Cuello AC, Henry JL (1992) Enkephalin-immunoreactive nociceptive neurons in the cat spinal cord. Neuroreport 3: 25–28

Ribeiro-da-Silva A, Mcleod AL, Ritchie J, Cuello AC, Julien J-P, Henry JL (1997) Thermal allodynia in transgenic mice over-expressing nerve growth factor: effects of morphine and naloxone. Soc Neurosci Abstr 23: 1806

Ruda MA, Bennett GJ, Dubner R (1986) Neurochemistry and neural circuitry in the dorsal horn. In: Emson PC, Rossor M, Tohyama M (eds) Peptides and Neurological Disease. Prog Brain Res, vol 66. Elsevier, Amsterdam, pp 219-268

Rusin KI, Ryu PD, Randic M (1992) Modulation of excitatory amino acid responses in rat dorsal horn neurons by tachykinins. J Neurophysiol 68: 265–286

Rusin KI, Bleakman D, Chard PS, Randic M, Miller RJ (1993a) Tachykinins potentiate *N*-methyl-D-aspartate responses in acutely isolated neurons from the dorsal horn. J Neurochem 60: 952–960

Rusin KI, Jiang MC, Cerne R, Randic M (1993b) Interactions between excitatory amino acids and tachykinins in the rat spinal dorsal horn. Brain Res Bull 30: 329–338

Rydh-Rinder M, Holmberg K, Elfvin LG, Wiesenfeld-Hallin Z, Hökfelt T (1996) Effects of peripheral axotomy on neuropeptides and nitric oxide synthase in dorsal root ganglia and spinal cord of the guinea pig: An immunohistochemical study. Brain Res 707: 180–188

Ryu PD, Randic M (1990) Low- and high-voltage-activated calcium currents in rat spinal dorsal horn neurons. J Neurophysiol 63: 273–285

Sagan S, Chassaing G, Pradier L, Lavielle S (1996) Tachykinin peptides affect differently the second messenger pathways after binding to CHO-expressed human NK-1 receptors. J Pharmacol Exp Ther 276: 1039–1048

Sakurada T, Sugiyama A, Sakurada C, Tanno K, Sakurada S, Kisara K, Hara A, Abiko Y (1996) Involvement of nitric oxide in spinally mediated capsaicin- and glutamate-induced behavioural responses in the mouse. Neurochem Int 29: 271–278

Salter MW, Henry JL (1987a) Evidence that adenosine mediates the depression of spinal dorsal horn neurons induced by peripheral vibration in the cat. Neuroscience 22: 631–650

Salter MW, Henry JL (1987b) Purine-induced depression of dorsal horn neurons in the cat spinal cord: enhancement by tachykinins. Neuroscience 23: 903–915

Salter MW, Henry JL (1988) Tachykinins enhance the depression of spinal nociceptive neurons caused by cutaneously applied vibration in the cat. Neuroscience 27: 243–249

Salter MW, Henry JL (1990) Physiological characteristics of responses of wide dynamic range spinal neurones to cutaneously applied vibration in the cat. Brain Res 507: 69–84

Salter MW, Henry JL (1991) Responses of functionally identified neurones in the dorsal horn of the cat spinal cord to substance P, neurokinin A and physalaemin. Neuroscience 43: 601–610

Shapiro MS, Hille B (1993) Substance P and somatostatin inhibit calcium channels in rat sympathetic neurons via different G protein pathways. Neuron 10: 11–20

Sharma HS, Nyberg F, Olsson Y, Dey PK (1990) Alteration of substance P after trauma to the spinal cord: An experimental study in the rat. Neuroscience 38: 205–212

Shen K-Z, Surprenant A (1993) Common ionic mechanisms of excitation by substance P and other transmitters in guinea-pig submucosal neurones. J Physiol (Lond) 462: 483–501
Skilling SR, Harkness DH, Larson AA (1992) Experimental peripheral neuropathy decreases the dose of substance P required to increase excitatory amino acid release in the CSF of the rat spinal cord. Neurosci Lett 139: 92–96
Sluka KA, Westlund KN (1993) Spinal cord amino acid release and content in an arthritis model: The effects of pretreatment with non-NMDA, NMDA, and NK1 receptor antagonists. Brain Res 627: 89–103
Soltoff SP, Toker A (1995) Carbachol, substance P, and phorbol ester promote the tyrosine phosphorylation of protein kinase Cδ in salivary gland epithelial cells. J Biol Chem 270: 13490–13495
Sugiura Y, Lee CL, Perl ER (1986) Central projections of identified, unmyelinated (C) afferent fibers innervating mammalian skin. Science 234: 358–361
Takahashi T, Otsuka M (1975) Regional distribution of substance P in the spinal cord and nerve roots of the cat and the effect of dorsal root section. Brain Res 87: 1–11
Tanabe T, Otani H, Bao LH, Mikami Y, Yasukura T, Ninomiya T, Ogawa R, Inagaki C (1996) Intracellular signaling pathway of substance P-induced superoxide production in human neutrophils. Eur J Pharmacol 299: 187–195
Taylor CW, Merritt JE, Putney JW, Rubin RP (1986) A guanine nucleotide-dependent regulatory protein couples substance P receptors to phospholipase C in rat parotid gland. Biochem Biophys Res Commun 136: 362–368
Terenghi G, Riveros-Moreno V, Hudson LD, Ibrahim NBN, Polak JM (1993) Immunohistochemistry of nitric oxide synthase demonstrates immunoreactive neurons in spinal cord and dorsal root ganglia of man and rat. J Neurol Sci 118: 34–37
Theriault E, Otsuka M, Jessell T (1979) Capsaicin-evoked release of substance P from primary sensory neurons. Brain Res 170: 209–213
Thompson SWN, Dray A, Urban L (1994) Injury-induced plasticity of spinal reflex activity: NK1 neurokinin receptor activation and enhanced A- and C-fiber mediated responses in the rat spinal cord in vitro. J Neurosci 14: 3672–3687
Urban L, Naeem S, Patel IA, Dray A (1994) Tachykinin induced regulation of excitatory amino acid responses in the rat spinal cord in vitro. Neurosci Lett 168: 185–188
Verge VMK, Richardson PM, Wiesenfeld-Hallin Z, Hökfelt T (1995) Differential influence of nerve growth factor on neuropeptide expression *in vivo*: A novel role in peptide suppression in adult sensory neurons. J Neurosci 15: 2081–2096
Womack MD, Macdermott AB, Jessell TM (1988) Sensory transmitters regulate intracellular calcium in dorsal horn neurons. Nature 334: 351–353
Yamaguchi K, Nakajima Y, Nakajima S, Stanfield PR (1990) Modulation of inwardly rectifying channels by substance P in cholinergic neurones from rat brain in culture. J Physiol (Lond) 426: 499–520
Yashpal K, Wright DM, Henry JL (1982) Substance P reduces tail-flick latency: implications for chronic pain syndromes. Pain 14: 155–167
Yashpal K, Dam T-V, Quirion R (1990) Quantitative autoradiographic distribution of multiple neurokinin binding sites in rat spinal cord. Brain Res 506: 259–266
Yashpal K, Dam T-V, Quirion R (1991a) Effects of dorsal rhizotomy on neurokinin receptor sub-types in the rat spinal cord: a quantitative autoradiographic study. Brain Res 552: 240–247
Yashpal K, Radhakrishnan V, Henry JL (1991b) NMDA receptor antagonist blocks the facilitation of the tail flick reflex in the rat induced by intrathecal administration of substance P and by noxious cutaneous stimulation. Neurosci Lett 128: 269–272
Yashpal K, Radhakrishnan V, Coderre TJ, Henry JL (1993) CP-96, 345, but not its stereoisomer, CP-96, 344, blocks the nociceptive responses to intrathecally administered substance P and to noxious thermal and chemical stimuli in the rat. Neuroscience 52: 1039–1047
Yashpal K, Pitcher GM, Parent A, Quirion R, Coderre TJ (1995) Noxious thermal and chemical stimulation induce increases in ^{3}H-phorbol 12,13-dibutyrate binding in spinal

cord dorsal horn as well as persistent pain and hyperalgesia, which is reduced by inhibition of protein kinase C. J Neurosci 15: 3263–3272
Yashpal K, Hui-Chan CWY, Henry JL (1996) SR 48968 specifically depresses neurokinin A- vs substance P-induced hyperalgesia in a nociceptive withdrawal reflex. Eur J Pharmacol 308: 41–48
Yashpal K, Henry JL (1983) Endorphins mediate overshoot of substance P-induced facilitation of a spinal nociceptive reflex. Can J Physiol Pharmacol 61: 303–307
Yashpal K, Henry JL (1984) Substance P analogue blocks SP-induced facilitation of a spinal nociceptive reflex. Brain Res Bull 13: 597–600
Zhang Q, Ji R-R, Lindsay R, Hökfelt T (1995) Effect of growth factors on substance P mRNA expression in axotomized dorsal root ganglia. Neuroreport 6: 1309–1312

Correspondence: Dr. J. L. Henry, Department of Physiology and Psychiatry, McGill University, 3655 Drummond St Montreal (Quebec), Canada H3G 1Y6.

Spinal Cord Monitoring
Eds: E. Stålberg, H. S. Sharma, Y. Olsson

Chapter 4

The role of glycine in pain and spasticity

R. K. Simpson, Jr.[1], **C. S. Robertson**[1], and **J. Clay Goodman**[2]

Departments of [1]Neurosurgery and [2]Pathology, Baylor College of Medicine, Houston, Texas, USA

Introduction

Neuropathic pain is an intense pain perceived without an obvious noxious stimulus. Spasticity is heightened stretch reflexes and excessive muscle tonus. Both are a common and severely debilitating sequela to injury of the nervous system and both conditions frequently occur together simultaneously. In order to support this glycine-based theory we tested the primary hypothesis with the assumption that a relationship exists between segmental glycine levels and neuropathic pain and spasticity. These tests included: 1) the detection of segmental glycine release after motor and sensory pathway stimulation in normal injured animals, 2) the reduction of pain and spasticity following administration of glycine, and its receptor agonists, and 3) the production of pain and spasticity following administration of glycine receptor antagonists. Once fulfilled, results from these experiments would support the goal directed toward enhancing glycine receptor mechanisms as treatment for neuropathic pain and spasticity. Two animal models were employed to test the hypothesis, 1) a classic model of neuropathic pain in the rat created by sciatic nerve constriction, and 2) an established model of spasticity in the rabbit created by spinal cord ischemia.

Topic background and development

Establishing a role for glycine in pain and spasticity

Neuropathic pain and spasticity are both highly complex expressions of neuronal activity following interruption of normal communications between cerebrum and peripheral nerves (Faganel *et al.*, 1982; Kuypers, 1981; Nathan *et al.*, 1959; Neilson, 1972; Shapovalov, 1975). A classic example of neuropathic pain is found in patients as reflex sympathetic dystrophy (RSD) (Schwartzman, 1992; Weil, 1992). A disorder associated with neuropathic pain is an abnormal motor condition commonly seen following any CNS injury, considered to be as spasticity (Herz *et al.*, 1990; Schwartzman, 1992). Current modalities available for treating neuropathic pain and spasticity are limited (Barolat *et al.*, 1989). These include some surgical procedures and few pharmacological agents (North *et al.*, 1991; Walta, 1982). The surgical procedures are, in general, ablative, not adjustable, and leave patients little to no capacity for recovery from associated deficits (Herz *et al.*, 1990). Few effective drugs are available and all those currently used have significant and often dangerous side effects (Albright *et al.*, 1991). Additional drugs that may add to the limited armamentarium of neuropathic pain and spasticity treatment would be highly desirable. To produce such an agent, a clearer understanding of the underlying neurochemical mechanisms responsible for these conditions is necessary.

Segmental interneuron activity plays a key role in mediating afferent input and smooth motor performance. These controls are lost as a consequence of a neurological injury (Barolat *et al.*, 1980; Dimitrijevic *et al.*, 1989; Masland, 1985). Loss of inhibitory interneuron activity contributes significantly to abnormal processing of sensory information and to a poorly coordinated muscle response (Sotgui, 1993; Sugimoto *et al.*, 1989). Glycine is an abundant postsynaptic inhibitory amino acid neurotransmitter associated with these inhibitory interneuron pools (Brugger *et al.*, 1992; Davidoff *et al.*, 1967; Hammerstad *et al.*, 1971; Hopkin *et al.*, 1971; Panter *et al.*, 1990; Rizzoli, 1968; Roberts *et al.*, 1972; Werman *et al.*, 1967; Werman *et al.*, 1968; Young *et al.*, 1987). Since glycine is likely to be involved in the mechanisms responsible for the expression of neuropathic pain and spasticity, glycine administration as a single agent or in combination with other glycinergic compounds should influence the severity of these conditions (Rao *et al.*, 1990; Scheardown *et al.*, 1989; Semba *et al.*, 1993; Simpson *et al.*, 1990; Simpson *et al.*, 1991; Simpson *et al.*, 1991; Simpson *et al.*, 1992; Simpson *et al.*, 1993; Simpson *et al.*, 1993). Since neuropathic pain and spasticity are frequently seen together after a neurological injury a common, basic segmental mechanism responsible is highly probable (Schwartzman, 1992; Weil 1992).

Glycine has been shown to be an abundant segmental amino acid which can be readily measured in the rat and rabbit animal model using microdialysis techniques (Araki *et al.*, 1988; Martiniak *et al.*, 1989). Recent reports from preliminary studies indicate that motor and sensory stimulation produce elevated segmental glycine levels in injured and control animals (Simpson *et al.*, 1990; Simpson *et al.*, 1991; Simpson *et al.*, 1991; Simpson *et al.*, 1991; Simpson *et al.*, 1992; Simpson *et al.*, 1993; Simpson *et al.*, 1993). The stimulation techniques used were designed to mimic those commonly applied to humans to treat neuropathic pain and spasticity (Barbeau, 1974; Barolat *et al.*, 1989; Herz *et al.*, 1990). Recent reports have also shown that neuropathic pain and spasticity are diminished with intrathecal glycine agonists and aggravated by glycine antagonists, respectively (Turski *et al.*, 1985; Turski *et al.*, 1987; Yamamoto *et al.*, 1992; Yamamoto *et al.*, 1993). Based on these observations, the close association of both neuropathic pain and spasticity after neurological injury, and the similarity in effective clinical treatment regimens, neuropathic pain and spasticity can be studied together.

Mechanisms of neuropathic pain and spasticity

Neuropathic pain, as a consequence of damage to peripheral or central neural structures, has been described since 1864 by Mitchell (Mitchell, 1864). Although many theories have been proposed to implicate peripheral mechanisms, the importance of spinal mechanisms, particularly the role of segmental interneurons, in the production of neuropathic pain is becoming increasingly clear. After the "Gate Theory" of pain was proposed by Melzack and Wall, several investigations showed that interneuron pools mediate nociceptive information from afferent sources (Melzack *et al.*, 1965). Superficial dorsal horn neurons (Rexed laminae I and II) respond primarily to Aδ and C fibers and are mediated by a variety of neurotransmitter substances and become hypersensitive when neuropathic pain is present (Basbaum, 1988; Kajander *et al.*, 1992; Laird *et al.*, 1992). Several interneuron pools contribute to control of dorsal horn neuron activity and appear to be selectively damaged under conditions where neuropathic pain exists (Mitchell *et al.*, 1993; Ren *et al.*, 1992; Schwartzman, 1992). Dorsal horn neurons of wide dynamic range that receive Aβ afferents are also likely to become sensitized without the modulation provided by interneurons and, therefore, contribute to the pain response (Rudomin *et al.*, 1986; Rudomin *et al.*, 1990; Sheen *et al.*, 1993). These pools include both small inhibitory interneurons in the dorsal horn as well as interneurons that are more ventrally located yet project dorsally (Campbell *et al.*, 1988). In support of this observation, immunohistochemical studies show that neurons in the dorsal horn regions involved in pain perception receive numerous postsynaptic glycinergic terminals (Sotgui, 1993; Todd, 1989; Wood *et al.*, 1989; Wood *et al.*, 1989; Yates *et al.*, 1982). In addition, dorsal horn cells in Rexed laminae I–V show reduced activity after ionophoretic application of glycinergic compounds and increased activity after application of glycine antagonists (Basbaum, 1988; Todd, 1989).

Several neurophysiologic mechanisms have been proposed to account for the increased muscle tone seen in spastic patients (Amassian *et al.*, 1987). Lack of descending inhibitory control or exaggerated afferent input leading to an imbalance of inhibitory and excitatory activity, as proposed by Sherrington, is accepted as the general mechanism of spasticity (Sherrington *et al.*, 1915). Discharge patterns of activated motor neurons in spastic patients suggest that a polyneuronal, plurisegmental network of "miscommunication" is involved (Faganel *et al.*, 1982; Nathan *et al.*, 1959). Facilitation of short latency stretch reflexes by cutaneous stimulation is evidence for a lack of adequate postsynaptic inhibitory control (Burke *et al.*, 1972). Segmental interneurons in the medial and ventral portions of segmental gray matter play a key role in mediating inhibitory control and govern the ultimate magnitude and quality of a motor response (Howell *et al.*, 1979; Kuypers, 1973). These cells include Renshaw cells, and type Ia, and type Ib interneurons and can be activated by corticospinal, rubrospinal, vestibulospinal, and reticulospinal pathways as well as segmental reflex loops incorporating peripheral afferent fibers, and the propriospinal network (Harrison *et al.*, 1985; Harrison *et al.*, 1985; Lundberg *et al.*, 1962; Magoun *et al.*, 1946; Rudomin *et al.*, 1986; Rudomin *et al.*, 1987; Rudomin *et al.*, 1990; Schreiner *et al.*, 1949; Young *et al.*, 1973). Excitatory postsynaptic potential (EPSP) records from motor neurons are enhanced and prolonged in spasticity due to altered interneuronal activity (Ashby *et al.*, 1987; Biscoe *et al.*, 1986; Mailis *et al.*, 1990). Likewise, inhibitory postsynaptic potentials (IPSPs) produced by activating these interneurons are reduced following the administration of glycine antagonists (Aprison, 1990; Curtis *et al.*, 1968; Rudomin *et al.*, 1986).

Association between glycine and neuropathic pain and spasticity

The role of glycine as an inhibitory neurotransmitter was initially suggested by Curtis in 1960, and since that time it has been established as the chief postsynaptic inhibitory neurotransmitter in the spinal cord (Aprison *et al.*, 1969; Aprison, 1990; Curtis *et al.*, 1968; Curtis *et al.*, 1971). Autoradiographic and immunohistochemical studies have shown glycine to be closely associated with medial and, in particular, ventral interneuron pools (Berger *et al.*, 1977; Fagg *et al.*, 1978; Fagg *et al.*, 1983). Glycine is liberated from cells corresponding in location to electrophysiologically defined Renshaw cells, and Ia and Ib interneurons (Aprison, 1990, Ashby *et al.*, 1987; Davidoff, 1989; Henneman, 1981). As with γ-amino butyric acid (GABA), glycine containing interneurons are also located within the dorsal horn (Farrant *et al.*, 1989; Farrant *et al.*, 1990; Hunt, 1983) and may act in concert with GABA to modify peripheral afferent activity.

Glycine is a simple, non-essential, ubiquitous amino acid that is found in higher concentration within the spinal cord than the brain, especially in the cervical and lumbosacral enlargements (Daly *et al.*, 1983; Hunt, 1983). The cerebrospinal fluid (CSF)/serum concentration ratio is lower than that of other amino acids and the influx of glycine into the CNS from external sources is negligible (Aprison, 1990; Shank *et al.*, 1970). Glycine is synthesized from serine by a mitochondrial enzyme, serine hydroxymethyltransferase (Shank *et al.*, 1970). The origin of serine is from glucose metabolism within the CNS (Farrant *et al.*, 1990). After synaptic release, the major system of glycine clearance is an energy dependent vesicular reuptake mechanism (Christensen *et al.*, 1990).

Glycine acts on postsynaptic neuronal membranes through two receptor subtypes, a strychnine sensitive chloride channel (Gly 1) producing membrane hyperpolarization and a much smaller population of *N*-methyl-D-aspartate (NMDA) sensitive channels (Gly 2) mediating calcium flux and producing membrane depolarization (Chizhmakov *et al.*, 1989; Larson, 1989; McNamara, 1990; Patel, 1990). Dorsal horn neurons subserving afferent inputs, particularly those involved in nociception, are readily activated by NMDA agonists (Basbaum, 1988, Beyer *et al.*, 1988; Beyer *et al.*, 1989). The primary effect of glycine on motor neuron activity is via activation of the hyperpolarizing chloride channel (Bruggencate *et al.*, 1968; Curtis *et al.*, 1968; Curtis *et al.*, 1971; Zeglgansberger *et al.*, 1979).

Neuropathic pain and spasticity can be produced by nonspecific NMDA channel stimulation whereas NMDA channel blockade by substances such as MK-801 can reduce neuropathic pain and spasticity (Cahusac *et al.*, 1984; Gerber *et al.*, 1991; Hao *et al.*, 1991; Turski *et al.*, 1985; Turski *et al.*, 1987). Blockade of glycine mediated chloride channels with the selective antagonist strychnine amplifies the pain response and produces spasticity. In contrast, glycine administration reduces sensitivity to noxious stimulation and muscle tone in neuropathic or spastic animals (Biscoe *et al.*, 1986; Erdo, 1990; Hall *et al.*, 1979; Larson, 1989; Smith *et al.*, 1979; Yaksh, 1989). The dual role of glycine in the spinal cord as it pertains to neuropathic pain and spasticity is only beginning to be deciphered using combinations of Gly 1 and Gly 2 agonists and antagonists (Lester *et al.*, 1993; Lynch *et al.*, 1993; Mao *et al.*, 1992; Mao *et al.*, 1992; Mao *et al.*, 1994; Nicola *et al.*, 1992; Nogaoka *et al.*, 1993; Sher *et al.*, 1990; Song *et al.*, 1993; Van den Pol *et al.*, 1988). Powerful glycine analogs such as D-serine have not been adequately tested to support for this therapeutic effect, nor have glycine-NMDA antagonists such as 5-7 dichlorokynurenic acid (5-7 DKA) (Cahusac *et al.*, 1984; Davar *et al.*, 1991; Thomson, 1991; Yoneda *et al.*, 1990).

Measurement of neuropathic pain and spasticity

A variety of methods have been developed to quantify pain (Brown *et al.*, 1975; D'Amour *et al.*, 1941; Hargreaves *et al.*, 1988; Randall *et al.*, 1957). Clinically, a numerical scale such as the visual analog score (VAS) is frequently used (Schwartzman, 1992; Weil, 1992). Experimentally, pain is generally measured using mechanical stimulation or thermal stimulation, each activating a relatively select group of nociceptors. The time to limb movement or vocalization is generally used to determine pain sensation in laboratory animals (Bennett *et al.*, 1988; Bennett *et al.*, 1989; Bennett *et al.*, 1991). Mechanical stimulation of pain receptors via calibrated pressure and thermal stimulation by a hot plate or a light source are the most accepted means of pain measurement, particularly in the laboratory setting (Attal *et al.*, 1990).

Spasticity is currently measured by a clinical grading scheme, most commonly the Ashworth score (Mathias *et al.*, 1989; Rudick *et al.*, 1987). The Ashworth score, although useful, is subjective and lacks quantitative precision. However, several types of electrophysiological recordings are generally accepted as providing quantitative measurements of spasticity (Dickenson *et al.*, 1990; Makitie *et al.*, 1977; Rymer *et al.*, 1989; Sheen *et al.*, 1993; Shimoji *et al.*, 1978). These include the raw or averaged electromyogram (EMG) and H- reflexes (Ashby *et al.*, 1974; Barolat *et al.*, 1980; Fasano *et al.*, 1982; Lindsley *et al.*, 1949). Amplitude and latency measurements from these tests are used to gauge the severity of spasticity. Although new approaches to quantifying spasticity are being developed such as the use of polyEMG activity evoked by transcranial motor cortex stimulation (Delwaide, 1985; Dimitrijevic *et al.*, 1989), H-reflex measurements have been shown to be the most reliable electrophysiologic measurement of spasticity. Ratios constructed from the amplitude of the H and M waves are frequently used to reflect the degree of muscle hypertonia (Heller *et al.*, 1982; Meinck, 1976).

Experimental procedure (neuropathic pain model)

Sprague-Dawley male rats between 350–400 grams were anesthetized with sodium pentobarbital (Nembutal, 40mg/kg IP). Body temperature was monitored with a rectal probe and a heating pad and light were used as needed. A cut-down to the sciatic nerve in the upper hind limb was performed. Chromic ligatures (4-O), four in number were loosely secured around the entire nerve diameter and separated 1 mm apart. The ligatures were then tightened just sufficient to minimally alter the diameter of the nerve, as viewed by an operating microscope. An incision was made in the atlanto-occipital membrane and an intrathecal catheter was placed into the cisterna magna and threaded to the lumbosacral region. A connector was secured to the catheter which was secured to the skull base with dental acrylic and the wound was closed. Daily flushing of the catheter was done until the chronic phase of study. Butorphanol tartrate (Torbutrol 0.1 mg/kg IP) and enrofloxcin (Baytril 0.1 mg/kg IP) was administered every 12 hours for 5 days. Each animal was maintained for 12 days prior to pain testing.

Experimental procedure (spasticity model)

Albino New Zealand male rabbits weighing between 3–5 kgs were used and were anesthetized with methohexital sodium (Brevital 10 mg/kg IV). The animals are ventilated with a volume-cycled rodent ventilator and maintained with 1–2% halothane (Fluothane) in 100% oxygen. Following surgery, the animals were immobilized with pancuronium bromide (Pavulon 0.2 mg/kg IV). Body temperature is monitored and maintained at 38°C using a heating pad and lamp. Blood pressure, blood gases, and pH were also monitored. A percutaneous cut-down was done in the left thigh through a lateral approach to transfemorally insert a sterile 4 French pediatric Swan-Ganz catheter. Approximately 0.6 cc of air was used to inflate the balloon just below the level of the renal arteries. A small vertex incision was made and sterile bipolar (4 mm Ag/AgCl cup) cortical stimulating electrodes were subcutaneously placed overlying the primary motor cortex for the left hindlimb. Two sterile stainless steel EMG needle recording electrodes, insulated except for the tips, were percutaneously inserted just lateral to the spinous process of the L5 and L6 vertebrae in contact with the lamina. Postoperatively, analgesics (butorphanol tartrate, Torbutrol 0.1 mg/kg SQ) and antibiotics (cephalothin sodium, Keflin 25 mg/kg IM) were administered every 8 hours. Daily assessments by H-reflex testing and Ashworth score determination were also done.

Electrophysiological analysis

Contributions to lumbar segmental activity by descending motor systems of cerebral origin were assessed by lamps CMEP analysis (Simpson *et al.*, 1989; Simpson *et al.*, 1993). The CMEPs were optimally produced by transcranial anodal stimulation of the motor cortex at an intensity of 5–10 mA, for a duration of 1.0 msec, and at a frequency of 3 Hz using a Grass S-88 stimulator. The signals were recorded from the paralaminar EMG electrodes at L5 (active) and L6 (inactive). All signals were amplified, filtered (5–3000 Hz), and averaged (n=100) using a Metraco Inc. evoked potential system. Signals were digitized and stored on floppy discs (Apple Computer IIe) and paper with a digital printer. CMEPs records were obtained before intraaortic balloon inflation, and frequently during ischemia. Records are also made immediately, during, and after reperfusion. Following the survival period, in injured or sham animals, CMEPs were recorded before and after insertion of the microdialysis probe. Signals were also obtained periodically during continuous epidural spinal cord or peripheral afferent nerve stimulation, microdialysis sampling, or drug infusion studies. Continuous cortical stimulation and CMEP records were made during investigations directed at the contributions to segmental glycine levels from primary cortex activation.

EMG records were obtained from surface disk and multiunit needle electrodes placed into the lateral muscle mass of the hindlimb ipsilateral to the microdialysis probe. Raw EMG records were recorded periodically before, during, and after continuous motor pathway stimulation. These records of 100 msec duration, sampled at a rate of 1/sec for 1 minute will be averaged, and stored using a TECA II electromyographic recording system. The number of spikes per second were tabulated from the raw EMG signals. Averaged EMG amplitude, and signal area were also analyzed.

H-reflexes were recorded from the plantar surface of the foot using surface disk electrodes in response to stimulation of the tibial nerve. M and H waves were measured in response to graded increments in stimulus intensity, from 0 mA to 20 mA using the TECA II system. H/M curves were constructed based on peak waveform amplitudes. Signals were recorded before ischemic spinal cord injury, and then daily until the infusion protocol began. H-reflexes were then recorded just before, and periodically after continuous infusion of the medication.

Amino acid sampling

The microdialysis probe used in the study is constructed by using a loop of dialysis tubing prepared from a saponified cellulose ester, 0.2 mm in diameter, with a molecular weight cut-off of 6000 (Simpson *et al.*, 1990). The probe is made rigid by inserting an extremely fine stainless steel wire. The construct is inserted into two stainless steel 25 gauge blunt needles for support. A seal between the dialysis tube and the needle is made with crazy glue, and the assembly is carefully bowed into a loop. Silastic tubing is connected to the needles for perfusion and collection. The probe is mounted on a spinal stereotaxic device, is inserted into the spinal cord, and positioned so that the permeable segment passes through the medial and ventral spinal cord gray matter via a L5-L6 laminectomy. Stereotaxic coordinates based on the dorsal midline septum at the L5-L6 region have been calculated and verified in several animals. Once the probe is secured with dental acrylic and the animal released from the frame, the probe is perfused with artificial CSF (pH-7.3, Na-140 meq/l, K-3 meq/l, Mg-1 meq/l) at a rate of 5 μl/min by a microinjection pump. Samples are collected in microvials mounted on the stereotaxic unit and handled with sterile technique. For the study, samples were obtained at frequent intervals prior to, during, and after termination of epidural spinal cord, peripheral afferent, or primary motor cortex stimulation, and placed in a freezer at −70°C.

Analysis of samples

Concentrations of glycine and related compounds were measured in the dialysate samples by precolumn phenylisothiocyanate (PITC) derivatization followed by reverse phase, gradient programmed, high pressure liquid chromatography with spectrophotometric detection (Waters PICOTAG system) (Bidlingmeyer *et al.*, 1984; Cohen *et al.*, 1986). Norleucine is used as an internal standard and quantitation is performed using area integration. The resolution of the system is 1.0 picomole per sample or 0.1 millimol per liter.

Pain measurements

Both mechanical nociception and thermal nociception were tested in response to glycinergic compound administration (Simpson *et al.*, 1995). On the twelfth day after creating the partial sciatic nerve ligation injury and placement of the catheter, the animals were subject to intrathecal drug infusions. Over the 12 days prior to infusion the animals were conditioned to the apparatus to prevent stress and addition injury. Mechanical stimulation of nociceptors was provided by the Randall-Sellito technique. A pneumatic pressure device increases the weight in grams applied to the footpad of the neuropathic and sham limb in a calibrated fashion. The sensitivity of the animal is based on the amount of pressure applied by the device that results in limb withdrawl, or the pain response. Both hindlimbs from each animal were tested prior to infusion. These tests continued during and after infusion. Thermal stimulation of nociceptors was provided by the hot plate technique. The time for limb withdrawl or elevation from the surface of a hot plate maintained at 42°C was measured. Again, both hindlimbs from each animal were tested prior to, during, and after infusion. Each animal was subjected to both mechanical and thermal nociceptor testing at alternating intervals.

Drug infusion

For the study of neuropathic pain, glycine, strychnine, and MK-801 and 5-7 DKA are infused at concentrations of 0.1 mmol in artificial CSF at 10 µl/min. The dose of each drug was delivered through an intrathecal catheter. For the study of spasticity, intrathecal infusion of glycine, strychnine, D-serine, β-alanine, MK-801 and 5-7 DKA was performed at a rate of 5 µl/min at a concentration of 100 mM in artificial CSF.

Statistical analysis

Data was analyzed using repeated measures analysis of variance (ANOVA) for experiments involving segmental amino acid concentrations, time to limb withdrawl after thermal or mechanical noxious stimulation, and H/M ratios.

Results from electrophysiological experiments (acute studies)

CMEPs have become a widely used clinical tool for evaluating spinal cord function (Boyd *et al.*, 1986; Dimitrijevic *et al.*, 1989; Grossi *et al.*, 1988; Kanek *et al.*, 1988; Kitagawa *et al.*, 1989; Levy *et al.*, 1984; Thompson *et al.*, 1987). These signals accurately reflect the outcome from ischemic spinal cord injury and reliably predict the degree of associated spasticity. Stimulus and recording parameters were chosen for optimum CMEP waveform records based on earlier studies by the applicant while developing this model using the methods described below (Simpson *et al.*, 1989; Simpson *et al.*, 1991; Simpson *et al.*, 1993). CMEPs recorded epidurally or at the level of the spinal laminae in response to transcranial stimulation reveal nine components that are consistently reproduced and are super-

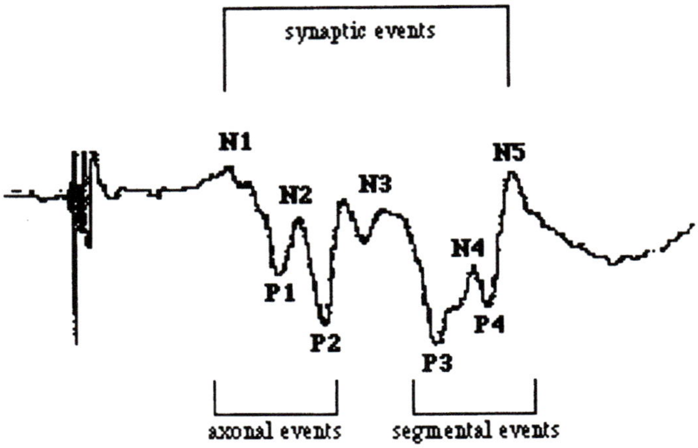

Fig. 1. Corticomotor evoked potential (*CMEP*). Long duration positive potential imposed on a positive potential of 5–7 msec in duration (Simpson *et al.*, 1989).

Alterations observed in CMEP components during timed periods of spinal cord ischemia reveal that after 10 minutes of ischemia, only the primary components or initial positive/negative deflection can be recorded. Reperfusion at this point results in return of all components. After 20 minutes of ischemia none of the components are clearly discernible. Reperfusion results in a rapid return of early latency components, however, late latency components remain significantly attenuated including the long duration, positive potential (Fig. 1). Reperfusion after 30 minutes of ischemia results in complete absence of the late latency components and the long duration positive potential along with significant loss of early latency response amplitudes (Simpson *et al.*, 1989; Simpson *et al.*, 1991; Simpson *et al.*, 1993).

The sequence of changes in the CMEP is consistent with classic studies of selective vulnerability of spinal cord structures to ischemia (Bernhard *et al.*, 1953; Brooks *et al.*, 1947; Cheng *et al.*, 1984; Davidoff, 1989; Ganes, 1982; Gelfan *et al.*, 1955; Lloyd, 1953; Martiniak *et al.*, 1989). Segmental neurons, particularly interneurons, synaptic elements, and axonal fibers will sequentially cease function after continuous ischemia and can be monitored for loss of electrical function by CMEP waveform analysis. These signals have been found to be helpful in predicting neurological deficits and the degree of spasticity in the animals, and compliment the H-reflex records but are less accurate as quantitative measures of spasticity. CMEPs primarily provide a method to assess the

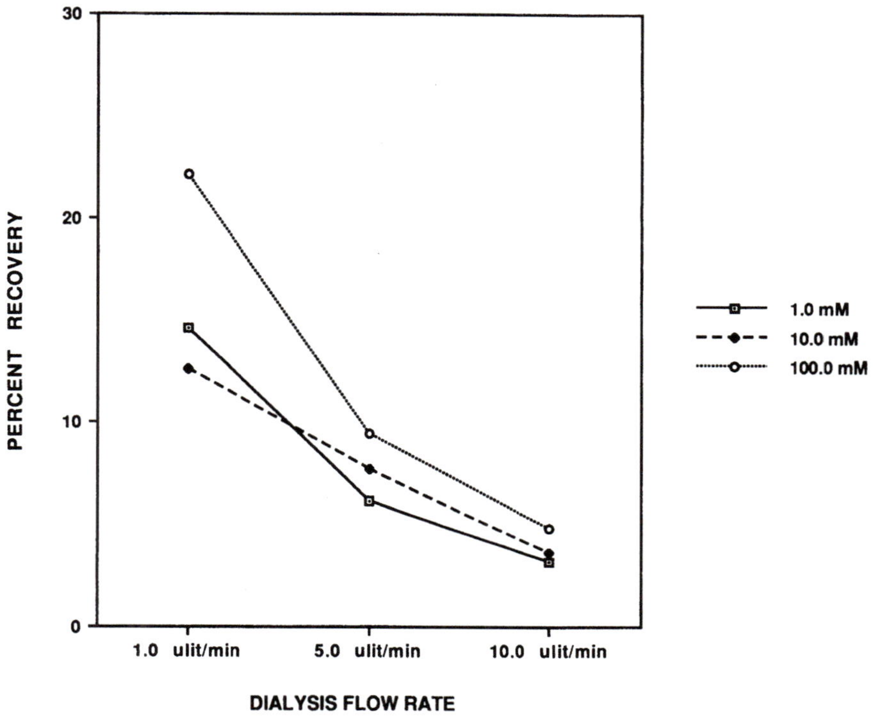

Fig. 2. A dialysis curve demonstrating the percent of glycine recovery from three standard solutions at three flow rates

integrity of the spinal cord during manipulation (Simpson *et al.*, 1989; Simpson *et al.*, 1993).

Microdialysis techniques permit continuous, accurate and repeatable estimates of concentrations of substances within the extracellular space with very little damage to surrounding tissue (Benveniste *et al.*, 1989; Juhasz *et al.*, 1989; Meyerson *et al.*, 1989; Quan *et al.*, 1989; Shuaib *et al.*, 1990). Preliminary studies revealed that sample aquisition approximately two hours following probe insertion was necessary to allow stabilization of the baseline extracellular environment (Benveniste 1989). Recovery rates for glycine determined using standard solutions dissolved in artificial CSF is consistently 6–8% of the standard solution when dialysis perfusion rate of 5 μl/min is used (Fig. 2) (Simpson *et al.*, 1990).

The segmental extracellular concentration of glycine was measured in response to spinal cord ischemia and stimulation of neural structures (Fig. 3). Although other amino acid neurotransmitter levels were measured,

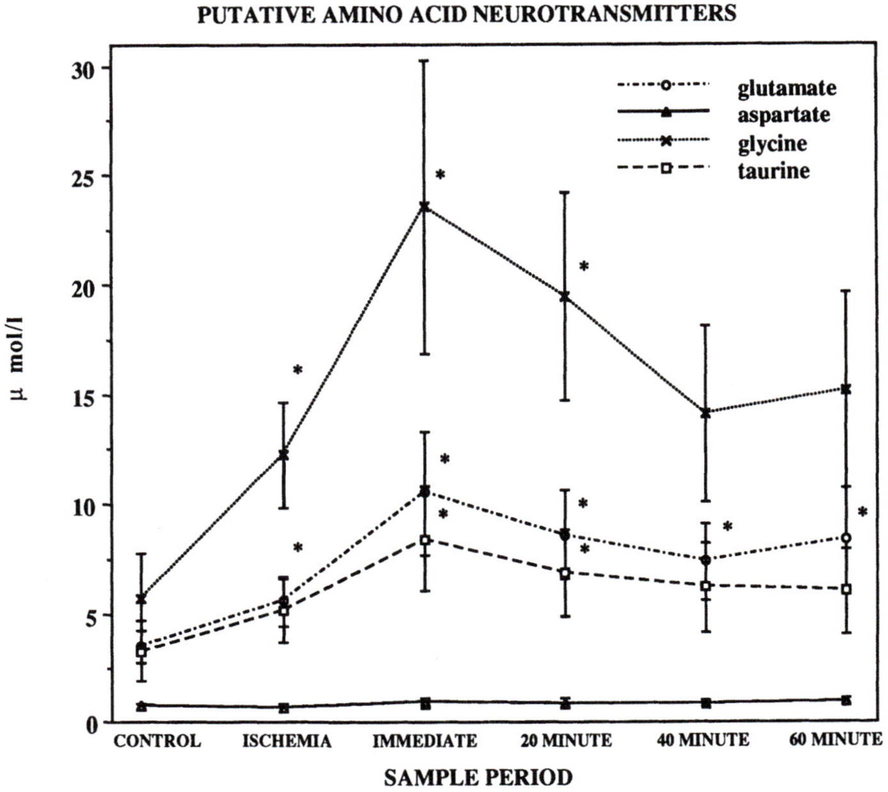

Fig. 3. Amino acid recovery before (control), during (ischemia), and after aortic occlusion or reperfusion (immediate, and 20, 40, and 60 minutes after balloon deflation). An asterix denotes a significant difference from control

glycine had the highest levels. Small elevations in the concentrations of non-neurotransmitter amino acids were measured after ischemia, including valine and leucine, which supports the lack of significant contamination of the segmental extracellular space by peripheral blood (Simpson *et al.*, 1990). In addition, GABA concentrations were exceedingly low in this model.

Results from electrophysiological experiments (chronic studies)

The degree of EMG activity recorded from hindlimb musculature in the injured animal reflects the degree of spasticity. The more complex and active the EMG record, the more spastic the animal. In addition, the amplitude and duration of the averaged EMG parallel spasticity. However, H-reflex records recorded from the plantar surface of the hind paw in response to tibial nerve stimulation were more sensitive to subtle changes in muscle hypertonia than either CMEPs or EMG. Stimulus intensities were increased in a sequential fashion from 0 to 20 mA and the amplitudes of the elicited H and M waves were plotted as a ratio (H/M) (Simpson *et al.*, 1995). This method has been found to be the most accurate reflection of the degree of clinically measured spasticity. The M waves were not significantly altered in amplitude regardless of injury. The H wave amplitudes were increased in proportion to the degree of clinically measured spasticity, thus producing a higher H/M ratio. Both clinical and electrophysiological measurements are used to determine the degree of spasticity throughout each experiment.

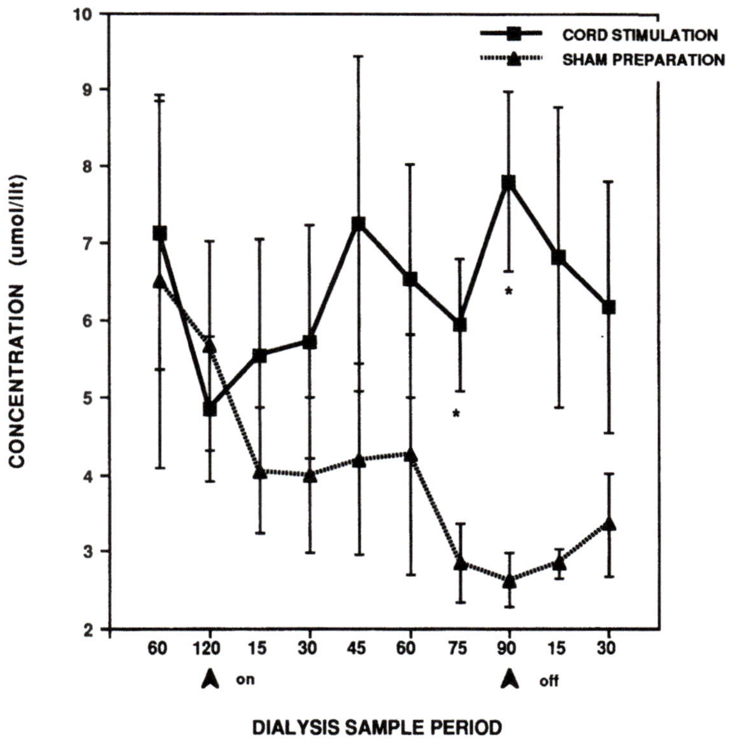

Fig. 4. Glycine recovery before, during, and after epidural spinal cord stimulation ("on" and "off") in uninjured animals. An asterix denotes a significant difference from control

Results from microdialysis experiments

Placement of the probe produced no measurable electrophysiological changes suggesting that little neuronal damage occurred. In uninjured animals, recovery of glycine was significantly elevated within 90 minutes of epidural spinal cord (Fig. 4) or peripheral afferent nerve stimulation (Fig. 5). A trend towards a significant increase was observed within 60 minutes after the onset of stimulation and was maintained in post-stimulation dialysis samples. Within 30 minutes after termination of stimulation, differences between the stimulated and sham groups were small and the variability in concentrations was large. Approximately two hours after stimulation, the concentration of glycine was near baseline values. These studies indicate that glycine is released into the segmental extracellular space in response to stimulation of neural structures that activate large numbers of interneuron pools.

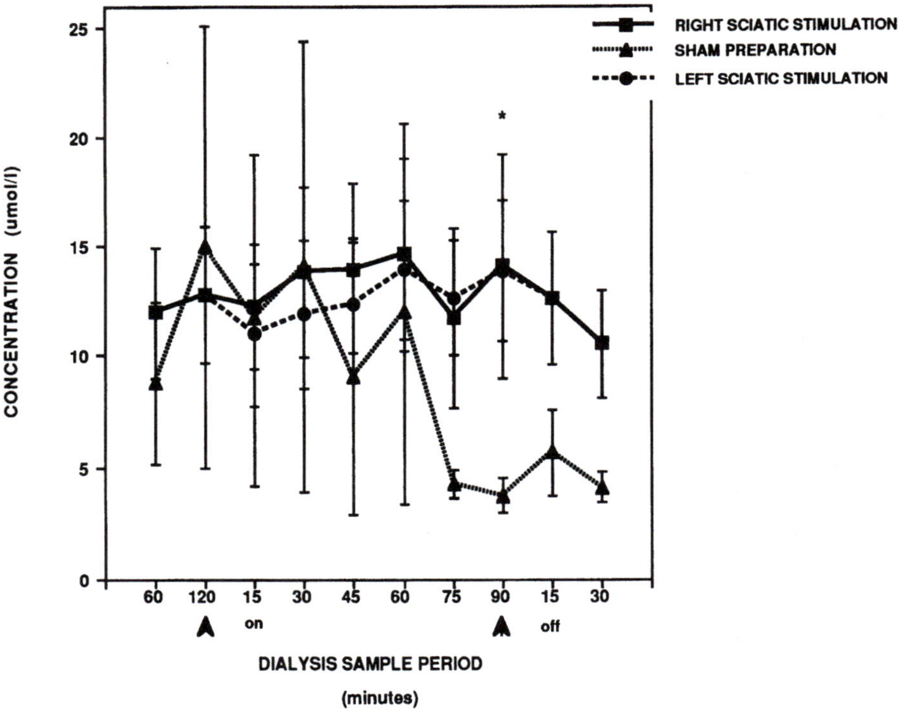

Fig. 5. Glycine recovery before, during, and after ipsilateral (left) or contralateral (right) sciatic nerve stimulation ("on" and "off") in uninjured animals. An asterix denotes a significant difference from control

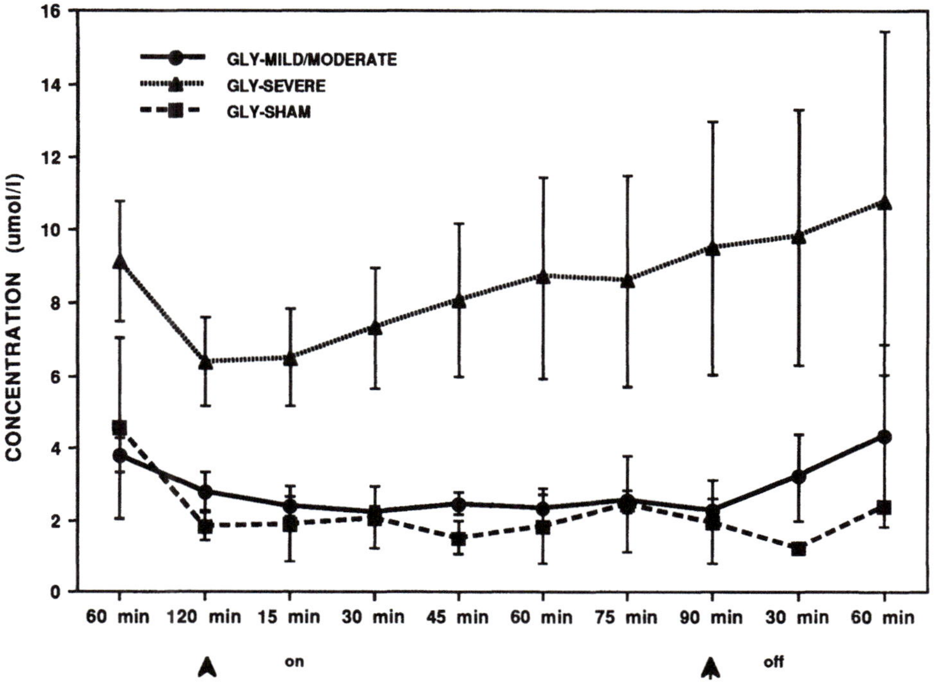

Fig. 6. Glycine recovery from mild/moderate or spastic animals, severe or flaccid animals, and sham animals. "On" and "off" indices in this figure refer to CMEP recording

Segmental microdialysis sampling was also performed in injured animals 3 days after spinal cord ischemia. Animals sustaining injuries that produced spasticity did not have a significant elevation in baseline glycine levels. However, animals sustaining severe, flaccid injuries had baseline glycine levels more than twice that of control pre-ischemia animals (Fig. 6). In addition, other injured animals underwent 90 minutes of continuous epidural spinal cord stimulation. Maintenance of glycine elevation was observed after 30 minutes of stimulation and was sustained throughout the 90 minute stimulation period irrespective of the degree of injury (Fig. 7). Likewise after approximately 75 minutes of peripheral afferent activation, glycine elevation was observed primarily in the ipsilateral (to the stimulus) section of the cord. Again, these results were obtained while using stimulation paradigms similar to those used in humans for control of neuropathic pain and spasticity (Simpson *et al.*, 1990; Simpson *et al.*, 1991; Simpson *et al.*, 1992; Simpson *et al.*, 1993; Simpson *et al.*, 1993). In addition, stimulation of the primary motor cortex produced a significant elevation of glycine. Such elevations were observed

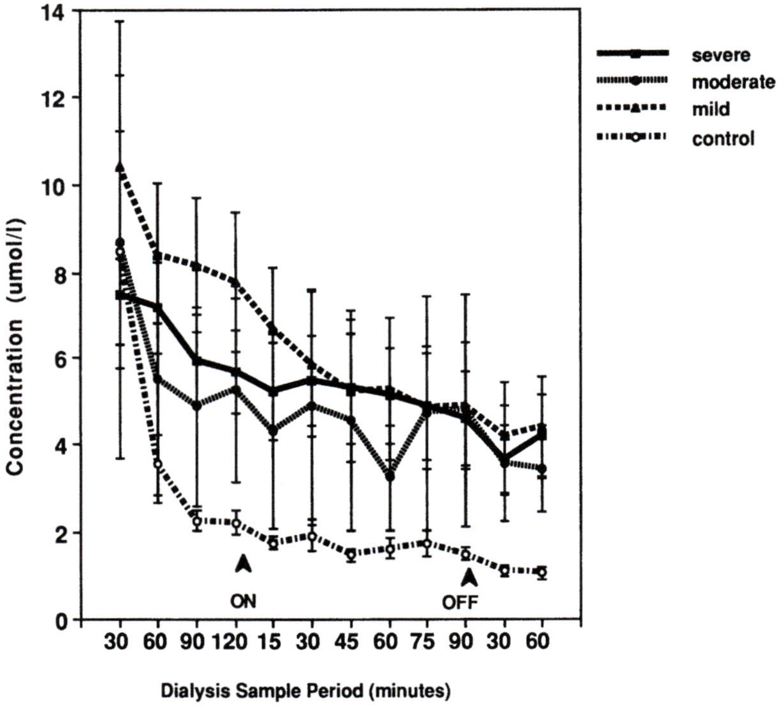

Fig. 7. Glycine recovery from mild/moderate or spastic animals, severe or flaccid animals, and sham animals during epidural spinal cord stimulation. "On" and "off" indices in this figure refer to application of epidural spinal cord stimulation

within 90 minutes and a trend towards significant differences was observed within 60 minutes of stimulation (Fig. 8).

Results from intrathecal drug delivery – pain

Neuropathic rats created by unilateral partial ligation of the sciatic nerve were treated twelve days after injury with intrathecal infusion of glycine, strychnine, MK-801, or 5–7 DKA, at a concentration of 0.1 μmol for 2 hours and at a rate of 5–10 μl/min (Simpson *et al.*, 1995; Simpson *et al.*, 1996; Simpson *et al.*, 1996). Pain scores were calculated using the Randall-Selitto pressure technique and the hot plate method and compared to rats receiving artificial CSF. The force necessary to produce a pain response, or movement of the neuropathic limb, was significantly increased after glycine administration. Strychnine, a specific glycine receptor (Gly 1) antagonist, significantly lowered the necessary force to evoke limb

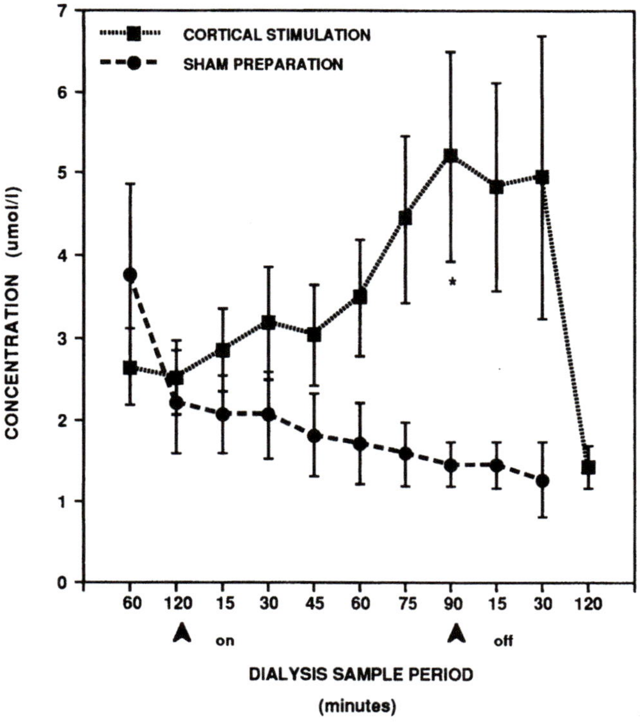

Fig. 8. Glycine recovery before, during, and after primary motor cortex stimulation ("on" and "off") in uninjured animals. An asterix denotes a significant difference from control

withdrawal and blocked the effect of glycine when administered simultaneously (Fig. 9 a-b). The influence of MK-801 or 5,7-DKA much less obvious (Fig. 9 c-d).

In contrast, strychnine failed to block the beneficial effects of glycine during hot plate testing yet the prolonged time for limb withdrawal in response to glycine administration was blocked when the non-specific NMDA receptor antagonist MK-801 was infused with glycine simultaneously (Fig. 10 a-b). A similar effect was observed although less dramatic after 5,7-DKA administration, a specific glycine-NMDA receptor (Gly 2) antagonist. These results are encouraging and provide a basis for additional studies related the differential effect of glycinergic (Gly 1 and Gly 2) compounds on pain, particularly with regard to mechanism of noxious stimulation. Mechanical stimulation via mechanonociceptors appears to be mediated through Gly 1 receptors whereas heat induced stimulation via thermonociceptors appears to be primarily mediated through Gly 2 receptors.

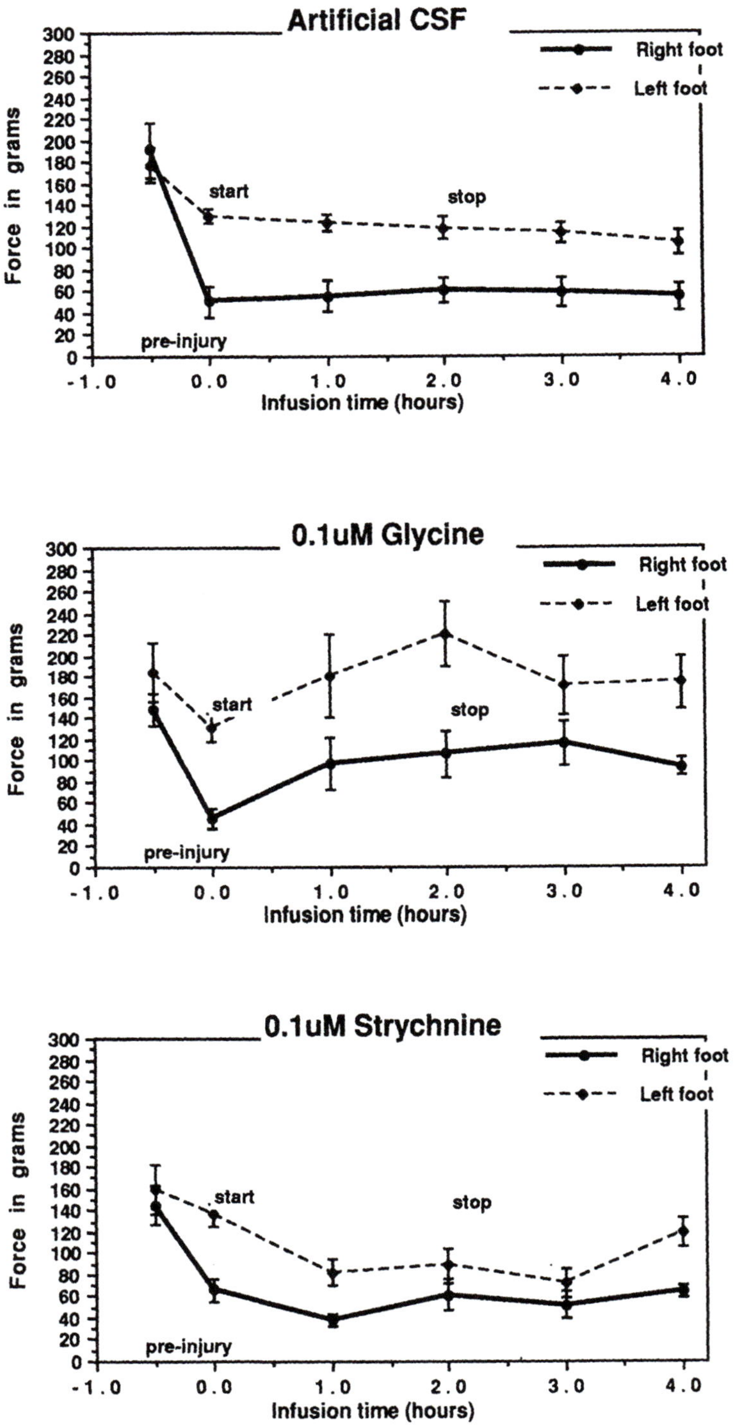

a

Fig. 9 a–d. Results of intrathecal drug infusion on the force necessary to produce foot withdrawal in the injured, neuropathic limb (right) and the uninjured limb (left)

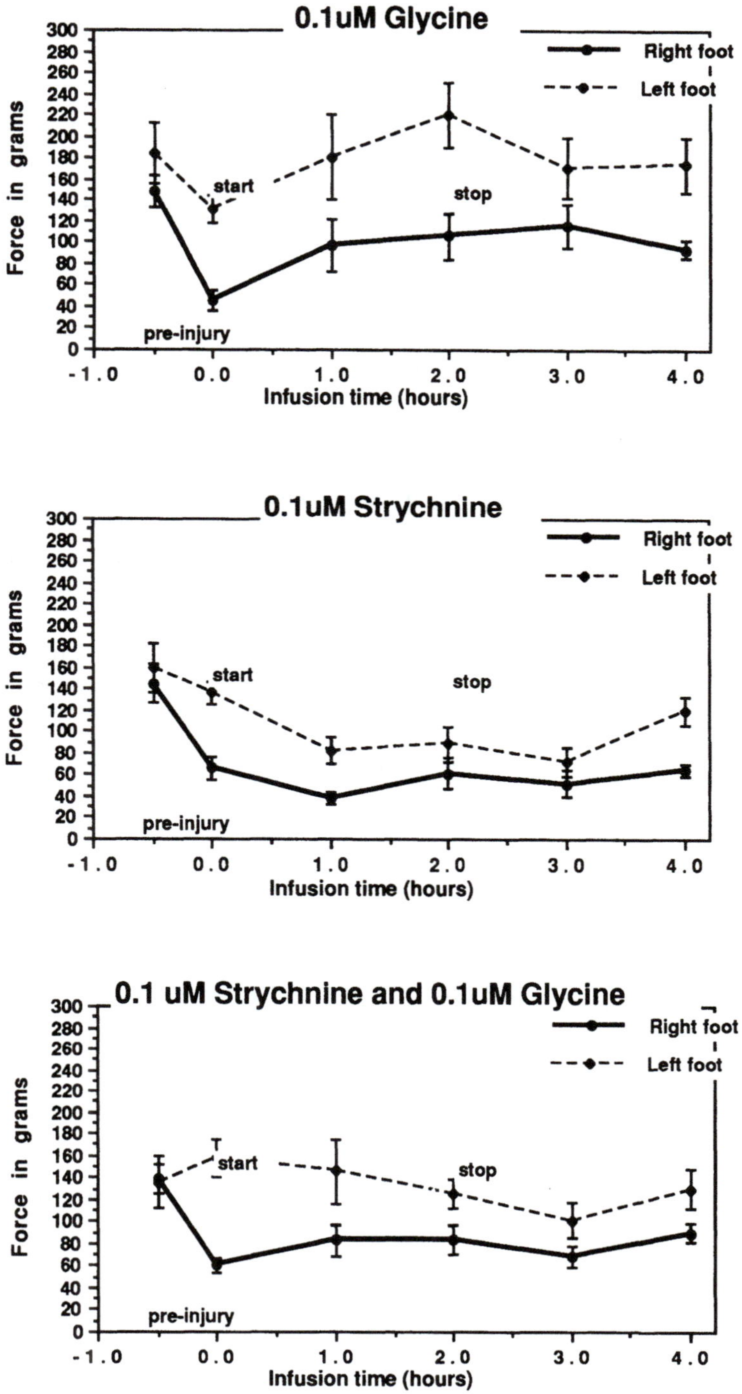
0.1uM Glycine
Right foot
Left foot
start
stop
pre-injury
Force in grams
Infusion time (hours)
0.1uM Strychnine
Right foot
Left foot
start
stop
pre-injury
Force in grams
Infusion time (hours)
0.1 uM Strychnine and 0.1uM Glycine
Right foot
Left foot
start
stop
pre-injury
Force in grams
Infusion time (hours)

b

Fig. 9b

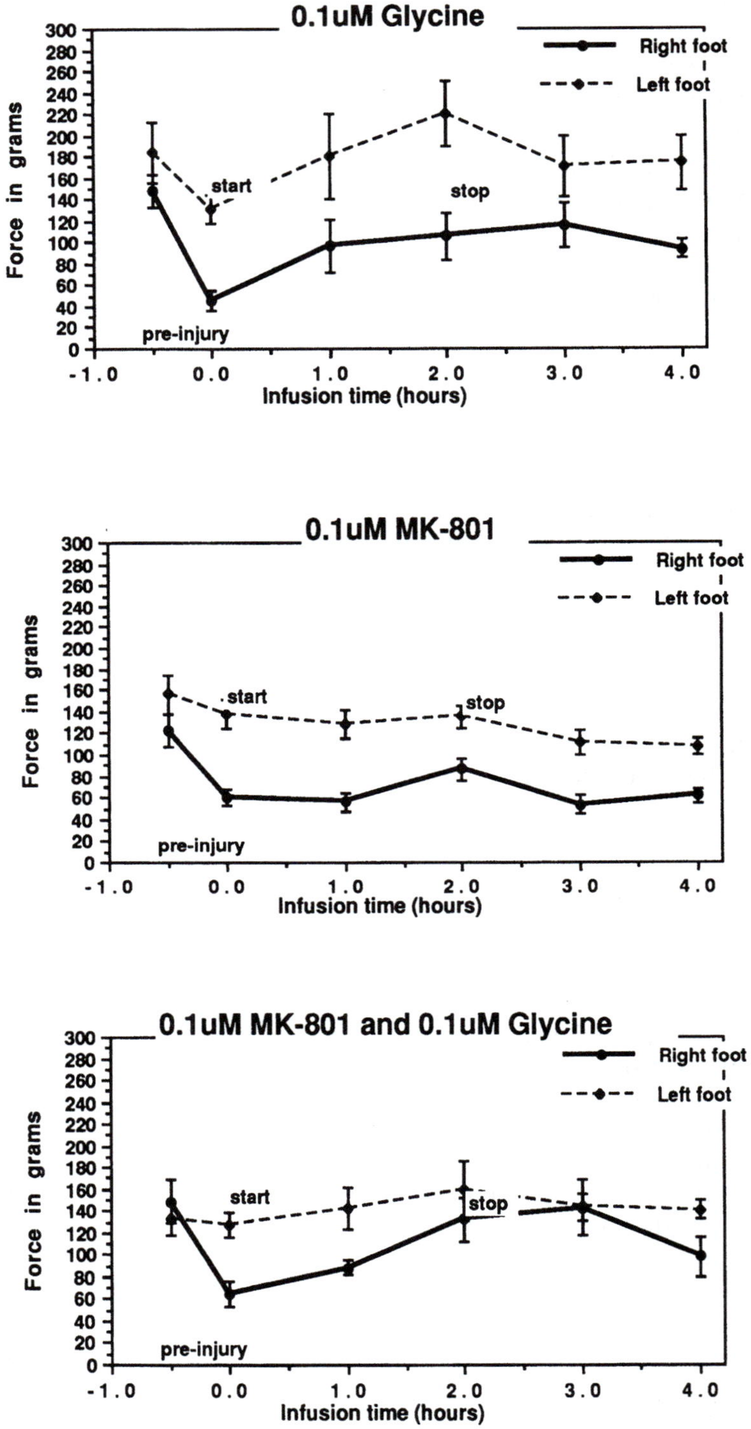
0.1uM Glycine
Right foot
Left foot
Force in grams
start
stop
pre-injury
Infusion time (hours)
0.1uM MK-801
Right foot
Left foot
Force in grams
start
stop
pre-injury
Infusion time (hours)
0.1uM MK-801 and 0.1uM Glycine
Right foot
Left foot
Force in grams
start
stop
pre-injury
Infusion time (hours)

c

Fig. 9c

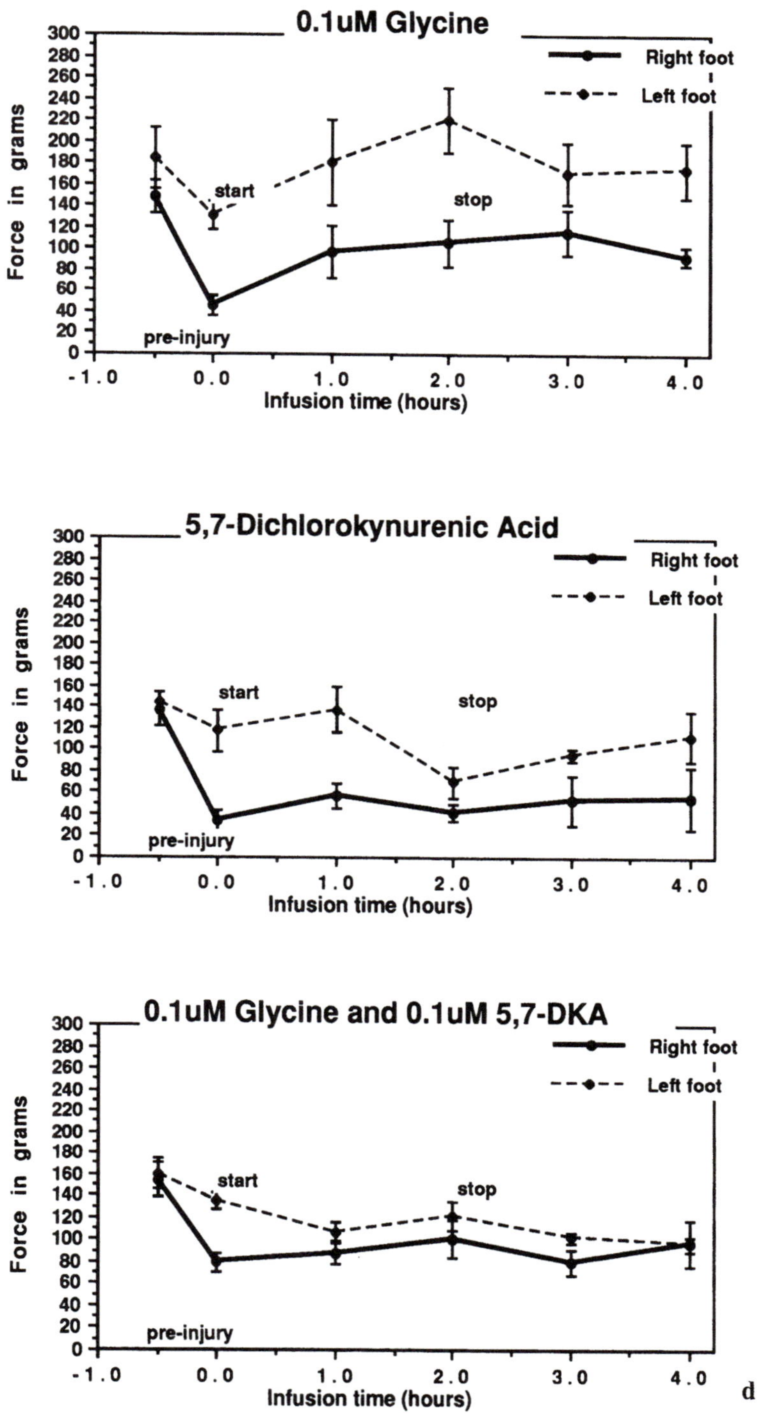
0.1uM Glycine
Right foot
Left foot
start
stop
pre-injury
Force in grams
300
280
260
240
220
200
180
160
140
120
100
80
60
40
20
0
-1.0
0.0
1.0
2.0
3.0
4.0
Infusion time (hours)
5,7-Dichlorokynurenic Acid
Right foot
Left foot
start
stop
pre-injury
Force in grams
Infusion time (hours)
0.1uM Glycine and 0.1uM 5,7-DKA
Right foot
Left foot
start
stop
pre-injury
Force in grams
Infusion time (hours)
d

Fig. 9d

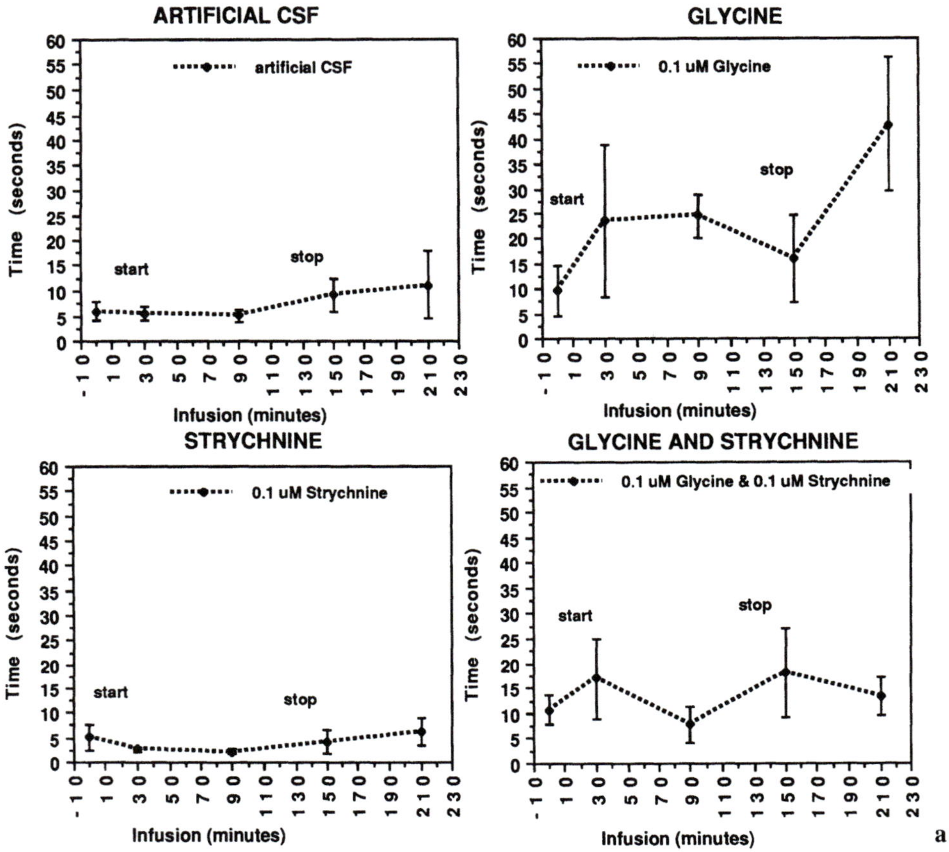

Fig. 10 a–b Results of intrathecal drug infusion on the time necessary for foot withdrawal from a hot plate in the injured, neuropathic limb

Results from intrathecal drug delivery – spasticity

Pharmacologically induced change in electrophysiologic measurements was also observed spastic animals examined with and H-reflexes (Simpson *et al.*, 1993; Simpson *et al.*, 1995; Simpson *et al.*, 1996). These data indicate that the model can be used with accuracy to study the effect of glycine and related compounds on spasticity. The animals were studied after stabilization of spasticity as determined clinically by the Ashworth score (average score was 3). A catheter inserted into the cisterna magna bathed the spinal cord with 100mmol/l solutions of glycine, strychnine, D-serine, β-alanine, MK-801, or artificial CSF for 4 hours at a rate of 5–10 ml/min. H-reflexes were periodically monitored before and during infusion

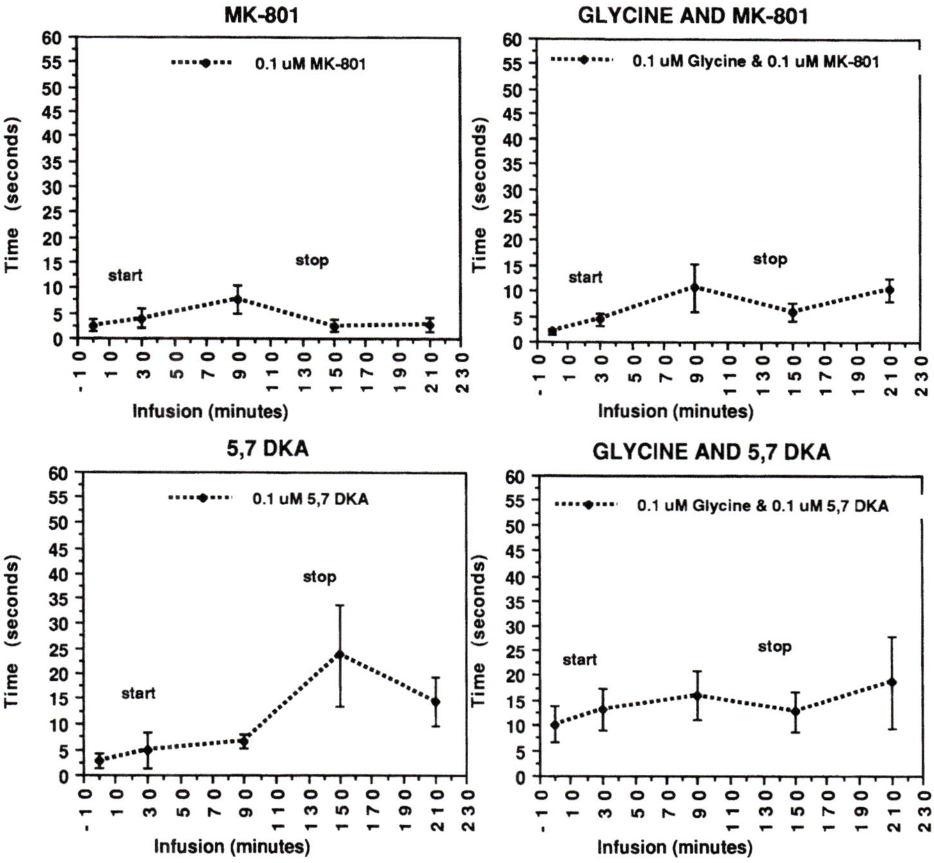

b

Fig. 10b

by stimulating the posterior tibial nerve at intensities ranging 0.5 to 20 mA and recording from the plantar surface of the foot.

Glycine and D-serine depressed the H wave in spastic animals after 90–120 minutes of infusion compared to artificial CSF (Fig. 11). Strychnine and β-alanine produced heightened H waves compared to artificial CSF infusion and MK-801 had no clear effect. The influence of each of these compounds on the H wave was most prominent using intensities ranging 0.5–4 mA. No changes in H wave amplitudes were seen by any infused compound when higher intensities were used, compared to artificial CSF. These results indicate that glycine and related compounds may influence spasticity but require further study.

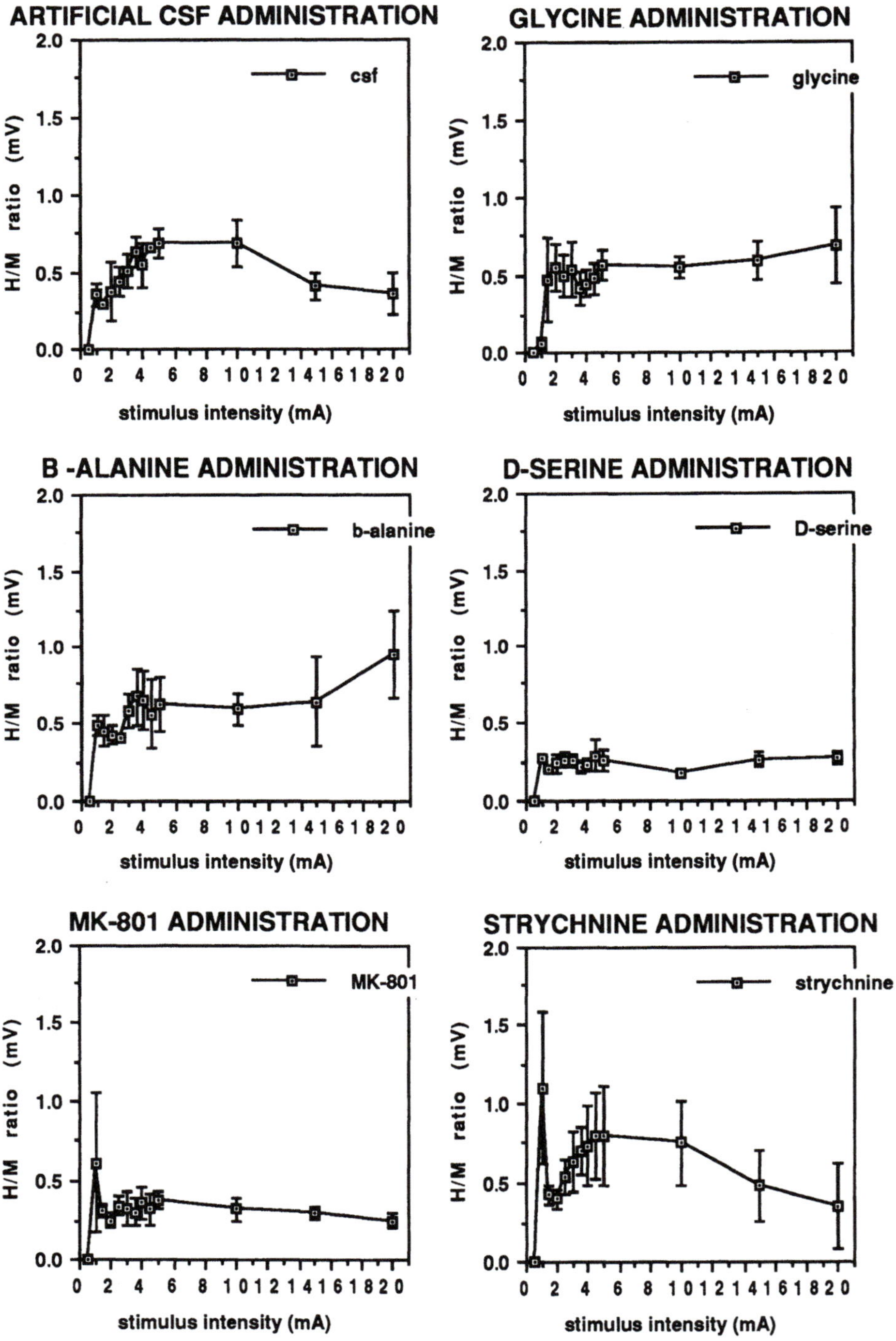

Fig. 11. Results of intrathecal drug infusion on H/M ratios, displayed as a curve, obtained from spastic animals

Conclusions

A complex relationship between the post synaptic inhibitory neurotransmitter glycine and the processing of afferent signals and motor control exists. Differentiation between the sensory and motor effects of the compound in experimental animals can be problematic. In addition, the influence of glycine on neurophysiologic activity appears to be mediated though at least two receptor subpopulations. However, several features regarding glycine and segmental neuron interaction can be discerned. These features include; 1) glycine is elevated within segmental gray matter during epidural spinal cord stimulation (ESCS), 2) glycine is elevated after primary motor cortex and ipsilateral large diameter peripheral nerve stimulation, 3) high levels of segmental glycine can be measured in non ESCS flaccid animals and these levels can be sustained in less injured animals with epidural spinal cord stimulation, 4) spasticity can be reduced by intrathecally infused glycine, and 5) neuropathic pain can be reduced by intrathecal administration of glycine, via Gly 1 if mechanically evoked and via Gly 2 if thermally evoked. The data yielded by the above experiments validates the assumptions made in our primary hypothesis. Furthermore, the data suggest that glycine or related compounds may provide a beneficial manner of therapy for spasticity and neuropathic pain and also provide laboratory tools to dissect the manner in which these conditions are expressed.

References

Albright AL, Cervi A, Singletary J (1991) Intrathecal baclofen for spasticity in cerebral palsy. JAMA 265: 1418–1422

Amassian VE, Stewart BS, Quirk GJ, Rosenthal JL (1987) Physiological basis of motor effects of a transient stimulation to cerebral cortex. Neurosurgery 20: 74–93

Aprison MH, Shank RP, Davidoff RA (1969) A comparison of the concentration of glycine, a transmitter suspect, in different areas of the brain and spinal cord in seven different vertebrates. Comp Biochem Physiol 28: 1345–1355

Aprison MH (1990) The discovery of the Neurotransmitter role of glycine. In: Ottersen OP, Storm-Mathisen J (eds) Glycine Neurotransmission. Wiley, New York, pp 1–23

Araki T, Yamano M, Murakami T, Wanaka A, Betz H, Tohyama M (1988) Localization of glycine receptors in the rat central nervous system: An immunocytochemical analysis using monclonal antibody. Neuroscience 25: 613–624

Ashby P, Verrier M, Lightfoot E (1974) Segmental reflex pathways in spinal shock and spinal spasticity in man. J Neurol Neurosurg Psychiatry 37: 1352–1360

Ashby P, McCrea DA (1987) Neurophysiology of spasticity. In: Davidoff RA (ed) Handbook of the Spinal Cord. Marcel Dekker, New York, pp 119–143

Attal N, Jazat F, Kayser V, Guilbaud G (1990) Further evidence for pain related behaviours in a model of unilateral peripheral mononeuropathy. Pain 41: 235–251

Barbeau A (1974) Preliminary study of glycine administration in patients with spasticity. Neurology 24: 392

Barolat-Romana G, Davis R (1980) Neurophysicological mechanisms in abnormal reflex activities in cerebral palsy and spinal spasticity. J Neurol Neurosurg Psychiatry 43: 333–342

Barolat G, Schwartzman RJ, Woo R (1989) Epidural spinal cord stimulation in the management of reflex syumpathetic dystrophy. Sterotac Funct Neurosurg 53: 29–39

Basbaum AI (1988) Distribution of glycine receptor immunoreactivity in the spinal cord of the rat: Cytochemical evidence for a differential glycinergic control of lamina I and V nociceptive neurons. J Comp Neurol 278: 330–336

Bennett GJ, Xie YK (1988) A peripheral mononeuropathy in rat that produces disorders of pain sensation like those seen in man. Pain 33: 87–107

Bennett GJ, Kajander KC, Sahara Y (1989) Neurochemical and anatomical changes in the dorsal horn of rats with an experimental painful peripheral neuropathy. In: Cervero F, Bennett GJ, Headley PM (eds) The Superficial Dorsal Horn of the Spinal Cord. Plenum, New York, pp 463–471

Bennett GJ (1991) Evidence from animal models on the pathogenesis of painful peripheral neuropathy: Relevance for pharmacotherapy. In: Basbaum AI, Beeson JM (eds) Toward a New Pharmacotherapy. Wiley, New York, pp 365–379

Benveniste H, Hansen AJ, Ottosen NS (1989) Determination of brain interstitial concentrations by microdialysis. J Neurochem 52: 1741–1750

Benveniste H (1989) Brain microdialysis: Short review. J Neurochem 52: 1667–1679

Berger SJ, Carter JG, Lowry OH (1977) The distribution of glycine, GABA, glutamate, and aspartate in rabbit spinal cord, cerebellum and hippocampus. J Neurochem 28: 149–158

Bernhard CG, Koll W (1953) On the effect of strychnine, asphyxia and dial on the spinal cord potentials. Acta Physiol Scand 29 [Suppl] 106: 30–41

Beyer C, Banas C, Gomora P, Komisaruk BR (1988) Prevention of the convulsant and hyperalgesic action of strychnine by intrathecal glycine and related amino acids. Pharmacol Biochem Behav 29: 73–78

Beyer C, Banas C, Gonzalez-Flores O, Komisaruk BR (1989) Blockage of substance P-induced scratching behavior in rats by the intrathecal adminstration of inhibitory amino acid agonists. Pharmacol Biochem Behav 34: 491–495

Bidlingmeyer BA, Cohen SA, Tarvin TL (1984) Rapid analysis of amino acids using pre-column derivatization. J Chromatogr 336: 93–104

Biscoe TJ, Duchen MR (1986) Synaptic physiology of spinal motoneurones of normal and spastic mice: an in vitro study. J Physiol 379: 275–292

Boyd SG, Tothwell JC, Cowan JMA, Webb PJ, Morley T, Asselman P, Marsden CD (1986) A method of monitoing function in corticospinal pathways during scoliosis surgery with a note on motor conduction velocities. J Neurol Neurosurg Psychiatry 49: 251–257

Brooks CM, Eccles JC (1947) A study of the effects of anaesthesia and asphyxia on the mono-synaptic pathway through the spinal cord. J Neurophysiol 10: 349–360

Brown AC, Headly PM, West (1975) A device for producing reproducible electronically-timed pinch to provide noxious mechanical input. Proc Royal Soc 2: 123–125

Bruggencate GT, Engberg I (1968) Analysis of glycine actions on spinal interneurones by intracellular recording. Brain Res 11: 446–450

Brugger F, Wicki U, Elton DN, Fagg GE, Olpe HR, Pozza NM (1992) Modulation of the NMDA receptor by D-serine in the cortex and the spinal cord, *in vitro*. Europ J Pharmacol 191: 29–38

Burke D, Ashby P (1972) Are spinal "presynaptic" inhibitory mechanisms supressed in spasticity? J Neurol Sci 15: 321–326

Cahusac PMB, Evans RH, Hill RG, Todriquez RE, Smait DAS (1984) The behavioural effects of an N-methylaspartate receptor antagonist following application to the lumbar spinal cord of conscious rats. Neuropharmacology 23: 719–724

Campbell JN, Raja SN, Melyer RA, Mackinnon SE (1988) Myelinated afferents signal the hyperalgesia associated with nerve injury. Pain 32: 89–94

Cheng MK, Robertson C, Grossman RG, Foltz R, Williams V (1984) Neurological outcome correlated with spinal evoked potentials in a spinal cord ischemia model. J Neurosurg 60: 786–795

Chizhmakov IV, Kiskin NI, Krishtal OA, Tsyndrenko AY (1989) Glycine action on N-methy-D-aspartate receptors in rat hippocampal neurons. Neurosci Lett 99: 131–136

Christensen H, Fykse EM, Fonnum F (1990) Uptake of glycine into synaptic vesicles isolated from rat spinal cord. J Neurochem 54: 1142–1147

Cohen SA, Bidlingmeyer BA, Tarvin TL (1986) PITC derivatives in amino acid analysis. Nature 320: 769–770
Curtis DR, Hosli L, Johnston GAR (1968) A pharmacological study of depression of spinal neurones by glycine and related amino acids. Exp Brain Res 6: 1–18
Curtis DR, Duggan AW, Johnston GAR (1971) The specificity of strychnine as a glycine antagonist in the mammalian spinal cord. Exp Brain Res 12: 547–565
D'Amour F, Smith D (1941) A method for determing loss of pain sensation. J Pharmacol Exp Therap 72: 74–79
Daly EC, Aprison MH (1983) Glycine. In: Lajtha A (ed) Handbook of Neurochemistry. Plenum Press, New York, pp 467–499
Davar G, Hama A, Deykin A, Vos B, Maciewicz R (1991) MK-801 blocks the development of thermal hyperalgesia in a rat model of experimental painful neuropathy. Brain Res 553: 327–330
Davidoff RA, Graham LT, Shank RP, Werman R, Aprison MH (1967) Changes in amino acid concentrations associated with loss of spinal interneurons. J Neurochem 14: 1025–1031
Davidoff RA (1989) Mode of action of antispasticity drugs. In: Park TS, Phillips LH, Peakock WJ (eds) Management of Spasticity in Cerebral Palsy and Spinal Cord Injury. Neurosurgery: State of the art review. Hanley & Belfus, Philadelphia, pp 315–324
Delwaide PJ (1985) Electrophysiological analysis of the mode of aciton of muscle relaxants in spasticity. Ann Neurol 17: 90–95
Dickenson AH, Sullivan AF (1990) Differential effects of excitatory amino acid antagonists on dorsal horn nociceptive neurones in the rat. Brain Res 506: 31–39
Dimitrijevic MM, Dimitrijevic MR, Sherwood AM, Vanderlinden C (1989) Clinical neurophysiological techniques in the assessment of spasticity. In: Park TS, Phillips LH, Peakock WJ (eds) Physical Medicine and Rehabilitation: State of the Art Reviews. Hanley & Belfus, Philadelphia, pp 64–83
Erdo SL (1990) Strycnine protection against excitotoxic cell death in primary cultures of rat cerebral cortex. Neurosci Lett 115: 341–344
Faganel J, Dimitrijevic MR (1982) Study of Propriospinal interneuron system in man. J Neurol Sci 56: 155–172
Fagg GE, Jordan CC, Webster RA (1978) Descending fibre-mediated release of endogenous glutamate and glycine from the perfused cat spinal cord *in vivo*. Brain Res 158: 159–170
Fagg GE, Foster AC (1983) Amino acid neurotransmitters and their pathways in the mammalian central nervous system. Neuroscience 9: 701–719
Farrant M, Webster RA (1989) Boulton A, Baker GB, Juorio AV (eds) Drugs as Tools in Neurotransmitter Research. Humana Press, Clifton, pp 161–219
Farrant M, Gibbs TT, Farb DH (1990) Molecular and cellular mechanisms of GABA/ Benzodiazepine-receptor regulation: Electrophysiological and biochemical studies. Neurochem Res 15: 175–191
Fasano VA, Barolat-Romana G, Zeme S, Sguazzi L (1979) Electrophysiological assessment of spinal circuits in spasticity by direct dorsal root stimulation. Neurosurgery 4: 146–151
Ganes T (1982) Synaptic and non-synaptic components of the human cervical evoked response. J Neurol Sci 4: 313–326
Gelfan S, Tarlov IM (1955) Differential vulnerability of spinal cord structures to anoxia. J Neurophysiol 18: 170–188
Gerber G, Cerne R, Randic M (1991) Participation of excitatory amino acid receptors in the slow excitatory synaptic transmission in rat spinal dorsal horn. Brain Res 561: 236–251
Graham LT, Shank RP, Werman R, Aprison MH (1967) Distribution of some synaptic transmitter suspects in cat spinal cord: glutamic acid, aspartic acid, γ-aminobutyric acid, glycine, and glutamine. J Neurochem 14: 465–472
Grossi EA, Laschinger JC, Krieger KH, Nathan IM, Colvin SB, Weiss MR, Baumann FG (1988) Epidural-evoked potentials: A more specific indicator of spinal cord ischemia. J Surg Res 44: 224–228

Hall PV, Smith JE, Lane J, Mote T, Campbell R (1979) Glycine and experimental spinal spasticity. Neurology 1979; 29: 262–267
Hammerstad JP, Murray JE, Cutler RWP (1971) Efflux of amino acid neurotransmitters from rat spinal cord slices. II. Factors influencing the electrically induced efflux of [^{14}C] glycine, and 3H-Gaba. Brain Res 35: 357–367
Hao JX, Xu XJ, Aldskogius H, Seiger A, Hallin ZW (1991) The excitatory amino acid receptor antagonist MK-801 prevents the hypersensitivity induced by spina cod ischemia in the rat. Exp Neurol 113: 182–191
Hargreaves K, Dubner R, Brown F, Flores C, Joris J (1988) A new and sensitive method for measuring thermal nociception in cutaneous hyperalgesia. Pain 32: 77–88
Harrison PJ, Jankowska E (1985) Organization of input to the interneurons mediating group I non-reciprocal inhibition of motorneurons in the cat. Part 1. J Physiol (Lond) 361: 379–401
Harrison PJ, Jankowska E (1985) Sources of input to interneurons mediating group I non-reciprocal inhibition of motorneurons in the cat. Part 2. J Physiol (Lond) 361: 403–418
Heller AH, Hallett M (1982) Electrophysioloigcal studies with the spastic mutant mouse. Brain Res 234: 299–308
Henneman E, Mendell LM (1981) Functional organization of motorneuron pool and its inputs. In: Brookhart JM, Mountcastle VB (eds) The Nervous System, Handbook of Physiology, American Physiological Society, Bethesda, pp 423–507
Herz DA, Looman JE, Tiberio A, Ketterling K, Kreitsch RK, Colwill HC, Grin OD (1990) The management of paralytic spasticity. Neurosurgery 26: 300–306
Hopkin J, Neal MJ (1971) Effect of electrical stimulation and high potassium concentrations on the efflux of [^{14}C] glycine from slices of spinal cord. Br J Pharmacol 42: 215–233
Howell DA, Lees AJ, Toghill PJ (1979) Spinal internuncial neurones in progressive encephalomyelitis with rigidity. J Neurol Neurosurg Psychiatry 42: 773–785
Hunt SP (1983) Cytochemistry of the spinal cord. In: Emson PC (ed) Chemical Neurochemistry. Raven Press, New York, pp 53–84
Juhasz G, Emri Z, Kekesi K, Pungor K (1989) Local perfusion of the thalamus with GABA increses sleep and induces long-lasting inhibition of somatosensory event-related potentials in cats. Neurosci Lett 103: 229–233
Kajander KC, Wakisaka S, Bennett GJ (1992) Spontaneous disharge originates in the dorsal root ganglion at the onset of a painful peripheral neuropathy in the rat. Neurosci Lett 138: 225–228
Kanek M, Fukamachi A, Sasaki H, Miyazawa N, Yagishita, Nukui H (1988) Intraoperative moinitoring of the motor function: Experimental and clinical study. Acta Neurochir Scand [Suppl] 42: 18–21
Kish PE, Fischer-Bovenkerk C, Ueda T (1989) Active transport of g-aminobutyric acid and glycine into synaptic vesicles. Proc Natl Acad Sci 86: 3877–3881
Kitagawa H, Itoh T, Takano H, Takakuwa K, Yamamoto N, Yamada H, Tsuji H (1989) Motor evoked potential monitoring during upper cervical spine surgery. Spine 14: 1078–1083
Kuypers HGJM (1973) In: Desmedt JE (ed) New Developments in Electromyography and Clinical Neurophysiology. Karger, Basel, pp 38–68
Kuypers HGJM (1981) Anatomy of the descending pathways. In: Brookhart JM, MountcastleVB (eds) Handbook of Physiology: The Nervous System. The American Physiological Society, Bethesda, pp 597–666
Laird JMA, Bennett GJ (1992) Dorsal root potentials and afferent input to the spinal cord in rats with an experimental peripheral neruopathy. Brain Res 584: 181–190
Larson AA (1989) Intrathecal GABA, glycine, taurine or beta-alanine elicits dyskinetic movements in mice. Pharmacol Biochem Behav 32: 505–509
Lester RAJ, Tong G, Jahr CE (1993) Interactions between the glycine and glutamate binding sites of the NMDA receptor. J Neurosci 13: 1088–1096.
Levy WJ, York DH, McCaffrey M, Tanzer F (1984) Motor evoked potentials from transcranial stimulation of the motor cortex in humans. Neurosurgery 15: 287–302

Lindsley DB, Schreiner LH, Magoun HW (1949) An electromyographic study of spasticity. J Neurophysiol 12: 197–205
Ljungdahl A, Hokfelt T (1973) Accumulation of ^{3}H-glycine in interneurons of the cat spinal cord. Histochemie 33: 277–280
Lloyd DPC (1953) Influence of asphyxia upon the responses of spinal motorneurons. J Gen Physiol 32: 409–443
Lundberg A, Norrsell U, Voorhoeve P (1962) Pyramidal effects on lumbosacral interneurones activated by somatic afferents. Acta Physiol Scand 56: 220–229
Lynch DR, Anegawa NJ, Verdoorn T, Pritchett DB (1993) *N*-methyl-D-aspartate receptors: different subunit requirements for binding of glutamate antagonists, glycine antagonists, and channel-blocking agents. Molec Pharmacol 45: 540–545
Magoun HW, Rhines R (1946) An inhibitory mechanism in the bulbar reticular formation. J Neurophysiol 9: 165–171
Mailis A, Ashby P (1990) Alterations in group Ia projections to motoneurons following spinal lesions in humans. J Neurophysiol 64: 637–647
Makitie J, Teravainen H (1977) Spinal cord dorsum potentials recorded in vivo after cold injury to the sural nerve in the rabbit. Cryobiology 14: 190–196
Mao J, Price DD, Mayer DJ, Lu J, Hayes RL (1992) Intrathecal MK-801 and local nerve anesthesia synergisticallyl reduce nociceptive behaviors in rats with experimental peripheral mononeuropathy. Brain Res 576: 254–262
Mao J, Price DD, Hayes RL, Lu J, Mayer DJ (1992) Differential roles of NMDA and non-NMDA receptor activation in induction and maintenance of thermal hyperalgesia in rats with painful peripheral mononeuropathy. Pain 49: 271–278
Mao J, Price DD, Mayer DJ (1994) Thermal hyperalgesia in association with the development of morphine tolerance in rats: roles of excitatory amino acid receptors and protein kinase. C J Neurosci 14: 2301–2312
Martiniak J, Chavko M, Danielisova V, Marsala J (1989) Regional concentrations of transmitter amino acids after spinal cord ischaemia in the rabbit. Physiol Bohemoslov 38: 275–281
Masland R (1985) Preface, In: Eccles J, Dimitrijevic MR (eds) Recent Achievements in Restorative Neurology. Upper Motor Neuron Functions and Dysfunctions. Karger, Basel, xv–xvi
Mathias CJ, Luckitt J, Desai P, Baker H, Masri WE, Frankel HL (1989) Pharmacodynamics and pharmacokinetics of the oral antispatic agent tizanidine in patients with spinal cord injury. J Rehab Res Devel 26: 9–16
McNamara D, Dingledine R (1990) Dual effect of glycine on NMDA-induced neurotoxicity in rat cortical cultures. J Neurosci 10: 3970–3976
Meinck HM (1976) Occurrence of the H reflex and the F wave in the rat. Electroencephalogr Clin Neurophysiol 41:530-533
Melzack R, Wall PD (1965) Pain mechanisms: A new theory. Science 150: 971–979
Meyerson BA, Linderoth B, Karlsson H, Ungerstedt U (1990) Microdialysis in the human brain: extracellular measurements in the thalamus of parkinsonian patients. Life Sci 46: 301–308
Mitchell K, Spike RC, Todd AJ (1993) An immunocytochemical study of glycine receptor and GABA in lamine 1-III of rat spinal dorsal horn. J Neurosci 13: 2371–2381
Mitchell SW, Morehouse GR, Keen WW (1864) Gunshot wounds and injuries of nerves. Lippincott, Philadelphia
Nathan PW, Smith MC (1959) Fasciculi proprii of the spinal cord in man. Review of present knowledge. Brain 82: 610–668
Neilson PD (1972) Interaction between voluntary contraction and tonic stretch reflex transmission in normal and spastic patients. J Neurol Neurosurg Psychiatry 35: 853–860
Nicola MA, Becker CM, Triller A (1992) Development of glycine receptor alpha subunit in cultivated rat spinal neurons: an immunocyochemical study. Neurosci Lett 138: 173–178

Nogaoka NH, Sadurada S, Sadurad T, Takeda S, Nakagawa Y, Kisara K, Arai Y (1993) Theophylline-induced nociceptive behavioral responses in mice: possible indirect interaction with spinal *N*-methyl-D-aspartate receptors. Neurochem Int 22: 69–74

North RB, Ewend MG, Lawton MT, Kidd DH, Piantadosi S (1991) Failed back surgery syndrome: 5-year follow-up after spinal cord stimulator implantation. Neurosurgery 28: 692–699

Panter SS, Yum SW, Faden AI (1990) Alteration in extracellular amino acids after traumatic spinal cord injury. Ann Neurol 27: 96–99

Patel J, Zinkland WC, Thompson C, Keith R, Salama A (1990) Role of glycine in the *N*-methyl-D-aspartate-mediated neuronal cytotoxicity. J Neurochem 54: 849–854

Quan N, Blatteis CM (1989) Microdialysis: A system for localized drug delivery into the brain. Brain Res Bull 22: 621–625

Randall LO, Selitto JJ (1957) A method for measurement of analgesic activity on inflamed tissue. Arch Int Pharmacodyn 111: 409–419

Rao TS, Cler JA, Emmett MR, Mick SJ, Iyengar S, Wood PL (1990) Glycine, glycinamide and D-serine act as positive modulators of signal transduction at the *N*-methyl-D-aspartate (NMDA) receptor in vivo: differential effects on mouse cerebellar gyclic guanosine monophosphate levels. Neuropharamacology 29: 1075–1080

Ren K, Hylden LK, Williams GM, Ruda MA, Dubner R (1992)The effects of a non-competative NMDA receptor antagonist, MK-801, on behavior hyperalgesia and dorsal horn neuronal activity in rats with unilateral inflammation. Pain 50: 331–344

Rizzoli AA (1968) Distribution of glutamic acid, aspartic acid, g-aminobutyric acid and glycine in six areas of cat spinal cord before and after transection. Brain Res 11: 11–18

Roberts PJ, Mitchell JF (1972) The release of amino acids from the hemisected spinal cord during stimulation. J Neurochem 19: 2473–2481

Rudick RA, Breton D, Krall RL (1987) The Gaba-agonist progabide for spasticity in multiple sclerosis. Arch Neurol 44: 1033–1036

Rudomin P, Solodkin M, Jimenz I (1986) PAD and PAH response patterns of group Ia and Ib fibers to cutaneous and descending inputs to the cat spinal cord. J Neurophysiol 56: 987–1006

Rudomin P, Solodkin M, Jimenz I (1987) Synaptic potentials of primary efferent fibers and motorneurons evoked by single intermediate nucleus interneurons in the cat spinal cord. J Neurophysiol 57: 1288–1313

Rudomin P, Jimenez I, Quevedo J, Solodkin M (1990) Pharmacologic analysis of inhibition produced by last-order intermediate neucleus interneurons mediating nonreciprocal inhibition of motoneurons in cat spinal cord. J Neurophysiol 63: 147–160

Rymer WZ, Powers RK (1989) Pathophysiology of muscular hypertonia in spasticity. Management of spasticity in cerebral palsly and spinal cord injury In: Park TS, Phillips LH, Peakock WJ (eds) Physical Medicine and Rehabilitation: State of the art reviews. Hanley & Belfus, Philadelphia, pp 291–301

Scheardown MJ, Drejer J, Jensen LH, Stidsen CE, Honore T (1989) A potent antagonist of the strychnine inbsensitive glycine receptor has anticonvulsant properties. Europ J Pharmacol 174: 197–204

Schreiner LH, Lindsley D, Magoun HW (1949) Role of brain stem facilitatory systems in maintenance of spasticity. J Neurophysiol 12: 207–216

Schwartzman RJ (1992) Reflex symmpathetic dystrophy and causalgia. The neurology of trauma. Neurologic Clin 10: 953–973

Semba J, Patsalos PN (1993) Milacemide effects on the temporal inter-relationship of amino acids and monoamine metabolites in rat cerebrospinal fluid. Europ J Pharmacol 230: 321–326

Shank RP, Aprison MH (1970) The metabolism in vivo of glycine and serine in eight areas of the rat central nervous system. J Neurochem 17: 1461–1475

Shapovalov AI (1975) Neuronal organization and synaptic mechanisms of supraspinal motor control in vertebrates. Rev Physiol Biochem Pharmacol 72: 1–54

Sheen K, Chung JM (1993) Signs of neuropathic pain depend on signals from injured nerve fibers in a rat model. Brain Res 610: 62–68
Sher GD, Mitchell D (1990) *N*-methyl-D-aspartate receptors mediate responses of rat dorsal hornneurones to hindlimb ischemia. Brain Res 522: 55–62
Sherrington CS, Sowton SCM (1915) Observations on reflex responses to single break shocks. J Physiol (Lond) 49: 331–348
Shimoji K, Shimizu H, Maruyama Y (1978) Origin of somatosensory evoked responses recorded from; the cervical skin surface. J Neurosurg 48: 980–984
Shuaib A, Xu K, Crain B, Siren AL, Feuerstein G, Hallenbeck J, Davis JN (1990) Assessment of damage from implantation of microdialysis probes in the rat hippocampus with silver degeneration staining. Neurosci Lett 112:149–154
Simpson RK Jr, Robertson CS, Goodman JC (1989) Alterations in the corticomotor evoked potential following spinal cord ischemia. J Neurosci Meth 28: 171–178
Simpson RK Jr, Robertson CS, Goodman JC (1990) Spinal cord ischemia-induced elevation of amino acids: extracellular measurement with microdialysis. Neurochem Res 15: 635–639
Simpson RK Jr, Robertson CS, Goodman JC (1991) Segmental release of amino acid neurotransmitters from transcranial stimulation. Neurochem Res 16: 89–94
Simpson RK Jr, Robertson CS, Goodman JC (1991) Release of segmental amino acid neurotransmitters in response to peripheral afferent and motor cortex stimulation: A pilot study. In Press, Life Sci (Pharm Lett) 49: 113–118
Simpson RK Jr, Robertson CS, Goodman JC, Halter JA (1991) Recovery of amino acid neurotransmitters from the spinal cord during posterior epidural stimulation: A preliminary report. J Am Paraplegia Soc 14: 4–9
Simpson RK Jr, Robertson CS, Goodman JC (1992) Segmental amino acid neurotransmitter recovery during posterior epidural stimulation after spinal cord injury. J Am Paraplegia Soc 16: 34–41
Simpson RK Jr, Robertson CS, Goodman JC (1993) Glycine: An important potential component of spinal shock. Neurochem Res 18: 887–892
Simpson RK Jr, Robertson CS, Goodman JC (1993) Glycine: A potential mediator of electrically induced pain modification. Biomed Lett 48: 193–207
Simpson RK Jr, Robertson CS, Goodman JC (1993) Spinal epidural corticomotor evoked potentials as a predictor of outcome from ischemic myelopathy. Neurol Res 15: 104–108
Simpson RK Jr, Gondo MM, Robertson CS, Goodman JC (1995) The influence of glycine and related compounds on spinal cord injury induced spasticity. Neurochem Res 20: 1203–1210
Simpson RK Jr, Robertson CS, Goodman JC (1995) Reduction of mechanonociceptive and thermonociceptive responses by intrathecal administration of glycine and related compounds. Surgical Forum 46: 581–583
Simpson RK Jr, Gondo MM, Robertson CS, Goodman JC (1996) Reduction in the mechanonociceptive response by intrathecal administration of glycine and related compounds. Neurochem Res 21:1221-1226
Simpson RK Jr, Gondo MM, Robertson CS, Goodman JC (1996) Reduction in thermal hyperalgesia by intrathecal administration of glycine and related compounds. Neurochem Res 22: 75–79
Simpson RK Jr, Robertson CS, Goodman JC (1996) The role of glycine in spinal shock. J Spinal Cord Med 19: 215–224
Smith JE, Hall PV, Galvin MR, Jones AR, Campbell RL (1979) Effects of glycine administration on canine experimental spinal spasticity and the levels of glycine, glutamate, and aspartate in the lumbar spinal cord. Neurosurgery 4: 152–156
Song XJ, Zhao ZQ (1993) Differential effects of NMDA and non-NMDA receptor antagonists on spinal cutaneous vs muscular nociception in the cat. Neuroreport 4: 17–20
Sotgui ML (1993) Descending influence on dorsal horn neuronal hyperactivity in a rat model of neuropathic pain. Neuroreport 4: 21–24

Sugimoto T, Bennett GJ, Kajander (1989) Strychnine-enhanced transsynaptic degeneration of dorsal horn neurons in rats with an experimental painful peripheral neuropathy. Neurosci Lett 98: 139–143

Thompson PD, Dick JPR, Asselman P, Griffin GB, Day BL, Rothwell JC, Sheehy MP, Marsden CD (1987) Examination of motor function in lesions of the spinal cord by stimulation of the motor cortex. Ann Neurol 21: 389–396

Thomson AM (1990) Glycine is a coagonist at the NMDA receptor/channel complex. Prog Neurobiol 35: 53–74

Todd AJ (1989) Cells in laminae III and IV of rat spinal dorsal horn receive monosynaptic primary afferent input in lamina II. J Comp Neurol 289: 676–686

Turski L, Schwarz M, Turski WA, Klockgether T, Sontag KH, Collins JF (1985) Muscle relaxant action of excitatory amino acid antagonists. Neurosci Lett 53: 321–326

Turski L, Klockgether T, Schwarz M, Sontag KH, Meldrum BS (1987) Neuroscience 20: 285–292

Van den Pol AN, Gorcs T (1988) Glycine and glycine receptor immunoreactivity in brain and spinal cord. J Neurosci 8: 472–492

Waltz JM (1982) Computerized percutaneous multilevel spinal cord stimulation in motor disorders. Appl Neurophysiol 45: 73–92

Weil S (1992) Reflex sympathetic dystrophy. In: Evans RW, Baskin DS, Yatsu FM (eds) Prognosis of Neurological Disorders. Oxford University Press, New York, pp 601–605

Werman R, Davidoff RA, Aprison MH (1967) Evidence for glycine as the principal transmitter mediating postsynaptic inhibition in the spinal cord of the cat. J Gen Physiol 50: 1093–1094

Werman R, Davidoff RA, Aprison MH (1968) Inhibitory action of glycine on spinal neurons in the cat. J Neurophysiol 31: 81–95

Wood PL, Emmett MR, Rao TS, Mick S, Cler J, Iyengar S (1989) In vivo modulation of the *N*-methyl-D-aspartate receptor complex by D-serine: potentiation of ongoing neuronal activity as evidenced by increased cerebellar cyclic GMP, Part 1. J Neurochem 53: 979–981

Wood PL, Emmett MR, Rao TS, Mick S, Cler J, Iyengar S (1989) In vivo modulation of the *N*-methyl-D-aspartate receptor complex by D-serine: potentiation of ongoing neuronal activity as evidenced by increased cerebellar cyclic GMP, Part 2. J Neurochem 53: 982–991

Yaksh TL (1989) Behavioral and autonomic correlates of the tactile evoked allodynia produced by spinal glycine inhibition: effects of modulatory receptor systems and excitatory amino acid antagonists. Pain 37: 111–123

Yamamoto T, Yaksh TL (1992) Spinal pharmacology of thermal hyperesthesia induced by constriction injury of sicatic nerve. Excitatoryo amino acid antagonists. Pain 49: 121–128

Yamamoto T, Yaksh TL (1993) Effects of intrathecal strychnine and bicuculline on nerve compression-induced thermal hyperalgesia and selective antagonism by MK-801. Pain 54: 79–84

Yates BJ, Thompson FJ, Mickle JP (1982) Origin and properties of spinal cord field potentials. Neurosurgery 11: 439–450

Yoneda Y, Ogita K, Suzuki T (1990) Interaction of strychnine-insensitive glycine binding with MK-801 binding in brain synaptic membranes. J Neurochem 55: 237–244

Young AB, Snyder SH (1973) Strychnine binding associated with glycine receptors of the central nervous system. Proc Nat Acad Sci 70: 2832–2836

Young AB, Macdonald RL (1987) Glycine as a spinal cord neurotransmitter. In: Davidoff RA (ed) Handbook of the spinal cord. Marcel Dekker, New York, pp 1–44

Zeglgansberger W, Champagnat J (1979) Cat spinal motorneurones exhibit topographic sensitivity to glutamate and glycine. Brain Res 160: 95–104

Correspondence: Dr. R. K. Simpson, Jr., Department of Neurosurgery, Baylor College of Medicine, One Baylor Plaza, Houston, Texas 77030, USA.

Spinal Cord Monitoring
Eds: E. Stålberg, H. S. Sharma, Y. Olsson

Chapter 5

Prostaglandin system in the spinal cord: a neuroanatomical study in the pathophysiological states

K. Matsumura[1,2], H. S. Sharma[3], C. Cao[2], Yu. Watanabe[1,2], K. Yamagata[5], M. Ozaki[2], K. Takeuchi[6], T. Gordh[4], J. Westman[3], and Y. Watanabe[1,2]

[1] Subfemtomol Biorecognition Project, Japan Science and Technology Corporation
[2] Department of Neuroscience, Osaka Bioscience Institute, Suita, Osaka, Japan
[3] Laboratory of Neuroanatomy, Department of Anatomy, Biomedical Centre
[4] Department of Anaesthesiology, University Hospital, Uppsala University, Uppsala, Sweden
[5] Deptartment of Molecular Neurobiology, Tokyo Metropolitan Institute for Neurosciences, Fuchu-shi, Tokyo, Japan, and
[6] The 2nd Department of Internal Medicine, Tohoku University, School of Medicine, Sendai, Miyagi, Japan

Introduction

Prostaglandins (PGs) exert a wide range of biological actions in various types of tissues in the body. The nervous system is not an exception. A large number of studies have implicated PGs as playing essential roles in a variety of physiological and pathological responses in the nervous system (Shimizu and Wolfe, 1990; Bazan *et al.*, 1995). In the brain, regional distributions of the sites of PG biosynthesis and PG actions have been clarified to some extent, although not completely (Watanabe *et al.*, 1983; Yamashita *et al.*, 1983; Watanabe *et al.*, 1988; Watanabe *et al.*, 1989; Matsumura *et al.*, 1990; Tsubokura *et al.*, 1991; Breder *et al.*, 1992; Matsumura *et al.*, 1992; Sugimoto *et al.*, 1994; Breder *et al.*, 1995; Cao *et al.*, 1995; Matsumura *et al.*, 1995; Breder and Saper, 1996; Cao *et al.*, 1996; Takechi *et al.*, 1996; Cao *et al.*, 1997; Matsumura *et al.*, 1997). In contrast, in the spinal cord, little is known as to where PGs are biosynthesized and where they act. Such information is of importance for a better understanding of the mechanism of PG-related disorders in the spinal cord and for proper clinical treatment. In this chapter, we would like to overview the functions and location of PG system in the spinal cord. In the first part, biochemical aspects of PG system are shortly summarized. The second part surveys the literature concerning the patho-physiological roles of PGs in the spinal cord. In the third part, our recent experimental results on the location of the PG system in the spinal cord are presented.

Biochemistry of prostaglandin system

Prostaglandins (PGs)

All PGs are unsaturated fatty acids with 20 carbon chains. Although there are three possible precursors for PGs, arachidonic acid is the major one from which PGs having two carbon-carbon double-bonds are biosynthesized. During the 1960s and 1970s, most of the bioactive PGs were identified and named in an alphabetical order from PGA to PGI and thromboxane A (TXA). Among them, the naturally occurring bioactive ones are PGs D_2, E_2, $F_{2\alpha}$, I_2, and TXA_2, the numeral "2" indicating the number of carbon-carbon double-bonds in the compound (Fig. 1). Enzymes for PG biosynthesis (Smith *et al.*, 1991) and receptors for their actions (Coleman *et al.*, 1994) are widely distributed in various tissues throughout the body; and, therefore, the biological actions of PGs are extremely diverse. There also seem to be yet unidentified functions of PGs especially in the nervous system.

Enzymes for PG biosynthesis

PGs are biosynthesized from phospholipids of the cell membrane through 3 enzymatic steps (Fig. 1) (see also review by Smith *et al.*, 1991). First, phospholipase A_2 (PLA_2) cleaves arachidonic acid from the membrane phospholipids. Arachidonic acid is then converted to PGH_2 by cyclooxygenase (COX). PGH_2 itself is biologically active in the blood vessels, and all other active PGs are biosynthesized from PGH_2 by PG isomerases. In this enzymatic cascade, PLA_2 and COX are common to all PGs and are considered to play rate-limiting roles in the biosynthesis of them. Three types of PLA_2s are well characterized in mammals. They are classified into

Table 1. Classification and nomenclature of PG receptors (Coleman *et al.*, 1994)

Receptor type	Endogenous ligand	Subtype	G-proteins	Second messengers
TP	TXA_2/PGH_2	–	Gs, Gq, Gi	Pl turn over ↑
EP	PGE_2	EP1	G(?)	Ca^{++} ↑
		EP2	Gs	cAMP ↑
		EP3	Gi, Gs, Gq	cAMP ↓↑, Pl turn over ↑
		EP4	Gs	cAMP ↑
FP	PGF_{2a}	–	Gq	Pl turn over ↑
IP	PGI_2	–	Gs, Gq	cAMP ↑, Pl turn over ↑
DP	PGD_2	–	Gs	cAMP ↑

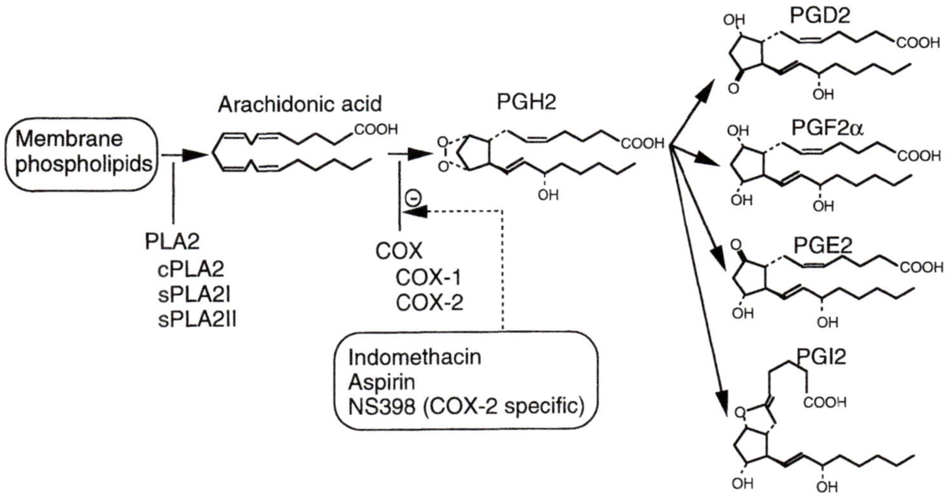

Fig. 1. Overview of the arachidonic acid cascade. *PLA_2* phospholipase A_2; *$cPLA_2$*, cytosolic PLA_2; *$sPLA_2I$* secretory PLA_2 type 1 (pancreatic type); *$sPLA_2II$* secretory PLA_2 type 2 (non-pancreatic type); *COX* cyclooxygenase

cytosolic PLA_2 ($cPLA_2$) and secretory PLA_2 ($sPLA_2$), the latter of which is further divided into type I (pancreatic type) and type II (nonpancreatic type) (Glaser *et al.*, 1993). The second enzyme, COX, catalyzes two chemical reactions that are from arachidonic acid to PGG_2 and, then, from PGG_2 to PGH_2. There are two types of COX, i.e., COX-1 and COX-2. COX-1 is constitutively expressed and seems to be involved in physiological functions under normal conditions. With a few exceptional cases, COX-2 is not constitutively expressed but strongly induced by several types of stimuli such as cytokines and is considered to act under pathological conditions (Goppelt-Struebe, 1995).

Receptors for PGs

Based on the results of bioassays, the existence of specific receptors for each PGs were postulated and designated DP, EP, FP, IP, and TP. The EP receptor was further classified into 4 subtypes (EP1, EP2, EP3, and EP4 in the case of the mouse) based on the preferences for synthetic agonists and antagonists and on the pattern of second-messenger recruitment (Coleman *et al.*, 1994). Molecular cloning of cDNAs for PG receptors was first reported in 1991, revealing the structure of the human thromboxane A_2 receptor (Hirata *et al.*, 1991). Based on the homology with the nucleotide sequence of TX receptor cDNA, cDNAs for all other PG receptors have been cloned and expressed *in vitro* in cell lines for further characterization (Ushikubi *et al.*, 1995). The receptor proteins are all coupled to G proteins and have 7 transmembrane domains as a common feature of G-protein-coupled receptors. PG signals at the receptor sites are, thus, converted to second messenger molecules, including cAMP and Ca^{++}.

Pathophysiological roles of PGs in the spinal cord

So far, in the spinal cord, roles of PGs have been intensively investigated in the nociceptive system and in the pathological changes associated with spinal cord injury (SCI). Both of these issues are not only of scientific interest but also of clinical importance.

Roles of PGs in nociception

Nonsteroidal anti-inflammatory drugs (NAIDs), which are potent and specific inhibitors of COX, are also potent analgesics (Vane, 1971; Ferreira, 1972; Ferreira *et al.*, 1973). This fact strongly suggests that endogenous PGs enhance the nociceptive function. Although there seem to be multiple sites within the nervous system where PGs act to modulate the pain signal transmission, accumulating evidence indicates that the primary nociceptive neurons are the major targets for the hyperalgesic actions of PGs. The primary sensory neurons are bipolar in shape, sending one process to peripheral tissue and the other one to the dorsal horn of the spinal cord; and the cell bodies are located in the dorsal root ganglia. PGs are likely to act on both terminals of these neuronal processes.

Action of PGs in the peripheral terminals of the primary nociceptive neurons. PGs when injected into the peripheral tissues evoke hyperalgesia. This was first demonstrated in humans in 1972 (Collier *et al.*, 1972) and was confirmed by a large number of studies on behavioral pain models using experimental animals, such as the paw-withdrawal reflex to a mechanical stimulus. The results are mostly consistent in the points that PGE_2 and PGI_2 are the most potent ones in augmenting behavioral pain when injected into the peripheral tissues (Ferreira *et al.*, 1978; Taiwo and Levine, 1990) and that these PGs themselves do not evoke pain but lower the threshold for the pain evoked by other stimuli, a phenomenon called sensitization. Hyperalgesia is often associated with peripheral inflammation. Application of NSAIDs, inhibitors of cyclooxygenase, to the inflamed tissue suppresses the hyperalgesia. Thus, endogenous PGs really seem to work as a mediator of hyperalgesia during inflammation.

Consistent with these finding from behavioral studies, neurophysiological experiments also showed that application of PGE_2 or PGI_2 or its analog to peripheral tissues increased the number of action potentials of the afferent nociceptive fibers (Chahl and Iggo, 1977; Mense, 1981; Birrell *et al.*, 1991; Birrell and McQueen, 1993). These results clearly indicate that peripheral terminals of the primary nociceptive neurons are the sites of hyperalgesic action of PGE_2 and PGI_2. Involvement of cAMP in the hyperalgesic action of PGs has been suggested (Taiwo *et al.*, 1989).

However, the ionic mechanism by which PGs lower the pain threshold remains to be clarified *in vivo*. This is mainly due to the difficulty in conducting electrophysiological experiments in the peripheral terminals of the primary sensory neurons.

Recently, however, electrophysiological studies employing cultured sensory neurons *in vitro* have provided important information on the mechanism of PGE_2-induced sensitization. A long-lasting hyperpolarization, which follows an action potential, was present in a subpopulation of primary sensory neurons. This hyperpolarization was driven by Ca^{++}-dependent K^+ current and thought to reduce the excitability of the neurons. PGE_2 suppressed this current and, therefore, increased the excitability of the neurons (Undem and Weinreich, 1993; Gold *et al*., 1996). This idea is well in agreement with the *in vivo* observation that PGE_2 itself does not induce pain but lowers the threshold for pain elicited by other stimuli. Another possible mechanism has also been proposed. A subpopulation of primary sensory neurons in culture was shown to possess tetrodotoxin-resistant Na^+ current. PGE_2 augmented this current in a cyclic-AMP-dependent manner (England *et al*., 1996; Gold *et al*., 1996), thus increasing neuronal excitability. On the other hand, no ionic mechanism of PGI_2-induced sensitization has been reported.

Action of PGs in the central terminals of the primary sensory neurons. As in the case of peripheral application of PGs, intrathecal (i.t.) infusions of PGs also modulate, mostly facilitate, nociception. PGE_2, PGD_2, and $PGF_{2\alpha}$ have been reported to induce hyperalgesia when infused intrathecally (Taiwo and Levince, 1988; Uda *et al*., 1990; Minami *et al*., 1992; Minami *et al*., 1994). Recently, effects of i.t. injection of PGs were classified into two categories, allodynia and hyperalgesia. Allodynia is a state of discomfort and pain evoked by innocuous stimuli such as a gentle touch or brushing, whereas hyperalgesia is a state of lowered threshold for pain to an eventually noxious stimulus. Minami et al. showed that PGE_2 and $PGF_{2\alpha}$ induced allodynia and that PGE_2 induced hyperalgesia also (Minami et al., 1992; Minami *et al*., 1994). Using subtype-specific agonists for PGE_2 receptors, they showed that PGE_2 induced allodynia through the EP1 receptor whereas hyperalgesia occurred through the EP3 receptor (Minami *et al*., 1994). Again, intrathecal injection of NSAIDs suppressed the pain elicited by injection of formalin to the peripheral tissue (Malmberg and Yaksh, 1992).

The central terminals of the primary sensory neurons release principally glutamate to transmit the pain signals. Neuropeptides, such as substance P (SP) and calcitonin gene-related peptide (CGRP) are also released and are considered to be involved in the modulation of pain signals. Using an i.t. microdialysis probe, Malmberg and Yaksh (Malmberg and Yaksh, 1995) showed that intradermal injection of formaldehyde into a hind paw of a rat, which treatment provokes a biphasic behavioral pain, increased the levels of excitatory amino acids and PGE_2 in the cerebrospinal fluid of the lumbar cord. Intrathecal as well as i.p. administration of an NSAID lowered the levels of PGE_2, excitatory amino acids, and behavioral pain. Similarly, using carrageenan/kaolin-induced knee joint inflammation as an experimental model, Yang *et al.* (Yang *et al.*, 1996) demonstrated increases in the levels of excitatory amino acids, PGE_2, and behavioral pain. I.t. infusion of capsaicin evoked the behavioral pain and amino acid release, and co-administration of PGE_2 enhanced these responses (Malmberg and Yaksh, 1994). The effects of PGs on neuropeptide release were mainly studied in cell culture systems. PGE_2 as well as PGI_2 and carbacyclin, a stable PGI_2 analog, enhanced the release of SP and CGRP by bradykinin from rat sensory neurons in culture (Hingtgen and Vasko, 1994; Vasko *et al.*, 1994).

These results imply that PGs also act on the central terminals of the sensory neurons to enhance the pain transmission by increasing the release of excitatory amino acids or neuropeptides.

PGs and spinal cord injury

Role of prostaglandins in the pathophysiology of CNS injury is not well known in spite of the fact that several workers examined PG content of the brain and spinal cord after trauma (see Chapter 13). The precursor of PGs such as arachidonic acid and some of its metabolites i.e., eicosanoids are involved in the secondary injury process. As mentioned above, these metabolites derived from two enzymatic pathways: cyclooxygenase which yields prostaglandins and thromboxane, and lipoxygenase which produce leukotrienes, lipoxins and hydroxy/hydroperoxy derivatives (for review see Barkai and Bazan 1989).

Contusion caused by closed head injury increases PGE_2 level in the traumatised hemisphere (Shohami *et al.*, 1988). In this study, the time course of PGE_2 elevation showed a close parallelism with edema development. This increased level of PGE_2 was also found within the remote brain regions following trauma with increase in the survival period ranging from few hours up to 10 days after primary insult. Increased levels of PG and eicosanoids are also evident following trauma to the spinal cord in experimental models. Thus, in rabbits and cats, increased levels of thromboxane B_2 occurs following impact injury to the cord, whereas elevated amounts of PGE_2 and $PGF_2\alpha$ can be seen within the spinal cord following compression injury in dogs and cats (Demeduik *et al.*, 1985; Hsu *et al.*, 1985). Faden *et al.*, (1988) showed elevation of thromboxane B_2 levels in rat after compression injury which correlates well with the severity of trauma. The increased levels of PGs and its metabolites can occur within 5 min after trauma which persists up to 7 to 10 days after the initial insult. These observations theoretically support the use of PG synthesis inhibitors in reducing the secondary tissue injury following brain or spinal cord trauma.

We are interested in the role of PGs as mediators of secondary lesions in spinal trauma because they are known modulators of microcirculation and microvascular permeability disturbances in the CNS. Secondly, injection of arachidonic acid into the cerebral cortex induces edema formation and PGs are known modulators of several other neurochemicals known to influence secondary injury processes following trauma (for details see Sharma *et al.*, 1993a,b: Winkler *et al.*, 1993). We used indomethacin, a potent inhibitor of cyclooxygenase enzyme to inhibit PG synthesis in the spinal cord before trauma. In this drug treated rats we examined microvascular permeability disturbances, spinal cord blood flow, edema and cell injury following a focal incision of the spinal cord (Sharma *et al.*, 1993a; 1995).

Our observations show that pretreatment with indomethacin significantly attenuated the breakdown of the blood-spinal cord barrier (BSCB) permeability to radioactive iodine and Evans blue tracers (Sharma *et al.*, 1993a, 1995). These observations suggest that PGs are somehow involved in the process of microvascular permeability disturbances. Obviously a reduction in BSCB permeability will reduce vasogenic edema formation and cell injury (for details see Winkler *et al.*, in this volume). Ultrastructural studies further confirm that blockade of PG synthesis before injury is capable of preventing the leakage of tracers across the endothelial cells of spinal cord microvessels. Thus using lanthanum, an electron dense tracers, we found that the tracer is mainly confined within the lumen of the endothelial cells (Fig. 2a) and the passage of the tracers is stopped at tight

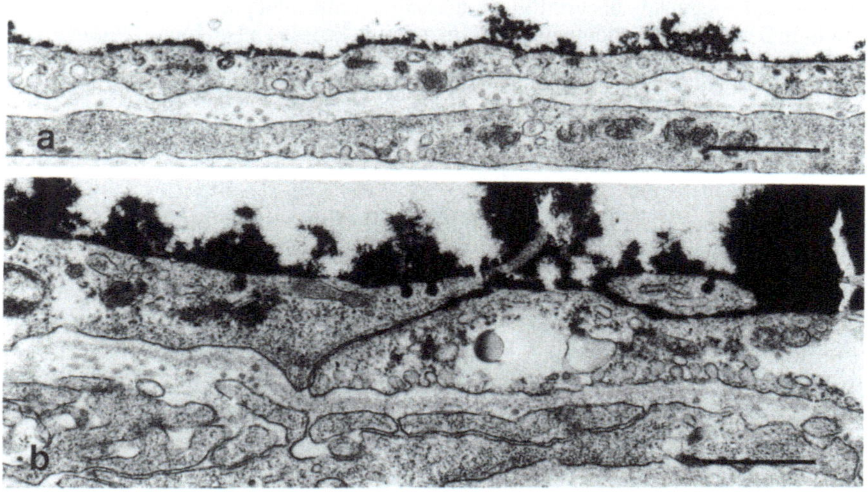

Fig. 2. Ultrastructural studies on the blood-spinal cord barrier using lanthanum as electron dense tracer in the rat spinal cord following trauma and its modification with indomethacin. Lanthanum seen as dark black particles is confined within the lumen of one indomethacin pretreated traumatised rat (**a**). The passage of lanthanum is stopped at one tight junction between two adjacent endothelial cells (**b**) (bar = 200 nm, Data from Sharma *et al.*, 1995 with permission)

Fig. 3. Ultrastructural studies on the blood-spinal cord barrier using lanthanum as electron dense tracer in the rat spinal cord following trauma. Lanthanum seen as dark black particles is infiltrated within the endothelial cell membrane (**a**) and also found within the basal lamina at other occasion (**b**). However, the passage of lanthanum is stopped at tight junction between two adjacent endothelial cells (**c**) (bar = 200 nm, Data from Sharma *et al.*, 1995 with permission)

junctions (Fig. 2b). On the other hand, in the untreated traumatised rats, lanthanum was seen within the endothelial cells infiltrated through the endothelial cell membrane (Fig. 3a). At some occasions, the tracer was also present in the basal lamina (Fig. 3b). However, the tight junctions were remained intact (Fig. 3c). These results strongly suggest that PGs are somehow involved in the endothelial cell membrane permeability. However it remains to be seen whether this effects of PGs on microvascular permeability is mediated by specific PG receptors.

Location of PG receptors in the spinal cord

The fact that intrathecal injections of PGs induce hyperalgesia or allodynia strongly suggests that the PG receptors are present in the spinal cord. Distributions of receptors for PGs were investigated autoradiographically either by binding of radio-labeled receptor ligands to frozen sections or by *in situ* hybridization of receptor mRNA. The former approach reveals the sites of receptor functioning, whereas the latter one shows the sites of biosynthesis of receptor protein.

Experimental procedure

For receptor binding studies, animals were anesthetized with thiopental and perfused with 20 mM PBS via the left ventricle to remove the blood. The spinal cord and lower brain stem were excised and frozen in dry ice powder. Frozen sections (10 µm) were cut in a cryostat and thaw-mounted on gelatin-coated glass slides. They were first preincubated in 50 mM Tris-HCl-buffered saline (pH 7.4) for one hour at 4°C and then incubated with ^{3}H-labeled PGE_2, PGD_2 or iloprost, a stable prostacyclin analog, at the concentration between 10 nM and 20 nM for 30 min. After having been washed, the sections were air-dried and exposed to ^{3}H-sensitive films for 4 to 12 weeks. After development of the films, the autoradiographic images were captured by a CCD camera, digitized to 8-bit values, and analyzed in a Macintosh computer using a software, NIH image. Distribution of EP3 receptor mRNA in both spinal cord and dorsal root ganglia was also studied by *in situ* hybridization with ^{35}S-labeled cRNA for rat EP3 mRNA used as a probe (Takeuchi *et al.*, 1993).

Results and discussion

In rats, a high density of iloprost binding sites and PGE_2 binding sites was found in the superficial layers of the dorsal horn and the spinal trigeminal nucleus (Fig. 4a) (Matsumura *et al.*, 1995). Both of these regions receive the somatosensory primary afferent inputs, the former from the dorsal root ganglia and the latter from the trigeminal ganglia. In addition, the nucleus tractus solitarius (NTS) also contained a high density of binding sites for iloprost and PGE_2 (Fig. 4a). The NTS receives viscerosensory afferent input via the vagus. Although the locations of iloprost binding sites and PGE_2 binding sites were very similar in all of these regions, the binding sites for the two ligands were definitely distinct because of the following two reasons: In a developmental study, iloprost binding appeared earlier than PGE_2 binding (Matsumura *et al.*, 1995). [^{3}H]Iloprost binding was little displaced by an excess amount of PGE_2 (Matsumura *et al.*, 1995).

Since all the above regions receive the primary sensory afferent, there arose a possibility that these PG receptors are located in the central terminals of the primary sensory afferent but not in the postsynaptic neurons in the dorsal horn or in the medulla. This possibility was examined in rats in which some of the dorsal roots were unilaterally sectioned between the dorsal root ganglia and spinal cord. Eight days after the dorsal roots had been sectioned, the rats were sacrificed; and their spinal cords were processed for the receptor binding study. The binding sites for iloprost and PGE_2 were significantly decreased, almost disappeared ipsilateral to the lesion (Fig. 4b) (Matsumura *et al.*, 1995). Similar reductions of iloprost and PGE_2 binding were also observed in the NTS after sectioning the vagus between the nodose ganglion and brain (Fig. 4c). In the case of the NTS, there still remained a low level of the binding sites ipsilateral to the lesion. This may be due to the bilateral innervation of the vagal afferent.

The superficial layers of the dorsal horn and spinal trigeminal nucleus are known to receive nociceptive and thermosensitive inputs via unmyelinated fibers (C-fiber) and fine myelinated fibers (Aδ-fiber). In order to further specify the types of afferent fibers possessing the PG receptors, we studied the effects of capsaicin, which specifically destroys C- and Aδ-fibers, on the receptor bindings. Neonatal rats (within 48 h after birth) were injected with capsaicin (50 mg/kg) or its vehicle subcutaneously. After eight weeks, hot plate test was performed to confirm the effect of capsaicin. Rats were put on a hot plate, and latency of the avoidance behavior was measured. Capsaicin-treated rats showed significantly longer latency than the vehicle-treated rats (mean $\pm$ s.e.; 10 ± 2.3 sec $(n=7)$ vs 4.2 ± 0.3, $(n=6)$; $p<0.01$). The rats were then sacrificed for the receptor binding study. In the dorsal horn, spinal trigeminal nucleus, and NTS, the densities of both binding iloprost (Fig. 4d) and PGE_2 (Fig. 4e) binding stites were significantly reduced in the capsaicin-treated rats, suggesting that at least some of the PG receptors are located in the C- or Aδ- fibers. The extent of the decrease, however, considerably varied among the locations and the types of receptor ligands. Especially, in the dorsal horn, the decrease in PGE_2 binding sites was much smaller (18%) than that in iloprost binding (86%). This result suggests that a large portion of PGE_2 receptors in the dorsal horn are expressed in primary sensory fibers other than C- or Aδ-fibers or that after the capsaicin treatment, PGE_2 receptors are newly expressed in fibers that do not express the receptor under normal conditions. Presence of PG receptors in the dorsal horn was also confirmed in monkeys (Fig. 4f, g).

The idea that the PG receptors are located in the primary sensory neurons was further supported by *in situ* hybridization studies. EP3 mRNA was not present in the dorsal horn, spinal trigeminal nucleus or NTS (data not shown). On the other hand, EP3 mRNA was abundantly expressed in the neurons of DRG, trigeminal ganglia, and nodose ganglia in rats (Fig. 5). Studies in mice by Sugimoto *et al.* (Sugimoto *et al.*, 1994) and by Oida *et al.* (Oida *et al.*, 1995) also showed abundant expression of mRNAs for EP1, EP3, EP4, and IP receptors in the neurons of these sensory ganglia but not in the spinal cord and the medullary regions.

These results clearly demonstrate that the receptors for PGs are biosynthesized in the sensory ganglia and exert their functions after being transported to their central terminals in the spinal cord and medulla oblongata. In addition, there is evidence that these receptors are transported to the peripheral terminals of the sensory fibers and exert functions there. When the vagus was ligated distal to the nodose ganglion, iloprost binding

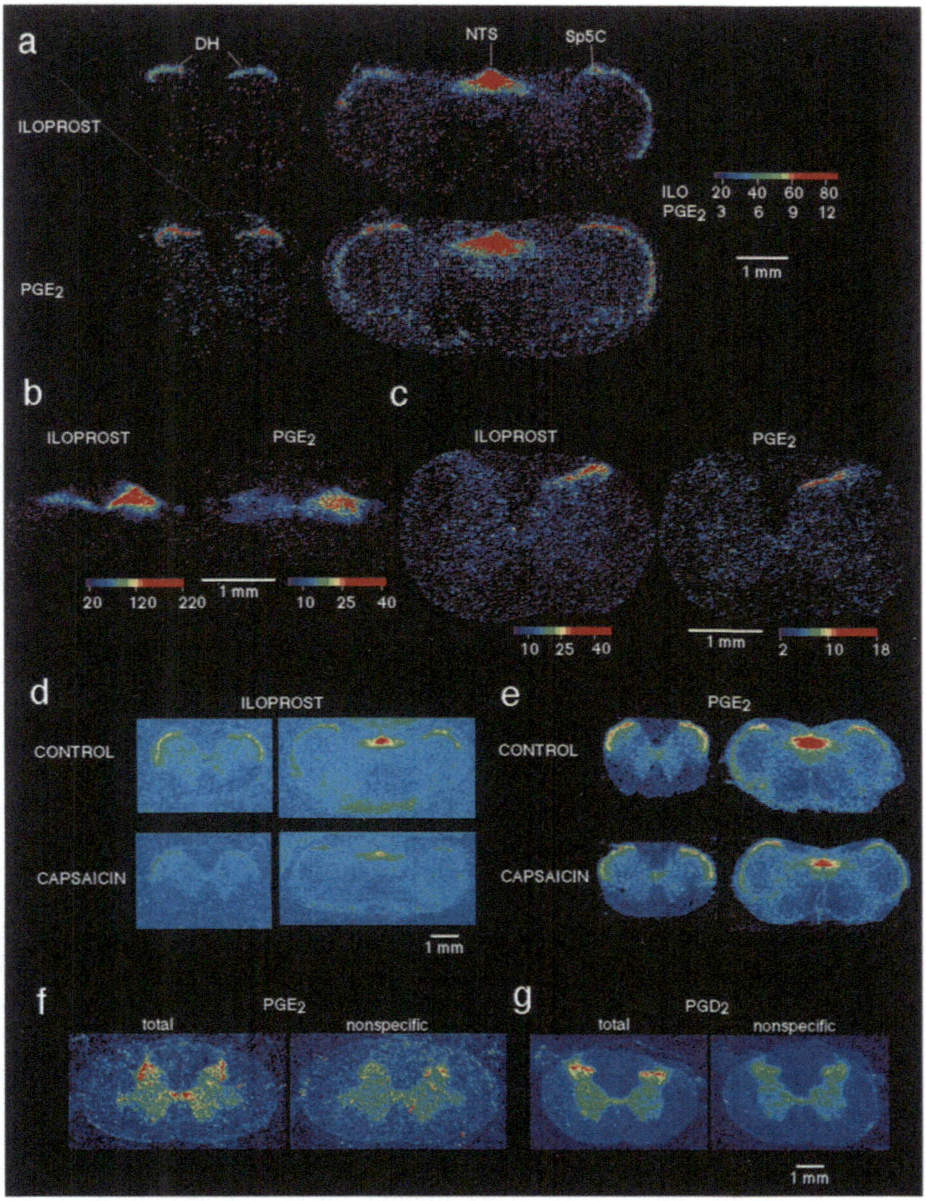

Fig. 4. Receptor autoradiography of PG receptors. (**a**) [^{3}H]iloprost and [^{3}H]PGE_2 binding sites in the spinal cord and medulla. *DH* dorsal horn; *NTS* nucleus tractus solitarius; *Sp5C* spinal trigeminal nucleus caudalis. Both binding sites were decreased after lesioning the dorsal root (**b**) or after nodose ganglionectomy (**c**). Neonatal treatment of capsaicin, which preferentially destroys C- and Aδ- fibers reduce iloprost binding (**d**) and PGE_2 binding (**e**) in the spinal cord and medulla. Receptors for PGE_2 (**f**) and PGD_2 (**g**) are also present in the spinal cord of the monkeys, however, PGI_2 receptors has not been studied

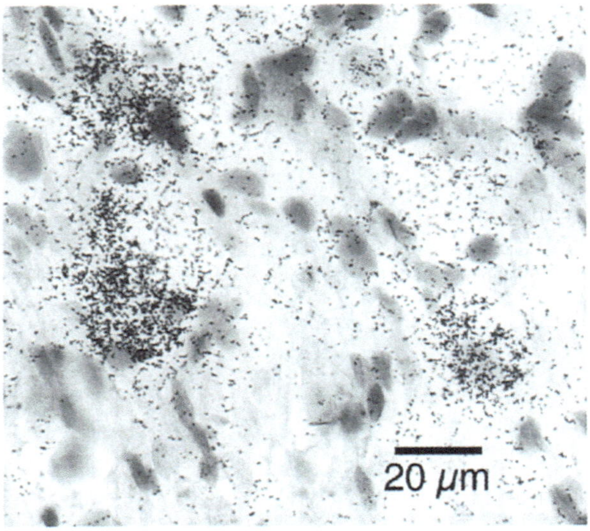

Fig. 5. EP3 receptor mRNA was highly expressed in the neurons of dorsal root ganglion in rats

sites accumulated in the central side of the ligature (Matsumura *et al.*, 1995).

A number of studies have demonstrated that PGE_2 induces hyperalgesia when injected either into the peripheral tissue or into the spinal cord. The presence of PGE_2 receptors in the dorsal horn and that of its mRNA in the DRG neurons well explain the hyperalgesic actions of PGE_2 applied at the two distinct sites. In the peripheral tissue, PGE_2 seems to enhance the excitability of nociceptors; and, in the central end, it seems to enhance the release of neurotransmitters or neuropeptides involved in the pain transmission.

PGI_2 is hyperalgesic when injected into the peripheral tissue (Ferreira *et al.*, 1978; Taiwo and Levine, 1990). However, i.t. injection of cicaprost, a stable PGI_2 anolog, induced only a weak allodynia and no hyperalgesia (Minami *et al.*, 1994). Nevertheless, there does exist a high level of PGI_2 receptors in the dorsal horn. This finding raises a question as to the physiological significance of PGI_2 receptors in the spinal cord. Another unexpected but intriguing result is that neonatal treatment with capsaicin decreased the density of PGE_2 receptors significantly but to a lesser extent compared with the decrease in PGI_2 receptor density in the spinal cord or with that in PGI_2 and PGE_2 receptor density in the NTS. On the other hand, neonatal treatment of capsaicin eliminates hyperalgesic actions of intradermally injected PGE_2 or PGE_1 (Khasar and Levine, 1996). The physiological relevance of this finding remains to be studied.

Location of enzymes for PG biosynthesis in the spinal cord

It is apparent that tissues in the spinal cord more or less possesses PLA_2 activity since the spinal cord does produce PGs in response to various stimuli, such as neuronal stimulation and tissue damage. Unfortunately, however, neither the type(s) of PLA_2 nor its location in the spinal cord has been elucidated. This was also the case for COXs under normal conditions.

Recently, we succeeded in demonstrating the location of COX-2 mRNA and COX-2 protein in the spinal cord under two pathological conditions in which PGs seem to play essential roles. These conditions are spinal cord injury and systemic injection of lipopolysaccharide (LPS), the latter being a model for infection.

Experimental procedure

Adult male rats were anesthetized by intraperitoneal injection with pentobarbital and were laminectomized at the level of the thoracic cord. Spinal cord injury (SCI) was produced by a longitudinal incision into the right dorsal horn (2 mm deep and 5 mm long) made with a sterile scalpel blade as described previously (Sharma *et al.*, 1993a; Sharma *et al.*, 1995). Sham-operated rats, which were laminectomized without SCI, were used as the control. Five hours after SCI or sham operation, the rats were perfused with saline followed by 4% paraformaldehyde (pfa, pH 6.5). The spinal cord was excised and immersed in 4% pfa containing 20% sucrose for 24 to 36 hours. The samples were frozen in dry ice powder, sectioned at a 16-μm thickness in a cryostat, and thaw-mounted on silane-coated glass slides. With ^{35}S-labeled cRNA for rat COX-2 mRNA used as a probe (Yamagata *et al.*, 1993), the sections were processed for *in situ* hybridization as described previously (Cao *et al.*, 1995).

For immunostaining of COX-2, rats with SCI were perfused with 20 mM PBS. The spinal cords were excised and frozen in dry ice powder. Frozen sections were made in a cryostat and thaw-mounted on silane-coated glass slides. The sections were preincubated with 3% normal goat serum with 0.03% TritonX-100 in 0.1 M PB for one hour at room temperature, and then incubated with the primary antibody (rabbit anti-murine COX-2 polyclonal antibody diluted 1200 times, Cayman Chemical) for at least 48 hours at 4°C. After having been washed, the sections were incubated with biotin-conjugated anti-rabbit IgG for 1 hour at 4°C. Avidin-biotin complex (Vectorstain) or avidin-conjugated Cy-3 (Amersham) were used for visualization of the antigen. In some sections, double immunostaining was made with anti-human factor VIII related antigen (anti-human von Willebrand factor) as a marker for endothelial cells. In the case of LPS treatment, rats were injected intraperitonealy with LPS (100 μg/kg) and were sacrificed 5 h after the injection. Rats injected with saline were used as the control. The spinal cords of the rats were processed for histological examinations as described above.

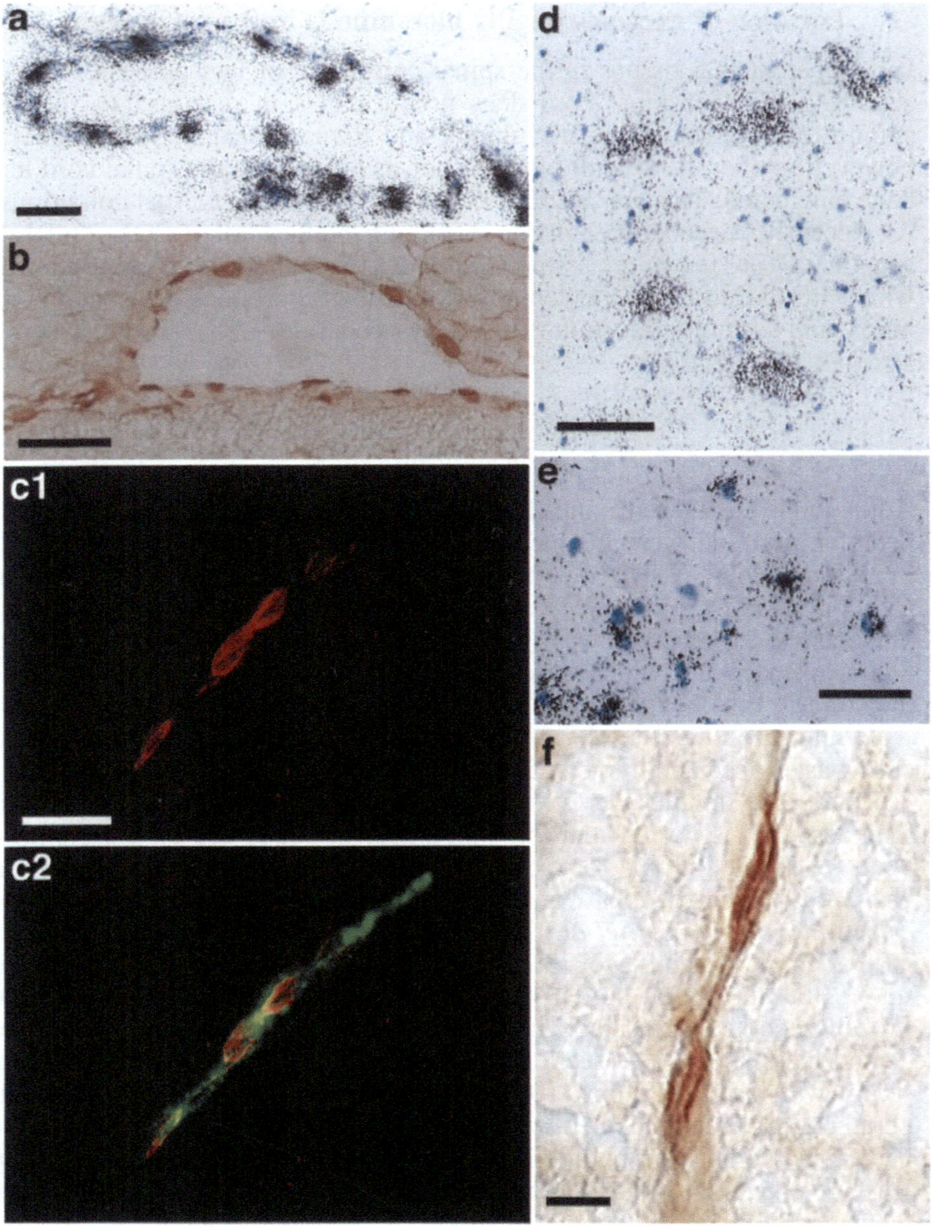

Fig. 6. Histochemical demonstration of COX-2 under pathological conditions. (**a**) Five hours after spinal cord injury (*SCI*), COX-2 mRNA was induced in the subarachnoidal and parenchymal blood vessels in the spinal cord. (**b**) COX-2-like immunoreactivity was also expressed in the blood vessels 5 h after SCI. (**c1**) COX-2-like immunoreactivity (red fluorescence) in the parenchymal blood vessels 5 h after SCI. (**c2**) A double immunostaining of COX-2 (red) and von Willbrand factor (green), an endothelial cell marker. (**d**) Neuron-like cells in the spinal cord also expressd a moderate level of COX-2 mRNA 5 h after SCI. (**e**) Small unidentified cells in the spinal cord also express COX-2 mRNA 5 h after SCI. (**f**) COX-2-like immunoreactivity in a parenchymal blood vessel 4 h after intraperitoneal injection of lipopolysaccharide (100 μg/kg). Scale bars represent 50 μm in a, b, d, 25 μm in c, e, and 10 μm in f

Results and discussion

Neither COX-2 mRNA nor COX-2-like immunoreactivity was observed in the spinal cord of sham-operated rats. Five hours after SCI, COX-2 mRNA was induced in three types of cells in the spinal cord. The most prominent induction was observed in the blood vessels of subarachinoidal space and of the parenchyma (Fig. 6a). COX-2-like immunoreactivity was also observed in the cells of blood vessels (Fig. 6b). Double immunostaining for COX-2 and von Willebrand factor revealed that COX-2 was induced in the nuclear membrane of the endothelial cells (Fig. 6c1, c2). A moderate level of COX-2 mRNA was also induced in relatively large cells in the gray matter of the spinal cord (Fig. 6d). Based on their size, location, and shape, they seemed to be neurons. As the third cell type, some small cells in the white matter also became positive for COX-2 mRNA after SCI (Fig. 6e). They appeared to be either glia or infiltrating leukocytes. The exact nature of the third group of cells remains to be identified.

As mentioned in the previous section, pretreatment with indomethacin, a potent inhibitor of COX, greatly rescued some pathological changes in the spinal cord. However, the induction of COX-2 after SCI was not attenuated by indomethacin pretreatment (data not shown). This result indicates that the induction of COX-2 was not the result of pathological responses after SCI but seemed to be the cause of the pathological responses. Further studies are necessary to understand the nature of the role of COX-2 in each type of cell in the pathogenesis of the spinal cord.

Intraperitoneal injection of LPS also resulted in COX-2 induction in the subarachinoidal and parenchymal blood vessels of the spinal cord (Fig. 6f). Double immunostaining for COX-2 and von Willebrand factor revealed the COX-2-positive cells to be the endothelial cells. In contrast to SCI, the LPS injection did not induce COX-2 in other types of cells. LPS injection is widely used as a model of bacterial infection. Infection is often associated with hyperalgesia in which PGs seem to play an essential role. The present result suggests that the blood vessels are the sites of PGs biosynthesis responsible for the hyperalgesia during infection.

Recently, Yamamoto and Nozaki-Taguchi (1996) demonstrated the involvement of COX-2 in formalin-induced hyperalgesia. In that study, rats were injected intrathecally with NS-398, a COX-2-specific inhibitor, or its vehicle either before or after injection of formalin into a hind paw. In the vehicle-treated rats, the subcutaneous injection of formalin resulted in a biphasic display of flinching behavior with the first peak just after the injection and the second one around 30 min after the formalin injection. In the rats treated with NS-398, the pain-related behavior was dose-dependently reduced in both phases. This result suggests that COX-2 is constitutively expressed in the spinal cord, since induction of COX-2 takes a much longer time. In our histological search for COX-2 mRNA and COX-2 protein, we could not demonstrate constitutively expressed COX-2 in the rat spinal cord. This seems to be due to an insufficiency in the sensitivity of the methods used. Indeed, an RT-PCR study on the rat spinal cord demonstrated that COX-2 mRNA was constitutively expressed and was upregulated by adjuvant-induced peripheral inflammation (Beiche *et al.*, 1996).

Concluding remarks

Although the PG system in the spinal cord is apparently involved in relatively frequent clinical problems, that is, pain and spinal cord injury, the neuroanatomical basis for these PG-related symptoms has been little clarified. Our current efforts have provided important but still fractional information on the PG system in the spinal cord under pathophysiological conditions. Recent biochemical studies, especially those that have employed molecular biological techniques, have been providing a great deal of information on the molecular nature of the PG system. Application of this knowledge to neuroanatomical research in the spinal cord would be of great value in this field.

Future direction

Our laboratory is currently engaged to map out various recepotrs of the prostaglandins in the spinal cord following injury and how these receptors regulate neuroprotection following treatment with several pharmacotherapeutic compounds known to exrt their beneficial effects on cell injury. In addition, an interaction of prostaglandins with nitric oxide production, a feature which recently attended wide recognition in the CNS, is also under investigation in our laboratory.

Acknowledgements

Part of this reserach is supported by funds from Swedish Medical Research Council grant nrs. 2710 (JW, HSS), 9077 (TG), Astra Pain Control, Sweden and University Grants Commission, New Delhi, India (HSS).

References

Barkai AI, Bazan NG (1989) Arachidonic acid metabolism in the nervous system. Physiological and Pathological significance. Ann NY Acad Sci 559: 1–353

Bazan NG, Rodriguez de Turco EB, Allan G (1995) Mediators of injury in neurotrauma: intracellular signal transduction and gene expression. J Neurotrauma 12: 791–814

Beiche F, Scheuerer S, Brune K, Geisslinger G, Goppelt-Struebe M (1996) Up-regulation of cyclooxygenase-2 mRNA in the rat spinal cord following peripheral inflammation. FEBS Lett 390: 165–169

Birrell GJ, McQueen DS (1993) The effects of capsaicin, bradykinin, PGE_2 and cicaprost on the discharge of articular sensory receptors *in vitro*. Brain Res 611: 103–107

Birrell GJ, McQueen DS, Iggo A, Coleman RA, Grubb BD (1991) PGI_2-induced activation and sensitization of articular mechanonociceptors. Neurosci Lett 124: 5–8

Breder CD, Dewitt D, Kraig RP (1995) Characterization of inducible cyclooxygenase in rat brain. J Comp Neurol 355: 296–315

Breder CD, Saper CB (1996) Expression of inducible cyclooxygenase mRNA in the mouse brain after systemic administration of bacterial lipopolysaccharide. Brain Res 713: 64–69

Breder CD, Smith WL, Raz A, Masferrer J, Seibert K, Needleman P, Saper CB (1992) Distribution and characterization of cyclooxygenase immunoreactivity in the ovine brain. J Comp Neurol 322: 409–438

Cao C, Matsumura K, Yamagata K, Watanabe Y (1995) Induction by lipopolysaccharide of cyclooxygenase-2 mRNA in rat brain; its possible role in the febrile response. Brain Res 697: 187–196

Cao C, Matsumura K, Yamagata K, Watanabe Y (1996) Endothelial cells of the rat brain vasculacture express cyclooxygenase-2 mRNA in response to systemic interleukin-1β: a possible site of prostaglandin synthesis responsible for fever. Brain Res 733: 263–272

Cao C, Matsumura K, Yamagata K, Watanabe Y (1997) Involvement of cyclooxygenase-2 in LPS-induced fever and its regulation in the brain. Am J Physiol 272: R1712–R1725

Chahl LA, Iggo A (1977) The effects of bradykinin and prostaglandin E_1 on rat cutaneous afferent nerve activity. Br J Pharmacol 59: 343–347

Coleman RA, Smith WL, Narumiya S (1994) International union of pharmacology classification of prostanoid receptors: properties, distribution, and structure of the receptors and their subtypes. Pharmacol Rev 46: 205–229

Collier JG, Karim SMM, Robinson B, Somers K (1972) Action of prostaglandin A_2, B_1, E_2 and $F_{2\alpha}$ on superficial hand veins of man. Br J Pharmacol 44: 374–375

Demeduik P, Saunders RD, Clendenon NR, Means ED, Anderson DK, Horrocks LA (1985) Changes in lipid metabolism in traumatised spinal cord. Prog Brain Res 63: 211–226

England S, Bevan S, Docherty RJ (1996) PGE_2 modulates the tetrodotoxin-resistant sodium current in neonatal rat dorsal root ganglion neurones via the cyclic AMP-protein kinase A cascade. J Physiol 495: 429–440

Faden AI, Lemke M, Demeduik P (1988) Effects of BW755C, a mixed cyclo-oxygenase-lipoxygenase inhibitor, following traumatic spinal cord injury in rats. Brain Res 463: 63–68

Ferreira SH (1972) Prostaglandins, aspirin-like drugs and analgesia. Nature 240: 200–203

Ferreira SH, Moncada S, Vane JR (1973) Prostaglandins and the mechanism of analggesia produced by aspirin-like drugs. Br J Pharmacol 49: 86–97

Ferreira SH, Nakamura M, Castro MSA (1978) The hyperalgesic effects of prostacyclin and prostaglandin E_2. Prostaglandin 16: 31–37

Glaser KB, Mobilio D, Chang JY, Senko N (1993) Phospholipase A_2 enzymes: regulation and inhibition. Trends Pharmacol Sci 14: 92–98

Gold MS, Reichling DB, Shuster MJ, Levine JD (1996) Hyperalgesic agents increase a tetrodotoxin-resistant Na^+ current in nociceptors. Proc Natl Acad Sci USA 93: 1108–1112

Gold MS, Shuster MJ, Levine JD (1996) Role of a Ca^{2+}-dependent slow afterhyperpolarization in prostaglandin E_2-induced sensitization of cultured rat sensory neurons. Neurosci Lett 205: 161–164

Goppelt-Struebe M (1995) Regulation of prostaglandin endoperoxide synthase (cyclooxygenase) isozyme expression. Prostaglandins Leukot. Essent Fatty Acids 52: 213–222

Hingtgen CM, Vasko MR (1994) Prostacyclin enhances the evoked-release of substance P and calcitonin gene-related peptide from rat sensory neurons. Brain Res 655: 51–60

Hirata M, Hayashi Y, Ushikubi F, Yokota Y, Kageyama R, Nakanishi S, Narumiya S (1991) Cloning and expression of cDNA for a human thromboxane A_2 receptor. Nature 349: 617–620

Hsu CY, Halushka PV, Hogan EL, Banik NL, Lee WA, Perot PL (1985) Alterations of thromboxane and prostacyclin levels in experimental spinal cord injury. Neurology 35: 1003–1009

Khasar SG, Levine JD (1996) Neonatal capsaicin attenuates mechanical nociception in the rat. Neurosci Lett 205: 141–143

Malmberg AB, Yaksh TL (1992) Antinociceptive actions of spinal nonsteroidal anti-inflammatory agents on the formalin test in the rat. J Pharmacol Exp Ther 263: 136–146

Malmberg AB, Yaksh TL (1994) Capsaicin-evoked prostaglandin E_2 release in spinal cord slices: relative effect of cyclooxygenase inhibitors. Eur J Pharmacol 271: 293–299

Malmberg AB, Yaksh TL (1995) Cyclooxygenase inhibition and the spinal release of prostaglandin E_2 and amino acids evoked by paw formalin injection: a microdialysis study in unanesthetized rats. J Neurosci 15: 2768–2776

Matsumura K, Cao C, Watanabe Yu, Watanabe Y (1998) Prostaglandin system in the brain: sites of biosynthesis and sites of action under normal and hyperthermic conditions. Prog Brain Res 115: 275–295

Matsumura K, Watanabe Yu, Imai-Matsumura K, Connolly M, Koyama Y, Onoe H, Watanabe Y (1992) Mapping of prostaglandin E_2 binding sites in rat brain using quantitative autoradiography Brain Res 581: 292–298

Matsumura K, Watanabe Yu, Onoe H, Watanabe Y (1995) Prostacyclin recpetor in the brain and central terminals of the primary sensory neurons: an autoradiographic study using a stable prostacyclin analogue [^{3}H]iloprost. Neuroscience 65: 493–503

Matsumura K, Watanabe Yu, Onoe H, Watanabe Y, Hayaishi O (1990) High density of prostaglandin E_2 binding sites in the anterior wall of the 3rd ventricle: a possible site of its hyperthermic action. Brain Res 533: 147–151

Mense S (1981) Sensitization of group IV muscle receptors to bradykinin by 5-hydroxy-tryptaminse and prostaglandin E_2. Brain Res 225: 95–105

Minami T, Nishihara I, Uda R, Ito S, Hyodo M, Hayaishi O (1994) Characterization of EP-receptor subtypes involved in allodynia and hyperalgesia induced by intrathecal administration of prostaglandin E_2 to mice. Br J Pharmacol 112: 735–740

Minami T, Uda R, Horiguchi S, Ito S, Hyodo M, Hayaishi O (1994) Allodynia evoked by intrathecal administration of prostaglandin E_2 to conscious mice. Pain 57: 217–223

Minami T, Uda R, Horiguchi S, Ito S, Hyodo M, Hayaishi O (1992) Allodynia evoked by intrathecal administration of prostaglandin $F_{2\alpha}$ to conscious mice. Pain 50: 223–229

Oida H, Namba T, Sugimoto Y, Ushikubi F, Ohishi H, Ichikawa A, Narumiya S (1995) *In situ* hybridization of prostacyclin receptor mRNA expression in various mouse organs. Br J Pharmacol 116: 2828–2837

Sharma HS, Olsson Y, Nyberg F, Dey PK (1993a) Prostaglandins modulate alterations of microvascular permeability, blood flow, edema and serotonin levels following spinal cord injury. An experimental study in the rat. Neuroscience 57: 443–449

Sharma HS, Olsson Y, Cervós-Navarro, J (1993b) Early perifocal cell changes and edema in traumatic injury of the spinal cord are reduced by indomethacin, an inhibitor of prostaglandin synthesis. Experimental study in the rat. Acta Neuropathol (Berl) 85: 145–153

Sharma HS, Olsson Y, Persson S, Nyberg F (1995) Trauma-induced opening of the the blood-spinal cord barrier is reduced by indomethacin, an inhibitor of prostaglandin biosynthesis. Experimental observations in the rat using [^{131}I]-sodium, Evans blue and lanthanum as tracers. Restorat Neurol Neurosci 7: 207–215

Shimizu T, Wolfe LS (1990) Arachidonic acid cascade and signal transduction. J Neurochem 55: 1–15

Shohami E, Shapiro Y, Cotev S (1988) Experimental closed head injury in rats: prostaglandin production in a noninjured zone. Neurosurgery 22: 859–863

Smith WL, Marnett LJ, DeWitt DL (1991) Prostaglandin and thromboxane biosynthesis. Pharmacol Ther 49: 153–179

Sugimoto Y, Shigemoto R, Namba T, Negishi M, Mizuno N, Narumiya S, Ichikawa A (1994) Distriburion of the mRNA for the prostaglandin E receptor subtype EP3 in the mouse nervous system. Neuroscience 62: 919–928

Taiwo YO, Bjerknes LK, Goetzl EJ, Levine JD (1989) Mediation of primary afferent peripheral hyperalgesia by the cAMP second messenger system. Neuroscience 32: 577–580

Taiwo YO, Levince JD (1988) Prostaglandin inhibit endogenous pain control mechanisms by blocking transmission at spinal noradrenergic synapses. J Neurosci 8: 1346–1349

Taiwo YO, Levine JD (1990) Effects of cyclooxygenase products of arachidonic acid metabolism on cutaneous nociceptive threshold in the rat. Brain Res 537: 372–374
Takechi H, Matsumura K, Watanabe Y, Kato K, Noyori R, Suzuki M, Watanabe Y (1996) A novel subtype of the prostaglandin receptor expressed in the central nervous system. J Biol Chem 271: 5901–5906
Takeuchi K, Abe T, Takahashi N, Abe K (1993) Molecular cloning and intrarenal localization of rat prostaglandin E_2 receptor EP3 subtype. Biochem Biophys Res Commun 194: 885–891
Tsubokura S, Watanabe Y, Ehara H, Imamura K, Sugimoto O, Kagamiyama H, Yamamoto S, Hayaishi O (1991) Localization of prostaglandin endoperoxide synthase in neurons and glia of monkey brain. Brain Res 543: 15–24
Uda R, Horiguchi S, Ito S, Hyodo M, Hayaishi O (1990) Nociceptive effects induced by intrathecal administration of prostaglandin D_2, E_2, or $F_{2\alpha}$ to conscious mice. Brain Res 510: 26–32
Undem BJ, Weinreich D (1993) Electrophysiological properties and chemosensitivity of guinea pig nodose ganglion neurons in vitro. J Auton Nerv Syst 44: 17–34
Ushikubi F, Hirata M, Narumiya S (1995) Molecular biology of prostanoid receptors; an overview. J Lipid Mediat 12: 343–359
Vane JR (1971) Inhibition of prostaglandin synthesis as a mechanism of action for aspirin-like drugs. Nature 231: 232–235
Vasko MR, Campbell WB, Waite KJ (1994) Prostaglandin E_2 enhances bradykinin-stimulated release of neuropeptides from rat sensory neurons in culture. J Neurosci 14: 4987–4997
Watanabe Yu, Watanabe Y, Hamada K, Bommelaer-Bayt M-C, Dray F, Kaneko T, Yumoto N, Hayaishi O (1989) Distinct localization of prostaglandin D_2, E_2, and $F_{2\alpha}$ binding sites in monkey brain. Brain Res 478: 143–148
Watanabe Yu, Watanabe Y, Hayaishi O (1988) Quantitative autoradiographic localization of prostaglandin E_2 binding sites in monkey diencephalon. J Neurosci 8: 2003–2010
Watanabe Y, Yamashita A, Tokumoto H, Hayaishi O (1983) Localization of prostaglandin D_2 binding protein and NADP-linked 15-hydroxyprostaglandin D_2 dehydrogenase in the Purkinje cells of miniature pig cerebellum. Proc Natl Acad Sci USA 80: 4542–4545
Winkler T, Sharma HS, Stålberg E, Olsson Y (1993) Indomethacin, an inhibitor of prostaglandin synthesis attenuates alteration in spinal cord evoked potentials and edema formation after trauma to the spinal cord. An experimental study in the rat. Neuroscience 52: 1057–1067
Yamagata K, Andreasson KI, Kaufmann WE, Barnes CA, Worley PF (1993) Expression of a mitogen-inducible cyclooxygenase in brain neurons: regulation by synaptic activity and glucocorticoids. Neuron 11: 371–386
Yamamoto T, Nozaki-Taguchi N (1996) Analysis of the effects of cyclooxygenase (COX)-1 and COX-2 in spinal nociceptive transmission using indomethacin, a non-selective COX inhibitor, and NS-398, a COX-2 selective inhibitor. Brain Res 739: 104–110
Yamashita A, Watanabe Y, Hayaishi O (1983) Autoradiographic localization of a binding protein(s) specific for prostaglandin D_2 in rat brain. Proc Natl Acad Sci USA 80: 6114–6118
Yang LC, Marsala M, Yaksh TL (1996) Characterization of time course of spinal amino acids, citruline and PGE_2 release after carrageenan/kaolin-induced knee joint inflammation: a chronic microdialysis study. Pain 67: 345–354

Correspondence: Dr. K. Matsumura, Department of Neuroscience, Osaka Bioscience Institute, 6-2-4 Furuedai, Suita 565-0874, Japan.

Section II

Spinal cord regeneration and repair

Spinal Cord Monitoring
Eds: E. Stålberg, H. S. Sharma, Y. Olsson

Chapter 6

Neurotrophic factors and CNS regeneration

T. Hagg[1] and **M. Oudega**[2]

[1]Department of Anatomy and Neurobiology, Faculty of Medicine, Dalhousie University, Halifax, Nova Scotia, Canada,
[2]Miami Project to Cure Paralysis, University of Miami School of Medicine, Miami, Florida, USA

Introduction

After a traumatic contusion or ischemic injury to the spinal cord many neurons, oligodendrocytes and other cells in the affected region die over the ensuing hours to days through a complex cascade of events. The neurons that contribute to the long projecting tracts generally do not die from such a distant lesion but their axons can be directly interrupted by the spinal cord injury or undergo subsequent degeneration. Ideally, treatments should be aimed at preventing these degenerative events, i.e., rescue local cells and axons from injury-induced death and degeneration. Such a treatment currently consists of infusions of high doses of methylprednisolon within 8 hours after injury which leads to some improvement in motor and sensory recovery (Bracken and Holford, 1993; Young *et al.*, 1994). However, many patients have existing spinal cord injuries and many receive injuries where tissue loss is inevitable. Thus, treatments also have to be developed to replace lost spinal cord tissue and to promote regeneration of long axonal projections. Spontaneous regeneration and recovery of function in the adult mammalian central nervous system (CNS) is very limited, especially in the spinal cord. This is not due to an inherent inability of the CNS neurons to regenerate since they are able to grow into peripheral nerve grafts. Evidently, the adult CNS environment, but not that of the peripheral nervous system (PNS), is restrictive to axonal regeneration. This is also exemplified by the observations that crushed central projections of primary sensory neurons can grow very well in the dorsal roots (PNS) but cannot cross the root-cord interface into the spinal cord (CNS). Similarly, of the CNS axons that grow into and across an implanted peripheral nerve graft, very few reenter the spinal cord tissue and only over very short distances. Recent advances in the understanding about the various elements of spinal cord degeneration and regeneration have been extensively reviewed elsewhere (Schwab and Bartholdi, 1996).

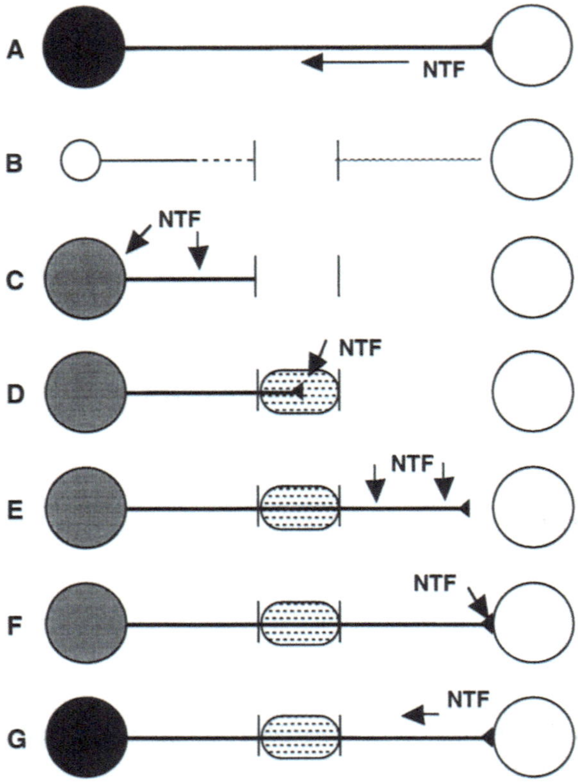

Fig. 1. Neurotrophic factors may facilitate different aspects of recovery from spinal cord injury. (**A**) In the normal adult CNS, endogenous neurotrophic factors (*NTF*) from the target or surrounding cells are essential for the maintenance of neuronal function. (**B**) After injury, neurons may atrophy or die, in part because of a relative (increased need) or absolute deficiency in neurotrophic factors. Neurotrophic factors could promote survival (**C**), sprouting into bridging materials (**D**), regeneration into the CNS and back to the original target (**E**), synapse formation (**F**) and transmitter-related functions (**G**)

This review will discuss a selected number of animal studies to illustrate the current understanding about neurotrophic factors and their therapeutic potential for spinal cord injury. It will focus on the protective and regeneration-promoting effects of neurotrophic factors on neurons. During the past decade we and others have found that neurotrophic factors can enhance several events that are crucial for functional recovery. These include, cell survival, axonal regeneration across bridging materials, regrowth into the CNS and regulation of transmitter-related function (Fig. 1). It has also become clear that different neurotrophic factors do not necessarily affect all these processes and that they have different sets of biological actions. The challenge is to understand how such different properties can be harnessed and combined to most efficiently affect the individual repair processes that are necessary for functional recovery after spinal cord and other CNS injuries.

General properties of neurotrophic factors

During development neurons become dependent for their survival and function on neurotrophic factors (Levi-Montalcini, 1987; Oppenheim, 1991). These 10–30 kD proteins are produced by surrounding cells and/or in the innervation territory. Nerve growth factor (NGF) was discovered more than 40 years ago and was recognized by its effects on developing sensory and sympathetic neurons that included neuronal survival, induction of neurite outgrowth and synthesis of transmitter enzymes (Levi-Montalcini, 1987). The general view is that neurotrophic factors promote growth of neurons (cell size or neurite length), which distinguishes them from growth factors that promote growth of a cell population (proliferation). This distinction is not always evident since typical growth factors (e.g., epidermal growth factor) also have neurotrophic actions and neurotrophic factors (e.g., NGF) can induce proliferation of certain cells (Hefti, 1993; Hagg, 1997). Many neurotrophic factors do not act only in the nervous system, e.g., NGF is involved in gamete production and B-cell function. Some neurotrophic factors appear to be selective for a few types of neurons, whereas others have a wide range of cell targets.

In the CNS, neurotrophic factors are synthesized by neurons and glial cells and in the PNS by Schwann cells and fibroblasts and are released through autocrine, paracrine or telecrine mechanisms (Fallon and Loughlin, 1993). Synthesis of neurotrophic factors appears to be regulated by neuronal activity and transmitter release and by hormones such as thyroid hormone and glucocorticoids (Lindholm *et al.*, 1994; Thoenen, 1995). After injury to the adult rat brain, the synthesis of several neurotrophic factors is greatly increased (Lindholm *et al.*, 1991; Ip *et al.*, 1993a; Mochetti and Wrathall, 1995; DeKosky *et al.*, 1996), a response that may be mediated by cytokines such as interleukin-1 (IL-1) that are released by microglia and immune cells (Fagan and Gage, 1990; Spranger *et al.*, 1990; DeKosky *et al.*, 1996). However, such increases in the level of neurotrophic factors are not large enough or do not occur at the right time to prevent degeneration and promote regeneration in the CNS. Potentially, endogenous sources of neurotrophic factors can be increased by pharmacological use of small molecules (e.g., dexamethasone, Saporito *et al.*, 1994), an approach that may prove to be useful in the treatment of spinal cord injury (see also section "Delivery of neurotrophic factors").

Neurotrophic factors can be internalized and then retrogradely transported from the innervation territory (presumably from axon terminals) to the cell body (DiStefano *et al.*, 1992; Korshing, 1993; Oppenheim, 1996) and some have been shown to be transported anterogradely (Von Bartheld *et al.*, 1996). It is still unclear whether neurotrophic factors need to be retrogradely transported to activate second message systems in the cell body or whether they can activate intracellular signaling at the distal portions of neuronal processes (Johanson *et al.*, 1996). Neurotrophic factors can most likely also activate receptors that are found on the cell bodies since they can affect neurons in vitro that do not have neurites. Understanding about where neurotrophic factors act will dictate where treatments would be best applied.

Neurotrophic factors and their receptors

NGF is a member of the neurotrophin family that includes brain-derived neurotrophic factor (BDNF), neurotrophin-3, -4/5 and -6 (NT-3, NT-4, NT-6) (Barbacid, 1995; Bothwell, 1995; Ip and Yancopoulos, 1996). These neurotrophins are structurally and functionally homologous and 50–60% identical in amino acid sequence. In their physiological state neurotrophins are homodimers of two approximately 120 amino acid proteins (~13 kD) that have specific binding domains for receptors (Ibanez, 1995). The neurotrophins bind to and activate high-affinity transmembrane tyrosine kinase (Trk) receptors that interact with several identified intracellular signaling pathways (Heumann, 1994; Bothwell, 1995; Chao and Hempstead, 1995; Greene and Kaplan, 1995; Segal and Greenberg, 1996). NGF specifically activates TrkA, BDNF and NT-4 activate TrkB and NT-3 activates TrkC, and in some cell types TrkB. Truncated forms of TrkB and TrkC without an intracellular domain have also been identified. These "receptors" do not transduce intracellular signals and their function is still unclear (Barbacid, 1995; Bothwell, 1995).

The low-affinity p75 NGF receptor ($p75^{NGFR}$) is a member of the tumor necrosis factor receptor superfamily and can bind all the neurotrophins (Heumann, 1994; Bothwell, 1995; Chao and Hempstead, 1995; Greene and Kaplan, 1995). The functions of $p75^{NGFR}$ are beginning to be elucidated and may include: providing ligand-binding specificity for NGF, enhancing Trk phosphorylation in the presence of ligand, decreasing Trk autophosphorylation in the absence of ligand and, in the PNS, facilitating retrograde transport of some neurotrophins (Curtis *et al.*, 1995). Through TrkA-independent activation of the ceramide pathway (Dobrowsky *et al.*, 1994; Chao, 1995; Hannun, 1996), $p75^{NGFR}$ can play a role in neuronal apoptosis (Rabizadeh *et al.*, 1993; Barrett and Bartlett, 1994; Frade *et al.*, 1996; Van der Zee *et al.*, 1996).

Ciliary neurotrophic factor (CNTF) is a member of the hematopoetic cytokine family, that includes leukemia-inhibitory factor (LIF) and interleukin-6 (IL-6)(Patterson, 1992; Manthorpe *et al.*, 1993; Sendtner *et al.*, 1994; Stahl *et al.*, 1994; Richardson, 1994; Ip and Yancopoulos, 1996). CNTF has ~200 amino acids (~22 kD) and under physiological conditions acts as a monomer. CNTF is expressed almost exclusively in the nervous system and is synthesized by astrocytes and Schwann cells. Whether and how CNTF is released under physiologic conditions is unclear since it lacks a signal sequence characteristic for classic secretion. After injury to the CNS, astrocytes produce more CNTF (Ip *et al.*, 1993a) and after injection of CNTF normal astrocytes become reactive (Kahn *et al.*, 1995; Winter *et al.*, 1995), suggesting that CNTF has an autocrine role for astrocytes. The CNTF α receptor is specific for CNTF but cannot directly activate intracellular signaling pathways because it lacks a transmembrane domain (Ip *et al.*, 1992; Stahl *et al.*, 1994; Ip and Yancopoulos, 1996). After binding, the CNTF- CNTF α receptor unit associates in a trimer receptor complex with a LIF β and a gp130 receptor that transduce the receptor signal to the intracellular Jak-STAT pathway (Ip and Yancopoulos, 1996; Segal and Greenberg, 1996). CNTF α receptor is soluble and can associate with the LIF β-gp130 dimer of other cells to render them responsive to CNTF (Davis *et al.*, 1993), an observation that may lead to new therapeutic approaches to influence cells that are not normally responsive to CNTF (Economides *et al.*, 1995). CNTF α receptor is found almost exclusively in the CNS where its mRNA and protein have been detected in a variety of neurons, many of them involved in motor function (Ip *et al.*, 1993b; MacLennan *et al.*, 1996).

Glial cell line-derived neurotrophic factor (GDNF) is a member of the transforming growth factor-β family (Lin *et al.*, 1993) and is an ~134 amino acid (~15–20 kD) protein that acts under physiologic conditions as a homodimer. During adulthood, GDNF is synthesized in various organs (e.g., lung, liver, ovary) at much higher levels than in the nervous system (Choi-Lundberg and Bohn, 1995; Suter-Crazzolara and Unsicker, 1994; Trupp *et al.*, 1995). GDNF binds to the non-transducing GDNF α receptor, which associates with a (possibly two) transducing Ret receptor (Jing *et al.*, 1996; Treanor *et al.*, 1996, see Chapter 8 in this volume).

NGF and cholinergic basal forebrain neurons

NGF prevents degeneration of cholinergic neurons

The cholinergic septohippocampal model in adult rats and the effects of NGF have been used extensively to formulate ideas about how neurotrophic factors function and how they may be utilized as therapeutic agents (Hagg *et al.*, 1993a; Henderson, 1996). Cholinergic neurons of the medial septum and vertical limb of the nucleus of the diagonal band of Broca project to the ipsilateral hippocampal formation through the fimbria-fornix and supracallosal stria (Fig. 2A) (Swanson *et al.*, 1987). NGF has a physiological role for these neurons *in vivo*, i.e., these neurons express TrkA and $p75^{NGFR}$ (Richardson *et al.*, 1986; Hagg *et al.*, 1989; Sobreviela *et al.*, 1994; Holtzman *et al.*, 1995), internalize and retrogradely transport NGF that is produced in the hippocampal formation and in the basal forebrain (Korshing *et al.*, 1985; Shelton and Reichardt, 1986; DiStefano *et al.*, 1992; Lauterborn *et al.*, 1993, 1995). Administration of exogenous NGF causes an increase (Mobley *et al.*, 1986; Fusco *et al.*, 1989) and

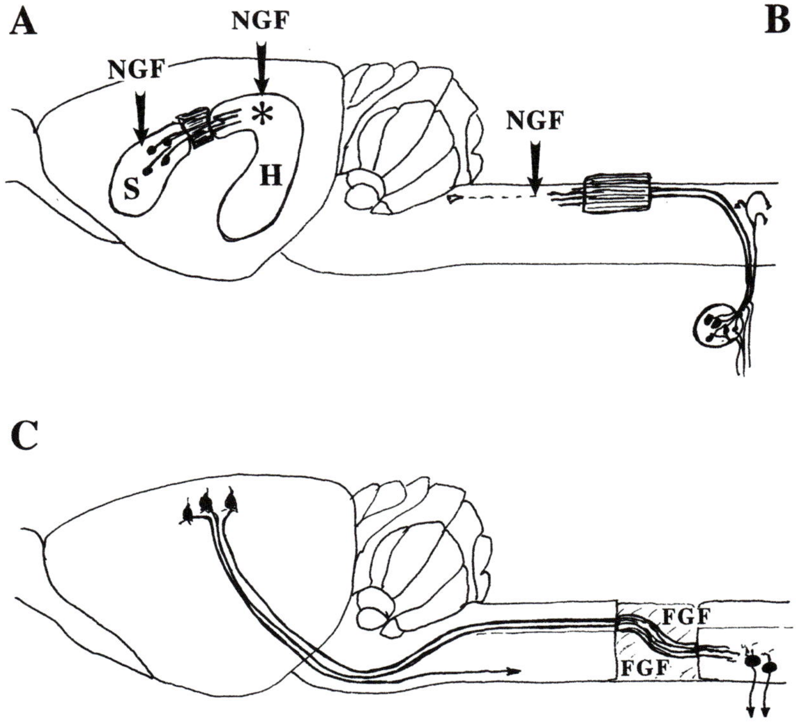

Fig. 2. Regeneration models for investigating the therapeutic potential of neurotrophic factors. (**A**) Cholinergic neurons in the septum (*S*) project through the fimbria-fornix to the hippocampal formation (*H*). NGF infusion close to the septum (arrow) prevents degeneration of the cholinergic neurons and NGF infusion into the hippocampal formation (arrow and asterisk) promotes reentry of cholinergic axons that have grown across a peripheral nerve bridge. (**B**) Primary sensory afferents terminate locally in the cord or ascend through the dorsal funiculus to the cuneatus and gracilis nuclei. NGF infusion (arrow) into the dorsal funiculus of the spinal cord promotes regeneration of ascending sensory axons that have grown across nerve bridges (**C**) In the rodent, most of the corticospinal motor axons descend through the dorsal corticospinal tract in the spinal cord to innervate lower motor neurons. Exogenous acidic fibroblast growth factor (*FGF*) plays a role in promoting ingrowth into redirecting thin nerve grafts and regeneration into spinal gray matter for recovery of motor functions

neutralizing endogenous NGF causes a decrease in choline acetyltransferase (ChAT)(Vantini *et al.*, 1989; Van der Zee *et al.*, 1995).

Over a 2 week period following transection of the septohippocampal pathway, ~70% of the axotomized cholinergic neurons lose their markers (ChAT, TrkA and $p75^{NGFR}$), their proximal axonal stumps degenerate and their cell bodies atrophy (Gage *et al.*, 1986; Armstrong *et al.*, 1987; Hagg *et al.*, 1989, 1992; Figueiredo *et al.*, 1995). Injection or chronic infusion of NGF close to the septal neurons over the same time can prevent these axotomy-induced degenerative changes (Hefti, 1986; Williams *et al.*, 1986;

Hagg *et al.*, 1989, 1992). Consistent with the finding that only cholinergic neurons express TrkA in the basal forebrain, NGF has no effect on the axotomized GABAergic septohippocampal neurons (Hagg *et al.*, 1992; Koliatsos *et al.*, 1994).

There is evidence that a proportion of the axotomized cholinergic neurons die between 2 and 4 weeks after axotomy (O'Brien *et al.*, 1990; Tuszynski *et al.*, 1990a; Wilcox *et al.*, 1995), although others have provided evidence to the contrary (Peterson *et al.*, 1990). More interesting is our finding that in 60% of these neurons, the cell body atrophy and loss of ChAT and $p75^{NGFR}$ can be rapidly reversed by an NGF treatment that starts more than 3 months after the transection (Hagg *et al.*, 1989). Apparently, these "dormant" neurons remain amenable to treatments, a finding that may have relevance for patients with existing injuries. The permanent maintenance of normal cell size and expression of ChAT after withdrawal of NGF treatment appears to depend on establishing stable synaptic contacts with the original targets (Montero and Hefti, 1988). The question whether and when neurotrophic factor treatments could be terminated without subsequent decreases in function is a clinically important one.

BDNF (Knüsel *et al.*, 1992; Morse *et al.*, 1993; Widmer *et al.*, 1993; Koliatsos *et al.*, 1994), but not NT-3, (Koliatsos *et al.*, 1994) can protect septal cholinergic (not GABAergic) neurons but is less effective and >10 times less potent than NGF (Williams *et al.*, 1996). This may be partially related to the poor diffusion characteristics of BDNF (Yan *et al.*, 1994; Anderson *et al.*, 1995). CNTF is as potent as NGF in rescuing all the axotomized cholinergic neurons but does not prevent the axotomy-induced loss of ChAT (Hagg *et al.*, 1992). In contrast to NGF, CNTF also rescues the lesioned GABAergic septohippocampal neurons (Hagg *et al.*, 1992) and activates CNS microglia (Hagg *et al.*, 1993b) and astrocytes (Kahn *et al.*, 1995; Winter *et al.*, 1995). Thus CNTF appears to have a more general injury-related and immune-mediator role in the CNS. The recently discovered GDNF reportedly is 10 times less potent than NGF and prevents the loss of $p75^{NGFR}$ but has less effect on the reduction in ChAT-staining (Williams *et al.*, 1996).

The finding that neurotrophic factors have different levels of effectiveness for these cholinergic neurons suggests that, although neurons are responsive to many neurotrophic factors, they respond optimally to a few. In addition, neurotrophic factors do not all have the same set of actions on these neurons, a characteristic that could be used to selectively affect individual components of the recovery process. The observed lack of effect of some neurotrophic factors on transmitter synthesis could be beneficial to regeneration. Regenerating neurons "switch" from synthesizing function-related proteins such as neurotransmitter enzymes to regeneration-related ones (Grafstein and McQuarrie, 1978), thereby possibly conserving cellular energies. BDNF has been shown to induce regeneration-associated genes in other axotomized CNS neurons (Tetzlaff *et al.*, 1994).

NGF promotes regeneration of cholinergic axons

Nerve growth factor was named for its neurite outgrowth-promoting effects (Levi-Montalcini, 1987). The neurites of developing peripheral neurons grow toward the source of NGF which indicates its chemoattractant or tropic properties (Menesini-Chen *et al.*, 1978; Gundersen and Barrett, 1980; Levi-Montalcini, 1987). In the adult rat, NGF induces sprouting of the lesioned, but not non-lesioned, cholinergic axons (Gage *et al.*, 1988; Hagg *et al.*, 1989, 1992; Vahlsing *et al.*, 1991; Hagg and Varon, 1993) and such axons grow toward the site of NGF infusion, even into brain regions that they did not occupy before (Rosenberg *et al.*, 1988; Vahlsing *et al.*, 1991; Hagg and Varon, 1993; Emerich *et al.*, 1994). GDNF induces only a moderate level of sprouting of the axotomized cholinergic axons (Williams *et al.*, 1996), BDNF has very limited effects (Morse *et al.*, 1993; Widmer *et al.*, 1993; Koliatsos *et al.*, 1994; Williams *et al.*, 1996), and CNTF has no effect (Hagg *et al.*, 1992). Of interest is our observation that NGF infusions close to the non-lesioned septum does not promote growth of sprouting axons from the lesioned septum into the non-lesioned septal terrain (Hagg and Varon, 1993). This suggests that non-lesioned CNS tissue does not permit entry of regenerating axons, a finding that may have relevance for the injured CNS where generally most regions and systems are intact.

Peripheral nerve grafts that are placed as bridges between the septum and hippocampus promote ingrowth of cholinergic axons and a substantial number of axons cross to the other end of such bridges between 2 and 4 weeks (Wendt *et al.*, 1983; Hagg *et al.*, 1990a). This regeneration-promoting effect of living peripheral nerve grafts may result from the increased synthesis of neurotrophic factors such as NGF by the nerve cells (Kromer and Cornbrooks, 1987; Olson *et al.*, 1990; Heumann *et al.*, 1987; Messersmith *et al.*, 1991; Meyer *et al.*, 1992). To investigate this possibility we implanted "acellular" nerves (Hagg *et al.*, 1991) which were devoid of myelin, axons, intact cells or their debris and consisted of longitudinally arranged basal lamina tubes that were rich in neurite-promoting molecules such as laminin (Manthorpe *et al.*, 1990). Very few cholinergic axons grew into such acellular nerve bridges (see also Berry *et al.*, 1988). In sharp contrast, when soaked in NGF before implantation the acellular nerves promoted almost the same amount of ingrowth as cell-containing nerve grafts (Hagg *et al.*, 1991). Similarly, implanted bridges of fibroblasts become permissive only when these cells are transfected to produce NGF (Kawaja *et al.*, 1992). These findings raise the possibility that cellular or artificial materials could be combined with specific neurotrophic factors to promote regeneration of selected types of neurons.

We attempted to further enhance regeneration into septohippocampal cell-containing nerve grafts by increasing the number of rescued cholinergic neurons with an NGF infusion close to the septum in adult rats (Hagg and Varon, 1993). The cholinergic axons remained in the region of the NGF infusion and did not grow into the grafts, most likely because of the chemoattractant properties of NGF. This suggests that NGF cannot induce axonal outgrowth into an environment with lower NGF levels, confirming previous findings in vitro (Campenot, 1995). Thus, such regeneration-promoting agents may always have to be presented in front of growing axons to induce regeneration into the appropriate target tissues. Others have shown that NGF infusion close to the septum during the first 2 weeks after graft implantation rescues more cholinergic neurons and stimulates regeneration of more cholinergic fibers into the hippocampal formation, presumably during the months after the transient NGF treatment (Tuszynski *et al.*, 1990b, 1995).

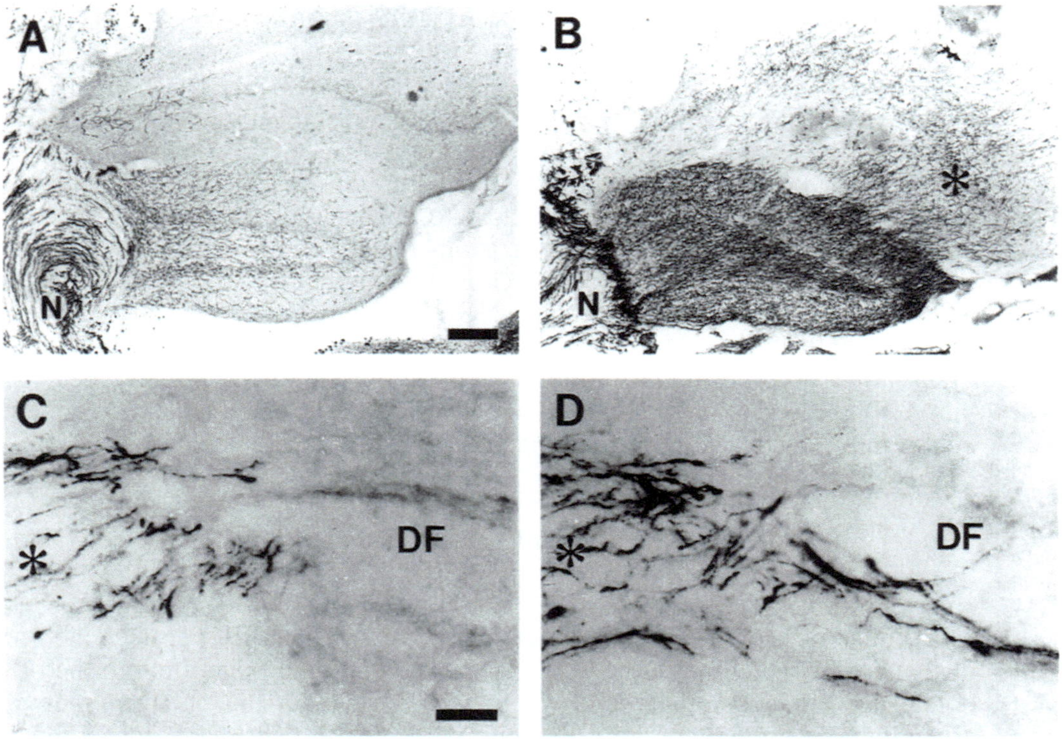

Fig. 3. NGF promotes regeneration of cholinergic septohippocampal and intraspinal sensory axons. (**A**) A modest number of cholinergic axons (stained for acetylcholinesterase) reenter the hippocampal formation from the ends (*N*) of septohippocampal nerve bridges after 1 month. Bar = 250 μm for **A** and **B**. (**B**) Infusion of NGF into the hippocampal formation (asterisk) promotes the reentry of a large number of cholinergic axons, leading to an almost normal fiber density in the dorsal hippocampal formation. (**C**) Essentially all ascending sensory axons (anterogradely labeled from the sciatic nerve) regenerate within 2 weeks into and across to the other end of a predegenerated nerve graft (asterisk), but essentially none reenter the rostral portions of the dorsal funiculus (*DF*). Bar = 250 μm for **C** and **D**. (**D**) Infusion of NGF over the same time into the rostral dorsal funiculus promotes regeneration of approximately half of these sensory axons back into the dorsal funiculus

In general, CNS and PNS axons can grow into and across long pieces of nerve graft but elongation into the CNS tissue is much slower or most often occurs only over very short distances (less than 1 mm)(Aguayo, 1985; Hagg *et al.*, 1990a). We had postulated that regeneration into the CNS tissue fails because of insufficient levels of growth-promoting substances such as NGF. Given the recognition of the chemoattracting effects of NGF, we infused NGF for 1 month into the dorsal hippocampal formation 2 mm from the end of a septohippocampal cell-containing peripheral nerve graft (Fig. 2A) (Hagg *et al.*, 1990b). This NGF treatment promoted the entry and further elongation in the hippocampal tissue of a large number of cholinergic axons (Fig. 3A,B). This effect was not due to the recruitment of more cholinergic axons into or growth across the nerve graft. Similar findings about the effects of NGF on regeneration of cholinergic axons into denervated CNS terrain have been reported by others (Butcher *et al.*, 1987; Ernfors *et al.*, 1989; Tuszynski *et al.*, 1990b; Farris *et al.*, 1995). An important aspect that has not been addressed yet, is whether NGF affects the normal pattern of reinnervation or the establishment of appropriate synapses.

Thus, the non-permissive or restrictive nature of the CNS environment can be modified by neurotrophic factors to enhance regeneration. In the septohippocampal system, cholinergic axons regenerate spontaneously and over longer times (9 months) grow back into the hippocampal formation to a similar extent as after a one month NGF infusion (Hagg *et al.*, 1990b). The NGF infusion may only have resulted in a faster regeneration and not in an larger extent of regeneration. This possibility led us to test the idea that CNS regeneration is limited due to an insufficient availability of neurotrophic factors in a more stringent system, i.e., the spinal cord.

Neurotrophic factors and spinal cord regeneration

Primary sensory afferents

The primary sensory neurons in the dorsal root ganglion (DRG) project one of their processes through the nerve to the periphery and one through the dorsal root to the spinal cord where they terminate and/or form the ascending sensory tract or dorsal funiculus (Figure 2B). The dorsal location of this tract makes it very accessible for experimental manipulations. In the adult DRG, the three Trk receptors are found in partially overlapping populations of sensory neurons (Verge *et al.*, 1989, 1992; Ernfors *et al.*, 1993; Mu *et al.*, 1993; McMahon *et al.*, 1994; Wright and Snider, 1995). Neurotrophins are retrogradely transported by these neurons when injected into the peripheral nerve (Richardson and Riopelle, 1984; DiStefano et al., 1992; Curtis *et al.*, 1995). The central processes of these neurons can also retrogradely transport NGF (Richardson and Riopelle, 1984). Approximately half of the sensory DRG neurons express TrkA and most of these neurons contain substance-P and CGRP (Verge *et al.*, 1989). Intrathecal infusion of NGF can affect these neurons and counteract sciatic nerve injury-induced reduction in NGF receptors (Verge *et al.*, 1992), and neurofilament (Verge *et al.*, 1990) and the loss of CGRP and substance-P (Fitzgerald *et al.*, 1985). These observations suggest that under physiological conditions the TrkA-positive sensory neurons are regulated by NGF that derives from the periphery and the spinal cord.

Most of the larger caliber ascending sensory axons originate from proprioceptive neurons that express TrkC and which constitute ~20% of the DRG neurons of the sciatic nerve (Mu *et al.*, 1993; McMahon *et al.*, 1994; Wright and Snider, 1995). Their responsiveness to and dependence on NT-3 is evidenced by the fact that these neurons are absent in mice that are homozygous for a null mutation of NT-3 (Ernfors *et al.*, 1995). Moreover, infusions of NT-3 around the nerve stump can prevent the increased levels of neuropeptide Y in these neurons and their central processes that is seen after peripheral injury (Ohara *et al.*, 1995). Approximately a quarter of the sensory DRG neuron population of the sciatic nerve is TrkB positive and almost all the TrkB-positive neurons also express either TrkA and/or TrkC (McMahon *et al.*, 1994). The BDNF mRNA that is expressed by cultured adult DRG neurons can be decreased by antisense oligonucleotides, a treatment that leads to cell death and is reversible by exogenous BDNF (Acheson *et al.*, 1995). This finding suggests that BDNF plays an autocrine role for these sensory neurons. A third of the adult DRG sensory neurons of the sciatic nerve does not express any of the three Trk's, and almost all of these cells are small diameter neurons (McMahon *et al.*, 1994). These cells are likely nonpeptidergic (Verge *et al.*, 1989, 1992) and express the enzyme fluoride-resistant acid phosphatase (Hunt *et al.*, 1992).

In adult rats, cats and monkeys, 20–40% of sensory DRG neurons apparently die beyond 2 weeks after nerve transection (Rich *et al.*, 1987; Baranowski *et al.*, 1993; Liss *et al.*, 1994, 1996). The sensory projections of the surviving neurons can regenerate very well in the peripheral nerve or dorsal root environment after a crush injury (Pollock, 1995). However, the adult dorsal root axons cannot grow across the dorsal root-spinal cord interface, some turn around and grow back along the root and others appear to terminate in the interface region (Carlstedt *et al.*, 1989, Oudega and Hagg, 1996). This inability to reenter the spinal cord may be caused by neurite-outgrowth inhibitors present on the oligodendrocytes and astrocytes of the adult spinal cord and/or because regenerating axons grow preferentially in a Schwann cell environment. Ever since the observations by Ramon Y Cajal and his colleague Tello the peripheral nerve graft has been recognized as a relatively good growth-promoting terrain compared to CNS tissue (Tello, 1911; Ramon Y Cajal, 1928; Richardson *et al.*, 1980; David and Aguayo, 1981; Aguayo, 1985). However, such nerve grafts as well as fetal tissue grafts (Tessler *et al.*, 1988) attract relatively few fibers, especially few from long projecting tracts. Moreover, those axons that grow across the grafts do not re-enter the spinal cord more than ~0.5–1 mm, again evidence of the non-permissive or restrictive nature of the CNS tissue (Schwab and Bartholdi, 1996, for details see Chapter 8 in this volume).

Of special significance is the observation that regeneration of central projecting sensory axons into peripheral nerve grafts is greatly enhanced by a previous conditioning lesion of the peripheral projections (Richardson and Issa, 1984; Richardson and Verge, 1986; Oudega *et al.*, 1994). We have found that 1 and 2 months after implantation, only 10% of transected ascending sensory axons will have grown into a freshly dissected peripheral nerve graft, but that with a conditioning lesion 7 days before implantation 45% of the fibers enter the graft. When the conditioning lesion of the peripheral projections was performed on the day of implantation, 22% entered the fresh nerve graft (Oudega *et al.*, 1994). The finding that a shorter time of conditioning results in less ingrowth, suggests that the axons can enter such grafts in a limited time-window. This window may be defined by the time it takes for the growth-response to develop and by the timing of the changes in the spinal cord-nerve graft interface that prevent axonal entry. The mechanism of the conditioning lesion may involve inflammation and satellite cell proliferation in the DRG (Lu and Richardson, 1991, 1993, 1995) possibly mediated in part by release of cytokines from the neurons (Murphy *et al.*, 1995). The conditioning response is a promising area of research that may lead to a better understanding of how to increase the regenerative capacity of CNS neurons.

Neurotrophic factors promote sensory regeneration. A distal portion of a peripheral nerve that is collected one week after nerve transection, promotes regeneration of PNS and CNS axons better than a freshly dissected one (Lewin-Kowalik *et al.*, 1992; Kerns *et al.*, 1993; Oudega *et al.*, 1994). The greater success of such predegenerated grafts may result from an increased proliferation and migration of Schwann cells at the time of implantation, which may enhance the fusion of the graft with the host (Bunge, 1993; Kerns *et al.*, 1993; Pollock, 1995). In addition, the cells in such predegenerated grafts would already have undergone the injury-induced increase in neurotrophic factor synthesis and the earlier release of these factors into the wound area and spinal cord may further facilitate regeneration (Kromer and Cornbrooks, 1987; Olson *et al.*, 1990; Heumann *et al.*, 1987; Messersmith *et al.*, 1991; Meyer *et al.*, 1992). Such neurotrophic factors may include NGF and BDNF. NGF can enhance neurite outgrowth of adult sensory neurons in culture (Lindsay, 1988) and regeneration of sensory processes in the peripheral nerve in vivo (DaSilva and Langone, 1989; Rich *et al.*, 1989). Moreover, an NGF infusion close to a peripheral nerve graft that is implanted into the spinal cord can promote the entry of intraspinal sensory axons (Fernandez *et al.*, 1990). Also, pieces of nitrocellulose paper coated with NGF promote ingrowth of sensory axons from dorsal roots into fetal spinal cord transplants and into the spinal cord itself, albeit over very short distances (Houle and Johanson, 1989;

Houle, 1992; Houle and Ziegler, 1994). The idea that neurotrophic factors can enhance the ability of grafts to promote sensory regeneration has also been supported by the findings with intraspinal grafts of fibroblasts that are genetically engineered to produce NGF (Tuszynski *et al.*, 1994; Nakahara *et al.*, 1996). Since many of the sensory axons in the spinal cord are responsive to NT-3, it is not surprising that grafts of fibroblasts that are engineered to overexpress NT-3 also cause a robust ingrowth of sensory axons into the graft (Senut *et al.*, 1995; Nakahara *et al.*, 1996). The regeneration-promoting ability of purified Schwann cells has recently been combined with the infusion of BDNF and NT-3 (into a Schwann cell graft), resulting in an increased ingrowth of propriospinal sensory and serotonergic supraspinal axons (Xu *et al.*, 1996).

We have found that by using a combination of a conditioning lesion and a predegeneration of the nerve graft, essentially all ascending (sciatic) sensory axons of the dorsal funiculus regenerate within one week into and across the grafts (Oudega *et al.*, 1994). Despite the enhanced regeneration with such methods, as well as with the others described above, the extent of ingrowth back into the spinal cord is very limited, especially into the white matter. This could very well result from the insufficient levels of neurotrophic factors in the spinal cord and/or from their chemoattractant properties that would limit growth away from the neurotrophic factor source. As described for the septohippocampal system, regenerating cholinergic axons grow towards the source of NGF and such infusions had to be given into the hippocampal formation for its enhanced reinnervation (Hagg *et al.*, 1990b). Thus, adult rats with a conditioning lesion and a predegenerated nerve bridge in the lesioned dorsal funiculus at T9–10 received an infusion of NGF for 2 weeks into the dorsal spinal cord 3–5 mm rostral from the end of the graft (Fig. 2B) (Oudega and Hagg, 1996). In response to the NGF infusion, 50% of the ascending sensory axons (anterogradely labeled from the sciatic nerve) reentered the dorsal funiculus (white matter) (Fig. 3C,D). The number of axons was smaller with increasing distance away from the graft and 10–20% reached the infusion site but not farther, perhaps because of the chemoattractant effects of NGF. It is possible that an NGF infusion over a time longer than 2 weeks would further enhance the axonal outgrowth, given the fact that it takes the axons approximately one week to grow across the nerve graft, i.e., the effective treatment time may have been only one week. The proportion of axons that had grown up to 2 and 3 mm back into the spinal cord was greater than we had observed in the septohippocampal model with NGF infusion. This is somewhat surprising since the cholinergic axons had been perceived as among the most readily regenerating ones in the CNS. This may be related to the fact that we used a predegenerated nerve graft and a conditioning lesion in the sensory system and not in the septohippocampal model.

It is still unresolved how NGF stimulates intraspinal regeneration of ascending sensory axons. Most of the neurons that give rise to the ascending sensory tract apparently express TrkC and not TrkA and are therefore expected to respond to NT-3 and not NGF. Our preliminary experiments have shown that BDNF and NT-3 are equally potent as NGF in promoting the regeneration of ascending sensory axons (Oudega and Hagg, unpublished). It is possible that the axons that had grown back into the spinal cord represent divergent sprouts of a small population of ascending axons that can respond to all three neurotrophins. Alternatively, the neurotrophic factors could have modified local cells (e.g., astrocytes, oligodendrocytes, microglia, white blood cells) that in turn release regeneration-promoting substances.

As mentioned before, central sensory processes regenerate very well in crushed dorsal roots, but in adults do not cross the dorsal root-spinal cord interface, even after a conditioning lesion of the peripheral nerve (Carlstedt *et al.*, 1989; Oudega and Hagg, 1996). Animals with a crushed L4 dorsal root (resected roots L3,5,6) and a one week conditioning lesion received an infusion of NGF into the lumbar spinal cord (Oudega and Hagg, 1996). This treatment promoted the reentry of ~20% of the sensory axons in the root back into the spinal cord and such axons grew toward the infusion site. This observation may be relevant for dorsal root injuries that have a poor prognosis for functional recovery in humans, in contrast to ventral root injuries.

Neurotrophic factors to prevent axonal and spinal degeneration. It is clear from all these studies that neurotrophins can promote regeneration of axons back into the adult spinal cord and other regions of the CNS. On the other hand, despite the progress that has been made so far, the improved regeneration in the spinal cord involves only a small number of axons. Morover, the elongation of the long projecting axons is limited to relatively short distances when considering the total length of the spinal cord. Ultimately, the process needed to achieve topographical and functionally appropriate connections may be very complex. As a potential successful approach to circumvent the difficulties associated with recovery strategies, we have started to investigate whether neurotrophic factors can prevent axonal degeneration after spinal cord injury. Among the early events after spinal cord injury is the disintegration of axons and the formation of swellings or terminal clubs on the proximal axonal stumps (Gilson and Stensaas, 1974; Kao, 1977). The rupture of these clubs and release of their hydrolytic enzymes may contribute to the secondary spinal cord damage. We have recently found that neurotrophins can prevent the formation of these terminal clubs in the ascending sensory tracts after spinal cord transection (Sayer *et al.*, 1996). Others have shown that BDNF can prevent axonal die-back or degeneration in the injured optic nerve of neonate rats (Weibel *et al.*, 1995). These observations may be relevant to humans where contusion is a common type of spinal cord injury. Immediately after such an injury most spinal axons are presumably still intact and the axonal degeneration occurs over the next hours to days. Thus, neurotrophic factors might be able to maintain the axonal connections and thereby prevent permanent neurological dysfunction (see Chapter 8 in this volume).

Corticospinal and rubrospinal tracts

Corticospinal motor neurons of layer V in the cerebral cortex project to the brainstem and spinal cord to innervate lower motor neurons (Fig. 2C). They express TrkC receptors and are responsive to NT-3 (Tetzlaff *et al.*, 1994). Motor neurons of the nucleus ruber give rise to the rubrospinal tract and they express TrkB receptors and respond to BDNF (Tetzlaff *et al.*, 1994). Both systems are critically involved in voluntary movement and have therefore received particular attention from various investigators. When the corticospinal or rubrospinal axons of adult rats are transected at the spinal cord level, these neurons do not die but also rarely regenerate into peripheral nerve or other types of grafts (Schwab and Bartholdi, 1996). This failure of regeneration may be related to the observation that such distant lesions do not induce regeneration-associated genes (Tetzlaff *et al.*, 1994). Lesions closer to the cell bodies of these neurons cause an increase in such regeneration-related molecules but also leads to cell death (corticospinal) or atrophy (rubrospinal) of the neurons. The neuronal degeneration of the corticospinal and rubrospinal neurons can be counteracted by NT-3 or BDNF, respectively (Tetzlaff *et al.*, 1994). In neonates, spinal cord transection causes loss of the rubrospinal neurons, an event that can be prevented by BDNF (Diener and Bregman, 1994). It has not been resolved whether an increased neuronal survival by neurotrophic factors after such a proximal lesion would translate into an improved regeneration. Studies aimed at understanding how these neurons can be conditioned after an injury to the spinal cord would be more directly relevant to spinal cord injury. Recently, a new treatment approach, consisting of spinal immobilization, implantation of multiple thin nerve grafts and acidic fibroblast growth factor-containing glue, was used very successfully to promote corticospinal regeneration back into the spinal gray matter (Fig. 2C) (Cheng *et al.*, 1996). It is possible that the acidic fibroblast growth factor induced a growth or conditioning-lesion response in these neurons, whose axons otherwise enter such grafts only in limited numbers.

Neurotrophins also have been used to promote corticospinal axonal outgrowth. Single injections of NT-3, but not BDNF, at the lesion site immediately after transection of the dorsal columns in adult rats promotes sprouting of the corticospinal fibers (Schnell *et al.*, 1994). The distance of outgrowth with such a treatment is very limited. In neonate rats, neutralization of the neurite-outgrowth inhibitor NI35 with specific antibodies results in the regrowth of a small number of corticospinal axons over long distances in the spinal cord (Schnell and Schwab, 1990). By combining this inhibitor neutralizing approach and the NT-3 injection, more of the transected corticospinal axons are able to grow over long distances in adult rats (Schnell *et al.*, 1994). It is conceivable that chronic infusions of NT-3 at the lesion site and more distal (caudal) to it would promote regeneration of even more corticospinal axons over longer distances than seen with the NT-3 injections.

In general, we expect that major advances in the understanding of the requirements for corticospinal and rubrospinal regeneration will be made if the successful approaches that have been discussed for the cholinergic CNS system and the ascending sensory tract are applied.

Future directions

Several areas of research are expected to lead to progress toward recovery strategies for spinal cord injury and in each one neurotrophic factors may prove to be beneficial. Firstly, the finding that a conditioning lesion greatly enhances the regenerative propensity of neurons is important and studies into the mechanisms of these regeneration-related cell responses are likely to provide information that is crucial for successful regeneration. A related and clinically relevant question is how long after an injury axotomized neurons remain responsive to regeneration-promoting treatments.

A second area of research where neurotrophic factors are expected to show promise is the understanding of how lengthy regeneration in the spinal cord can be achieved through guidance or chemoattraction of regrowing axonal growth cones back to the original targets.

A third fruitful research area that is attracting a great deal of attention is fueled by the findings that the CNS contains various specific inhibitors for axonal outgrowth (Pindzola *et al.*, 1993; Schwab *et al.*, 1993; McKerracher *et al.*, 1994; Mukhopadhyay *et al.*, 1994; Keynes and Cook, 1995). Neutralization of these inhibitors or removal of the cells that produce them leads to lengthy regeneration of a limited number of axons (Schnell *et al.*, 1990; Cadelli and Schwab, 1991). Neurotrophic factors, when combined with this approach, further enhance the regeneration by recruiting more axons into the regeneration process (Schnell *et al.*, 1994). This approach of removing chemical obstacles for regenerating axons and adding specific stimulants is a logical approach and could very well become a component of the treatment and recovery of spinal cord injury.

Neurotrophic factors are also likely to play a role in preventing degeneration of axons after spinal cord contusion and thus maintain the integrity of the long ascending and descending connections between the upper CNS nuclei and the spinal cord. If such treatments are successful and applied early enough after an acute injury, they might eliminate the need for the much more complex task of promoting long-distance regeneration and topographically and functionally appropriate reinnervation. Another area of research that warrants investigation is the possibility that neurotrophic factors can prevent degeneration of neurons and other cells in the surrounding area of the spinal cord injury. Neurotrophic factors might also play a role in promoting remyelination, since oligodendrocytes are responsive to some of them in vitro (Louis *et al.*, 1993; Barres *et al.*, 1994).

Delivery of neurotrophic factors

Since neurotrophic factors are medium sized proteins they cannot cross the blood-brain-barrier and their systemic administration does not lead to significant concentrations in the CNS. Therefore, various other methods of delivery are currently being developed. An important consideration for such treatments is related to the location and extent of the injured region since, ideally, the treatment should only affect this region. The spinal cord may be relatively unique in this respect since it is readily accessible through intrathecal administration and because it is relatively isolated from the rest of the CNS. In most cases, the injured spinal region has a reduced blood-brain-barrier for a time after the injury, and neurotrophins may then be able to enter the cord from the cerebrospinal fluid. As with any pharmacological intervention, it would be important to be able to control the dose and the time-course of administration, including the termination of the neurotrophic factor treatment.

A very direct administration method would consist of injections or infusions of neurotrophic factors directly into the affected CNS region (see Chapter 8 in this volume). This method would result in local effects and would permit the regulation of the dose and timing of the treatment. This requires a surgical approach that is not appropriate for the spinal cord since it would result in further damage. However, as mentioned, neurotrophic factors may be able to reach the injured spinal cord from the intrathecal space. Chronic administration could be also achieved by using slow-release polymers that contain the neurotrophic factor(s) (Hoffman *et al.*, 1990). A second approach could consist of systemic injection of neurotrophic factors combined with a temporary interruption of the blood-brain-barrier by substances that are injected into the intrathecal space. This may not be without problems if the treatment is lengthy, given the fact that the barrier plays a protective role for the normal non-injured CNS regions. Systemically injected neurotrophic factors may be able to cross the barrier by using carrier molecules such as antibodies to transferrin receptor that can shuttle neurotrophic factors across the intact barrier (Granholm *et al.*, 1994; Kordower *et al.*, 1994). A third strategy for local and chronic effects may consist of implants of neurotrophic factor-producing cells such as genetically engineered fibroblasts (Rosenberg *et al.*, 1988; Kawaja *et al.*, 1992; Tuszynski *et al.*, 1994). These cells may also secrete other proteins that are not generally found in the spinal cord and by their nature are a risk of forming neoplasia. The latter complication is circumvented by introducing such cells in semi-permeable containers that are ideally removable (Winn *et al.*, 1994; Aebischer *et al.*, 1996). A fourth approach may consist of viral transfection of the spinal cord with genes for new and continuous synthesis of neurotrophic factors by the local spinal cells in the injured region. Although this would ensure a chronic and local effect, the new neurotrophic factor expression may interfere with the normal functions of the host cells and it would be difficulty to regulate the doses and time-course of delivery. A fifth potential delivery approach is based on the fact that almost all neurotrophic factors are produced in the CNS. Thus, agents may be developed that can induce the synthesis of these endogenous sources (Saporito *et al.*, 1994). Such compounds are ideally small enough for systemic administration. An increase in endogenous neurotrophic factor levels might also be achieved by manipulation of the inflammatory response and the associated neurotrophic factor-inducing cytokines. Finally, efforts are underway to design small molecules that mimic the binding sites of neurotrophic factors to their receptors and these could include agonists as well as antagonists (Ibanez, 1995; Van der Zee *et al.*, 1996). As with all pharmacological agents, systemic administration of neurotrophic factors, their derivatives or agents that increase endogenous levels may result in unwanted side effects, as was observed in the recent clinical trials with CNTF for amyotrophic lateral sclerosis (ALS-CNTF-Treatment-Study-Group, 1995). This is to be expected since many neurotrophic factors also play physiological roles in non-nervous tissues.

Acknowledgements

Part of the authors' studies reviewed here was supported by grants from the International Research Institute for Paraplegia, Rick Hansen Man in Motion Foundation, and Scholarships from the Medical Research Council of Canada (TH) and the Miami Project (MO). In memory of Dr. Richard P. Bunge, Scientific Director, The Miami Project to Cure Paralysis, Professor of Neurological Surgery, Cell Biology and Anatomy, and Neurology, University of Miami School of Medicine.

References

Acheson A, Conover JC, Fandl JP, DeChiara TM, Russell M, Thadani A, Squinto SP, Yancopoulos GD, Lindsay RM (1995) A BDNF autocrine loop in adult sensory neurons prevents cell death. Nature 374: 450–453

Aebischer P, Schluep M, Deglon N, Joseph JM, Hirt L, Heyd B, Goddard M, Hammang JP, Zurn AD, Kato AC, Regli F, Baetge EE (1996) Intrathecal delivery of CNTF using encapsulated genetically modified xenogeneic cells in amyotrophic lateral sclerosis patients. Nat Med 2: 696–699

Aguayo AJ (1985) Capacity for renewed axonal growth in the mammalian central nervous system. In: Bignami A, Bloom FE, Bolis CL, Adeloye A (eds) Central Nervous System Plasticity and Repair. Raven Press, New York, pp 31–40

ALS-CNTF-Treatment-Study-Group (1996) A double-blind placebo-controlled clinical trial of subcutaneous recombinant human ciliary neurotrophic factor (rHCNTF) in amyotrophic lateral sclerosis. Neurology 46: 1244–1249

Anderson KD, Alderson RF, Altar CA, DiStefano PS, Corcoran TL, Lindsay RM, Wiegand SJ (1995) Differential distribution of exogenous BDNF, NGF, and NT-3 in the brain corresponds to the relative abundance and distribution of high-affinity and low-affinity neurotrophin receptors. J Comp Neurol 357: 296–317

Armstrong DM, Terry RD, Deteresa RM, Bruce G, Hersh LB, Gage FH (1987) Response of septal cholinergic neurons to axotomy. J Comp Neurol 264: 421–436

Baranowski AP, Priestley JV, McMahon S (1993) Substance P in cutaneous primary sensory neurons–a comparison of models of nerve injury that allow varying degrees of regeneration. Neuroscience 55: 1025–1036

Barbacid M (1995) Neurotrophic factors and their receptors. Curr Opin Cell Biol 7: 148–155
Barres BA, Raff MC, Gaese F, Bartke I, Dechant G, Barde YA, Department of Biology UCLU (1994) A crucial role for neurotrophin-3 in oligodendrocyte development. Nature 367: 371–375
Barrett GL, Bartlett PF (1994) The p75 nerve growth factor receptor mediates survival or death depending on the stage of sensory neuron development. Proc Natl Acad Sci USA 91: 6501–6505
Berry M, Rees L, Hall S, Yiu P, Sievers J (1988) Optic axons regenerate into sciatic nerve isografts only in the presence of Schwann cells. Brain Res Bull 20: 223-231
Bothwell M (1995) Functional interactions of neurotrophins and neurotrophin receptors. Annu Rev Neurosci 18: 223–253
Bracken MB, Holford TR (1993) Effects of timing of methylprednisolone or naloxone administration on recovery of segmental and long-tract neurological function in NASCIS 2. J Neurosurg 79: 500–507
Bunge RP (1993) Expanding roles for the Schwann cell: ensheathment, myelination, trophism and regeneration. Curr Opin Neurobiol 3: 805–809
Butcher LL, Woolf NJ (1987) Cholinergic neuronal regeneration can be modified by growth factors. In: Dowdall MJ, Hawthorne JN (eds) Cellular and molecular basis of cholinergic function. Ellis Horwood, Chichester, pp 395–402
Cadelli D, Schwab ME (1991) Regeneration of lesioned septo-hippocampal acetylcholinesterase-positive axons is improved by antibodies against the myelin-assocaited neurite growth inhibitors NI-35/250. Eur J Neurosci 3: 825–832
Campenot RB (1995) NGF and the local control of nerve terminal growth. J Neurobiol 20: 599–611
Carlstedt T, Cullheim S, Risling M, Ulfhake B (1989) Nerve fibre regeneration across the PNS-CNS interface at the root-spinal cord junction. Brain Res Bull 22: 93–102
Chao MV (1995) Ceramide: a potential second messenger in the nervous system. Mol Cell Neurosci 6: 91–96
Chao MV, Hempstead BL (1995) p75 and Trk: a two-receptor system. Trends Neurosci 18: 321–326
Cheng H, Cao Y, Olson L (1996) Spinal cord repair in adult paraplegic rats: partial restoration of hind limb function. Science 273: 510–513
Choi-Lundberg DL, Bohn MC (1995) Ontogeny and distribution of glial cell line-derived neurotrophic factor (GDNF) mRNA in rat. Dev Brain Res 85: 80–88
Curtis R, Adryan KM, Stark JL, Park JS, Compton DL, Weskamp G, Huber LJ, Chao MV, Jaenisch R, Lee KF, Lindsay RM, DiStefano PS (1995) Differential role of the low affinity neurotrophin receptor (p75) in retrograde axonal transport of the neurotrophins. Neuron 14: 1201–1211
DaSilva CF, Langone F (1989) Addition of nerve growth factor to the interior of a tubular prosthesis increases sensory neuron regeneration in vivo. Braz J Med Biol Res 22: 691–694
David S, Aguayo AJ (1981) Axonal elongation into peripheral nervous system "bridges" after central nervous system injury in adult rats. Science 214: 931–933
Davis S, Aldrich TH, Ip NY, Stahl N, Scherer S, Farruggella T, DiStefano PS, Curtis R, Panayotatos N, Gascan H, Chevalier S, Yancopoulos GD (1993) Released form of CNTF receptor alpha component as a soluble mediator of CNTF responses. Science 259: 1736–1739
DeKosky ST, Styren SD, O'Malley ME, Goss JR, Kochanek P, Marion D, Evans CH, Robbins PD (1996) Interleukin-1 receptor antagonist suppresses neurotrophin response in injured rat brain. Ann Neurol 39: 123–127
Diener PS, Bregman BS (1994) Neurotrophic factors prevent the death of CNS neurons after spinal cord lesions in newborn rats. Neuroreport 5: 1913–1917
DiStefano PS, Friedman B, Radziejewski C, Alexander C, Boland P, Schick CM, Lindsay RM, Wiegand SJ (1992) The neurotrophins BDNF, NT-3, and NGF display distinct

patterns of retrograde axonal transport in peripheral and central neurons. Neuron 8: 983–993

Dobrowsky RT, Werner MH, Castellino AM, Chao MV, Hannun YA (1994) Activation of the sphingomyelin cycle through the low-affinity neurotrophin receptor. Science 265: 1596–1599

Economides AN, Ravetch JV, Yancopoulos GD, Stahl N (1995) Designer cytokines: targeting actions to cells of choice. Science 270: 1351–1353

Emerich DF, Winn SR, Harper J, Hammang JP, Baetge EE, Kordower JH (1994) Implants of polymer-encapsulated human NGF-secreting cells in the nonhuman primate: rescue and sprouting of degenerating cholinergic basal forebrain neurons. J Comp Neurol 349: 148–164

Ernfors P, Ebendal T, Olson L, Mouton P, Stromberg I, Persson H (1989) A cell line producing recombinant nerve growth factor evokes growth responses in intrinsic and grafted central cholinergic neurons. Proc Natl Acad Sci USA 86: 4756–4760

Ernfors P, Kucera J, Lee KF, Loring J, Jaenisch R (1995) Studies on the physiological role of brain-derived neurotrophic factor and neurotrophin-3 in knockout mice. Int J Dev Biol 39: 799–807

Ernfors P, Rosario CM, Merlio JP, Grant G, Aldskogius H, Persson H (1993) Expression of mRNAs for neurotrophin receptors in the dorsal root ganglion and spinal cord during development and following peripheral or central axotomy. Mol Brain Res 17: 217–226

Fagan AM, Gage FH (1990) Cholinergic sprouting in the hippocampus: a proposed role for IL-1. Exp Neurol 110: 105–120

Fallon JH, Loughlin SE (1993) Functional implications of the anatomical localization of neurotrophic factors. In: Fallon JH, Loughlin SE (eds) Neurotrophic factors. Academic Press, San Diego, pp 1–24

Farris TW, Butcher LL, Oh JD, Woolf NJ (1995) Trophic-factor modulation of cortical acetylcholinesterase reappearance following transection of the medial cholinergic pathway in the adult rat. Exp Neurol 131: 180–192

Fernandez E, Pallini R, Mercanti D (1990) Effects of topically administered nerve growth factor on axonal regeneration in peripheral nerve autografts implanted in the spinal cord of rats. Neurosurgery 26: 37–42

Figueiredo BC, Skup M, Bedard AM, Tetzlaff W, Cuello AC (1995) Differential expression of p140trk, p75NGFR and growth-associated phosphoprotein-43 genes in nucleus basalis magnocellularis, thalamus and adjacent cortex following neocortical infarction and nerve growth factor treatment. Neurosci 68: 29–45

Fitzgerald M, Wall PD, Goedert M, Emson C (1985) Nerve growth factor counteracts the neuro-physiological and neurochemical effects of chronic sciatic nerve section. Brain Res 332: 131–141

Frade JM, Rodriguez-Tebar A, Barde YA (1996) Induction of cell death by endogenous nerve growth factor through its p75 receptor. Nature 383: 166–168

Fusco M, Oderfeld NB, Vantini G, Schiavo N, Gradkowska M, Zaremba M, Leon A (1989) Nerve growth factor affects uninjured, adult rat septohippocampal cholinergic neurons. Neuroscience 33: 47–52

Gage FH, Armstrong DM, Williams LR, Varon S (1988) Morphological response of axotomized septal neurons to nerve growth factor. J Comp Neurol 269: 147–155

Gage FH, Wictorin K, Fischer W, Williams LR, Varon S, Björklund A (1986) Retrograde cell changes in medial septum and diagonal band following fimbria-fornix transection: quantitative temporal analysis. Neuroscience 19: 241–255

Gilson BC, Stensaas LJ (1974) Early axonal changes following lesions of the dorsal columns in rats. Cell Tiss Res 149: 1–20

Grafstein B, McQuarrie IG (1978) Role of the nerve cell body in axonal regeneration. In: Cotman CW (ed) Neuronal plasticity. Raven Press , New York, pp 155–195

Granholm AC, Backman C, Bloom F, Ebendal T, Gerhardt GA, Hoffer B, Mackerlova L, Olson L, Soderstrom S, Walus LR, *et al.* (1994) NGF and anti-transferrin receptor

antibody conjugate: short and long-term effects on survival of cholinergic neurons in intraocular septal transplants. J Pharmacol Exp Ther 268: 448–459
Greene LA, Kaplan DR (1995) Early events in neurotrophin signalling via Trk abd p75 receptors. Curr Opin Neurobiol 5: 579–587
Gundersen RW, Barrett JN (1980) Characterization of the turning response of dorsal root neurites toward nerve growth factor. J Cell Biol 87: 546–554
Hagg T (1997) Neurotrophic factors. In: Dulbecco R (ed) Encyclopedia of Human Biology, Vol 6. Academic Press, San Diego, pp 223–233
Hagg T, Fass HB, Vahlsing HL, Manthorpe M, Conner JM, Varon S (1989) Nerve growth factor (NGF) reverses axotomy-induced decreases in choline acetyltransferase, NGF receptor and size of medial septum cholinergic neurons. Brain Res 505: 29–38
Hagg T, Gulati AK, Behzadian MA, Vahlsing HL, Varon S, Manthorpe M (1991) Nerve growth factor promotes CNS axonal regeneration into acellular peripheral nerve grafts. Exp Neurol 112: 79–88
Hagg T, Louis JC, Varon S (1993a) Neurotrophic factors and CNS regeneration. In: Gorio A (ed) Neuroregeneration. Raven Press, New York, pp 265–287
Hagg T, Quon D, Higaki J, Varon S (1992) Ciliary neurotrophic factor prevents neuronal degeneration and promotes low affinity NGF receptor expression in the adult rat CNS. Neuron 8: 145–158
Hagg T, Vahlsing HL, Manthorpe M, Varon S (1990a) Septohippocampal cholinergic axonal regeneration through peripheral nerve bridges: quantification and temporal development. Exp Neurol 109: 153–163
Hagg T, Vahlsing HL, Manthorpe M, Varon S (1990b) Nerve growth factor infusion into the denervated adult rat hippocampal formation promotes its cholinergic reinnervation. J Neurosci 10: 3087–3092
Hagg T, Varon S (1993) Neurotropism of nerve growth factor for adult rat septal cholinergic axons in vivo. Exp Neurol 119: 37–45
Hagg T, Varon S, Louis JC (1993b) Ciliary neurotrophic factor (CNTF) promotes low-affinity nerve growth factor receptor and CD4 expression by rat CNS microglia. J Neuroimmunol 48: 177–187
Hannun YA (1996) Functions of ceramide in coordinating cellular responses to stress. Science 274: 1855–1859
Hefti F (1986) Nerve growth factor promotes survival of septal cholinergic neurons after fimbrial transections. J Neurosci 6: 2155–2162
Hefti F (1993) Neurotrophic factors: what are they and what are they doing ? In: Fallon JH, Loughlin SE (eds) Neurotrophic factors. Academic Press, San Diego, pp 25–49
Henderson Z (1996) Responses of basal forebrain cholinergic neurons to damage in the adult brain. Prog Brain Res 48: 219–254
Heumann R (1994) Neurotrophin signalling. Curr Opin Neurobiol 4: 668–679
Heumann R, Korsching S, Bandtlow C, Thoenen H (1987) Changes of nerve growth factor synthesis in nonneuronal cells in response to sciatic nerve transection. J Cell Biol 104: 1623–1631
Hoffman D, Wahlberg L, Aebischer P (1990) NGF released from a polymer matrix prevents loss of ChAT expression in basal forebrain neurons following a fimbria-fornix lesion. Exp Neurol 110: 39–44
Holtzman DM, Kilbridge J, Li Y, Cunningham E Jr, Lenn NJ, Clary DO, Reichardt LF, Mobley WC (1995) TrkA expression in the CNS: evidence for the existence of several novel NGF-responsive CNS neurons. J Neurosci 15: 1567–1576
Houle JD (1992) Regeneration of dorsal root axons is related to specific non-neuronal cells lining NGF-treated intraspinal nitrocellulose implants. Exp Neurol 118: 133–142
Houle JD, Johnson JE (1989) Nerve growth factor (NGF)-treated nitrocellulose enhances and directs the regeneration of adult rat dorsal root axons through intraspinal neural tissue transplants. Neurosci Lett 103: 17–23
Houle JD, Ziegler MK (1994) Bridging a complete transection lesion of adult rat spinal cord with growth factor-treated nitrocellulose implants. J Neural Transplant Plast 5: 115–124

Hunt S, Mantyh PW, Priestley JV (1992) Patterns of termination of biochemically characterized sensory neurons. In: Scott SA (ed) Sensory Neurons. Diversity, Development and Plasticity. Oxford University Press, New York, pp 60–76

Ibanez CF (1995) Neurotrophic factors: from structure-function studies to designing effective therapeutics. Trends Biotechnol 13: 217–227

Ip NY, McClain J, Barrezueta NX, Aldrich TH, Pan L, Li Y, Wiegand SJ, Friedman B, Davis S, Yancopoulos GD (1993b) The alpha component of the CNTF receptor is required for signaling and defines potential CNTF targets in the adult and during development. Neuron 10: 89–102

Ip NY, Nye SH, Boulton TG, Davis S, Taga T, Li Y, Birren SJ, Yasukawa K, Kishimoto T, Anderson DJ, Stahl N, Yancopoulos GD (1992) CNTF and LIF act on neuronal cells via shared signaling pathways that involve the IL-6 signal transducing receptor component gp130. Cell 69: 1121–1132

Ip NY, Wiegand SJ, Morse J, Rudge JS (1993a) Injury-induced regulation of ciliary neurotrophic factor mRNA in the adult rat brain. Eur J Neurosci 5: 25–33

Ip NY, Yancopoulos GD (1996) The neurotrophins and CNTF : Two families of collaborative neurotrophic factors. Ann Rev Neurosci 19: 491–515

Jing S, Wen D, Yu Y, Holst PL, Luo Y, Fang M, Tamir R, Antonio L, Hu Z, Cupples R, Louis JC, Hu S, Altrock BW, Fox GM (1996) GDNF-induced activation of the ret protein tyrosine kinase is mediated by GDNFR-alpha, a novel receptor for GDNF. Cell 85: 1113–1124

Johanson SO, Crouch MF, Hendry IA (1996) Signal transduction from membrane to nucleus: the special case for neurons. Neurochem Res 21: 779–785

Kahn MA, Ellison JA, Speight GJ, de-Vellis J (1995) CNTF regulation of astrogliosis and the activation of microglia in the developing rat central nervous system. Brain Res 685: 55–67

Kao CC, Chang LW, Bloodworth JMB (1977) The mechanism of spinal cord cavitation following spinal cord transection. J Neurosurg 46: 197–209, 745–756, 757–766

Kawaja MD, Rosenberg MB, Yoshida K, Gage FH (1992) Somatic gene transfer of nerve growth factor promotes the survival of axotomized septal neurons and the regeneration of their axons in adult rats. J Neurosci 12: 2849–2864

Kerns JM, Danielsen N, Holmquist B, Kanje M, Lundborg G (1993) The influence of predegeneration on regeneration through peripheral nerve grafts in the rat. Exp Neurol 122: 28–36

Keynes RJ, Cook GM (1995) Repulsive and inhibitory signals. Curr Opin Neurobiol 5: 75–82

Knüsel B, Beck KD, Winslow JW, Rosenthal A, Burton LE, Widmer HR, Nikolics K, Hefti F (1992) Brain-derived neurotrophic factor administration protects basal forebrain cholinergic but not nigral dopaminergic neurons from degenerative changes after axotomy in the adult rat brain. J Neurosci 12: 4391–4402

Koliatsos VE, Price DL, Gouras GK, Cayouette MH, Burton LE, Winslow JW (1994) Highly selective effects of nerve growth factor, brain-derived neurotrophic factor, and neurotrophin-3 on intact and injured basal forebrain magnocellular neurons. J Comp Neurol 343: 247–262

Kordower JH, Charles V, Bayer R, Bartus RT, Putney S, Walus LR, Friden PM (1994) Intravenous administration of a transferrin receptor antibody-nerve growth factor conjugate prevents the degeneration of cholinergic striatal neurons in a model of Huntington disease. Proc Natl Acad Sci USA 91: 9077–9080

Korsching S (1993) The neurotrophic factor concept: a reexamination. J Neurosci 13: 2739–2748

Korsching S, Auburger G, Heumann R, Scott J, Thoenen H (1985) Levels of nerve growth factor and its mRNA in the central nervous system of the rat correlate with cholinergic innervation. Embo J 4: 1389–1393

Kromer LF, Cornbrooks CJ (1987) Identification of trophic factors and transplanted cellular environments that promote CNS axonal regeneration. Ann N Y Acad Sci 495: 207–224

Lauterborn JC, Bizon JL, Tran TM, Gall CM (1995) NGF mRNA is expressed by GABAergic but not cholinergic neurons in rat basal forebrain. J Comp Neurol 360: 454–462
Lauterborn JC, Tran TM, Isackson PJ, Gall CM (1993) Nerve growth factor mRNA is expressed by GABAergic neurons in rat hippocampus. Neuroreport 5: 273–276
Levi-Montalcini R (1987) The nerve growth factor 35 years later. Science 237: 1154–1162
Lewin-Kowalik J, Sieron AL, Krause M, Barski JJ, Gorka D (1992) Time-dependent regenerative influence of predegenerated nerve grafts on hippocampus. Brain Res Bull 29: 831–835
Lin LF, Doherty DH, Lile JD, Bektesh S, Collins F (1993) GDNF: a glial cell line-derived neurotrophic factor for midbrain dopaminergic neurons. Science 260: 1130–1132
Lindholm D, Castren E, Berzaghi M, Blochl A, Thoenen H (1994) Activity-dependent and hormonal regulation of neurotrophin mRNA levels in the brain–implications for neuronal plasticity. J Neurobiol 25: 1362–1372
Lindholm D, Castren E, Kiefer R, Zafra F, Thoenen H (1992) Transforming growth factor-beta 1 in the rat brain: increase after injury and inhibition of astrocyte proliferation. J Cell Biol 117: 395–400
Lindsay RM (1988) Nerve growth factors (NGF, BDNF) enhance axonal regeneration but are not required for survival of adult sensory neurons. J Neurosci 8: 2394–2405
Liss AG, af-Ekenstam FW, Wiberg M (1994) Cell loss in sensory ganglia after peripheral nerve injury. An anatomical tracer study using lectin-coupled horseradish peroxidase in cats. Scand J Plast Reconstr Surg Hand Surg 28: 177–188
Liss AG, af-Ekenstam FW, Wiberg M (1996) Loss of neurons in the dorsal root ganglia after transection of a peripheral sensory nerve. An anatomical study in monkeys. Scand J Plast Reconstr Surg Hand Surg 30: 1–6
Louis JC, Magal E, Takayama S, Varon S (1993) CNTF protection of oligodendrocytes against natural and tumor necrosis factor-induced death. Science 259: 689–692
Lu X, Richardson PM (1991) Inflammation near the nerve cell body enhances axonal regeneration. J Neurosci 11: 972–978
Lu X, Richardson PM (1993) Responses of macrophages in rat dorsal root ganglia following peripheral nerve injury. J Neurocytol 22: 334–341
Lu X, Richardson PM (1995) Changes in neuronal mRNAs induced by a local inflammatory reaction. J Neurosci Res 41: 8–14
MacLennan AJ, Vinson EN, Marks L, McLaurin DL, Pfeifer M, Lee N (1996) Immunohistochemical localization of ciliary neurotrophic factor receptor alpha expression in the rat nervous system. J Neurosci 16: 621–630
Manthorpe M, Louis JC, Hagg T, Varon S (1993) The ciliary neuronotrophic factor. In: Fallon J, Loughlin S (eds) Neurotrophic factors. Academic Press , San Diego, pp 443–473
Manthorpe M, Muir D, Hagg T, Varon S (1990) Growth promoting and inhibiting factors for neurons. In: Seil F (ed) Advances in neural regeneration research : The third international neural regeneration research symposium. Liss, New York, pp 87–102
McKerracher L, David S, Jackson DL, Kottis V, Dunn RJ, Braun PE (1994) Identification of myelin-associated glycoprotein as a major myelin-derived inhibitor of neurite growth. Neuron 13: 805–811
McMahon SB, Armanini MP, Ling LH, Phillips HS (1994) Expression and coexpression of Trk receptors in subpopulations of adult primary sensory neurons projecting to identified peripheral targets. Neuron 12: 1161–1171
Menesini-Chen MG, Chen JS, Levi-Montalcini R (1978) Sympathetic nerve fibers ingrowth in the central nervous system of neonatal rodent upon intracerebral NGF injections. Arch Ital Biol 116: 53–84
Messersmith DJ, Fabrazzo M, Mocchetti I, Kromer LF (1991) Effects of sciatic nerve transplants after fimbria-fornix lesion: examination of the role of nerve growth factor. Brain Res 557: 293–297

Meyer M, Matsuoka I, Wetmore C, Olson L, Thoenen H (1992) Enhanced synthesis of brain-derived neurotrophic factor in the lesioned peripheral nerve: different mechanisms are responsible for the regulation of BDNF and NGF mRNA. J Cell Biol 119: 45–54
Mobley WC, Rutkowski JL, Tennekoon GI, Gemski J, Buchanan K, Johnston MV (1986) Nerve growth factor increases choline acetyltransferase activity in developing basal forebrain neurons. Brain Res 387: 53–62
Mocchetti I, Wrathall JR (1995) Neurotrophic factors in central nervous system trauma. J Neurotrauma 12: 853–870
Montero CN, Hefti F (1988) Rescue of lesioned septal cholinergic neurons by nerve growth factor: specificity and requirement for chronic treatment. J Neurosci 8: 2986–2999
Morse JK, Wiegand SJ, Anderson K, You Y, Cai N, Carnahan J, Miller J, DiStefano PS, Altar CA, Lindsay RM, Alderson RF (1993) Brain-derived neurotrophic factor (BDNF) prevents the degeneration of medial septal cholinergic neurons following fimbria transection. J Neurosci 13: 4146–4156
Mu X, Silos-Santiago I, Carroll SL, Snider WD (1993) Neurotrophin receptor genes are expressed in distinct patterns in developing dorsal root ganglia. J Neurosci 13: 4029–4041
Mukhopadhyay G, Doherty P, Walsh FS, Crocker PR, Filbin MT (1994) A novel role for myelin-associated glycoprotein as an inhibitor of axonal regeneration. Neuron 13: 757–767
Murphy PG, Grondin J, Altares M, Richardson PM (1995) Induction of interleukin-6 in axotomized sensory neurons. J Neurosci 15: 5130–5138
Nakahara Y, Gage FH, Tuszynski MH (1996) Grafts of fibroblasts genetically modified to secrete NGF, BDNF, NT-3, or basic FGF elicit differential responses in the adult spinal cord. Cell Transplant 5: 191–204
O'Brien TS, Svendsen CN, Isacson O, Sofroniew MV (1990) Loss of true blue labelling from the medial septum following transection of the fimbria-fornix: evidence for the death of cholinergic and non-cholinergic neurons. Brain Res 508: 249–256
Ohara S, Tantuwaya V, DiStefano PS, Schmidt RE (1995) Exogenous NT-3 mitigates the transganglionic neuropeptide Y response to sciatic nerve injury. Brain Res 699: 143–148
Olson L, Ayer LC, Ebendal T, Eriksdotter NM, Ernfors P, Henschen A, Hoffer B, Giacobini M, Mouton P, Palmer M, Persson H, Sara V, Strömberg I, Wetmore C (1990) Grafts, growth factors and grafts that make growth factors. Progr Brain Res 82: 55–66
Oppenheim RW (1991) Cell death during development of the nervous system. Annu Rev Neurosci 14: 453–501
Oppenheim RW (1996) The concept of uptake and retrograde transport of neurotrophic molecuales during development: history and present status. Neurochem Res 21: 769–777
Oudega M, Hagg T (1996) Nerve growth factor promotes regeneration of sensory axons into adult rat spinal cord. Exp Neurol 140: 218–229
Oudega M, Varon S, Hagg T (1994) Regeneration of adult rat sensory axons into intraspinal nerve grafts: promoting effects of conditioning lesion and graft predegeneration. Exp Neurol 129: 194–206
Patterson PH (1992) The emerging neuropoietic cytokine family: first CDF/LIF, CNTF and IL-6; next ONC, MGF, GCSF? Curr Opin Neurobiol 2: 94–97
Peterson GM, Lanford GW, Powell EW (1990) Fate of septohippocampal neurons following fimbria-fornix transection: a time course analysis. Brain Res Bull 25: 129–137
Pindzola RR, Doller C, Silver J (1993) Putative inhibitory extracellular matrix molecules at the dorsal root entry zone of the spinal cord during development and after root and sciatic nerve lesions. Dev Biol 156: 34–48
Pollock M (1995) Nerve regeneration. Curr Opin Neurol 8: 354–358
Rabizadeh S, Oh J, Zhong LT, Yang J, Bitler CM, Butcher LL, Bredesen DE (1993) Induction of apoptosis by the low-affinity NGF receptor. Science 261: 345–348
Ramon Y Cajal S (1928) Degeneration and regeneration of the nervous system. (May RM, translator.) Oxford University Press, London
Rich KM, Alexander TD, Pryor JC, Hollowell JP (1989) Nerve growth factor enhances regeneration through silicone chambers. Exp Neurol 105: 162–170

Rich KM, Luszczynski JR, Osborne PA, Johnson E Jr. (1987) Nerve growth factor protects adult sensory neurons from cell death and atrophy caused by nerve injury. J Neurocytol 16: 261–268

Richardson PM (1994) Ciliary neurotrophic factor: a review. Pharmacol Ther 63: 187–198

Richardson PM, Issa VMK (1984) Peripheral injury enhances central regeneration of primary sensory neurones. Nature 309: 791–793

Richardson PM, Issa VM, Riopelle RJ (1986) Distribution of neuronal receptors for nerve growth factor in the rat. J Neurosci 6: 2312–2321

Richardson PM, McGuinness UM, Aguayo AJ (1980) Axons from CNS neurons regenerate into PNS grafts. Nature 284: 264–265

Richardson PM, Riopelle RJ (1984) Uptake of nerve growth factor along peripheral and spinal axons of primary sensory neurons. J Neurosci 4: 1683–1689

Richardson PM, Verge VM (1986) The induction of a regenerative propensity in sensory neurons following peripheral axonal injury. J Neurocytol 15: 585–594

Rosenberg MB, Friedmann T, Robertson RC, Tuszynski M, Wolff JA, Breakefield XO, Gage FH (1988) Grafting genetically modified cells to the damaged brain: restorative effects of NGF expression. Science 242: 1575–1578

Saporito MS, Brown ER, Hartpence KC, Wilcox HM, Robbins E, Vaught JL, Carswell S (1994) Systemic dexamethasone administration increases septal Trk autophosphorylation in adult rats via an induction of nerve growth factor. Mol Pharmacol 45: 395–401

Sayer F, Oudega M, Hagg T (1996) Neurotrophins prevent degeneration of proximal stumps of transected sensory axons in adult rat spinal cord. Soc Neurosci Abstr 22, p 745

Schnell L, Schneider R, Kolbeck R, Barde YA, Schwab ME (1994) Neurotrophin-3 enhances sprouting of corticospinal tract during development and after adult spinal cord lesion. Nature 367: 170–173

Schnell L, Schwab ME (1990) Axonal regeneration in the rat spinal cord produced by an antibody against myelin-associated neurite growth inhibitors. Nature 343: 269–272

Schwab ME, Bartholdi D (1996) Degeneration and regeneration of axons in the lesioned spinal cord. Physiol Rev 76: 319–370

Schwab ME, Kapfhammer JP, Bandtlow CE (1993) Inhibitors of neurite growth. Ann Rev Neurosci 16: 565–595

Segal RA, Greenberg ME (1996) Intracellular signaling pathways activated by neurotrophic factors. Ann Rev Neurosci 19: 463–489

Sendtner M, Carroll P, Holtmann B, Hughes RA, Thoenen H (1994) Ciliary neurotrophic factor. J Neurobiol 25: 1436–1453

Senut MC, Tuszynski MH, Raymon HK, Suhr ST, Liou NH, Jones KR, Reichardt LF, Gage FH (1995) Regional differences in responsiveness of adult CNS axons to grafts of cells expressing human neurotrophin 3. Exp Neurol 135: 36–55

Shelton DL, Reichardt LF (1986) Studies on the expression of the beta nerve growth factor (NGF) gene in the central nervous system: level and regional distribution of NGF mRNA suggest that NGF functions as a trophic factor for several distinct populations of neurons. Proc Natl Acad Sci USA 83: 2714–2718

Sobreviela T, Clary DO, Reichardt LF, Brandabur MM, Kordower JH, Mufson EJ (1994) TrkA-immunoreactive profiles in the central nervous system: colocalization with neurons containing p75 nerve growth factor receptor, choline acetyltransferase, and serotonin. J Comp Neurol 350: 587–611

Spranger M, Lindholm D, Bandtlow C, Heumann R, Gnahn H, Naher-Noe M, Thoenen H (1990) Regulation of nerve growth factor (NGF) synthesis in the central nervous system: comparison between the effects of interleukin-1 and various growth factors in astrocyte cultures and in vivo. Eur J Neurosci 2: 69–76

Stahl N, Boulton TG, Farruggella T, Ip NY, Davis S, Witthuhn BA, Quelle FW, Silvennoinen O, Barbieri G, Pellegrini S, *et al.* (1994) Association and activation of Jak-Tyk kinases by CNTF-LIF-OSM-IL-6 beta receptor components. Science 263: 92–95

Suter-Crazzolara C, Unsicker K (1994) GDNF is expressed in two forms in many tissues outside the CNS. Neuroreport 5: 2486–2488

Swanson LW, Köhler C, Björklund A (1987) The limbic region. I: The septohippocampal system. In: Björklund A, Hökfelt T, Swanson LW (eds) Handbook of chemical neuroanatomy, vol 5: Integrated systems of the CNS, Part I. Elsevier , Amsterdam, pp 125–277
Tello F (1911) La influencia del neurotropismo en la generación de los centros nerviosos. Trab Lab Invest Biol 9: 123–159
Tessler A, Himes BT, Houle J, Reier PJ (1988) Regeneration of adult dorsal root axons into transplants of embryonic spinal cord. J Comp Neurol 270: 537–548
Tetzlaff W, Kobayashi NR, Giehl KM, Tsui BJ, Cassar SL, Bedard AM (1994) Response of rubrospinal and corticospinal neurons to injury and neurotrophins. Prog Brain Res 103: 271–286
Thoenen H (1995) Neurotrophins and neuronal plasticity. Science 270: 593–598
Treanor JJ, Goodman L, de-Sauvage F, Stone DM, Poulsen KT, Beck CD, Gray C, Armanini MP, Pollock RA, Hefti F, Phillips HS, Goddard A, Moore MW, Buj-Bello A, Davies AM, Asai N, Takahashi M, Vandlen R, Henderson CE, Rosenthal A (1996) Characterization of a multicomponent receptor for GDNF. Nature 382: 80–83
Trupp M, Ryden M, Jornvall H, Funakoshi H, Timmusk T, Arenas E, Ibanez CF (1995) Peripheral expression and biological activities of GDNF, a new neurotrophic factor for avian and mammalian peripheral neurons. J Cell Biol 130: 137–148
Tuszynski MH, Armstrong DM, Gage FH (1990) Basal forebrain cell loss following fimbria/ fornix transection. Brain Res 508: 241–248
Tuszynski MH, Buzsaki G, Gage FH (1990b) Nerve growth factor infusions combined with fetal hippocampal grafts enhance reconstruction of the lesioned septohippocampal projection. Neurosci 36: 33–44
Tuszynski MH, Gage FH (1995) Bridging grafts and transient nerve growth factor infusions promote long-term central nervous system neuronal rescue and partial functional recovery. Proc Natl Acad Sci USA 92: 4621–4625
Tuszynski MH, Peterson DA, Ray J, Baird A, Nakahara Y, Gage FH (1994) Fibroblasts genetically modified to produce nerve growth factor induce robust neuritic ingrowth after grafting to the spinal cord. Exp Neurol 126: 1–14
Vahlsing HL, Hagg T, Spencer M, Conner JM, Manthorpe M, Varon S (1991) Dose-dependent responses to nerve growth factor by adult rat cholinergic medial septum and neostriatum neurons. Brain Res 552: 320–329
Van der Zee CE, Lourenssen S, Stanisz J, Diamond J (1995) NGF deprivation of adult rat brain results in cholinergic hypofunction and selective impairments in spatial learning. Eur J Neurosci 7: 160–168
Van der Zee CEEM, Ross GM, Riopelle RJ, Hagg T (1996) Survival of cholinergic forebrain neurons in developing $p75^{NGFR}$-null mice. Science 274: 1729–1732
Vantini G, Schiavo N, Di MA, Polato P, Triban C, Callegaro L, Toffano G, Leon A (1989) Evidence for a physiological role of nerve growth factor in the central nervous system of neonatal rats. Neuron 3: 267–273
Verge VM, Merlio JP, Grondin J, Ernfors P, Persson H, Riopelle RJ, Hökfelt T, Richardson PM (1992) Colocalization of NGF binding sites, trk mRNA, and low-affinity NGF receptor mRNA in primary sensory neurons: responses to injury and infusion of NGF. J Neurosci Res 33: 82–90
Verge VM, Richardson PM, Benoit R, Riopelle RJ (1989) Histochemical characterization of sensory neurons with high-affinity receptors for nerve growth factor. J Neurocytol 18: 583–591
Verge VM, Tetzlaff W, Bisby MA, Richardson PM (1990) Influence of nerve growth factor on neurofilament gene expression in mature primary sensory neurons. J Neurosci 10: 2018–2025
von-Bartheld CS, Byers MR, Williams R, Bothwell M (1996) Anterograde transport of neurotrophins and axodendritic transfer in the developing visual system. Nature 379: 830–833
Weibel D, Kreutzberg GW, Schwab ME (1995) Brain-derived neurotrophic factor (BDNF) prevents lesion-induced axonal die-back in young rat optic nerve. Brain Res 679: 249–254

Wendt JS, Fagg GE, Cotman CW (1983) Regeneration of rat hippocampal fimbria fibers after fimbria transection and peripheral nerve or fetal hippocampal implantation. Exp Neurol 79: 452–461

Widmer HR, Knusel B, Hefti F (1993) BDNF protection of basal forebrain cholinergic neurons after axotomy: complete protection of p75NGFR-positive cells. Neuroreport 4: 363–366

Wilcox BJ, Applegate MD, Portera-Cailliau C, Koliatsos VE (1995) Nerve growth factor prevents apoptotic cell death in injured central cholinergic neurons. J Comp Neurol 359: 573–585

Williams LR, Inouye G, Cummins V, Pelleymounter MA (1996) Glial cell line-derived neurotrophic factor sustains axotomized basal forebrain cholinergic neurons in vivo: dose-response comparison to nerve growth factor and brain-derived neurotrophic factor. J Pharmacol Exp Ther 277: 1140–1151

Williams LR, Varon S, Peterson GM, Wictorin K, Fischer W, Björklund A, Gage FH (1986) Continuous infusion of nerve growth factor prevents basal forebrain neuronal death after fimbria fornix transection. Proc Natl Acad Sci USA 83: 9231–9235

Winn SR, Hammang JP, Emerich DF, Lee A, Palmiter RD, Baetge EE (1994) Polymer-encapsulated cells genetically modified to secrete human nerve growth factor promote the survival of axotomized septal cholinergic neurons. Proc Natl Acad Sci USA 91: 2324–2328

Winter CG, Saotome Y, Levison SW, Hirsh D (1995) A role for ciliary neurotrophic factor as an inducer of reactive gliosis, the glial response to central nervous system injury. Proc Natl Acad Sci USA 92: 5865–5869

Wright DE, Snider WD (1995) Neurotrophin receptor mRNA expression defines distinct populations of neurons in rat dorsal root ganglia. J Comp Neurol 351: 329–338

Xu XM, Guenard V, Kleitman N, Aebischer P, Bunge MB (1995) A combination of BDNF and NT-3 promotes supraspinal axonal regeneration into Schwann cell grafts in adult rat thoracic spinal cord. Exp Neurol 134: 261–272

Yan Q, Matheson C, Sun J, Radeke MJ, Feinstein SC, Miller JA (1994) Distribution of intracerebral ventricularly administered neurotrophins in rat brain and its correlation with trk receptor expression. Exp Neurol 127: 23–36

Young W, Kume-Kick J, Constantini S (1994) Glucocorticoid therapy of spinal cord injury. Ann N Y Acad Sci 743: 241–265

Correspondence: Dr. T. Hagg, Department of Anatomy and Neurobiology, Faculty of Medicine, Dalhousie University, Halifax, Nova Scotia, 1B3H 4H7 Canada.

Spinal Cord Monitoring
Eds: E. Stålberg, H. S. Sharma, Y. Olsson

Chapter 7

Regeneration in the axotomized cord: influence of cyclosporine A and neonatal immune desensitization in mammals

G. Palladini and **B. Caronti**

Dipartimento di Scienze Neurologiche, Università 'La Sapienza', Roma, Italy

Central axonal regeneration

Principles of axonal regeneration

Regeneration is a normal regulative process by which an organism re-establishes its lost equilibrium and restores, more or less completely, body parts that have been lost or severely injured. *Repetitive regeneration* indicates the renewal of cells, tissue and organs that have a shorter life then the whole organism itself, whereas the term *reparative regeneration* applies to the repair of damage after injury (Filoni, 1981).

Central axonal regeneration in the zoological scale

Due to its unique morphology, the nerve cell is practically the only cytotype that can be used to study the regeneration of cell portions and not just reproduction. Regeneration of peripheral axons is maintained along the zoological scale. On the contrary, the ability to regenerate central nervous system (CNS) axons after injury has been lost progressively in the zoological scale. Central axonal regeneration is, in fact, extremely rapid in simplest animals (Coelenterates), and in any case rapid in the Nemertines and Anellids. Considerable axon, but not cell, regenerating power is observed in the nerve ganglia of Arthropods (Bullock and Horridge, 1965). In lower Vertebrates, such as Chordata, Cyclostomata, Amphibians and Teleosteans, central axonal regeneration spontaneously occurs, both at spinal cord level and, to a lesser extent, at encephalic level (Filoni, 1981). Regeneration of spinal cord axons is observed to a fairly high degree in Reptiles (Lacertidae and Gekkonidae). CNS regenerating activity is present at spinal cord level also in Birds (Margotta *et al.*, 1989b), whereas it seems absent at encephalic level (for review see Lauro *et al.*, 1992). The dichotomy between axonal regeneration in peripheral nerves and regeneration of the central axons is, on the contrary, very marked in Mammals (Polezhaev, 1972; Kiernan, 1979). CNS regeneration in this phylum, in fact, merely becomes abortive, i.e., limited to the onset of growth cones and a few micrometers of axons, whereas the regenerating ability of peripheral axons has been morphologically and functionally demonstrated (Ramon y Cajal, 1907, 1928).

Hypotheses to explain the lack of regeneration in mammalian CNS

Considerable scientific attention has been focused to explain the inability of mammalian CNS axons to regenerate after injury (Berry, 1979). Early studies by Aguayo and collaborators (David and Aguayo, 1981; Benfey and Aguayo, 1982; So and Aguayo, 1985) furnished evidence that CNS axons are indeed able to regenerate once they are removed from their original environment and allowed to grow into peripheral nerve transplants. Conversely, insertion of a segment of the optic tract between the two cut ends of a peripheral nerve interrupts spontaneous axonal growth (Marx, 1980). Based on these data it has been hypothesised that microenvironmental factors can influence axonal regenerative ability. In other words, the failure of CNS axon regeneration has been attributed to the CNS environment rather than to an intrinsic lack of regenerative power of the neurons (Collins and West, 1989; Chen *et al.*, 1995). CNS extracellular matrix components have been first hypothesised and later identified as non-permissive for regeneration (Bahr and Bonhoeffer, 1994). Non-permissive properties have been also attributed to myelinic proteins of oligodendrocyte origin (Rutka *et al.*, 1988; Caroni and Schwab, 1988a,b; Schwab and Caroni, 1988; Schnell and Schwab, 1990), but the same molecules have been successively identified in the CNS of Teleosteans, whose central axons regenerate spontaneously (Sivron and Schwartz, 1995; Stichel and Muller, 1995; Nona, 1995). Other authors (Kiernan, 1979) interpreted the abortive regeneration of mammalian CNS axons as due to the lack of neurotrophic factors acting on neuronal survival and axonal guidance elongation, such as proteins, endogenous opioids, nerve growth factor (NGF), brain derived neurotrophic factor (BDNF), fibroblast growth factor (FGF), and neurotrophin-3 (NT3) (for review see Lauro *et al.*, 1992). A mechanical action of a *glial barrier* has been also proposed to explain the regenerative failure of mammalian CNS axons (Reier and Holle, 1988); however, the inhibitory role of microglia and/or astrocytes in the axonal regenerative process has been recently discussed (Aubert *et al.*, 1995).

The autoimmune hypothesis

Data from the literature and results from our laboratory

One further hypothesis to explain the abortive regeneration in mammalian CNS, to which our laboratory has given considerable experimental support, is that autoimmune reactions occur following traumatic lesion of the CNS in Mammals, that contribute to inhibit regeneration. In 1965 we demonstrated that treatment with antimitotic drugs markedly enhances axonal regeneration in rats submitted to spinal cord transection (Palladini and Alfei, 1965). Those data gave the first indirect evidence that inflammatory and immunological phenomena may contribute to block axonal regeneration in the rat CNS. The autoimmune hypothesis was originally suggested by Berry and Riches (1974) who proposed that antibodies directed against CNS components would directly damage the axonal membrane, thus inhibiting spontaneous regeneration. Soon before the formulation of this hypothesis, Matinian and Andreasian (1973) showed that treatment with Pyromen (a bacterial lipopolysaccharydes extract) and proteases has marked effects on CNS axonal regeneration in the rat. Further reports by several authors consistently demonstrated that traumatic lesion of the CNS in Mammals elicits an autoimmune reaction in serum and in situ lesionis (Masada, 1965; Prochazka *et al.*, 1971; Stefan *et al.*, 1971; Mizrachi and Orhy, 1983; Bowen, 1986; Gannuschkina *et al.*, 1982). The autoimmune hypothesis, and particularly the results of the study by Matinian and Andreasian have been strongly adversed by several authors (Aarli *et al.* 1975; Willenborg *et al.*, 1977; Aarli, 1983) who doubted the specificity of the autoantibodies and, consequently, their inhibitory role in CNS regeneration, and the surgical approach used by the Russian authors to lesion the animals. The autoimmune hypothesis was, however, indirectly supported by the evidence that a variety of treatments potentially acting on the immune system, such as puromicine (Bernstein and Wells, 1980), ciclophosphamide-azathioprine (Feringa *et al.*, 1975), corticosteroids or Pyromen (Littrel, 1955; Stuart, 1955; Feringa *et al.*, 1974, 1976; Kiernan, 1979; Gelderd *et al.*, 1980; Pettegrew, 1980; Kelly, 1988; Bracken *et al.*, 1994), x-ray irradiation (Kalderon *et al.*, 1990), colchicine and cloroquine (Giulian and Robertson, 1989), transglutaminase (Eitan *et al.*, 1994), and desensitization (Feringa *et al.*, 1979) have regeneration-enhancing properties. Moreover, the regenerative enhancement was observed in CNS lesioned animals in condition of natural immunodeficiency, such as in rat foetus (Gerard and Koppany, 1962), in newborn opossum (Martin *et al.*, 1988) and hamster (Kalil and Reh, 1982; Reh and Kalil, 1982), and in hibernating animals (Guth *et al.*, 1981).

Autoantibodies following spinal cord lesion in mammals

Serum autoantibodies belonging to the IgG family, directed against proteic antigens derived from spinal cord homogenate (prepared according Willard and Simon, 1981), have been demonstrated following traumatic spinal cord injury in rabbits and rats (Palladini *et al.*, 1987).

Briefly, approximatively 5 g of rat or rabbit spinal cord, fresh or frozen at −70°C, were homogenized at 4°C in 30 ml buffer pH 8 (5 mM EDTA, 2 mM dithiothreitol, 1 mM phenylmethylsulfonylfluoride, 10 mM Tris) and then diluted to 360 ml with the same buffer. The homogenate was stirred 1 hour, then centrifuged 1 hour at 100,000 g. Proteins were precipited from the supernatant by addiction of NaCl (final concentration 0.3 M) followed by stirring for 25 minutes, then centrifuging 1 hour at 100,000 g. The pellet was resuspended at room temperature in 4 ml of modified buffer (10 mM dithiothreitol, 3% Na-dodecylsulphate, without phenylmethylsulfonylfluoride). The suspension was heated 5 minutes at 90°C and centrifuged for 45 minutes at 100,000 g at 20°C. The pellet was discared and the supernatant was stored at −70°C.

In the rabbit, serum IgG appear 2–3 days after spinal cord lesion and persist at least 21 days (the maximal survival time of animals used in those experiments). In the rat, a faint positivity of antibodies is present before spinal cord lesion, that disappears 1 hour after lesion and reappears 3 days after spinal cord transection. The time-course of the immune reaction observed in our study suggests that the autoimmune reaction directed against CNS antigens is anamnestic in nature. Immunohistochemical methods to detect antibodies in situ lesionis revealed the presence of IgG uniformly deposited on the axon since the third day after lesion. Antigen-antibody binding occurred at least until day 12 post-lesion. A similar pattern of IgG positivity has been observed on intact spinal cord sections incubated with sera from spinal cord lesioned animals. The use of anti-IgG antibodies to detect immunoglobulins in situ lesionis allowed to exclude non specific binding of IgM to lipid antigens (Hofstetter *et al.*, 1985). The binding of IgG to axonal antigens has been further confirmed by the disappearance of positivity after incubation of lesioned spinal cord sections, previously incubated with anti-IgG antibody, with acetic acid pH 2.4, a method commonly used to break the antigen-antibody linkage (Gilden *et al.*, 1978). Using this method, similar result have been obtained on intact spinal cord sections previously incubated with sera from lesioned animals; a successive incubation with these latter sera restored the positivity and the pattern of immune reaction. Technique controls were performed both on lesioned and intact spinal cord sections replacing anti-IgG or serum from lesioned animal with dilution buffer alone or normal serum, respectively. In both cases, we did not observed the immune reaction. Moreover, in order to exclude the possibility of aspecific positivity due to the binding of anti-IgG antibodies to immunoglobulins Fc-receptors on axonal membrane (Aarli *et al.*, 1975), spinal cord sections were incubated with non-immune IgG to saturate Fc-receptors, and, successively, with anti-IgG conjugated with fluorescein isothyocianate (FITC). Results obtained using this control procedure were negative.

Autoantibodies production following spinal cord lesion in anatomocomparative researches

A series of anatomocomparative researches by our group produced definitive evidence for a relationship between abortive regeneration of central axons and appearance of autoimmune response and, vice versa, between possible central axonal regeneration and absence of autoimmune response, although there was not a direct demonstration of a cause-effect relationship. Results of these researches showed a marked morphological regeneration of spinal descending axons (*fasciculus longitudinalis medianus*) in adult birds submitted to transverse spinal cord transection, although the regenerative phenomenon seemed not permanent and it was not followed by functional recovery. Immunofluorescence techniques and immunoelettrophoresis using spinal cord homogenate, showed the absence of IgG both in situ lesionis and in serum. Although it is well known that rapid and intense synthesis of antibody occurs in birds, brain is considered an *immunological privileged* organ, like in mammals, as demonstrated by the possibility of experimental allergic encephalitis (EAE) induction (Palladini and Lauro, 1988). The lack of autoimmune reaction against the axonal antigens contained in spinal cord homogenate, can be assumed to be related with the extreme slowness and graduality of the development of blood brain barrier (3 months) (for review see Palladini and Papalia, 1970) and with the dubious immunological memory of birds (Marchalonis, 1977). The marked morphofunctional similarities between birds and reptiles (both phyla belong to superphylum of Sauropsida), strongly suggest the lack of immune reaction following spinal cord transection also in reptiles. In fact, our previous study revealed a high regenerative ability in *Lacerta*, despite its high systematic position. In *Lacerta*, in which spinal tracts regeneration is functionally complete (Raffaelli and Palladini, 1969), immune reaction was not present in situ lesionis nor in serum; this occurrence is probably related to the extreme slowness of the antibody synthesis and to the rapidity of spinal cord regeneration (Palladini *et al.*, 1987). Complete morphofunctional regeneration of spinal cord is possible in Urodele Amphibians (*Triturus*) (Butler and Ward, 1967); in this species we did not observed immune response in serum, using an anti-triturus Ig serum obtained in our laboratory (Margotta *et al.*, 1989a), although Urodele Amphibians are able of humoral and cellular immune response (Marchalonis *et al.*, 1977). Based on these issues, we designed a series of experiments on spinal cord transected Urodeles (*Triturus*) in order to investigate whether the induction of immune phenomena may influence the spontaneous spinal cord regeneration (Margotta *et al.*, 1989a). After remotion of 2–3 mm long segment of spinal cord, a homo- or heteroimplant of spinal cord covered with a potent mineral adjuvant (venetian talcum) was inserted into a subcutaneous dorsolateral pouch. Both treatment inhibited spinal cord regeneration, and immunocomplexes in situ lesionis were observed. On the contrary, the remaining two treatments included in the experimental design (i.e., insertion of the adjuvant alone or spinal cord graft alone, respectively), not influence the spontaneous spinal cord regeneration; in these control-animals we did not observed immunocomplexes in situ lesionis.

Experimental models confirm the autoimmune hypothesis

Other experimental protocols recently performed in our laboratory allow to further determine the possible cause-effect relationship between autoimmune phenomena and abortive regeneration in mammalian CNS.

Neonatal desensitization with CNS extract

Specific immunosuppression against the CNS extract was achieved in rats according to Hokfield (1987). Newborn Wistar rats were injected subcutaneously 5 times every other day with 0.1ml of rat CNS protein extract (prepared as described by Willard and Simon, 1981). Controls were injected with phosphate buffer saline (PBS) according to the same protocol. All animals underwent complete transverse section of the spinal cord at the level of T7/T8 3 months after treatment with CNS extract or vehicle (Palladini *et al.*, in press). Surgery and visual inspection of the spinal cord section were performed under operative microscope. Accuracy of the lesion was verified by drawing a probe through the spinal cord stumps, as recommended by Guth *et al.* (1980). Bilateral hind limb paralysis occurred immediately after surgery in all animals. Pre-lesion serum antibody levels, measured by enzyme immunoassay (EIA) were not significantly different in controls and desensitized animals. Serum IgG levels were unchanged in desensitized-transected animals after surgery, whereas significant increases were measured in control-transected animals 7 days after surgery. The immunoenzymatic reaction for IgG performed 17 days after spinal cord transection demonstrated that in desensitized-transected animals immuno-

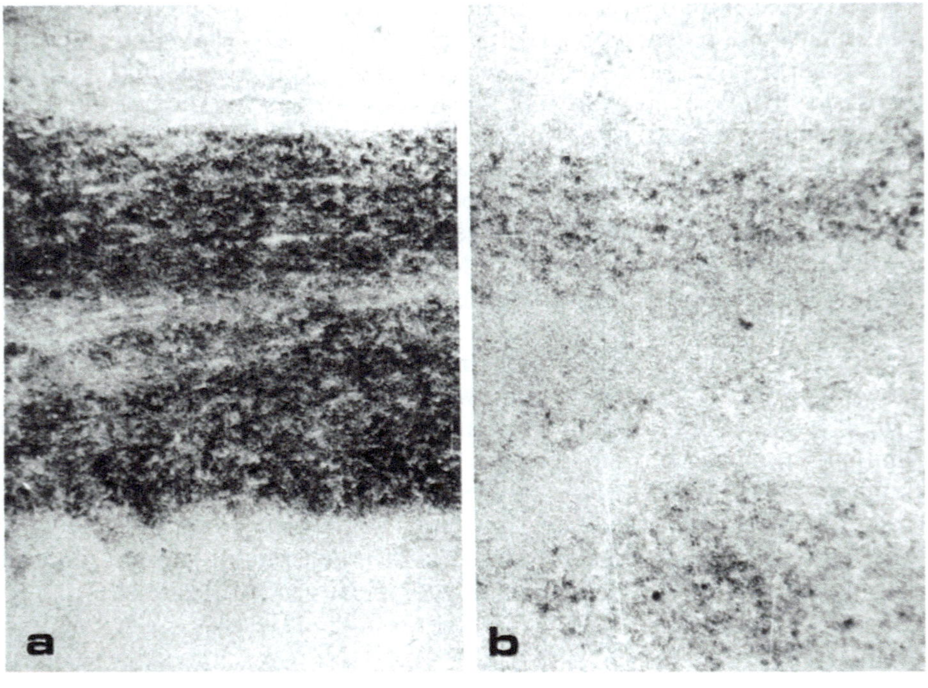

Fig. 1. Immunoenzymatic localization of IgG in rat spinal cord. Shown are longitudinal sections from a control-transected (**a**) and desensitized-transected (**b**) rat 17 days after surgery. Density of IgG is high in the control-transected rat, whereas no signal is observed in desensitized-transected rat. Magnification 60×

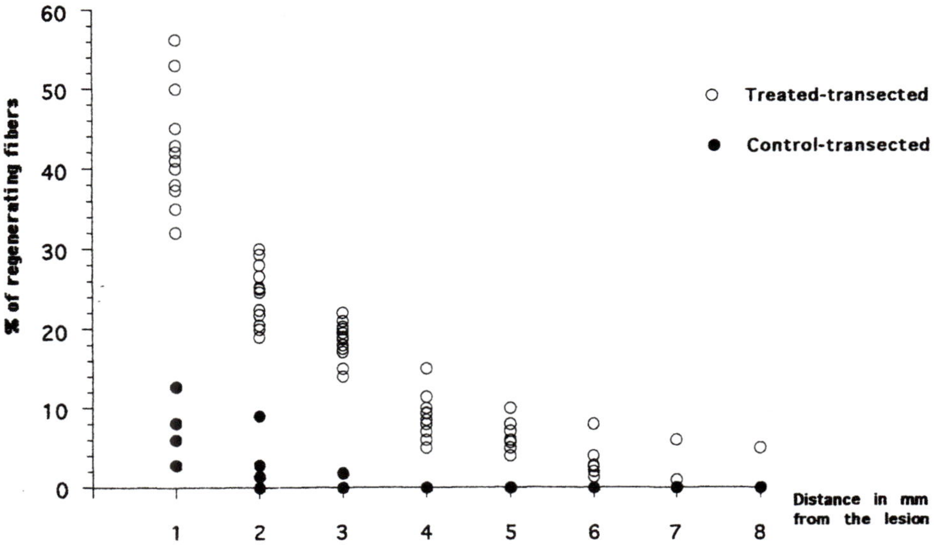

Fig. 2. Quantitative evaluation of regenerating fibres in control-transected and desensitized rats. Measurement was performed by comparing the number of fibres in the caudal and rostral stumps. For each animal, the number of fibres observed in the rostral stump was considered as 100%. The number of fibres in the distal stump was recorded at 1 mm interval from 1 to 8 mm caudally to the lesion, and expressed as percentage of the number of fibres in the rostral stump

complexes precipitation in situ lesionis was markedly lower than in control-transected rats (Fig. 1). Regeneration of descending fibres, organized in thick bundles which transversed the lesion and penetrated into the caudal stump, was revealed in longitudinal section of spinal cord of desensitized-transected rats, stained for Bodian's silver impregnation. Thin, dark, irregular fibres with frequent branching, similar to those previously described by other authors in transected-NT3 treated rats (Schnell *et al.*, 1994) were observed. Quantitative evaluation of regenerated fibres in desensitized-transected rats demonstrated a higher number of regenerated fibres in spinal cord *pyramidal field*, which in most cases reached the length of 5–6 mm in the distal stump (Fig. 2). No clinical improvement was, however, observed. In control transected animals only few fibres were found at 1 mm caudally to the lesion. These experiments provide evidence that down-regulation of the autoimmune response in newborn Mammals enhances axonal regeneration in the adult life.

Immunosuppressive treatment with cyclosporine A (CyA)

CyA is an immunosuppressant drug with peculiar properties. It has been demonstrated that CyA acts specifically on T-lymphocytes, thus blocking T-mediated antibody production (Borel *et al.*, 1976). Systemic administration of CyA induces antigen-specific tolerance during a persistent intracerebral virus infection in rats (Schreiber and Crabtree, 1992; Stits, 1992). For its properties, CyA differs from other drugs previously used as immunosuppressors, that inhibit protein synthesis and produce numerous collateral effects. To further investigate whether the block of the autoimmune response is able to facilitate central axonal regeneration in Mammals, we studied the effects of CyA in female adult rats submitted to transverse section of the spinal cord at mid-thoracic level, as described above. One group of transected animals was treated with CyA (2.5–5 mg/Kg/day, subcutaneously), while another group of transected rats received saline. In the first series of experiments, treatment with CyA or vehicle was begun immediately after surgery and carried out for 14 days (Teichner *et al.*, 1993), while in the second series of experiments treatment was continued for 35 days postoperatively (Palladini *et al.*, 1996). In order to analyze circulating antibodies, blood samples were collected from all animals, at the time of surgery and soon before sacrifice, respectively. The immunodetection of antibodies in situ lesionis was carried out both in vehicle-treated and in CyA-treated rats at days 3, 7, 14, 21, 30 after spinal cord section. Immunohistochemical studies were performed on spinal cord sections obtained from the animals. Morpho-functional recovery of lesioned axons was analyzed by retrograde Fast Blue tracing procedure. Clinical recovery was observed and examined following different approaches and tests.

Transected vehicle-treated rats. Spinal cord of rats treated with vehicle for 14 days showed the typical pattern of alterations previously reported (Palladini *et al.*, 1987) (Fig. 3). Wide areas of necrosis were observed in the two cut ends of the spinal cord; within these areas, numerous foamy cells and reactive astrocytes were identified. Degeneration of descending fibres in the dorsal columns could be followed for as long as 2–3 mm cranially to the lesion. The scar appeared strongly incomplete and presented wide arachnoidal cystis. In the scar region, a few fibres running horizontally were observed. Sometimes axonal growth cones were present, while there was not evidence of nerve fibres bridging the scar and penetrating the opposite portion of the spinal cord. A similar pattern of alterations was observed in spinal cord of rats treated with vehicle for 35 days; however, in these animals the few fibres in the scar region observed 14 days after surgery disappeared. Numerous microglial-like cells and macrophages, with evident positivity for MHC class II antigens, were present at level of the cut end region. Immunostaining for neurofilaments, which appeared still complete 3 days after surgery, decreased progressively in vehicle-treated rats and eventually disappeared 14 days after spinal cord transection. The axonal degeneration evidenced by histological and immunohistochemical examinations was confirmed by results obtained following retrograde tracing procedure performed in a subgroup of rats treated with vehicle or CyA for 35 days. Seven days before sacrifice, the fluorescent retrograde tracer Fast Blue was injected 4 mm below the

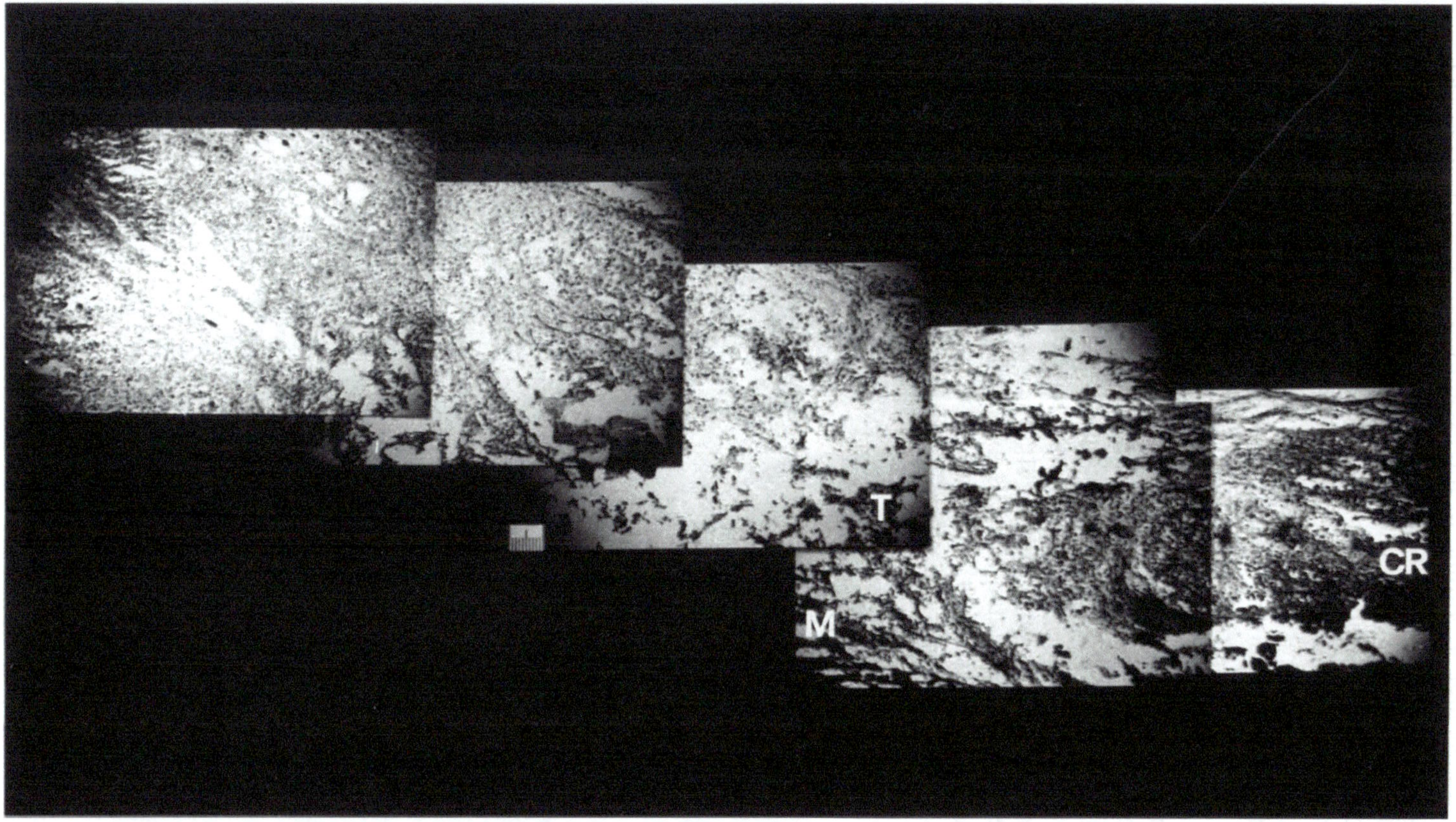

Fig. 3. Longitudinal section of spinal cord from a control-transected rat, 17 days after surgery. Silver impregnation according to Bodian's method, mosaic of 6 frames. Cephalic cortico-spinal fibres stop at the level of the lesion. *CR* rostral stump; *T* transection site; *M* meninges. Calibration bar 200 µm. Magnification 40×

original transection, according to Bentivoglio *et al.* (1980). After sacrifice, spinal cord was removed and histological sections were observed at UV-rays microscope. In spinal vehicle-treated rats there wasn't evidence of fluorescent pyramidal fibres cranially or rostrally to the lesion. In addition, there was not evidence of fluorescent neuronal columns at level of ventral horns; aspecific fluorescence was present at level of meninges and reactive astrocytes. In situ lesionis, 3 days after surgery IgG were weakly identifiable in both vehicle-treated and CyA-treated animals. Seven days after surgery, IgG were identifiable at level of axoplasm and axonal membranes (Fig. 4a), but no decoration was observed at level of the spinal roots nor in the pirenophora of the gray matter. Positivity for IgG in situ lesionis was markedly reduced in CyA-treated animals, with respect of vehicle-treated rats; positivity for IgG progressively decreased in all animals and not signal was identifiable 30 days after surgery. No functional behaviour was observed in vehicle-treated animals: bilateral hind limb paralysis occurred immediately after surgery, and lasted throughout all the experimental period. Spinal walking was never observed, according to Kajihara *et al.* (1973) and De La Torre *et al.* (1975); in fact, spinal walking has been

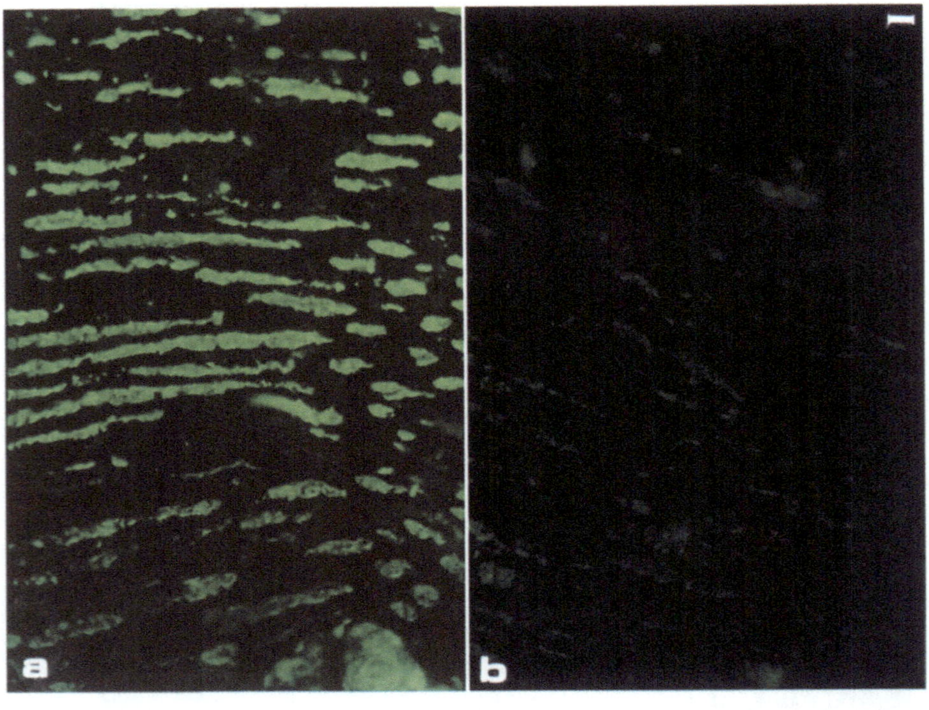

Fig. 4. Longitudinal spinal cord of transected rat, untreated (**a**) and treated with CyA (**b**), 300 µm cranially to the lesion 14 days after surgery; direct immunofluorescence signal for IgG. High signal is present in the untreated animal; CyA treatment markedly reduces the presence of IgG *in situ* lesionis. Calibration bar 10 µm

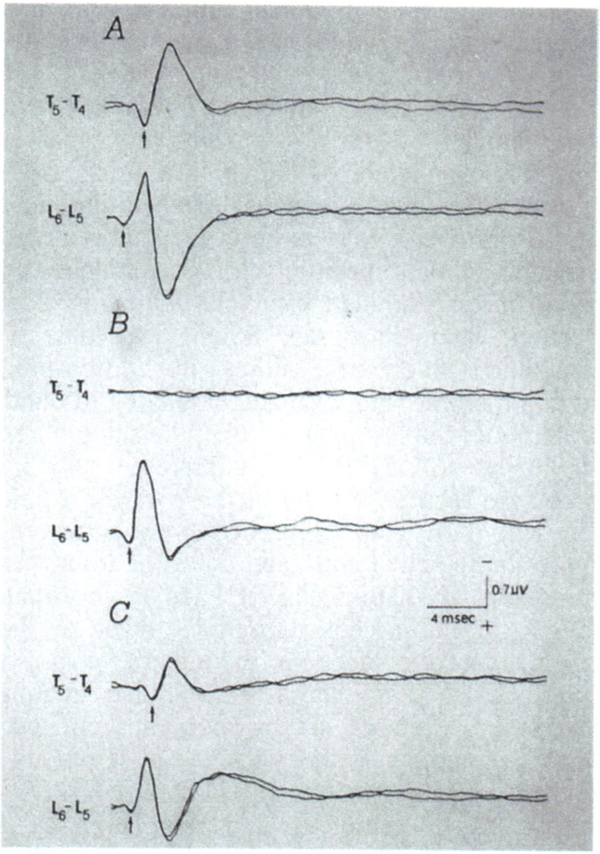

Fig. 5. SSEP in rats with a transected spinal cord showing in (**A**) tracing of normal rat, in (**B**) tracing of transected, vehicle-treated rat and in (**C**) tracing of transected rat treated with CyA. The signal is recorded caudally to the lesion (L6) in all cases; immunosuppression by means of CyA promotes recovery of the triphasic wave rostrally to the lesion (T5), thus demonstrating that regeneration also occurs along the sensitive ascending long pathways

observed in cat and dog, but not in rat. Pain stimulation produces weak reflex activity in some animals (2/13). Somato-sensory evoked potentials (SSEP), recorded by sciatic nerve electrical stimulation according to Pozzessere *et al.* (1987) at day 25 after surgery, never showed reappearance of the triphasic wave at the dorsal level (Fig. 5a).

CyA-treated transected rats. Histological pictures of spinal cord of CyA treated animals (regardless of the dosage of CyA) (Figs. 6A and B) were completely different from vehicle-treated rats. Fifteen days after surgery, a strong glial-connectival scar was observed, bridging the cut ends of the spinal cord. Within the scar, bundles of thin, argirophil and convoluted fibres, that are characteristic of newly developed fibres (Cohen *et al.*, 1988) were present; most of the fibres in dorsal columns of the cranial stumps of the cord appeared normal, while just a few degenerated fibres were observed. Areas of cyst were rare or absent. At day 35 after surgery, an increase of the glial-connectival component in scar was evident. The scar contained strong bundles of fibres that extensively penetrated the caudal stump. This histological picture of the spinal cord of CyA-treated rat is similar to that observed in the spontaneous regeneration in *Lacerta* (Fig. 7). Few MHC class II-positive cells were observed 3 days after surgery, but no later in CyA-treated animals; positivity for neurofilaments was present in almost all rostral fibres examined up to 30 days after transection. In all CyA-treated rats in which retrograde tracing procedure was performed, numerous convoluted fluorescent fibres in corticospinal tract were observed, bridging the scar (Fig. 6C) and extending in bundles for several centimetres cranially to the lesion (Fig. 6E); the same fibres were positive to Bodian's stain (Figs. 6F and B). Many fluorescent fibres were observed in cervical transverse section at level of the pyramidal field. Internuncial neurons traced with Fast Blue (Fig. 6G) were observed rostrally and caudally to the lesion in spinal cord, and traced neurons were observed up the brain stem reticular formation (Fig. 6H) and sensory-motor cortex (Fig. 6I). The presence of fluorescent retrograde tracer in the interneurons around the lesion site gave evidence of the integrity of these cells in CyA-treated rats. Serological examination showed that treatment with CyA reduces the increased levels of serum antibodies elicited by lesion. In particular, following treatment with CyA levels of serum antibodies against proteic components weighing between 45 kD and 65 kD were markedly lower than measured in vehicle-treated rats (Fig. 8). As previously described, IgG positivity was strongly reduced also in situ lesionis.

Behavioural study. Animals were monitored for sensory-motor recovery of the hind limbs twice weekly after surgery. Locomotion and sensibility

Fig. 6. **(A)** Cryostatic longitudinal sections around the lesion area (mosaic of 8 frames) of rat spinal cord, CyA-treated, 35 days after surgery. As shown in figure 10, placing contact reaction and incomplete spontaneous motor activity are observed in this rat. Spinal cord was injected with Fast Blue 4000 μm caudally to the scar 30 days after surgery. The slide was observed and photographed in fluorescence and subsequently refixed and impregnated according to Bodian's method. *Cs* cranial stump; *T...t* section line. **(B)** Enlarged details of **(A)**. Calibration bar 100 μm. Nervous fibres bridge the transected area, silver impregnation according to Bodian's method. **(C)** Fast Blue tracing. Regenerated fibres, that extend at least as far the injection site, 4 mm further down, **(D)** bridge the transected area (the same field photographed in Fig. **B**) and give a fluorescent signal for retrograde tracing procedure from the injection site. **(E)** Longitudinal section of descending long fibres (pyramidal field) photographed after Fast Blue tracing procedure, and **(F)** after silver impregnation according Bodian's method. **(G)** Fast Blue tracing. Transverse section. Pyramidal field at the level of the cervical tract regenerated fluorescent by retrograde tracing procedure from the site of lesion (c.s.). **(H)** Transverse section of red nucleus in the mesencephalic area; fluorescent signal due to Fast Blue tracer is present at level of the lesion site. **(I)** Ventral horn in caudal stump. Fast Blue tracer evidenced the integrity of ventral motor cells and internuncial neurons

H
G
E
A
F
B
C
D

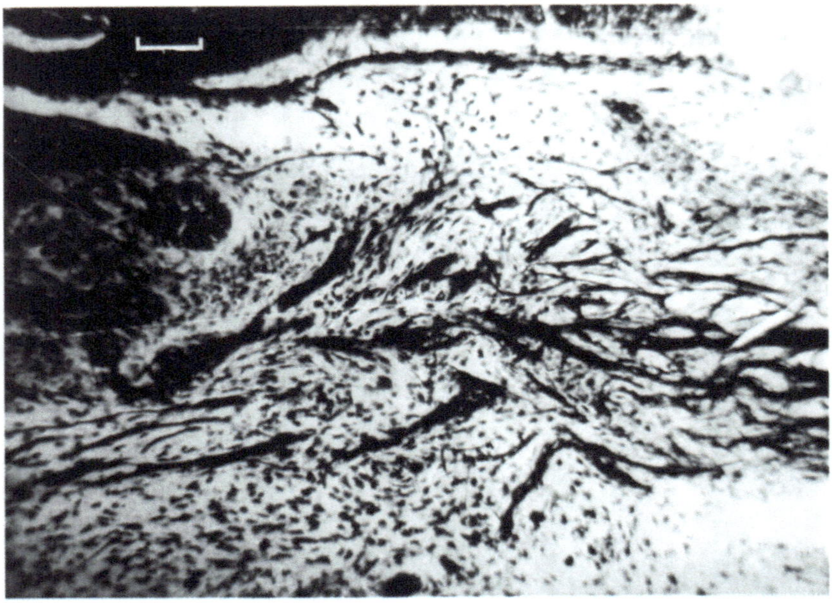

Fig. 7. Longitudinal spinal cord section of transected *Lacerta*, at day 17 after surgery. Glees stain. Calibration bar 100 μm; *CR* cranial stump

were evaluated by an unbiased observer according to Kunkel-Bagden *et al.* (1993). *Locomotion*: *Climb*: the ability of the animals suspended by its tail to climb onto a platform was observed. A normal animal lifts both hind limbs together with its hips and rapidly places his hind limbs onto the platform. This behaviour was observed in 32% of CyA-treated animals, while was never observed in vehicle-treated animals. *Runway*: the ability of the animal to deambulate over a plane or inclined surface was examined. Hind limb movements were observed and videotaped. Slow motion analysis was then performed in order to identify the coordination of movements and the animal's individual limbs movements. The time to cross, the number of fore limb steps and the errors (footslips off the side) were compared between normal, saline-treated and CyA-treated rats. In 27% of CyA-treated rats recovery of spontaneous hind limb motor activity was observed starting approximatively 20 days after surgery. The time required for this recovery is in agreement with literature data concerning the growth rate of adult spinal cord axons following treatment with NGF or NT3 (Fernandez *et al.*, 1993; Schnell *et al.*, 1994). Recovery of function was incomplete in so far as coordinated left hind limb – left fore limb – right hind limb – right fore limb, as described by Iwashita *et al.* (1994), in normal rat, was maintained only for a few steps, then interrupted by spastic hind limb movements before being restored for another short series of steps. During the correct runway, the position of hind limbs was very

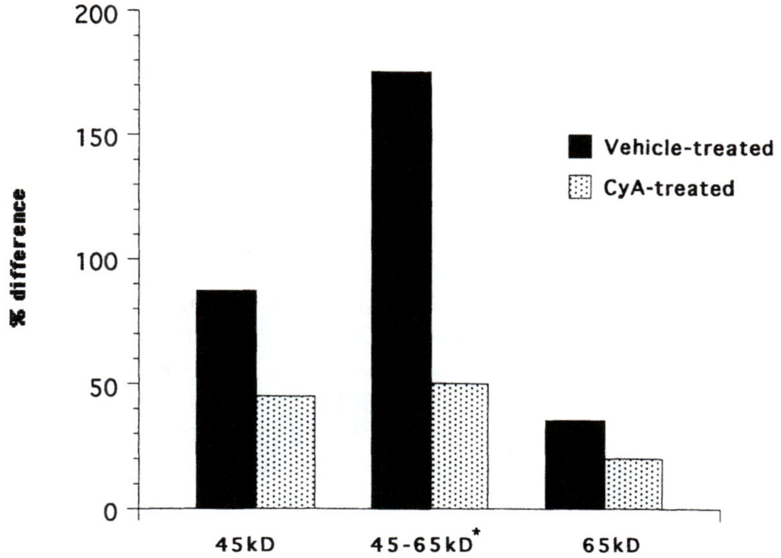

Fig. 8. Percentage differences of antibody level between the post-lesion sera and the pre-lesion ones in CyA-treated rats and vehicle-treated rats. Western blots were analyzed using a computerized image processing system by means of a dedicated software. Differences in optical density of identifiable bands were analyzed for significancy by means of *t* test. Levels of seric antibodies against several proteic components of the CNS extract are markedly increased by sectioning the spinal cord. Treatment with CyA partially blocks the increase in antibody levels elicited by the lesion. $^{*}p < 0.01$, *t* test

similar to that of an intact rat. *Placing*: the ability of an animal to place the hind limbs onto a surface for support in response to proprioceptive (joint bending) and to contact (light touch) stimuli was observed. Contact and proprioceptive placing can be distinguished visually by the joint displacement required to elicit the response (Kunkel-Bagden *et al.*, 1993), i.e., a flexion of the limb followed by placement of the limb onto the surface. After transverse section and hemisection of spinal cord in adult animals, proprioceptive placing response recovery, but contact placing reaction is abolished; in fact, contact placing reaction is considered as strictly indicative of supraspinal motor control (Donatelle, 1977; Bregman and Goldberger, 1982, 1983; Bregman *et al.*, 1993; Kunkel-Bagden *et al.*, 1993). The 77% of CyA-treated animals displayed proprioceptive reflex and positive contact placing reaction, similarly to intact animals (Fig. 9a and 9b). Contact placing reaction was never observed in saline-treated animals. *Sensibility*: in 91% of CyA-treated animals, strong reflex activity to pain stimulation was observed. Triphasic wave rostrally to the lesion, as shown by SSEPs, was recorded in all CyA-treated tested animals, 30 days after surgery, but never in saline-treated rats. SSEPs were used to measure the capacity of different sensory pathways to transmit impulses from L6 (caudally to the lesion) to T5 (rostrally to the lesion) along the spinal cord:

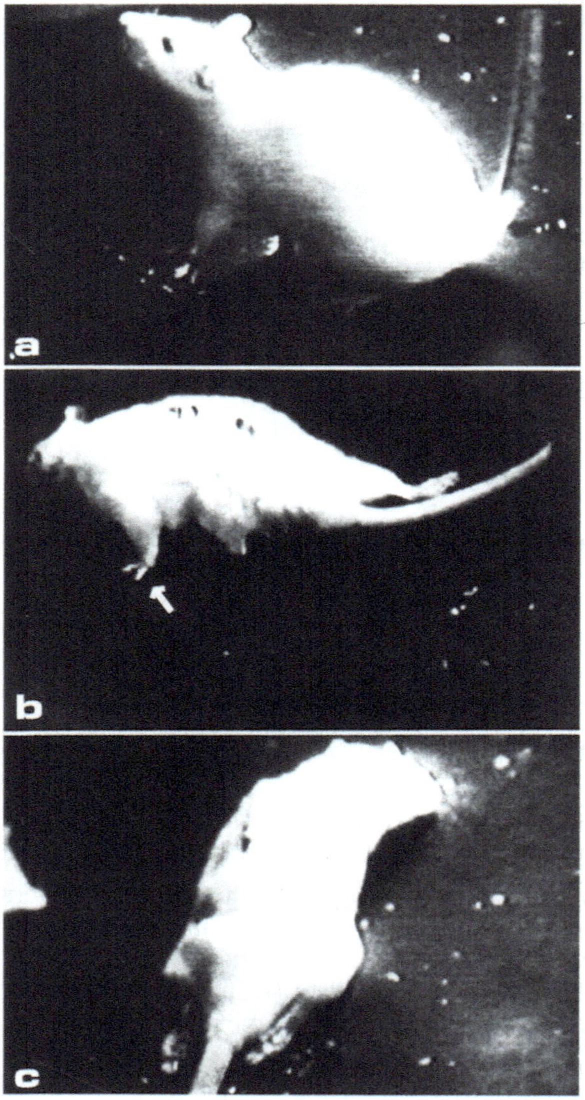

Fig. 9. Frames taken from videotape of (**a**) a normal rat, (**b**) one transected and CyA-treated rat 24 days after surgery, (**c**) one transected and vehicle-treated rat 24 days after surgery. Treatment with CyA promotes reappearance of contact placing reaction in the left hind leg (arrow). Histological examination revealed marked asymmetry in axonal regeneration in favour of the left side of the spinal cord. Similar results were obtained in 80% of the rats treated with CyA, 25% of the treated rats also presented incomplete spontaneous locomotion (see text for details)

then, they furnish further proof of the functional recovery of long tracts of CyA-treated animals (Fig. 5c).

Conclusions

As discussed earlier, the autoimmune hypothesis proposed by Berry and Riches in 1974, although supported by numerous experimental reports, was strongly doubted by several scientists for different reasons, such as the completeness of spinal cord transection, and the methodological approaches to the detection of the immune reaction in situ lesionis. Our studies were, then, performed taking in account these critical reviewing.

Anatomocomparative data demonstrate a relationship between autoimmunity and abortive central axonal regeneration, although there isn't definite demonstration of a direct cause-effect relationship between the two phenomena. Our results obtained with specific immunosuppressive protocol demonstrate the existence of this type of relationship. In fact, we showed serum immunoglobulins that bind to CNS antigens following transection, target molecules of axoplasm and axonal membrane. Furthermore, our data show that the autoimmune reaction takes place a few days after spinal cord transection and persists in situ lesionis until the axon is completely destroyed. The specificity of the autoimmune reaction is confirmed by the absence of autoimmune reaction following lesions in the peripheral nervous system (PNS) and by the absence of cross-reactivity between CNS and PNS. CyA (and, partially, neonatal desensitization) treatment markedly reduces the presence of antibodies in situ lesionis, thus permitting survival and growth, as shown by the persisting positive neurofilaments immunoreaction. These data demonstrate that autoimmune phenomena contribute to the abortive central axonal regeneration in Mammals. In fact, the results obtained in spinal cord lesioned rats following treatment with CyA provide evidence of a functional recovery according to the criteria proposed by Guth *et al.* (1980) and De La Torre (1984). Moreover, recent clinical studies demonstrate that treatment with CyA of human spinal cord lesions significantly reduce lesion's volume (Davis *et al.*, 1995). Although a direct non-immunomediated positive action of CyA on neuritogenesis seems to be excluded by our in vitro experiments (Fabrizi *et al.*, 1995), it is, however, conceivable that some other effects of the drug may exert a positive action on functional recovery on spinal cord. In particular, it has been recently reported that traumatic lesions are able to activate neuronal nitric oxide synthase (NOS), which significantly contributes to the neuronal damage (Wu and Scott, 1993; Wu *et al.*, 1994; Sharma *et al.*, 1995). As a NOS inhibitory action has been attributed to CyA (Fast *et al.*, 1993; Muhl *et al.*, 1993), this property can contribute to the integrity of perilesional interneurons observed in spinal cord of CyA-treated rats. In the rat, in fact, survival of the spinal cord intrinsic neurons is a potent factor of functional recovery in the regeneration of pyramidal ways, as it is well known that in rat all corticospinal innervation to motor neurons in the ventral horn of the spinal cord is through interneurons (Bernstein and Goldberg, 1995).

The results of the studies reviewed herein definitively demonstrate that CNS lesions elicit the production of antibodies directed against central axonal antigens. Although the biochemical nature of these antigens has not been identified yet, our recent study furnished informations regarding CNS molecules target for antibodies that appeared in serum after CNS lesion (Teichner *et al.*, 1993). As demonstrated via immunoblotting on SDS-page, two peaks of serum antibodies binding proteic components of CNS homogenate were measured after surgical complete section of spinal cord: these proteic components weight 45 kD and 65 kD, respectively. The serum immune response markedly increases at day 15 after lesion. Preliminary observations by our laboratory show that increased levels of circulating immunoglobulins against human CNS protein extract were measured in patients affected by traumatic lesion of the spinal cord, in comparison with healthy subjects (Palladini *et al.*, manuscript in preparation); in particular, these antibodies bind CNS molecules weighing 30 kD and 110 kD (Fig. 10).

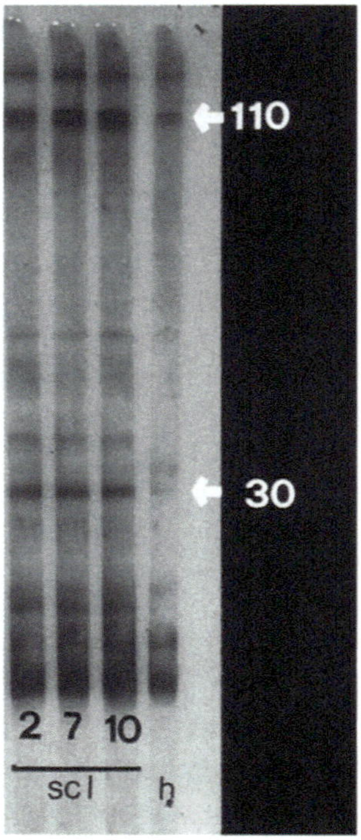

Fig. 10. Analysis of circulating antibodies performed via immunoblotting on SDS-page resolved on human CNS extract. Sera samples were collected from a healthy subject (*h*) and from a patient with spinal cord traumatic lesion (*scl*), at days 2, 7, 10 after lesion. Shown is a significant increase of circulating antibodies against CNS proteic components weighing 30 kD and 110 kD in sera samples from the spinal cord lesioned patient, with respect of the normal serum

The immune response observed in serum and in situ lesionis following CNS lesion is absent following PNS lesions.

The mechanism by which antibodies would block spontaneous axonal regeneration in CNS is unknown at present. Based on reported data showing that antibodies decorate axonal membrane after traumatic lesion of spinal cord, a hypothesis can be drawn, according to which binding of antibodies to specific receptors of the axon membrane prevent enhancing factors to act on neurons. We therefore propose a mechanism alternative to the original theory by Berry and Riches. According to our hypothesis,

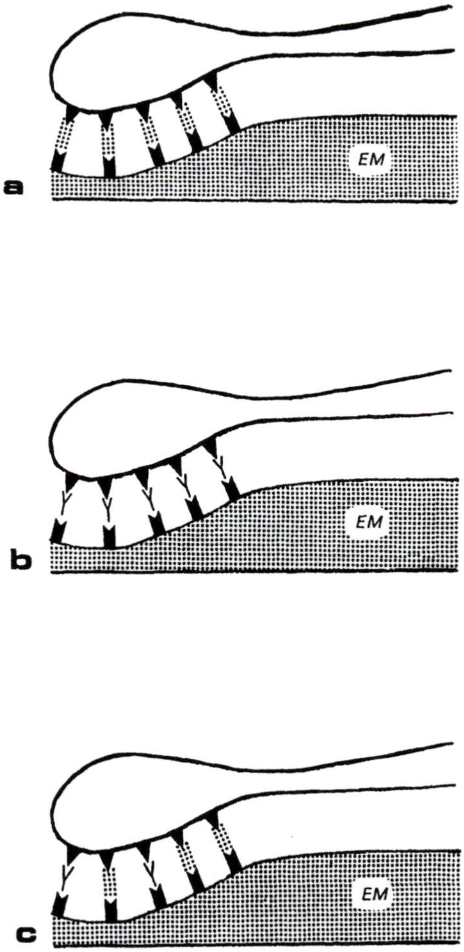

Fig. 11. Diagram summarizing the autoimmune hypothesis of the abortive regeneration in the mammalian CNS. (**a**) It is assumed that the interaction of molecules present in the extracellular matrix EM and axonal membrane receptors acts permissively on axonal growth. (**b**) Following traumatic exposure of CNS antigens, antibodies directed to membrane receptors are present in situ lesionis, that block the interaction. (**c**) Removal of the immune response would allow the interaction, thus promoting axonal regeneration

trophic agents are present in the CNS extracellular matrix, that acts permissively on axonal regeneration by binding with specific "receptors" on the neuronal membrane: antibodies direct against CNS antigens would bind to axonal receptors, thus blocking the interaction between axonal membrane and the trophic agents; removal of the immune reaction would, therefore, allow the interaction between trophic agents and the axons, and enable regeneration (Fig. 11). Further studies are necessary in order to characterize the molecular targets of this immune reaction.

References

Aarli JA (1983) The immune system and nervous system. J Neurol 229: 137–154
Aarli JA, Aparicio SR, Lumdsden CE, Tonder O (1975) Binding of normal human IgG to myelin sheats, glia and neurons. Immunology 28: 171–185
Aubert I, Ridet JL, Gage FH (1995) Regeneration in the adult mammalian CNS: guided by development. Curr Opinion Neurobiol 5: 625–635
Bahr M, Bonhoeffer F (1994) Perspectives on axonal regeneration in the mammalian CNS. TINS 17: 473–479
Benfey M, Aguayo AJ (1982) Extensive elongation of axon from rat brain into peripheral nerve graft. Nature (Lond) 296: 150–152
Bentivoglio M, Kuypers HGJM, Catsman-Berrevoets CE, Loewe H, Dann O (1980) Two new fluorescent retrograde neuronal tracers whichare transported over long distances. Neurosci Lett 18: 25–30
Bernstein JJ, Goldberg WJ (1995) Experimental spinal cord transplantation as a mechanism of spinal cord regeneration. Paraplegia 33: 250–253
Bernstein JJ, Wells MR (1980) Puromycine induction of transient regeneration in mammalian spinal cord. In: McConnel PS, Boer GJ, Romjin HJ, van de Pioll NE, Corner MA (eds) Adaptive capabilities of the nervous system. Prog Brain Res 53: 21–38
Berry M (1979) Regeneration in central nervous system. In: Thomas Smith WT, Cavanagh JB (eds) Recent Advances in Neuropathology. Churchill-Livingstone, Edinburgh, pp 67–111
Berry M, Riches AC (1974) An immunological approach to regeneration in central nervous system. Br Med Bull 30: 135–140
Bjorklund A (1994) A question of making it work. Nature (Lond) 367: 112
Borel JM, Feurer C, Guble HU, Stahelin H (1976) Biological effect of Cyclosporine A: a new anti-lymphocytic agent. Agents Actions 6: 468–475
Bowen FP (1986) Immunological reactions after cortical lesions in rabbits. Arch Neurol 19: 398–402
Bracken MB, Shepard MJ, Collins WF jr, Holford TR, Baskin DS, Eisenberg HM, Flamm E, Leo-Summers L, Maroon JC, Marshall LF, Perot PL Jr, Piemeier J, Sonntag VKH, Wagner FC, Wildelger JL, Winn HR, Young W (1992) Methylprednisolone or naloxone treatment after acute spinal cord injury: 1 years follow-up data. J Neurosurg 76: 23–31
Bregman BS, Goldberger ME (1982) Anatomical plasticity and sparing of function after spinal cord damage in neonatal rats. Science 217: 553–555
Bregman BS, Goldberger ME (1983) Infant lesion effects. II. Sparing and recovery of function after spinal cord damage in newborn and adult rats. Dev Brain Res 9: 119–135
Bregman BS, Kunkel-Badgen E, Reier PJ, Hai HD, McAtee M, Da G (1993) Recovery of function after spinal cord injury: mechanism underlying transplant-mediated recovery of function differ after spinal cord injury in newborn and adult rat. Exp Neurol 123: 3–16
Bullock TH, Horridge GA (1965) Structure and Function in the Nervous System of Invertebrates, Vol 1. Freeman, San Francisco, pp 108–109
Butler EG, Ward MB (1967) Reconstitution of the spinal cord after ablation in adult Triturus. Dev Biol 15: 464–486

Caroni P, Schwab M (1988a) Two membrane protein fractions from rat central myelin with inhibitory properties from neurite growth and fibroblast spreading. J Cell Biol 106: 1281–1288

Caroni P, Schwab M (1988b) Antibody against myelin-associated inhibitor of neurite growth neutralized nonpermissive substrates properties of CNS white matter. Neuron 1: 85–96

Chen D, Jhaveri S, Schneider GE (1995) Intrinsic changes in developing retinal neurons results in regenerative failure of their axons. Proc Natl Acad Sci USA 92: 7287–7291

Cohen AH, Baker MT, Dobrov TA (1988) Functional regeneration demonstrated in the adult spinal cord of lampreys. Soc Neurosci Abstract 14: 482–485

Collins GH, West NR (1989) Prospects for axonal regrowth in spinal cord injury. Brain Res Bull 22: 89–92

David S, Aguayo AJ (1981) Axonal elongation into peripheral nervous system bridges after central nervous system injury in adult rat. Science 214: 931–933

Davies S, Illis LS, Raisman G (1995) Regeneration in the central nervous system and related factors. Summary of the Bermuda Paraplegia Conference, 1994. Paraplegia 33: 10–17

DeLaTorre JC, Johnson CM, Goode DJ, Mullan S (1975) Pharmacologic treatment and evaluation of permanent experimental spinal trauma. Neurology 25: 508–511

DeLaTorre JC (1984) Spinal cord injury models. Prog Neurobiol 4: 289–344

Donatelle JM (1977) Growth of the corticospinal tracts and the development of placing reactionsin the postnatal life. J Comp Neurol 175: 207–232

Eitan S, Solomon A, Lavie V, Yoles E, Hirschberg DL, Belkin M, Schwartz M (1994) Recovery of visual response of injured adult rat optic nerves treated with transglutaminase. Science 264: 1764–1768

Fabrizi C, Caronti B, Palladini G (1995) Autoimmunity, central axonal regeneration and Cyclosporine A. *In vitro* observations on the action mechanism. Rend Fis Acc Lincei 6: 87–93

Fast DJ, Lynch R, Lau RW (1993) Cyclosporine A inhibit nitric oxide production by L529 cells in response to tumor necrosis factor and interferon. J Interferon Res 13: 235–240

Feringa EF, Wendt JS, Randall DJ (1974) Immunosuppressive treatment to enhance spinal cord regeneration in rats. Neurology 263: 287–293

Feringa ER, Johnson RD, Wendt JS (1975) Spinal cord regeneration in rats after immunosuppressive treatment. Arch Neurol 32: 676–683

Feringa ER, Kinning WK, Britten AG, Vahlsing HL (1976) Recovery in rats after spinal cord injury. Neurology 26: 839–843

Feringa EF, Nelson KR, Vahlsing HL, Dauser RC (1979) Spinal cord regeneration in rats made immunologically unresponsive to CNS antigens. J Neurol Neurosurg Psychiatry 42: 642–648

Fernandez E, Pallini R, Lauretti L, Mercanti D, Serra A, Calissano P (1993) Spinal cord transection in adult rats: effect of local infusion of nerve growth factor on the corticospinal tract axons. Neurosurgery 33: 889–893

Filoni S (1981) La rigenerazione nei Vertebrati. Quad Morfol Funz Comp, Piccin, Padova

Gannuschkina IV, Johansson BB, Person LI (1982) Increased blood brain barrier dysfunction around cerebral stab wounds in rat immunized to brain antigens. Acta Neurol Scand 66: 482–487

Gelderd JB, Matthews MA, StOnge MF, Faciane CL (1980) Qualitative and quantitative effects of ACHT, Pyromen, Cytoxan and isobutyl-2-cyanoacrylate treatment following spinal cord transection in rats. Acta Neurobiol Exp 40: 439–500

Gerard RW, Koppany T (1962) Studies on spinal cord regeneration in rat. Am J Physiol 76: 211–212

Gilden D, Delvin M, Wromblewska L (1978) A technique for the elution of cell-surface antibody from human brain tissue. Ann Neurol 3: 403–405

Giulian D, Robertson C (1990) Inhibition of mononuclear phagocytes reduces ischemic injury in the spinal cord. Ann Neurol 27: 33–42

Guth L, Albuquerque EX, Deshpande SS, Barret CP, Donati EJ, Warnick JE (1980) Ineffectiveness of enzyme therapy on regeneration in the transected spinal cord of the rat. J Neurosurg 52: 73–86
Guth L, Barret CP, Donati EJ, Deshapande SS, Albuquerque EX (1981) Histopathological reactions and axonal regeneration in the transected spinal cord of hibernating squirrel. J Comp Neurol 203: 297–308
Hockfield S (1987) A mab to unique cerebellar neuron generated by immunosuppression and rapid immunisation. Science 237: 67–70
Hofstetter W, Heusser CH, Blaser K (1985) Nonspecific binding of mouse IgM antibodies to lipid antigen. J Neuroimmunol 7: 207–214
Iwashita Y, Kawaguchi S, Murata M (1994) Restoration of function by replacement of spinal cord segments in the rat. Nature (Lond) 367: 167–169
Kajihara K, Kawanaga HM, DeLaTorre JC, Mullan S (1973) Dimethylsulfoxyde in the treatment of experimental acute spinal cord injury. Surg Neurol 1: 16–22
Kalderon N, Alfieri AA, Fuka Z (1990) Beneficial effects of x-irradiation on recovery of lesioned mammalian central nervous system. Proc Natl Acad Sci USA 87: 10058–10062
Kalil K, Reh TH (1982) A light and electron microscopic study of the regrowing pyramidal tract fibres. J Comp Neurol 211: 265–275
Kelly TA (1988) The role of the immune system in central nervous system regeneration (theoretical considerations). Med Hypoth 26: 13–15
Kiernan JA (1979) Hypothesis concerned with axonal regeneration in the mammalian nervous system. Biol Rev 54: 155–197
Kunkel-Bagden E, Dai HN, Bregman BS (1993) Method to asses the development and recovery of locomotor function after spinal cord injury in rats. Exp Neurol 119: 153–164
Lauro G, Margotta V, Venturini G, Teichner A, Caronti B, Palladini G (1992) Correlations between immune response and CNS regeneration in vertebrate phylogenesis. Boll Zool 59: 215–220
Littrel JL (1955) Apparent functional restitution in pyromen treated spinal cord. In: Windle WF (ed) Regeneration in CNS. Thomas, Springfield, pp 219–228
Marchalonis JJ (1977) Immunity in Evolution. Arnold, London.
Margotta V, Filoni S, Venturini G, Lauro GM, Scorsini D, Palladini G (1989 a) Autoimmunity and central nervous system regeneration in Urodele Amphibians. J Hirnforsch 30: 99–106
Margotta V, Lauro GM, Di Lorenzo N, Grossi M, Scorsini D, Grifone N, Palladini G (1989 b) Central axonal regeneration and autoimmunity in adult birds. J Hirnforsch 5: 595–602
Martin GF, Xiao Ming X (1988) Evidence of developmental plasticity of the neurospinal tract. Studies using the North American opossum. Develop Brain Res 39: 308–309
Marx L (1980) Regeneration in central nervous system. Science 209: 379–380
Masada S (1965) Immunological studies of brain injuries. Clin Neurol 5: 547–555
Matinian LA, Andreasian AS (1973) [Enzyme therapy in organic lesions of the spinal cord] Akademia Nauk Armenian SSR, 1973, pp 94 [englisch transl.: Los Angeles: Brain Information Service, University of California, 1976, pp 156]
Mizrachi J, Ohry J (1983) Systemic humoral factors partecipating in the course of spinal cord injuries. Paraplegia 21: 287–293
Muhl H, Kunz D, Rob P, Pfeilschiter J (1993) Cyclosporin derivatives inhibit interleukin 1 (induction of nitric oxide synthase in renal mesangial cells. Eur J Pharmacol 249: 95–100
Nona SN (1995) Regenerative failure in the mammalian CNS. TiNS 18: 128
Owens T, Renno T, Taupin V, Krakowski M (1994) Inflammatory cytokines in the brain: does the CNS shape immune response? Immunol Today 15: 566–571
Palladini G, Alfei L (1965) Observations concerning the regeneration of the spinal cord of the adult rat during treatment with antiblastic substances. In: Kiortsis V, Trapusch HAL (eds) Regeneration in Animals and Related Problems. North-Holland, Amsterdam, pp 515–520
Palladini G, Lauro GM (1988) Neuroimmunologia: stato attuale delle conoscenze. Quaderni di Neuropatologia 3: 19–30

Palladini G, Papalia C (1970) La barriera ematoencefalica per il ferrocianuro nella ontogenesi del pollo. Acta Neurol 25: 137–140
Palladini G, Grossi M, Maleci A, Lauro GM, Guidetti B (1987) Immunocomplexes in rat and rabbit spinal cord after injury. Exp Neurol 95: 639-651
Palladini G, Caronti B, Pozzessere G, Teichner A, Buttarelli FR, Morselli E, Valle E, Venturini G, Fortuna A, Pontieri FE (1996) Treatment with Cyclosporine A promotes axonal regeneration in rats submitted to transverse section of the spinal cord – II – Recovery of function. J Hirnforsch 37: 145–153
Palladini G, Caronti B, Buttarelli FR, Teichner A, Morselli E, Maione D, Venturini G, Pontieri FE (1998) Involvement of the immune system in the abortive regeneration in mammalia central nervous system. Int J Neurol (in press)
Pettgrew RK (1980) Evaluation of the use of enzymes for functional restitution after spinal cord severance in rat. Exp Neurol 68: 284–294
Pozzessere G, Valle E, Santoro A, Delfini R, Rino PA, Cantore G, Morocutti C (1987) Prognostic value of early somatosensory evoked potentials during carotid surgery: relationship with electroencephalogram, stump pressure and clinical outcome. Acta Neurochir (Wien) 89: 28–33
Prochazka , Voltherova M, Stephan J (1971) Studies on immunological reaction after brain injuries. II. Int J Surg 55: 322–326
Raffaelli E, Palladini G (1969) Rigenerazione delle cellule e degli assoni del midollo spinale dorsale di Lacerta sicula. Boll Zool 36: 105–110
Ramon y Cajal S (1907) Notes sur la dégénérescence traumatique des fibres du cervelet et du cerveau. Tr Lab Rech Biol (Madrid) 5: 105–115
Ramon y Cajal S (1928) Degeneration and Regeneration of Nervous System. University Press, London
Reh T, Kalil K (1982) Functional role of regrowing pyramidal tract fibres. J Comp Neurol 211: 276–283
Reier DJ, Holle JD (1988) The glial scar: its bearing axonal elongation and transplantation approaches to CNS. In: Watzan SG (ed) Advances in Neurology: Functional Recovery in Neurological Disease. Raven Press, New York, pp 87–138
Rutka JT, Apodaca G, Stern R, Rosemblum M (1988) The extracellular matrix in the central and peripheral nervous system: structure and function. J Neurosurg 69: 155–170
Schnell L, Schwab ME (1990) Axonal regeneration in the rat spinal cord produced by an antibody against myelin-associated neurite growth inhibitors. Nature (Lond) 343: 269–272
Schnell L, Schneider R, Kolbeck R, Sardo YA, Schwab ME (1994) Neurotrophine-3 enhances sprouting of corticospinal tract during development and after adult spinal cord lesion. Nature (Lond) 367: 170–173
Schreiber SL, Crabtree GR (1992) The mechanism of action of cyclosporine A and FK506. Immunol Today 13: 136–139
Schwab ME, Caroni P (1988) Oligodendrocytes and CNS myelin are nonpermissive substrate for neurite growth and fibroblast spreading in vitro. J Neurosci 8: 2381–2393
Sharma HS, Lindholm D, Alm P, Gordh T, Olsson Y, Westman J (1996) Topical application of brain-derived neurotrophic factor nitric oxide synthase following focal trauma to the rat spinal cord. XVI International Winter Meeting, Swiss Society of Neuropathology. Degeneration and Regeneration in the Nervous System. Abstract Booklet, p 16
Sivron T, Schwartz M (1995) Glial cell types, lineages and response to injury in rat and fish: implication for regeneration. Glia 13: 157–165
So F, Aguayo AJ (1985) Lenghty regrowth of cut axons from ganglion cells after peripheral nerve transplantation into the retina of adult rat. Brain Res 328: 349–354
Stefan J, Prochazka M, Volterova M (1971) Studies of immunological reactions after brain injury. I. Antibodies against brain tissue lipids after experimental injury of the brain in rabbits. Int Surgery 55: 316–321
Stichel CC, Muller HW (1995) Regenerative failure in the mammalian CNS. TiNS 18, p 128

Stitz L (1992) Induction of antigen-specific tolerance by ciclosporine A. Eur J Immunol 22: 1995–2001

Stuart EG (1955) Time reaction and possible mechanism of pyromen and desoxycorticosterone acetate in CNS regeneration. In: Windle FW (ed) Regeneration in CNS. Thomas, Springfield, pp 162–170

Teichner A, Morselli E, Buttarelli FR, Caronti B, Pontieri FE, Venturini G, Palladini G (1993) Treatment with cyclosporine A promotes axonal regeneration in rats submitted to transverse section of the spinal cord. J Hirnforsch 34: 343–349

Willard M, Simon C (1981) Antibody decoration of neurofilaments. J Cell Biol 89: 198–205

Willenborg DO, Staten EA, Eidelberg E (1977) Studies on cell-mediated hypersensitivity to neural antigens after experimental spinal cord injury. Exp Neurol 54: 383–392

Wu W, Scott DE (1993) Increased expression of nitric oxyde synthase in hypothalamic neuronal regeneration. Exp Neurol 121: 279–283

Wu W, Liuzzi FJ, Schinco FP, Depto AS, Li Y, Mong JA, Dawson TM, Snyder SH (1994) Neuronal nitric oxide synthase is induced in spinal neurons by traumatic injury. Neuroscience 61: 719–726

Correspondence: Dr. G. Palladini, Dipartimento di Scienze Neurologiche, Università degli Studi, 'La Sapienza', Viale dell'Università, 30-00185 Roma, Italy.

Spinal Cord Monitoring
Eds: E. Stålberg, H. S. Sharma, Y. Olsson

Chapter 8

Neurotrophic factors attenuate neuronal nitric oxide synthase upregulation, microvascular permeability disturbances, edema formation and cell injury in the spinal cord following trauma: an experimental study using topical application of BDNF and IGF-1 in the rat

H. S. Sharma[1], **F. Nyberg**[2], **T. Gordh**[3], **P. Alm**[4], and **J. Westman**[1]

[1] Laboratory of Neuroanatomy, Department of Medical Cell Biology, [2] Department of Pharmaceutical Biosciences, Biomedical Centre, [3] Department of Anaesthesiology, University Hospital, Uppsala University, Uppsala, Sweden, and [4] Department of Pathology, University Hospital, Lund University, Lund, Sweden

Introduction

Injury to the spinal cord is one of the important causes of disability in human population suffering from trauma and no success to treat this medical catastrophe has yet been achieved (Faden and Salzman, 1992; Faden, 1993). Paralysis, paraplegia and life-time rehabilitation are the main problems of spinal cord injury victims throughout the world (Schwab and Bartholdi, 1996). Most of the incomplete spinal cord damage are at least partially reversible, if treated immediately following the primary insult. On the other hand, a delay in treatment schedule beyond 8 h following the primary insult, no significant neuroprotection is achieved irrespective of therapy (for review see Faden, 1993; Schwab and Bartholdi, 1996; Winkler *et al.*, 1998, for details see Chapters 14, 16, 17, 19 in this volume). Thus efforts should be made to treat trauma patients as soon as possible following initial traumatic events.

Destructive changes in the spinal cord tissue occur within hours following trauma which is progressive with time leading to permanent disability (Sharma and Olsson, 1990; Olsson *et al.*, 1990; Sharma *et al.*, 1990a, Olsson *et al.*, 1995; Sharma *et al.*, 1995a; Schwab and Bartholdi, 1996). Pathophysiological studies of spinal cord and its reaction to cell injury in the past was characterised only following days or weeks after the initial insult (Schwab and Bartholdi, 1996). Thus, it is not certain whether these late pathological changes represent a reversible or irreversible damage to the cord.

The early consequences of spinal trauma and its pathological reaction in relation with neuronal degeneration are not well known (Lipton, 1993; Faden and Salzman, 1992; Winkler *et al.*, 1998). A detailed knowledge of early cell reactions following trauma is necessary to explore suitable therapeutic measures to minimise cell and tissue injury (Sharma and Olsson, 1990; Sharma *et al.*, 1998c). A promising approach to reduce paraplegia or paralysis following spinal cord injury to enhance axonal regeneration has emerged recently (Schwab and Bartholdi, 1996). Regeneration and repair mechanisms may constitute one of the most important aspects of potential therapy in future for trauma victims. To enhance axonal regeneration use of neurotrophins, foetal tissue implants, or graft of peripheral nerves following trauma is suggested by several workers which is still preliminary and currently used at the level of experimental models (for review see Schwab and Bartholdi, 1996). Further studies are thus highly needed to increase our understanding of molecular and cellular mechanisms of neurotrophin function in the CNS with regard to axonal regeneration.

The ability of neurotrophins to promote survival and to rescue nerve cells from death following traumatic, ischemic or hypoxic insults has encouraged great interest in neurotrauma research recently (Frisen *et al.*, 1992; Yan *et al.*, 1992; Risling *et al.*, 1993; Burke *et al.*, 1994; Cheng and Mattson, 1994; Frisen *et al.*, 1994; Hefti 1994; Li *et al.*, 1994; Lindholm 1994; Lindvall *et al.*, 1994; Mattson *et al.*, 1995; Nakao *et al.*, 1995; Andersson *et al.*, 1996; Samdani *et al.*, 1997). Neurotrophins are a family of growth factors consisting of nerve growth factor (NGF), brain derived neurotrophic factor (BDNF), neurotrophin-3 (NT-3) and neurotrophin-4/5 (NT-4/5) (Barde, 1994; Davies, 1994; Huber *et al.*, 1995). Besides these neurotrophic factors, other growth factors include glial-derived neurotrophic factor (GDNF), ciliary neurotrophic factor (CNTF), and transforming growth factor-β (TGF-β) (Barde, 1994; Thoenen, 1995). In general, neurotrophic factors promote neurite extension, neuronal survival and differentiation (Barbacid, 1995; Thoenen, 1995; Schwab and Bartholdi, 1996).

Recently, Choi and his collaborators provided experimental evidences which suggest that some neurotrophins may in fact enhance excitotoxic cell injury (Koh *et al.*, 1995). One possibility behind the neurotrophin-induced cell injury is their ability to stimulate nitric oxide synthase (NOS), an enzyme responsible for synthesis of nitric oxide (NO) from its precursor L-ariginine, in cell culture studies (Holtzman *et al.*, 1994; Huber *et al.*, 1995; Peunova and Enikolopov, 1995). Several reports suggest that neurotrophins are able to enhance NO production in cell cultures (see Novikov *et al.*, 1995; Samdani *et al.*, 1997). Thus, neuronal NOS (nNOS) expression is increased by NGF treatment in PC-12 cell lines (Peunova and Enikolopov, 1995). Likewise, application of BDNF, NT-3 and NT-4 on 16 days old rat embryonic spinal cord enhances nNOS expression (Huber *et al.*, 1995). These observations led these authors to conclude that nNOS levels are maintained by neurotrophins during development. However, the role of neurotrophins on NOS upregulation in adult spinal cord is still a new subject which require further investigation. Since increased expression of nNOS seems to be associated with NO production (Chiueh *et al.*, 1994; Dawson and Dawson, 1996; Kimura and Steinbusch, 1996; Yun *et al.*, 1996), it appears that neurotrophins can enhance cell injury by increased NO production (Samdani *et al.*, 1997). Since most of the studies are done in cell culture, the effect of neurotrophins on NO production *in vivo* is not yet clearly known.

NO is involved in tissue damage following ischemia, stroke and infarction (Dawson and Dawson, 1996; Yun *et al.*, 1996). It seems quite likely that NO is involved in pathological mechanisms of spinal cord injury. Experiments from our laboratory show that a focal trauma to the spinal cord induces profound upregulation of nNOS which correlates well with the edema formation and cell injury (Sharma *et al.*, 1995e, 1996a). Application of nNOS antiserum following trauma significantly attenuated the NOS upregulation and prevented edematous expansion and cellular reactions in the cord (Sharma *et al.*, 1996a). These observations were the first to provide a direct evidence that NOS upregulation is injurious to the spinal cord following trauma, a feature which is now well documented in ischemic brain injury (see Dawson and Dawson, 1996; Kimura and Steinbusch, 1996).

Our present knowledge regarding the involvement of NO in regeneration and repair mechanisms following trauma is rudimentary. To address this question, we have initiated a series of investigations in spinal trauma to examine the potential efficacy of BDNF and IGF-1 to attenuate NO production and/or cell damage. It appears that the neuroprotective effects of BDNF and IGF-1 are somehow related with their ability to downregulate the NO activity as evident from a reduced NOS expression in the spinal cord following trauma (Sharma *et al.*, 1997a; 1998b).

This review is based on current research dealing with neurotrophins in the CNS with particular reference to trauma. Furthermore, based on our own investigations in the spinal cord injury, influence of neurotrophins on NOS regulation as one possible mechanism of neuroprotection, is discussed.

Axotomy and regeneration in adult spinal cord

After axotomy adult spinal cord neurones exhibit regenerating response such as, upregulation of immediate early genes, cytoskeletal proteins and growth-associated proteins (43-kD, GAP-43) (Davies, 1994; Kapfhammer and Schwab, 1994). The lesioned axons can sprout but do not normally elongate beyond 2 millimetres (Risling *et al.*, 1993; Schwab and Bartholdi, 1996). Application of neurotrophic factors following axotomy enhances axonal sprouting and elongation several millimetres beyond the lesion scar indicating that growth factors are important in regeneration (Caroni and Schwab, 1988, for details see chapter 6 in this volume). This is further evident from the fact that application of NGF-antibodies in cell culture induced pronounced cell death which is significantly attenuated by application of either single or a combination of neurotrophic factors (Caroni and Schwab, 1988; Schwab and Bartholdi, 1996).

GAP-43 is the major phosphoprotein of neuronal growth cone found in dorsal horn of the adult spinal cord, particularly in substantia gelatinosa (Kapfhammer and Schwab, 1994). A pronounced increase in GAP-43 mRNA or GAP-43 protein occurs following spinal cord lesion or compression injury in perifocal segments which persists even 2–3 weeks after the initial insult. Following axotomy, upregulation of GAP-43 occurs in dorsal root ganglion and motoneurons (Kapfhammer and Schwab, 1994). The expression of GAP-43 is much higher in dorsal root ganglion following peripheral nerve lesion compared to dorsal root lesion (see Schwab and Bartholdi, 1996).

Elevation of GAP-43 levels depends on the local microenvironment following lesion (Kapfhammer and Schwab, 1994). Sciatic nerve lesion induces a rapid and pronounced upregulation of GAP-43 in the peripheral nerve branch and in dorsal nerve roots of all dorsal root ganglion neurones, whereas, only a small and less intense increase in GAP-43 can be seen in dorsal root axon branches forming the dorsal funiculi of the spinal cord (Schreyer and Skene, 1991).

The factors influencing spinal cord regeneration in the CNS microenvironment are not well understood. Glial cells, myelin and many other neurochemicals seem to influence the regeneration phenomena (Barde, 1994; Davies, 1994). Meticulous experiments by Schwab and his group show that myelin has a suppressor role on GAP-43 expression in the corresponding parts of the axons following axotomy (Kapfhammer and Schwab, 1994; Schwab and Bartholdi, 1996). New evidences suggest that glial cell population and biochemical changes associated with axotomy also influence the molecular mechanism of sprouting and regeneration (Samdani *et al.*, 1997).

The detailed molecular mechanisms of axotomy induced regeneration or cell death are not yet clarified. Available evidences show that a number of genes that are normally expressed during development become reactivated in the lesioned adult neurons (for reviews see Schwab and Bartholdi, 1996; Haag and Oudega in this volume). Thus, immediate early genes such as c-*fos* and c-*Jun* which are transcription factors are activated in specific regions within hours following axotomy. Expression of c-*Jun* and c-*fos* in rubrospinal neurons occurs following lesion at cervical level but not at low thoracic level of the spinal cord (Jenkins *et al.*, 1993). Likewise lesion or ligation of peripheral nerves or a dorsal root lesion induces an upregulation of c-*Jun* in dorsal root ganglion (Jenkins *et al.*, 1993). The functional significance of these findings, however, are still obscure.

Myelin inhibiting factor and regeneration of lesioned spinal cord

Schwab and his group were the first to show that myelin can inhibit regeneration in the spinal cord following lesion (Caroni and Schwab, 1988; see Schwab and Bartholdi, 1996). Bilateral transection of the dorsal half of the spinal cord in myelin-free spinal cord tissue produced by repeated X-irradiation of newborn rats (Gilmore, 1963; Savio and Schwab, 1990) exhibits sprouting and elongation of axons beyond several mm of the lesion site compared to less than 1 mm growth in control rats (Caroni and Schwab, 1988; for details see Schwab and Bartholdi, 1996). Since, X-irradiation will result in immunologic removal of myelin by killing of the oligodendrocyte precursors (Gilmore, 1963), the authors suggested that myelin has some inhibitory influence on regeneration (Schwab and Bartholdi, 1996; Hagg and Oudega in this volume). Immunological removal of myelin may have some consequences on astrocytes, microglia and neuronal function which in turn will influence regeneration as well. However, this is a new subject which require further investigation.

The idea that myelin has some influence on regeneration is further explored by Schwab and his group by utilising a monoclonal antibody raised against oligodendrocyte-associated neurite growth inhibitory proteins (IN-1). This antibody neutralises myelin- and oligodendrocyte-associated neurite growth inhibitory protein NI-250, NI-35 as well as whole CNS myelin (Caroni and Schwab, 1988). Application of IN-1 antibody following spinal cord lesion in young adult rats resulted in sprouting of axons several mm beyond the lesion compared to the control group in which the sprouting was limited to 1 mm (Schnell and Schwab, 1990; 1993). This observation clearly suggest that myelin has significant inhibitory influence on regeneration in the lesioned spinal cord.

Regeneration is also enhanced by application of neurotrophins or implantation of foetal tissue grafts. A combination of neurotrophins or tissue graft with antibody application gives superior results (Schnell *et al.*, 1994). Local NT-3 application but not with BDNF treatment results in a remarkable increase in regenerative sprouting in the lesioned spinal cord (Schnell and Schwab, 1993). A similar increase in sprouting was also observed in lesioned spinal cord after implanting foetal spinal cord tissue graft (embryonic day 14) (for details see Schwab and Bartholdi, 1996). When the treatment is combined with IN-1 antibody the elongation of sprouting is even more marked compared to the neurotrophins or embryonic tissue graft alone indicating that myelin inhibiting factors play a dominant role in spinal cord regeneration (see Schwab and Bartholdi, 1996; Haag and Oudega in this volume).

Neurotrophic factors and their receptors in the spinal cord

Almost all the known neurotrophic factors and their receptors are present in the developing or adult spinal cord (see Tables 1 and 2). A focal lesion to

Table 1. Neurotrophins/receptors and mRNA present in developing or adult spinal cord

Neurotrophins/growth factors	mRNA	Receptors
Normal spinal cord		
NGF	NT-3	trk-B
BDNF	BDNF	trk-C
NT-3	FGF	p 75
NT-4/5		
FGF		
CNTF		
GDNF		
HBGH		

Compiled from various sources.
NGF Nerve growth factor; *FGF* fibroblast growth factor; *BDNF* brain derived neurotrophic factor; *NT* neurotrophin; *trk* tyrosine kinase; *CNTF* ciliary neurotrophic factor; *GDNF* glial derived neurotrophic factor; *HBGF* heparin binding growth factor.

Table 2. Neurotrophic factors in the spinal cord and their possible functions

mRNA for neurotrophins	
NGF, BDNF, NT-3, NT-4/5	
Neurotrophic receptors	
trk-B, trk-C	
Motoneurons express	
trk-B, trk-C receptors	
mRNA for NT-3, BDNF	
low affinity NGF receptors (p 75)	
neurons, motoneurons, fibre tracts	
FGF mRNA	
FGF	proliferation, cell survival, neurite outgrowth, transmitter synthesis (*in vitro*)
CNTF	
GDNF	
HBGF	
Physiological role in spinal cord?	
Influence of neurotransmitters/neuromodulators/neuropeptides?	
Neurotrophins may act as	
soluble messengers between neurons, glial cells,	
vascular endothelium during development	
adult CNS?	

Compiled from various sources.
NGF Nerve growth factor; *FGF* fibroblast growth factor; *BDNF* brain derived neurotrophic factor, *NT* neurotrophin; *trk* tyrosine kinase, *CNTF* ciliary neurotrophic factor, *GDNF* glial derived neurotrophic factor; *HBGF* heparin binding growth factor.

the cord influences these neurotrophic factors and their receptors (Table 3). The functional significance of these findings are not known in all its details (see Schwab and Bartholdi, 1996; Hagg and Oudega in this volume). Previous studies are based on *in vitro* experiments and thus further studies using *in vivo* approaches are needed to establish a role of neurotrophic factors and their receptors in spinal cord injury and regeneration.

Most of the experimental data on neurotrophic factors and their receptors are based on investigations in normal spinal cord during development (Tables 1 and 2). Alterations in neurotrophic factors and their receptors following spinal cord lesion and the influence of neurotrophic factors on the injured cord is described in few sporadic reports (Tables 3 and 4). A brief survey of these findings is given below.

Normal distribution

During development mRNA for NGF, BDNF, NT-3 and NT-4/5 are expressed in different segments of the normal spinal cord (Table 1) (Ernfors and Persson, 1991; Maissonpierre *et al.*, 1990; Schecterson and Bothwell, 1992; Timmusk *et al.*, 1993). BDNF and NGF content of the spinal cord is very low (Ernfors and Persson, 1991). Low level of FGF mRNA and protein is present in developing motoneurons. NT-3 content of the spinal cord is high during development which decreases after birth and maturity (Ernfors and Persson, 1991; Maissonpierre *et al.*, 1990). In conformity with the spinal cord, GDNF, HBGF, and CNTF content in the brain is also very low (see Schwab and Bartholdi, 1996) (Tables 1–2).

The receptors for the neurotrophins are present in the brain and spinal cord (Table 1). High affinity neurotrophin receptors trkB and trkC are expressed during development and maturity on motoneurons of the spinal cord (Weise *et al.*, 1993). The motoneurons also express NGF receptors (Frautschty *et al.*, 1991). These neurotrophin receptors influence neuronal survival by modulation of neurotransmitters, neuropeptidess synthesis and/or release in the spinal cord (Lindsay and Harmar, 1989; Nawa *et al.*, 1993; Barde, 1994; Burke *et al.*, 1994; Folli *et al.*, 1994; Hefti, 1994; Holtzman *et al.*, 1994; Lindsay, 1994; Mulderry, 1994; Nawa *et al.*, 1994; Huber *et al.*, 1995; Mattson *et al.*, 1995; Schwab and Bartholdi, 1996).

Lesion of the spinal cord

Hemisection or complete transection of the spinal cord significantly influences the distribution of neurotrophic factors and their receptors (Table 3). Contusion injury in rat resulted in increased expression of bFGF mRNA in motoneurons and astrocytes 6 h after trauma which continued until 2 days after the initial injury (Follesa *et al.*, 1994; Koshinaga *et al.*, 1993). An increase in FGF, CNTF and TGFβ expression also occurs in brain injury indicating that neurotrophins are sensitive to CNS insults (Frautschy *et al.*, 1991; Lindholm *et al.*, 1992; Logan *et al.*, 1992; Ip *et al.*, 1993).

Incision of the cat and rat spinal cord induces an upregulation of trkB mRNA and protein in the lesion scar several days after the primary lesion (Frisen *et al.*, 1992). An increased immunoreactivity of trkC was observed 3 weeks after spinal transection in rats (Frisen *et al.*, 1994). Upregulation of

Table 3. Neurotrophins in the lesioned spinal cord

Lesion	Secies	Survival period	Neurotrophins/growth factors
Knife wounds Spinal cord	young adult cat/rat	several days to weeks	upregulation of trk-B mRNA, trk-B protein in lesion scar
			NGF/BDNF/NT-3/NT-4 (slight upregulation)
		3 weeks	trk-C upregulation in lateral funiculus fibres
	cat/rat		low affinity FGF receptor p 75 immunoreactivity in scar tissues, motoneurons, blood vessels, ventral funiculus
		6 h	β-FGF mRNA (3-fold increase)
		2 days	β-FGF protein in motoneurons, gray matter, neurons
		5 days	β-FGF protein in astrocytes
Brain injury	rat	2–5 days	BDNF, NGF, FGF, CNTF, TGF-β

Compiled from various sources.
NGF Nerve growth factor; *FGF* fibroblast growth factor; *BDNF* brain derived neurotrophic factor; *NT* neurotrophin; *trk* tyrosine kinase; *CNTF* ciliary neurotrophic factor; *GDNF* glial derived neurotrophic factor; *HBGF* heparin binding growth factor; *TGF-β* transforming growth factor beta.

low-affinity NGF receptors following contusion injury was observed in scar tissue, motoneurons, microvessels and in ventral funiculus of cats and rats (Reynolds *et al.*, 1991; Risling *et al.*, 1993). However, mRNA for NGF, BDNF, NT-3 and trkA did not show such an upregulation (Schwab and Bartholdi, 1996). These observations suggest that injury to the spinal cord is associated with alterations in neurotrophin receptors (Table 3).

Effect of exogenous neurotrophic factors

Previously, the effects of neurotrophic factors on adult spinal cord lesion *in vivo* are quite few (Table 4). With the advent of new techniques, highly purified or recombinant neurotrophic factors are now available in large quantities. This has accentuated some research in this field. Schwab and his group injected NGF, BDNF or NT-3 into the spinal cord of adult rats following bilateral transection and studied the sprouting axons using anterograde tracing (Schnell *et al.*, 1994). Their results show a marked increase in axonal sprouting following NT-3 application, whereas NGF exerted a weak effect and BDNF was without any significant effect on axonal regeneration (Schwab and Bartholdi, 1996). Interestingly, when these authors combined neurotrophins with the antibody to myelin associated neurite growth inhibitory proteins (NI-1) the increase in sprout-

Table 4. Effects of neurotrophin factors on lesioned spinal cord

Treatment	Result
Recombinant neurotrophic factors (injection of NGF/BDNF/NT-3 into the lesion site in adult spinal cord dorsal bilateral transaction)	spontaneous sprouting enhanced by NT-3 (about 1 mm) NGF (weak effect) BDNF (without effect)
NT-3 + antibody against myelin associated neurite growth inhibitory proteins	sprouting enhanced in large portions of denervated spinal cord
Transfection of NGF gene (i.e., NGF producing fibroblasts, but not bFGF producing graft)	sprouting enhanced in dorsal root fibres
BDNF/NT-3 treatment	survival of axotomised newborn rubrospinal neurons
Neurotrophins in perinatal/adult cord?	regrowth/regeneration/sprouting?
Implantation of embryonic tissues/ Schwann cells in the spinal cord	increased synthesis of neurotrophic factors by these transplanted cells (direct demonstration is lacking)
Cytokine receptors CNTF IL-6 LIF	cross talk between neurotrophic factors, cytokines and their receptors on neurons, glial cells, inflammatory cells, meningeal cells, blood vessels, endothelial cells in the lesion site?

Compiled from various sources, for details see Schwaab and Bartholdi (1996).
NGF Nerve growth factor; *FGF* fibroblast growth factor; *BDNF* brain derived neurotrophic factor; *NT* neurotrophin, *trk* tyrosine kinase; *CNTF* ciliary neurotrophic factor; *GDNF* glial derived neurotrophic factor; *HBGF* heparin binding growth factor; *TGF-β* transforming growth factor beta; *IL-6* interleukin 6; *LIF* leukaemia inhibitory factor.

ing axons was much more remarkable compared to rats treated without antibody (Schwab and Bartholdi, 1996).

Using another approach, Schwab and his group found that transplants of NGF-producing fibroblasts transfected with the NGF gene, but not with bFGF-producing grafts in the spinal cord, resulted in a massive increase in axonal sprouting in dorsal root fibres (see Schwab and Bartholdi, 1996). Similarly, axotomised neurons are rescued by application of BDNF and NT-3 in new born rats (Schwab and Bartholdi, 1996). These results suggest that neurotrophins influence regeneration of the lesioned neurons in spinal cord. However, it is not certain whether neurotrophins will enhance regneration or rescue axotomised neurons in the adult spinal cord as well. This is a new subject which require additional investigation.

Apart from application of neurotrophins with or without NI-1 antibody treatment, transplantation of Schwan cells or embryonic tissue in the spinal cord enhances axonal regrowth. An increased synthesis of neurotrophins by the transplanted cells will influence regeneration (Schwab and Bartholdi, 1996). However, a direct demonstration of this phenomena is still lacking.

The molecular mechanisms underlying the action of neurotrophins in spinal cord regeneration are not fully understood. One of the recently discovered new aspects of neurotrophin effect on other signalling molecules such as cytokines seems to be quite exciting in this direction (Pettersson and Nawa, 1993). CNTF belongs to the subfamily of cytokine receptors IL-6 and leukaemia inhibiting factor (LIF) (Ip *et al.*, 1993; Pettersson and Nawa, 1993; Lindsay, 1994). There are evidences that NGF receptors are present on macrophages and lymphocytes (Pettersson and Nawa, 1993; Lindvall *et al.*, 1994). Thus, it seems quite likely that there is a cross talk between neurotrophins and cytokines with their receptors located on neurons, glial cells, inflammatory cells, meninges, blood vessels in scar tissue or perifocal edematous tissue (see Schwab and Bartholdi, 1996). Obviously, some signals may be beneficial following an upregulation of certain kinds of neurotrophins, whereas upregulation of some other types of neurotrophins may induce tissue loss (Koh *et al.*, 1995; Samdani *et al.*, 1997). This area of research is quite complex at this moment and further new insights are highly needed to explore this emerging new concept of CNS injury.

Other functions of neurotrophins during development

Apart from having trophic effects in the CNS, the neurotrophins can exert powerful instructive signals by inducing expression of certain neurotransmitter and neuropeptides (Hökfelt *et al.*, 1994; Huber *et al.*, 1995). Thus, there are evidences that BDNF induces somatostatin and neuropeptide Y in cerebral cortical neurons both *in vitro* and *in vivo* (Nawa *et al.*, 1993). These observations indicate that neurotrophins modulate neuropeptide transmission in the CNS (Mulderry, 1994). Most of these neuropeptides participates in sensory information processing system of the CNS. Thus, a possibility exists that neurotrophins may influence sensory functions in the CNS, a feature which require further investigation.

Neurotrophins and nitric oxide function during development

Neurotrophins are able to enhance NO production in cell culture which appears to be mediated by activation of specific neurotrophin receptors (Hökfelt *et al.*, 1994; Holtzman *et al.*, 1994; Garthwaite and Boulton, 1995; Huber *et al.*, 1995; Koh *et al.*, 1995). Thus, using 16 day old rat embryo spinal cord cells in culture, treatment with BDNF, NT-3 or NT-4 in a dose of 10 ng/ml for 6 days significantly increased the numbers of NADPH-diaphorase reactive neurons compared to control group. On the other hand, treatment with NGF, GDNF (10 ng/ml) or TGF-β (2 ng/ml) were not effective in enhancing NADPH-diaphorase reactive neurons (Huber *et al.*, 1995). Both trkB and trkC neurotrophin receptors activation is responsible for high increases in NOS containing neurons. This is evident from the fact that BDNF and NT-4 which activate trkB receptor were able to increase NOS reactive neurons 2 fold, whereas, NT-3 which has a high affinity for trkC receptors increased NOS activity by 1.5-fold (Huber *et al.*, 1995). These data suggest that development of NOS neurons in the embryonic spinal cord is influenced by BDNF and NT-3/4 probably via trkB receptor. However, whether neurotrophins exert similar effect on NOS neurons in adult spinal cord *in vivo* is not known.

There are several other examples of axotomy or root avulsions upregulating NADPH-diaphorase, NOS protein or mRNA (Choi, 1993; Dun *et al.*, 1993; Chiueh *et al.*, 1994; Hökfelt *et al.*, 1994; Yu, 1994; Novikov *et al.*, 1995; see Dawson and Dawson, 1996; Yamada *et al.*, 1996; Sharma, 1998; Sharma *et al.*, 1998a). Recently, Novikov and co-workers showed that long-term treatment with BDNF significantly attenuated up-regulation of NOS neurons in the spinal cord following spinal root avulsions *in vivo* (Novikov *et al.*, 1995). These observations point out that in adult spinal cord neurotrophins may act in opposite way on NOS neurons. The probable mechanisms of such discrepancy between *in vivo* and *in vitro* studies are still unclear.

Pathophysiology of a focal spinal cord injury

Pathophysiology of spinal cord injury is quite complex and many factors are involved in tissue destruction following trauma (see Sharma *et al.*, 1990a–c; 1991; 1992a,b; Faden, 1993; Sharma *et al.*, 1993a–d; 1995a–d; Schwab and Bartholdi, 1996; Sharma *et al.*, 1996a–c; Sharma and Westman, 1997; Sharma *et al.*, 1997a,b; 1998b; Winkler *et al.*, 1998). Experiments carried out in our laboratory in the past showed that many neurochemicals such as serotonin, prostaglandins, opioids, histamine, catecholamines are involved in edema formation and cell injury following spinal cord trauma (Sharma and Olsson, 1990; Sharma *et al.*, 1990a–c; 1991; Olsson *et al.*, 1992; Sharma *et al.*, 1992a,b; 1993a–d; Olsson *et al.*, 1995; Sharma *et al.*, 1995a–c). These observations suggest that no single chemical compound alone is responsible for all the pathological disturbances following spinal trauma. The final outcome will be determined by a combination of all the elements responsible for tissue destruction and tissue protection (Winkler *et al.*, 1998). Many new factors or chemicals in future will thus join the list contributing to cell injury following trauma to the spinal cord.

Nitric oxide in the CNS

The recently identified gaseous molecule NO has influenced our basic concept of neurotransmission in the central nervous system (Bredt *et al.*, 1991; Choi, 1993; Chiueh *et al.*, 1994; Yun *et al.*, 1996; Sharma *et al.*, 1998a). NO is a free radical gas which does not require any synaptic machinery for storage or releases into the synaptic cleft for its action

(Dawson and Snyder, 1994, Garthwaite and Boulton, 1995; Sharma, 1998; Sharma *et al.*, 1998a–c). NO diffuses quite rapidly among the cell membranes in order to induce its biological action (Garthwaite and Boulton, 1995). Since its half life is extremely short (about 5 sec) most studies in the CNS are concentrated to examine NOS, the enzyme which is responsible for NO synthesis from the amino acid L-arginine (Dun *et al.*, 1993; Hökfelt *et al.*, 1994; see Dawson and Dawson, 1996). NOS exists in various isoforms (Kimura and Steinbusch, 1996; Yamada *et al.*, 1996) and recently specific antibodies to these isoforms are available commercially (Sharma *et al.*, 1997c–e; Sharma, 1998), which provide new tools in the NO research with great specificity and precision (for review see Sharma *et al.*, 1998a).

Physiological functions of NO

There are evidences that NO is involved in the control of blood flow, heart rate, blood pressure, long-term potentiation, depression and release of neurotransmitters (Schuman and Madison, 1994). Recently, NOS is found to be co-localised with several neuropeptides (Yamada *et al.*, 1996) in various parts of the CNS, which seems to be important for the modulation of neurotransmission, ion channels, receptor function and signal transduction mechanisms (Yun *et al.*, 1996). There are indications that NO is involved in neuronal plasticity as well in neurotoxicity in the CNS (Hökfelt *et al.*, 1994; Dawson and Dawson, 1996; Yun *et al.*, 1996: Sharma *et al.*, 1998a). Thus, it is still not clear whether upregulation of NOS and/or NO production seen in the CNS during various neurological diseases or in experimental models involving brain pathology such as, stroke, infarction, ischemia and trauma is contributing to cell injury (Dawson and Snyder, 1994; Schuman and Madison, 1994; Dawson and Dawson, 1996).

NO and cell injury

Recent studies suggest that upregulation of NOS is associated with increased NO production which is responsible for cell injury in several models of CNS toxicity (Choi, 1993; see Chiueh *et al.*, 1994; Kimura and Steinbusch, 1996; Dawson and Dawson, 1996; Sharma, 1998; Sharma *et al.*, 1998a,b). Experiments based on neuronal NOS null mice provides one of the most strong evidence supporting the role of NO in neurotoxicity in ischemia (see Dawson and Dawson, 1996). Thus, the NOS null mice are quite resistant to ischemia induced cell damage in the brain. On the other hand, use of NOS inhibitors in ischemia, stroke or infarction provide contradictory results making the hypothesis that NO is involved in cell injury and cell death a bit controversial (Sharma *et al.*, 1998a). There are indications that blockade of the endothelial isoform of NOS (eNOS) aggravates ischemic injury, whereas the blockade of neuronal NOS (nNOS) in pathological conditions is beneficial for cell survival (see Dawson and

Dawson, 1996). It seems likely that inhibition of particular isoforms of NOS is important in cell injury or survival, a feature which require additional investigation (see Chapter 19 in this volume).

Nitric oxide in spinal cord injury

Involvement of NO in the pathophysiology of spinal cord injury is evident from our observations that topical application of antibody raised against neuronal isoform of NOS (nNOS) is capable to attenuate edema formation and cell injury of the traumatised cord (Figs. 1 and 2) (Sharma *et al.*, 1995e, 1996a, 1998b). This beneficial effect of the antiserum was related with its ability to attenuate upregulation of nNOS expression following trauma (Fig. 2) (Sharma *et al.*, 1996a). This observation opens up a new line of scientific research in this field implicating NO as an injurious factor responsible for cell injury in the spinal cord following trauma. This idea is recently well supported by many laboratories using other approaches (Fukuda *et al.*, 1996; Novikov *et al.*, 1995, for details see chapter 19 in this volume).

Neurotrophins, NOS and spinal cord injury

BDNF and insulin like growth factor-1 (IGF-I) are neuroprotective in several animal models of ischemic brain injury indicating that neurotrophins are neuroprotective in the adult CNS as well (Davies, 1994; Folli *et al.*, 1994; Lindsay, 1994; Lindvall *et al.*, 1994; Yao *et al.*, 1995). NGF and BDNF has the capacity to induce somatostatin and neuropeptide Y synthesis in cultured cortical neurons (Nawa *et al.*, 1993). Intravenous administration of BDNF in newborn rats induces synthesis of several neuropeptides in the CNS (Nawa *et al.*, 1994) and NGF can selectively upregulate expression of neuropeptides in the sensory neurons (Mulderry, 1994).

IGF has numerous actions on neurons and glia cell functions *in vitro* (Folli *et al.*, 1994). However its role in the CNS *in vivo* situation is still largely unknown (Logan *et al.*, 1992). Brain and spinal cord contains high levels of IGF which is reduced significantly following spinal cord injury (Tsitouras *et al.*, 1995). On the other hand human cases of amyotrophic lateral sclerosis (ALS) exhibit upregulation of IGF binding in the spinal cord gray matter (Adem *et al.*, 1994; Barbecid, 1995). These observations suggest that IGF is involved in several disease processes of the CNS. Recent reports show that growth factors such as IGF, NGF and CNTF can rescue motor neuron atrophy in multiple scleroses, Alzheimer's, Parkinson's and Huntington's diseases (Barde, 1994; Barbecid, 1995; Andersson *et al.*, 1996; Samdani *et al.*, 1997) indicating a profound neuroprotective role of neurotrophins in neurodegeneration.

Nerve cells in the brain and in spinal cord express a constitutive isoform of the enzyme NOS which is originally isolated and cloned by Bredt,

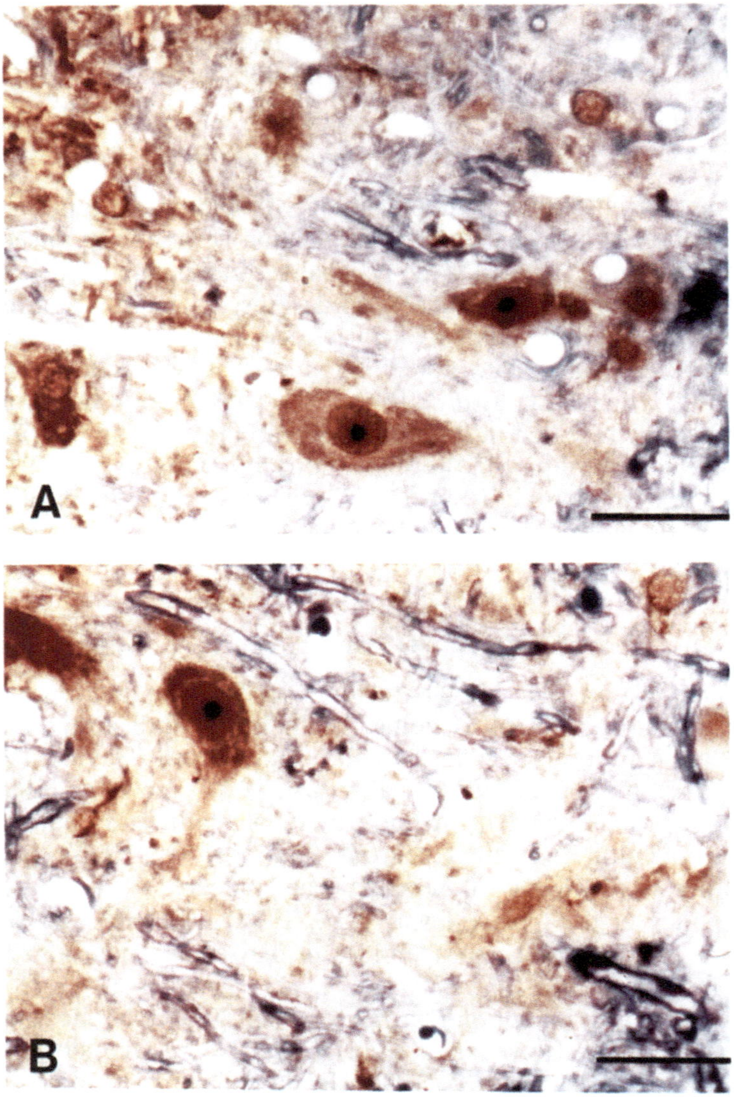

Fig. 1. nNOS upregulation in the spinal cord ventral horn of the T9 segment following trauma. (bar = 25 µm, Data modified from Sharma *et al.*, 1996a with permission)

Snyder and their colleagues (Bredt *et al.*, 1991). In adult spinal cord, NOS positive neurons are mainly located in the superficial dorsal horn and in lamina X (Dun *et al.*, 1993). Upregulation of NOS following spinal cord injury is one of the key factors in inducing cell damage (Sharma *et al.*, 1996a, 1998a). Since, neurotrophins influence NO production (Holtzman

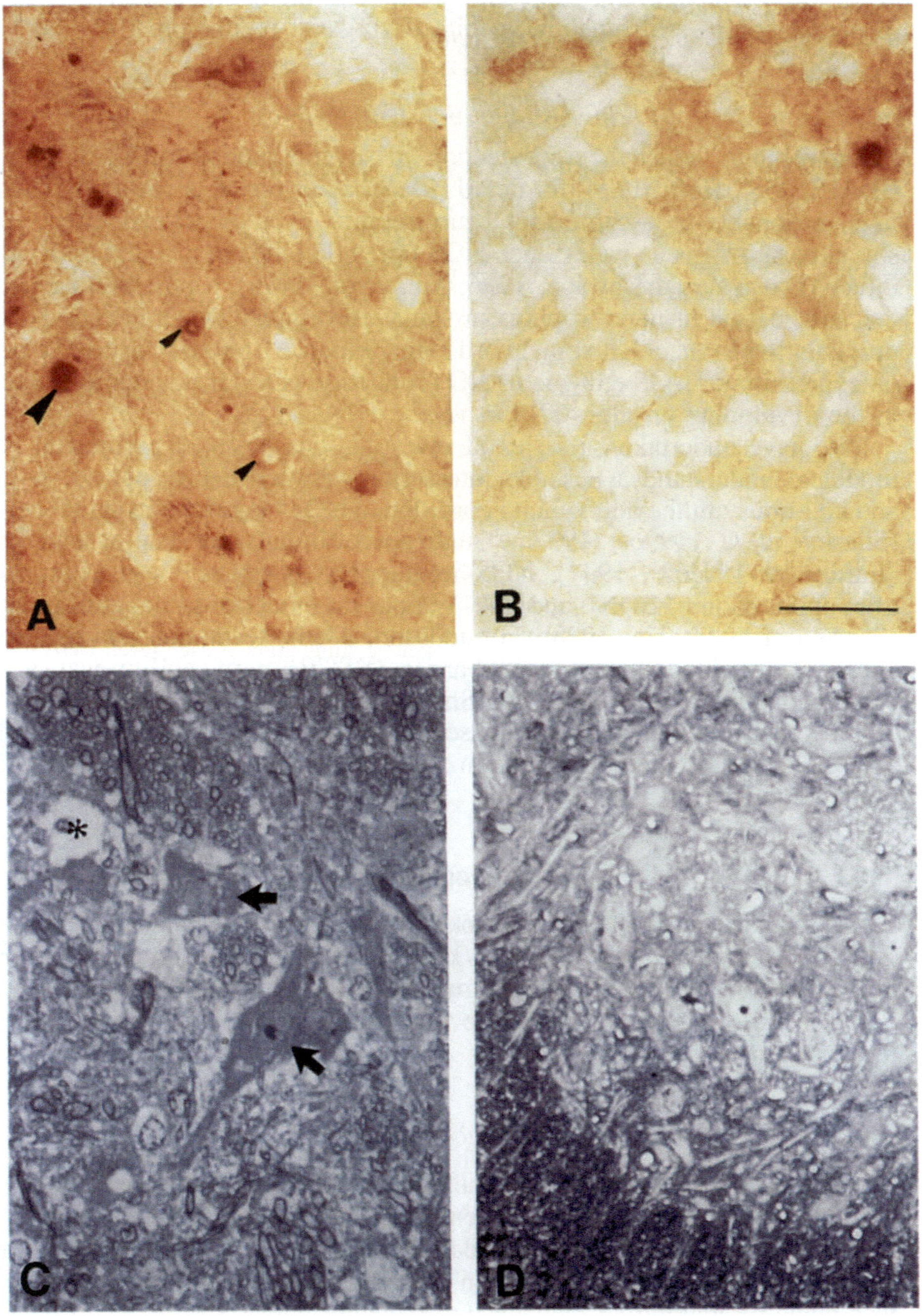

Fig. 2. Pretreatment with nNOS antiserum significantly attenuated nNOS upregulation (**B**) compared to saline treated (**A**) traumatised rat. Application of nNOS antiserum also attenuated edema and cell changes (**D**) compared to untreated traumatised rat (**C**) which exhibited many dark and distorted nerve cells (arrows) and edema (*) in the spinal cord (bar: a,b = 50 μm, c,d = 40 μm; Data modified from Sharma *et al.*, 1996a with permission)

et al., 1994; Huber *et al.*, 1995), it appears that the neuroprotective effects of neurotrophins are somehow mediated via NOS regulation.

Background and hypotheses

Neurotrophins induce neuroprotection in several models of CNS injury (Barde, 1994; Hefti, 1994, for details see Chapter 6 in this volume). Thus, it is reasonable to assume that neurotrophins exert beneficial effects in spinal cord trauma as well. Since NOS upregulation is associated with cell injury (Dawson and Dawson, 1996; Yun *et al.*, 1996), a possibility exists that application of neurotrophins in spinal cord injury will influence NOS upregulation in the adult spinal cord. The influence of neurotrophic factors on expression of NOS enzyme or NO production is so far not studied either in the developing or adult CNS *in vivo*.

We have undertaken a series of investigations in our experimental rat model to understand a possible interaction between neurotrophins, NOS and cell injury in the adult spinal cord. Since NO is injurious in spinal cord (Sharma *et al.*, 1996a; 1997c,e) and neurotrophins are neuroprotective (Barde, 1994; Hefti, 1994), NOS immunohistochemistry was examined after application of BDNF and IGF-1 in our rat model of spinal cord injury (Sharma *et al.*, 1997a; 1998b). In addition, edema formation and disturbances in the blood-spinal cord barrier (BSCB) permeability was also examined in relation to cell injury (Sharma and Olsson, 1990; Sharma *et al.*, 1995b). The following section is based on our own investigations on BDNF and IGF induced neuroprotection in relation with changes in NOS upregulation.

Methodological aspects

Experiments were carried out on male Sprague Dawley rats (body weight 250–300 g) under equithesin (3 ml/kg, i.p.) anaesthesia. The animals were housed at controlled room temperature ($21 \pm 1°C$) with 12 h light and 12 h dark schedule. The rat food and tap water were supplied *ad libitum* before the experiment.

Spinal cord injury

One segment laminectomy was done on the T10–11 segment and the spinal cord was exposed by carefully removing the dura mater. The exposed spinal cord was covered with surgical cotton soaked in physiological saline at room temperature. After 30 min of stabilisation period, a focal injury was made by making an incision into the right dorsal horn of the T10–11 segments using a scalpel blade (Sharma and Olsson, 1990). The deepest part of the lesion was close to the lamina VII–VIII. This experimental condition was approved by the Ethical Committee of Banaras Hindu University, Varanasi, India and Uppsala University, Uppsala, Sweden.

Treatment with neurotrophins

In separate group of animals BDNF or IGF-1 (0.1 μg/10 μl in phosphate buffer saline) was applied topically 30 min before injury on the exposed spinal cord (20 μl) followed by repeated doses of the growth factors immediately before and 30 min after injury. Thereafter, application of BDNF or IGF-1 was carried out at every 1 h interval until sacrifice (Sharma *et al.*, 1997a; 1998b). Five hours after injury, the tissue pieces from the T9 and T12 segments were processed for nNOS immunostaining. Normal animals served as intact controls. For comparison with the growth factor treated group untreated traumatised rats were used as experimental controls.

The following parameters were examined in control, spinal cord injured and neurotrophins treated injured animals.

NOS immunohistochemistry

Five h after spinal cord injury, the animals were perfused with about 200 ml of 4 % paraformaldehyde based fixative in 0.1 M phosphate buffer pH 7.4 at 4°C preceded with a brief saline rinse (Sharma *et al.*, 1996a). After perfusion, the animals were wrapped in an aluminium foil and kept overnight at 4°C in a refrigerator. On the next day, the spinal cord samples were dissected out and placed in the same fixative at 4°C. About 60 μm thick multiple vibratome sections were cut from the T9 and T12 segments and processed for NOS immunoreactivity using monoclonal antiserum to constitutive isoform of neuronal NOS as described earlier (Sharma *et al.*, 1996a–c). Some tissue pieces were embedded in epon for routine light and electron microscopy for morphological examination using standard procedures (Sharma *et al.*, 1990a–c; Sharma and Westman, 1997; Sharma *et al.*, 1998c).

Microvascular permeability and edema

In separate groups of control, traumatised or neurotrophins treated injured rats spinal cord water content and microvascular permeability to $^{[131]}$I-sodium was measured in the perifocal segments (Sharma *et al.*, 1990a). For this purpose radioactive iodine (about $0.5\,\text{ml} \times 10^6$ cpm) in 0.9 % saline was injected into the right femoral vein 5 min before termination of the experiment (Olsson *et al.*, 1990). The intravascular tracer was washed out by a brief saline rinse through heart (Sharma *et al.*, 1990a; 1995b). Tissue pieces from T9, T10–11 and T12 were dissected out, weighed immediately and counted in a 3-in well type gamma counter (energy window 500–800 KeV). After counting the radioactivity the tissue samples were dried in an incubator maintained at 90°C for 72 h (Sharma *et al.*, 1990a; Sharma and Cervós-Navarro, 1990). The water content was calculated from the differences between wet and dry weight of the cord (Sharma and Olsson 1990).

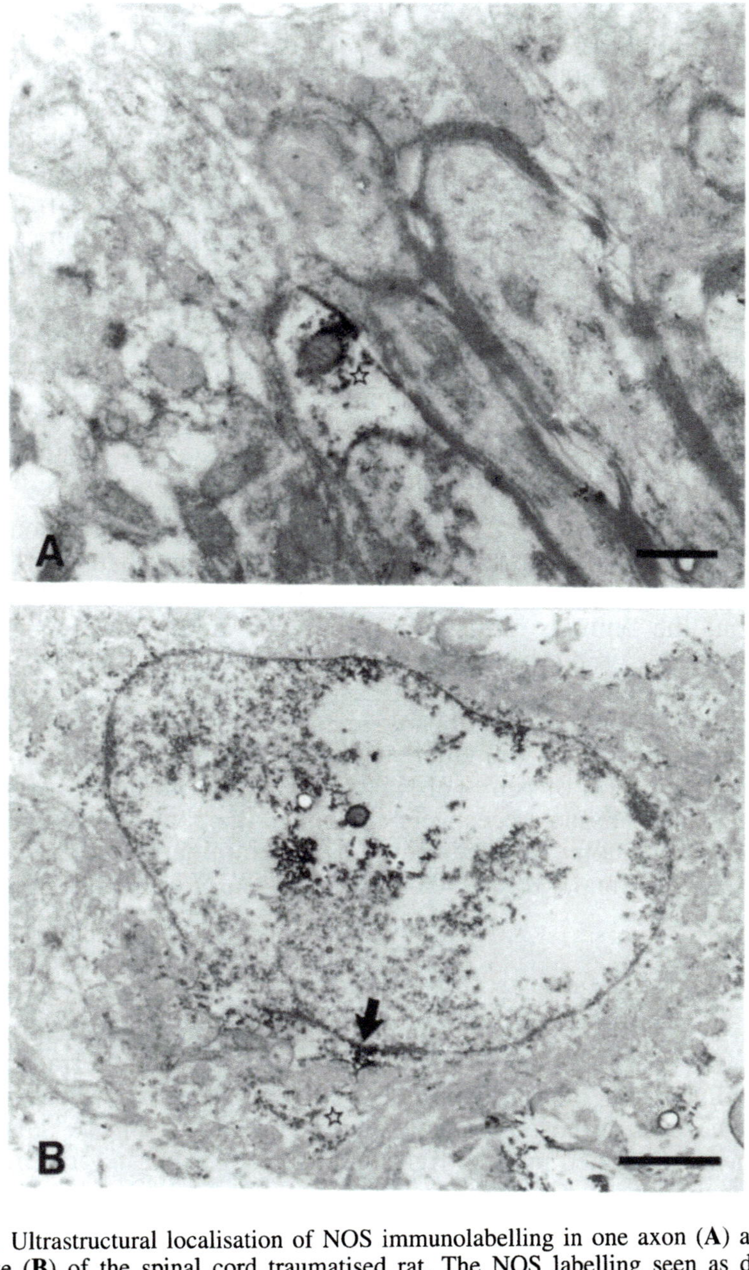

Fig. 3. Ultrastructural localisation of NOS immunolabelling in one axon (**A**) and in one dendrite (**B**) of the spinal cord traumatised rat. The NOS labelling seen as dark black particles is confined within the cell border. This section is not contrasted with lead citrate or uranyl acetate (bar = 1 µm, Data modified from Sharma *et al.*, 1996a)

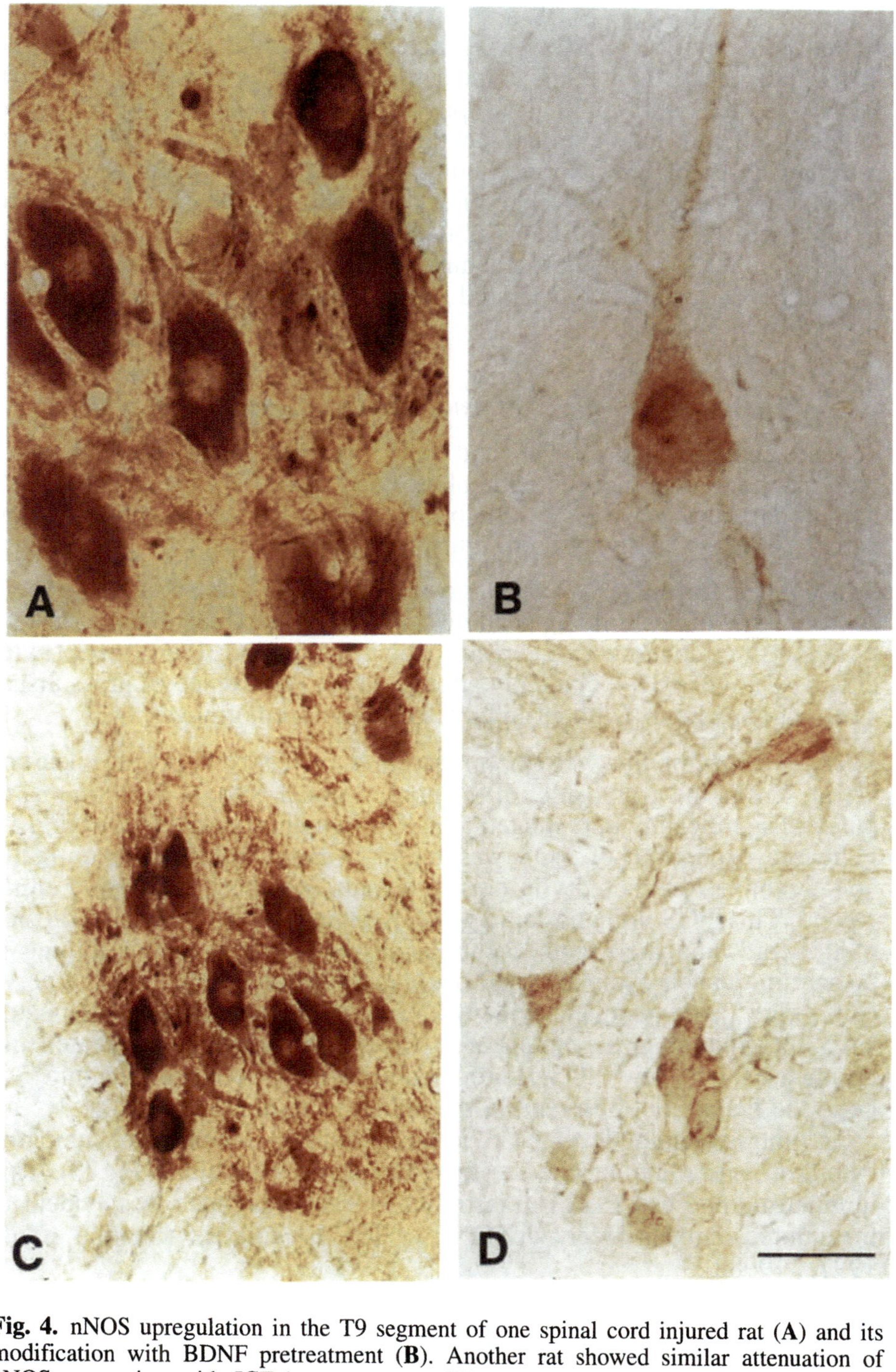

Fig. 4. nNOS upregulation in the T9 segment of one spinal cord injured rat (**A**) and its modification with BDNF pretreatment (**B**). Another rat showed similar attenuation of nNOS expression with IGF-1 pretreatment (**D**) compared to untreated injured rat (**C**) (bar = 30 μm) (Data modified from Sharma *et al.*, 1997a and 1998b with permission)

Experimental protocol

Spinal cord injury (n = 20) was inflicted on the right dorsal horn of the T10–11 segments by making an incision. In one group of animals (n = 10) BDNF or IGF (0.1 µg/10 µl in phosphate buffer saline) was applied topically 30 min before injury on the exposed spinal cord followed by repeated doses of IGF. Normal animals (n = 10) were used as controls. Spinal cord edema was examined in control (n = 5), spinal cord traumatised (n = 5) and BDNF or IGF-1 treated and traumatised (n = 5) rats while NOS immunohistochemistry was done in separate groups of normal (n = 5), spinal cord injured (n = 5) and BDNF or IGF treated and injured (n = 5) rats.

Neuroprotective effects of neurotrophins in spinal cord injury

Topical application of BDNF and IGF-1 in high quantity significantly reduced the trauma induced pathophysiology of cell changes in the spinal cord (Sharma *et al.*, 1997a; 1998b). The effect of neurotrophins on the pathophysiology of spinal cord injury is described below.

Effect of neurotrophins on visual swelling. Spinal cord traumatised rats showed signs of haemorrhage and visual swelling immediately after lesion. This swelling continued to expand over a 5 h period. Pretreatment with either BDNF or IGF-1 markedly attenuated the occurrence of gross visual swelling after injury without influencing microhaemorrhages.

Effect of neurotrophins on NOS upregulation. A few NOS positive neurons are present in the gray matter of the normal spinal cord. A focal trauma to the cord significantly upregulated the NOS immunostaining at 5 h (Fig. 1). Increased expression of NOS was found in nerve cells which normally donot exhibit NOS activity in normal spinal cord (Fig. 2) (Sharma *et al.*, 1996a). This NOS immunostaining was most marked in the ipsilateral side of the cord compared to the contralateral half. NOS immunostaining is present in the cell cytoplasm and occasionally in the nucleus of the nerve cells (Fig. 1). Pretreatment with BDNF or IGF-1 significantly attenuated NOS upregulation in the motoneuron (Fig. 4). In neurotrophins treated and traumatised rats, few NOS neurons showed elongated axons within 5 h period, a feature not observed in the untreated injured group (Fig. 4).

At ultrastructural level NOS immunostaining is confined within the cytoplasm and often attached to the endoplasmic reticulum (Fig. 3). This immunostaining of NOS at ultrastructural level was much less evident in the injured animals received BDNF or IGF-1 treatment (Sharma *et al.*, 1997a; 1998b).

Effect of neurotrophins on edema and microvascular permeability. Spinal cord injured rats exhibited a profound increase in the BSCB permeability to ^{131}I-albumin and water content in T9 and T12 perifocal segments (Sharma *et al.*, 1990a, 1995b). Pretreatment with neurotrophic factors significantly

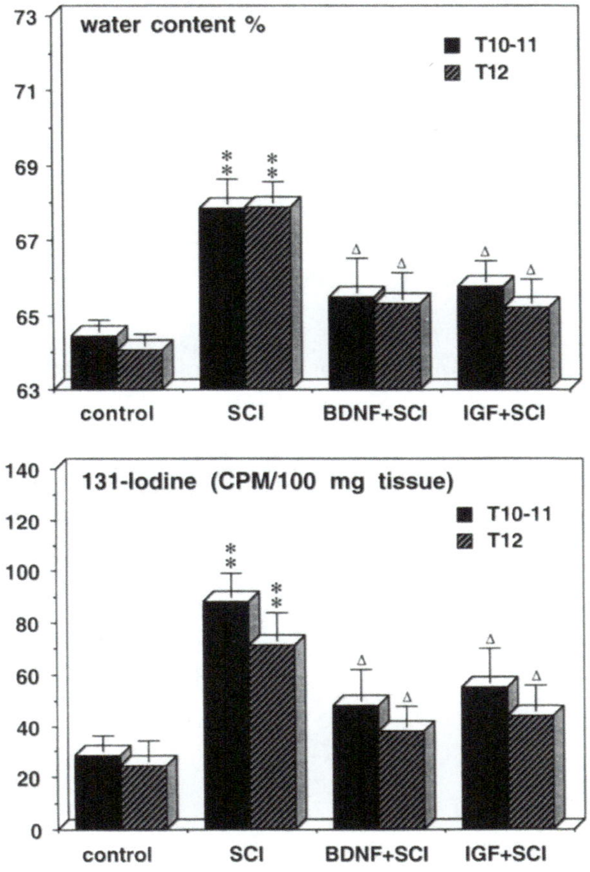

Fig. 5. Water content (above) and blood-spinal cord barrier (*BSCB*) permeability (below) in the untreated injured (T10–11) and adjacent caudal (T12) spinal cord segments and their modification with BDNF or IGF-1 treatment. * $P < 0.001$ compared from control; Δ $P < 0.05$ compared from SCI, ANOVA followed by Dunnett test (*SCI* spinal cord injury; *BDNF* brain derived neurotrophic factor; *IGF* insulin like growth factor) (Data modified from Sharma *et al.*, 1997a, 1998b)

attenuated the extravasation of the BSCB permeability to iodine (Fig. 5). The water content of the T9 and T12 segment was increased significantly from the control group. Pretreatment with BDNF or IGF significantly thwarted the edema development in the spinal cord 5 h after injury (Sharma *et al.*, 1997a; 1998b). Thus, the water content in the T9 and T12 segments was significantly less compared to the untreated injured group (Fig. 5).

Effect of neurotrophins on cell changes. Samples of T9 and T12 segments obtained from untreated rats showed profound cell injury to the nerve cells, glial cells and myelin (Sharma *et al.*, 1990a, 1993a, 1995a, 1997a; Winkler *et al.*, 1998). Thus, spinal cord injured rats showed profound swelling and

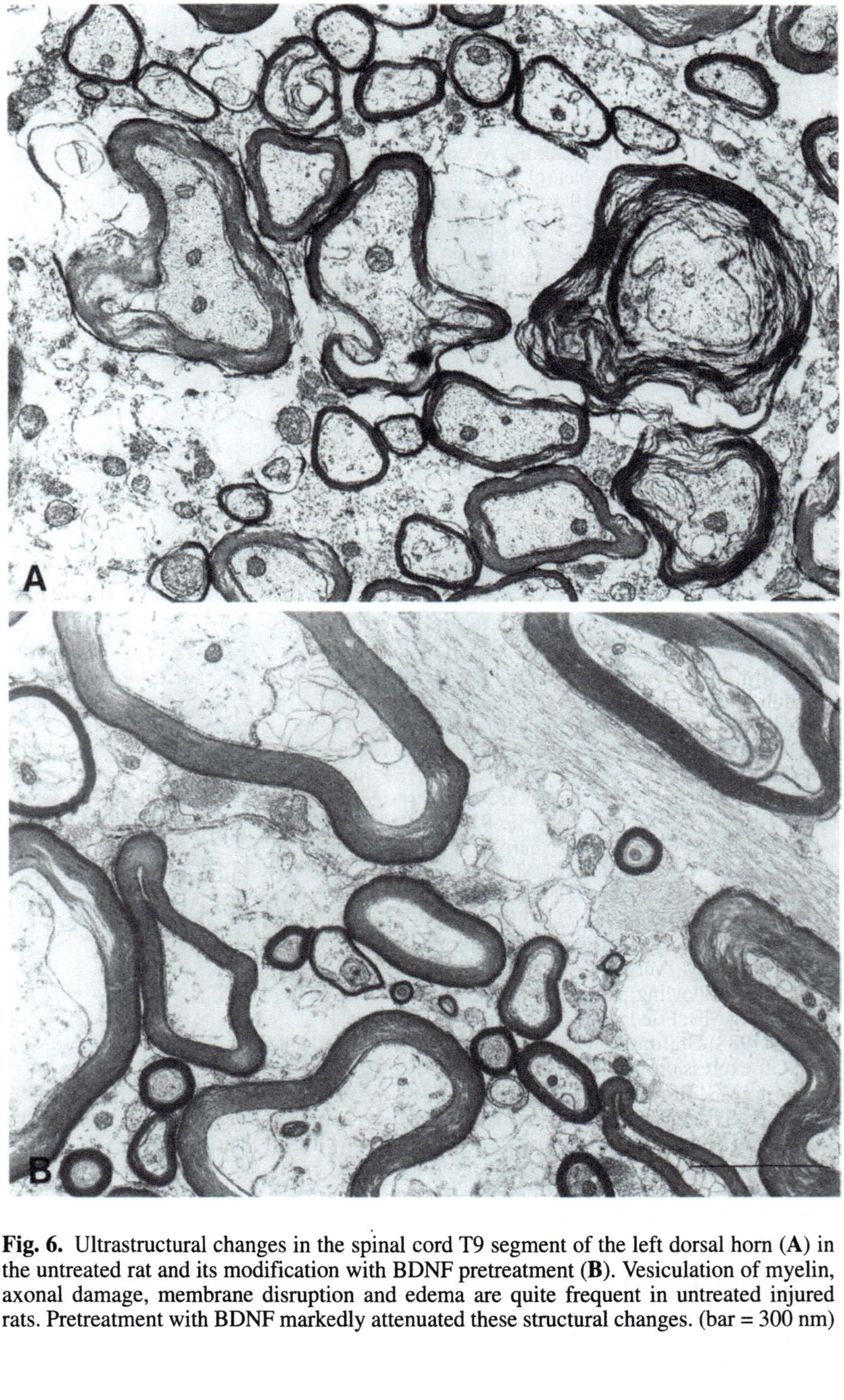

Fig. 6. Ultrastructural changes in the spinal cord T9 segment of the left dorsal horn (**A**) in the untreated rat and its modification with BDNF pretreatment (**B**). Vesiculation of myelin, axonal damage, membrane disruption and edema are quite frequent in untreated injured rats. Pretreatment with BDNF markedly attenuated these structural changes. (bar = 300 nm)

edematous expansion of the cord in the T9 and the T12 segments of the cord (Fig. 6). These cell changes were considerably reduced by pretreatment with neurotrophins. Thus, application of BDNF (Fig. 6) or IGF significantly attenuated the cell injury following trauma (H S Sharma unpublished observations). The spinal cord nerve cells look normal in appearance, and damage to the glial cells or myelin was less pronounced (results not shown).

Mechanisms of neuroprotection by neurotrophins

Topical application of BDNF or IGF-1 reduced the visual swelling of traumatised cord in this study. This clearly demonstrate that neurotrophins can induce significant neuroprotection in the injured spinal cord of adult rats. These observations are in agreement with reports which show a significant reduction in cell damage in ischemia by neurotrophins treatment. Our study further added a new dimension in the neuroprotective action of neurotrophins regarding their ability to reduce microvascular permeability disturbances and edema formation following trauma. These findings strongly support the idea that neurotrophins can influence the secondary injury processes probably by alterations in the microfluid environment of the cord.

A significant attenuation of BSCB permeability by neurotrophins suggests that secondary injury mechanisms following trauma are thwarted by neurotrophins. A reduction in BSCB permeability will attenuate leakage of plasma proteins and thereby reducing vasogenic edema, which is clearly evident from our findings of a reduction in water content (for review see Sharma *et al.*, 1998c). Obviously, a less vascular or cellular reaction in the cord following injury in neurotrophins treated group is likely to exhibit less cell and tissue damage (for review see Winkler *et al.*, 1998).

Neurotrophins promote survival of the embryonic motoneurons in vitro and prevent loss of motoneurons following axotomy in vivo (Yan *et al.*, 1992; Yin *et al.*, 1994). Peripheral axotomy induces NOS upregulation in the spinal cord motoneurons (Yu, 1994) indicating that NO is involved in the mechanisms of motoneurons damage (Dawson and Dawson, 1996). BDNF pretreatment significantly reduced the upregulation of NOS following axotomy in a rat model (Novikov *et al.*, 1995). This confirms the hypothesis that neurotrophins can influence NOS expression (Hökfelt *et al.*, 1994; Holtzman *et al.*, 1994; Huber *et al.*, 1995).

Our observation further extend this idea in spinal trauma and demonstrate that pretreatment with BDNF and IGF-1 prevented the NOS upregulation around the lesion following injury. These results suggest that one of the important mechanisms of neuroprotection by neurotrophins is to inhibit NOS upregulation (Nivikov *et al.*, 1995; Sharma *et al.*, 1996a; Sharma, 1998; Sharma *et al.*, 1998a–c). Activation of NOS will result in production of NO (Chieuh *et al.*, 1994; Yun *et al.*, 1996) which exerts its neurotoxic effect on cells probably via activation of NMDA receptor and

Ca^{2+} influx (Choi, 1993; Chiueh *et al.*, 1994). Prevention of NMDA receptor activation by MK-801 or inhibition of NOS is neuroprotective (Garthwaite and Boulton, 1995; Yun *et al.*, 1996). Thus, it seems quite likely that neurotrophins which attenuated NOS activity following trauma may have strong neuroprotective effects.

Spinal cord is very rich in IGF and spinal cord injury reduces the IGF content (Hefti, 1994; Folli *et al.*, 1994; Tsitours *et al.*, 1995). A decrease in the IGF level following spinal cord injury will deprive neurons to trophic support resulting in atrophy or cell death. Exogenous supplement of IGF-1 rescue the nerve cell damage which suggests that growth factors can induce neuroprotection following trauma (Yao *et al.*, 1992; Yin *et al.*, 1994). Another possibility of these growth factors is to attenuate cellular stress reaction by influencing fluid microenvironment.

Motoneurons and other neuronal populations in the spinal cord are important targets of neurotrophins (Ernfors and Persson, 1991; Barde, 1994). BDNF mRNA is produced in dorsal root ganglion which is anterogradely transported to the neurons in the spinal cord dorsal horn (Frisen *et al.*, 1994; Hefti, 1994). Most of the neuronal population in the spinal cord express receptors for neurotrophins such as trk-B, trk-C and low-affinity neurotrophin receptor p75 (Frisen *et al.*, 1992; 1994; Schwab and Bartholdi, 1996). Thus, neurotrophins via receptor regulations can influence spinal cord fluid microenvironment.

We applied neurotrophins topically on the spinal cord which was repeated several times after injury. This may have given access of neurotrophins into the tissues in order to have an effect. That neurotrophins gained access into the spinal cord tissue is evident from our results with prolonged axons of the neurons stained with NOS positive immunostaining, a finding not seen in the rats without neurotrophins treatment. These observations suggest that neurotrophins when applied topically can gain access within the cord tissue in a short duration. Previously, using immunohistochemical techniques, we demostrated penetration of serotonin or nNOS antisera within the spinal cord following their topical application (Sharma *et al.*, 1996a, 1997d). Thus, serotonin and NOS antisera were capable to neutralise tissue antigens *in vivo* which is confirmed by prevention of trauma induced increased serotonin and NOS immunoreactivity within the spinal cord (Sharma *et al.*, 1990b; 1996a, 1997b).

The possible mechanisms by which neurotrophins attenuate NOS upregulation is not clear from this study. Neurotrophins influence intracellular cascade following trauma by modifying signal transduction and Ca^{2+} permeability (Mattsson *et al.*, 1995). These events can influence NOS upregulation. Another possibility of neuroprotection caused by neurotrophins is to attenuates the intensity of cellular stress or oxidative stress caused by trauma. A reduction in cellular stress due to modulation of sensory information process by neurotrophins is also likely (Davies, 1994; Hökfelt *et al.*, 1994; Mulderry, 1994). In order to testify this hypothesis, upregulation of heat shock protein response in neurotrophins treated traumatised rats are needed.

Conclusion

In conclusion, our results suggest that BDNF and IGF pretreatment is neuroprotective in spinal cord injury. This beneficial effect seems to be related with the capacity of the neurotrophins to downregulate NOS expression following spinal trauma. Our results, thus provide new experimental evidences which suggest that BDNF and IGF-1 may exert their potential neuroprotective effect probably via NOS regulation.

Future directions

Studies are in progress in our laboratory to understand a common mechanisms of neuroprotective actions of neurotrophins in spinal cord trauma. It remains to be seen whether topical application of other neurotrophins such as NGF, NT-3/4, CNTF, GDNF, LIF, TGF, FGF can also influence NOS upregulation in spinal cord injury. Another approach in this direction would be to map several neurotrophins and their receptors in the spinal cord following injury in our model.

Acknowledgements

Due to space limitation, only key references are used in this review. Original references can be found in the major reviews listed in this chapter. Authors are grateful to Dr. Dan Lindholm for providing BDNF as a gift. This investigation is supported by grants from Swedish Medical Research Council nr 2710, 9459, 9077, 11205; Astra Pain Control, Sweden; The University Grants Commission, New Delhi, India; Alexander von Humboldt Foundation, Bonn, Germany. Authors are thankful to Joachim Burman for help in library work and photocopying. The technical assistance of Kärstin Flink, Ingamarie Olsson, Inga Hörte, Ulla Johansson is highly appreciated. The secretarial assistance of Gun-Britt Lind and Aruna Sharma is acknowledged with thanks.

References

Adem A, Ekblom J, Gillberg PG (1994) Growth factor receptors in amyotrophic lateral sclerosis. Mol Neurobiol 9: 225–231

Anderson KD, Panayotatos N, Corcoran TL, Lindsay RM, Wiegand SJ (1996) Ciliary neurotrophic factor protects striatal output neurons in an animal model of Huntington disease. Proc Natl Acad Sci USA 93: 7346–7351

Barbacid M (1995) Neurotrophic factors and their receptors. Curr Opin Cell Biol 7: 148–155

Barde YA (1994) Neurotrophins: a family of proteins supporting the survival of neurons. Prog Clin Biol Res 390: 45–56

Bredt DS, Hwang PM, Glatt CE, Lowenstein C, Reed RR, Snyder SH (1991) Cloned and expressed nitric oxide synthase structurally resembles P-450 reductase. Nature 351: 714–718

Burke MA, Mobley WC, Cho J, Wiegand SJ, Lindsay RM, Mufson EJ, Kordower JH (1994) Loss of developing cholinergic basal forebrain neurons following excitotoxic lesions of the hippocampus: rescue by neurotrophins. Exp Neurol 130: 178–195

Caroni P, Schwab ME (1988) Antibody against myelin-associated inhibitor of neurite outgrowth neutralizes nonpermissive substrate properties of CNS white matter. Neuron 1: 85–96
Cheng B, Mattson MP (1994) NT-3 and BDNF protects CNS neurons against metabolic/excitotoxic insults. Brain Res 640: 56–67
Chiueh CC, Gilbert DL, Colton CA (1994) The neurobiology of NO˙ and ˙OH. Ann NY Acad Sci 738: 1–471
Choi DW (1993) Nitric oxide: foe or friend to the injured brain. Proc Natl Acad Sci USA 90: 9741–9743
Davies AM (1994) The role of neurotrophins in the developing nervous system. J Neurobiol 25: 1334–1348
Dawson VL, Dawson TM (1996) Nitric oxide neurotoxicity. J Chem Neuroanat 10: 179–190
Dawson TM, Snyder SH (1994) Gases as biological messengers: nitric oxide and carbon monoxide in the brain. J Neurosci 14: 5147–5159
Dun NJ, Dun SL, Wu SY, Förstermann U, Schmidt HHW, Tseng LF (1993) Nitric oxide synthase-immunoreactivity in the rat, mouse, cat and squirrel monkey spinal cord. Neuroscience 54: 845–857
Ernfors P, Persson H (1991) Developmentally regulated expression of HDNF/NT-3 mRNA in rat spinal cord motoneurons and expression of BDNF mRNA in embryonic dorsal root ganglion. Eur J Neurosci 3: 953–961
Faden AI (1993) Experimental neurobiology of central nervous system trauma. Crit Rev Neurobiol 7: 175–186
Faden AI, Salzman SK (1992) Pharmacological strategies in CNS trauma. Trends Pharmacol Sci 13: 29–35
Follesa P, Wrathal JR, Mocchetti I (1994) Increased basic fibroblast growth factor mRNA following contusive spinal cord injury. Mol Brain Res 22: 1–8
Folli F, Bonfanti L, Renard E, Kahn CR, Merighi A (1994) Insulin receptor substrate-1 (IRS-1) distribution in the rat central nervous system. J Neurosci 14: 6412–6422
Frautschy SA, Walicke PA, Baird A (1991) Localization of basic fibroblast growth factor and its mRNA after CNS injury. Brain Res 553: 291–299
Frisen J, Verge VMK, Cullheim S, Persson H, Fried K, Middlemas DS, Hunter T, Hökfelt T, Risling M (1992) Increased levels of trkB mRNA and trkB protein-like immunoreactivity in the injured rat and cat spinal cord. Proc Natl Acad Sci USA 89: 11282–11286
Frisen J, Arvidson U, Lindholm D, Fried K, Verges VMK, Cullheim S, Hökfelt T, Risling M (1994) trkC expression in the injured rat spinal cord. Neuroreport 5: 349–352
Fukuda K, Richman JD, Sato M, Sharp FR, Panter SS, Noble LJ (1996) Induction of heme oxygenase-1 (HO-1) in glia after traumatic brain injury. Brain Res 736: 68–75
Garthwaite J, Boulton CL (1995) Nitric oxide signalling in the central nervous system. Ann Rev Physiol 57: 683–706
Gilmore SA (1963) The effects of X-irradiation on the spinal cords of neonatal rats. II. Histological observations. J Neuropathol Exp Neurol 22: 294–301
Hefti F (1994) Neurotrophic factor therpay for nervous system degenrative diseases. J Neurobiol 25: 1418–1435
Hokfelt T, Ceccatelli S, Gustafsson L, Hulting A-L, Verge V, Villar M, Xu X-J, Xu Z-Q, Wiessenfeld-Hallin Z (1994) Plasticity of NO synthase expression in the nervous and endocrine systems. Neuropharmacology 33: 1221–1227
Höltzman DM, Kilbridge J, Bredt DS, Black SM, Li Y, Clary DO, Reichardt LF, Mobley WC (1994) NOS induction by NGF in basal forebrain cholinergic neurons: evidence for regulation of brain NOS by a neurotrophin. Neurobiol Dis 1: 51–60
Huber KA, Krieglstein K, Unsicker K (1995) The neurotrophins BDNF, NT-3 and NT-4 but not NGF, TGF-β1 and GDNF, increases the number of NADPH-diaphorase-reactive neurons in rat spinal cord cultures. Neuroscience 69: 771–779
Ip NY, Weigand SJ, Morse J, Rudge JS (1993) Injury induced regulation of ciliary neurotrophic factor mRNA in the adult rat brain. Eur J Neurosci 5: 25–33

Jenkins RW, Tetzlaff W, Hunt SP (1993) Differential expression of immediate early genes in rubrospinal neurons following axotomy in rat. Eur J Neurosci 5: 203–209
Kapfhammer JP, Schwab ME (1994) Increased expression of the growth-associated protein GAP-43 in the myelin-free rat spinal cord. Eur J Neurosci 6: 403–411
Kimura H, Steinbusch HWM (1996) The role of nitric oxide in the central nervous system. From basic to therapeutic aspect. J Chem Neuroanat 10: 179–322
Koh JY, Gwag BJ, Lobner D, Choi DW (1995) Potentiated necrosis of cultured cortical neurons by neurotrophins. Science 268: 573–575
Koshinaga M, Sanon HR, Whitemore SR (1993) Altered acidic and basic fibroblast growth factor expression following spinal cord injury. Exp Neurol 120: 32–48
Li L, Oppenheim RW, Lei M, Houenou LJ (1994) Neurotrophic agents prevent motoneuron death following scieatic nerve sections in the neonatal mouse. J Neurobiol 25: 759–766
Lindholm D (1994) Role of neurotrophins in preventing glutamate induced neuronal cell death. J Neurol 242: S16–S18
Lindholm D, Castrén E, Kiefer R, Zafra F, Thoenen H (1992) Transforming growth factor-b1 in the rat brain: increase after injury and inhibition of astrocyte proliferation. J Cell Biol 117: 395–400
Lindsay RM (1994) Neurotrophic growth factors and neurodegenerative diseases: therapeutic potential of the neurotrophins and ciliary neurotrophic factor. Neurobiol Aging 15: 249–251
Lindsay RM, Harmar AJ (1989) Nerve growth factor regulates expression of neuropeptide genes in adult sensory neurons. Nature 337: 362–364
Lindvall O, Kokaia Z, Bengzon J, Elmer E, Kokaia M (1994) Neurotrophins and brain insults. Trends Neurosci 17: 490–496
Lipton SA (1993) Molecular mechanisms of trauma-induced neuronal degeneration. Curr Opin Neurol Neurosurg 6: 588–596
Logan A, Frautschy SA, Gonzalez A-M, Sporn MB, Baird A (1992) Enhanced expression of transforming growth factor beta-1 in the rat brain after a localized cerebellar injury. Brain Res 587: 216–225
Maissonpierre PC, Belluscio L, Friedman B, Alderson RF, Weigand SJ, Fruth ME, Lindsay RM Yancopoulos GD (1990) NT-3, BDNF, and NGF in the developing rat nervous system: parallel as well as reciprocal patterns of expression. Neuron 5: 501–509
Mattson MP, Lovell MA, Furukawa K, Markesbery WR (1995) Neurotrophic factors attenuate glutamate-induced accumulation of peroxides, elevation of intracellular Ca^{2+} concentration, and neurotoxicity and increase antioxidants enzyme activities in hippocampal neurons. J Neurochem 65: 1740–1751
Mulderry PK (1994) Neuropeptide expression by newborn and adult rat sensory neurons in culture. Effects of nerve growth factor and other neurotrophic factor. Neuroscience 59: 673–688
Nakao N, Kokaia Z, Odin P, Lindvall O (1995) Protective effects of BDNF and NT-3 but not PGDF against hypoglycemic injury to cultured striatal neurons. Exp Neurol 13: 1–10
Nawa H, Bassho Y, Carnahan J, Nakanishi S, Mizuno K (1993) Regulation of neuropeptide expression in cultured cerebral cortical neurons by brain-derived neurotrophic factor. J Neurochem 60: 772–775
Nawa H, Peleymounter MA, Carnahan J (1994) Intraventricular administration of BDNF increases neuropeptide expression in newborn rat brain. J Neurosci 14: 3751–3765
Novikov L, Novikova L, Kellerth J-O (1995) Brain-derived neurotrophic factor promotes survival and blocks nitric oxide synthase expression in adult rat spinal motoneurons after ventral nerve root avulsion. Neurosci Lett 200: 45–48
Olsson Y, Sharma HS, Pettersson CÅV (1990) Effects of p-chlorophenylalanine on microvascular permeability changes in spinal cord trauma. An experimental study in the rat using ^{131}I-sodium and lanthanum tracers. Acta Neuropathol (Berl) 79: 595–603
Olsson Y, Sharma HS, Pettersson Å, Cervós-Navarro J (1992) Endogenous release of neurochemicals may increase Vvscular permeability, induce edema and influence on cell changes in trauma to the spinal cord. Prog Brain Res 91: 197–203

Olsson Y, Sharma HS, Nyberg F, Westman J (1995) The opioid receptor antagonist naloxone influences the pathophysiology of spinal cord injury. In: Nyberg F, Sharma HS, Wissenfeld-Hallin Z (eds) Progress in Brain Research 104. Elsevier, Amsterdam, pp 381–399
Pattersson PH, Nawa H (1993) Neuronal differentiation factors/cytokines and synaptic plasticity. Cell 72, Neuron 10 [Suppl] 123–137
Peunova N, Enikolopov G (1995) Nitric oxide triggers a switch to growth arrest during differentiation of neuronal cells. Nature 375: 68–73
Reynolds ME, Brunello N, Mocchetti I, Wrathall JR (1991) Localization of nerve growth factor receptor mRNA in contused rat spinal cord by in situ hybridisation. Brain Res 559: 149–153
Risling M, Fried K, Linda H, Carlstedt T, Cullheim S (1993) Regrowth of motor axons following spinal cord lesions: distribution of laminin and collagen in CNS scar tissue. Brain Res Bull 30: 405–411
Samdani AF, Newcamp C, Resnik A, Facchinetti F, Hoffman BE, Dawson VL, Dawson TM (1997) Differential susceptibility to neurotoxicity mediated by neurotrophins and neuronal nitric oxide synthase. J Neurosci 17: 4633–4641
Savio T, Schwab ME (1990) Lesioned corticospinal tract axons regenerate in myelin-free rat spinal cord. Proc Natl Acad Sci USA 87: 4130–4133
Schecterson LC, Bothwell M (1992) Novel roles for neurotrophins are suggested by BDNF and NT-3 mRNA expression in developing neurons. Neuron 9: 449–463
Schnell L, Schneider R, Kolbeck R, Barde Y-A, Schwab ME (1994) Neurotrophin-3 enhances sprouting of corticospinal tract during development and after adult spinal cord lesion. Nature 367: 170–173
Schnell L, Schwab ME (1990) Axonal regeneration in the rat spinal cord produced by an antibody against myelin-associated neurite growth inhibitors. Nature 343: 269–272
Schnell L, Schwab ME (1993) Sprouting and regeneration of lesioned corticospinal tract fibers in the adult rat spinal cord. Eur J Neurosci 5: 1156–1171
Schreyer DJ, Skene JHP (1991) Fate of GAP-43 in ascending spinal axons of DRG neurons after peripheral nerve injury: delayed accumulation and correlation with regenerative potential. J Neurosci 11: 3738–3751
Schwab ME, Bartholdi D (1966) Degeneration and regeneration of axons in the lesioned spinal cord. Physiol Rev 76: 319–370
Schuman EM, Madison DV (1994) Nitric oxide and synaptic function. Ann Rev Neurosci 17: 153–183
Sharma HS (1998) Neurobiology of the nitric oxide in the nervous system. Basic and clinical perspectives. Editorial. Amino Acids 14: 83–85
Sharma HS, Cervós-Navarro J (1990) Nimodipine improves cerebral blood flow and reduces brain edema, cellular damage and blood-brain barrier permeability following heat stress in young rats. In: Krieglstein J, Oberpichler H (eds) Pharmacology of Cerebral Ischemia 1990. CRC Press, Boca Raton, pp 303–310
Sharma HS, Olsson Y (1990) Edema formation and cellular alteration in spinal cord injury in the rat and their modification with p-chlorophenylalanine. Acta Neuropathol (Berl) 79: 604–610
Sharma HS, Westman J (1997) Prostaglandins modulate constitutive isoform of heat shock protein (72 kD) response following trauma to the rat spinal cord. Acta Neurochir (Wien) [Suppl] 70: 134–137
Sharma HS, Olsson Y, Dey PK (1990a) Early accumulation of serotonin in rat spinal cord subjected to traumatic injury. Relation to edema and blood flow changes. Neuroscience 36: 725–730
Sharma HS, Westman J, Olsson Y, Johansson O, Dey PK (1990b) Increased 5-HT immunoreactivity in traumatized spinal cord. An experimental study in the rat. Acta Neuropathol (Berl) 80: 12–17
Sharma HS, Nyberg F, Olsson Y, Dey PK (1990c) Alteration in substance P in brain and spinal cord following spinal cord injury. An experimental study in the rat. Neuroscience 38: 205–212

Sharma HS, Winkler T, Stålberg E, Olsson Y, Dey PK (1991) Evaluation of traumatic spinal cord edema using evoked potentials recorded from the spinal epidural space. An experimental study in the rat. J Neurol Sci 102: 150–162

Sharma HS, Nyberg F, Olsson Y (1992a) Dynorphin A content in the rat brain and spinal cord after a localized trauma to the spinal cord and its modification with p-chlorophenylalanine. An experimental study using radioimmunoassay technique. Neuroscience Res 14: 195–203

Sharma HS, Westman J, Olsson Y (1992b) Increased heat shock protein (HSP-70 kD) immunoreactivity following acute spinal cord trauma in the rat. Clin Neuropathol 11 (4) p 174

Sharma HS, Olsson Y, Cervós-Navarro J (1993a) Early perifocal cell changes and edema in traumatic injury of the spinal cord are reduced by indomethacin, an inhibitor of prostaglandin synthesis. Acta Neuropathol (Berl) 85: 145–153

Sharma HS, Olsson Y, Nyberg F, Dey PK (1993b) Prostaglandins modulate alterations of microvascular permeability, blood flow, edema and serotonin levels following spinal cord injury. An experimental study in the rat. Neuroscience 57: 443–449

Sharma HS, Olsson Y, Cervós-Navarro J (1993c) p-Chlorophenylalanine, a serotonin synthesis inhibitor, reduces the response of glial fibrillary acidic protein induced by trauma to the spinal cord. Acta Neuropathol (Berl) 86: 422–427

Sharma HS, Nyberg F, Thörnwall M, Olsson Y (1993d) Met-Enkephalin-Arg^6-Phe^7 in spinal cord and brain following traumatic injury of the spinal cord: Influence of p-chlorophenylalanine. An experimental study in the rat using radioimmunoassay technique. Neuropharmacology 32: 711–717

Sharma HS, Olsson Y, Nyberg F (1995a) Influence of dynorphin-A antibodies on the formation of edema and cell changes in spinal cord trauma. Prog Brain Res 104: 401–416

Sharma HS, Olsson Y, Pearsson S, Nyberg F (1995b) Trauma induced opening of the blood-spinal cord barrier is reduced by indomethacin, an inhibitor of prostaglandin synthesis. Experimental observations in the rat using ^{131}I-sodium, Evans blue and lanthanum as tracers. Restor Neurol Neurosci 7: 207–215

Sharma HS, Olsson Y, Westman J (1995c) A serotonin synthesis inhibitor, p-chlorophenylalanine reduces the heat shock protein response following trauma to the spinal cord. An immunohistochemical and ultrastructural study in the rat. Neurosci Res 21: 241–249

Sharma HS, Fonnum F, Olsson Y, Gordh T, Westman J (1995d) p-Chlorophenylalanine, a serotonin synthesis inhibitor reduces the alterations of GABA and glutamate levels following a focal spinal cord injury. An experimental study in the rat using immunohistochemical and HPLC technique. Amino Acids 9: 38

Sharma HS, Alm P, Gordh T, Nyberg F, Olsson Y, Westman J (1995e) Involvement of nitric oxide in the pathophysiology of spinal cord injury in the rat. J Neurotrauma 12, p 370

Sharma HS, Westman J, Olsson Y, Alm P (1996a) Involvement of nitric oxide in acute spinal cord injury: an immunohistochemical study using light and electron microscopy in the rat. Neurosci Res 24: 373–384

Sharma HS, Westman J, Olsson Y (1996b) Spinal cord injury induced *c-fos* expression is reduced by p-CPA, a serotonin synthesis inhibitor. Neuropathol Appl Neurobiol 22: 15–16

Sharma HS, Lindholm D, Alm P, Gordh T, Olsson Y, Westman J (1996c) Topical application of brain-derived neurotrophic factor reduces upregulation of neuronal nitric oxide synthase following focal trauma to the rat spinal cord. XVI Swiss International Winter Meeting, St. Moritz, pp 16–17

Sharma HS, Nyberg F, Gordh T, Alm P, Westman J (1997a) Topical application of insulin like growth factor-1 reduces edema and upregulation of neuronal nitric oxide synthase following trauma to the rat spinal cord. Acta Neurochir (Wien) [Suppl] 70: 130–133

Sharma HS, Westman J, Nyberg F (1997b) Topical application of 5-HT antibodies reduces edema and cell changes following trauma to the rat spinal cord. Acta Neurochir (Wien) [Suppl] 70: 155–158

Sharma HS, Westman J, Alm P, Sjöquist P-O, Cervós-Navarro J, Nyberg F (1997c) Involvement of nitric oxide in the pathophysiology of acute heat stress in the rat. Influence of a new antioxidant compound H-290/51. Ann NY Acad Sci 813: 581–590

Sharma HS, Alm P, Westman J (1997d) Upregulation of heme oxygenase-II in the rat spinal cord following heat stress. In: Nielsen Johannsen B and Nielsen R (eds) Thermal Physiology. August Krogh Institute, Copenhagen, pp 135–138

Sharma HS, Westman J, Cervós-Navarro J, Dey PK, Nyberg F (1997e) Opioid receptor antagonists attenuate heat stress-induced reduction in cerebral blood flow, increased blood-brain barrier permeability, vasogenic edema and cell changes in the rat. Ann NY Acad Sci 813: 559–571

Sharma HS, Alm P, Westman (1998a) Nitric oxide and carbon monoxide in the pathophysiology of brain functions in heat stress. In: Sharma HS, Westman J (eds) Brain Function in Hot Environment. Prog Brain Res. Elsevier, Amsterdam 115: 297–336

Sharma HS, Nyberg F, Westman J, Alm P, Gordh T, Lindholm D (1998b) Brain derived neurotrophic factor and insulin like growth factor-1 attenuate upregulation of nitric oxide synthase and cell injury following trauma to the spinal cord. An immunohistochemical study in the rat. Amino Acids 14: 121–129

Sharma HS, Westman J, Nyberg F (1998c) Pathophysiology of brain edema and cell changes following hyperthermic brain injury. In: Sharma HS, Westman J (eds) Brain Function in Hot Environment. Prog Brain Res. Elsevier, Amsterdam 115: 351–412

Timmusk TN, Belluardo N, Metsis M, Persson H (1993) Widespread and developmentally regulated expression of neurotrophin-4 mRNA in rat brain and peripheral tissue. Eur J Neurosci 5: 605–613

Thoenen H (1995) Neurotrophins and neuronal plasticity. Science 270: 593–598

Tsitouras PD, Zhong YG, Spungen AM, Bauman WA (1995) Serum testosterone and growth hormone/insulin-like growth factor-I in adults with spinal cord injury. Horm Metab Res 27: 287–292

Yan Q, Elliott J, Snider WD (1992) Brain-derived neurotrophic factor rescues spinal cord motor neurons from axotomy induced cell death. Nature 360: 753–755

Yamada K, Emson P, Hökfelt T (1996) Immunohistochemical mapping of nitric oxide synthase in the rat hypothalamus and colocalization with neuropeptides. J Chem Neuroanat 10: 295–316

Yao DL, Liu X, Hudson LD, Webster HD (1995) Insulin-like growth factor I treatment reduces demyealination and up-regulates gene expression of myelin-related proteins in experimental autoimmune encephalomyelitis. Proc Natl Acad Sci USA 92: 6190–6194

Yin QW, Johnson J, Prevette D, Oppenheim RW (1994) Cell death of spinal motoneurons in the chick embryo following deafferentation: rescue effects of tissue extracts, soluble proteins, and neurotrophic agents. J Neurosci 14: 7629–7640

Yu W-HA (1994) Nitric oxide synthase in motor neurons after axotomy. J Histochem Cytochem 42: 451–457

Yun H-Y, Dawson VL, Dawson TM (1996) Neurobiology of nitric oxide. Crit Rev Neurobiol 10: 291–316

Weise B, Janet T, Grothe C (1993) Localization of bFGF and FGF-receptor in the developing nervous system of the embryonic and newborn rat. J Neurosci Res 34: 442–453

Winkler T, Sharma HS, Stålberg E, Westman J (1998) Spinal cord bioelectrical activity, edema and cell injury following a focal trauma to the rat spinal cord. An experimental study using pharmacological and morphological approach. In: Stålberg E, Sharma HS, Olsson Y (eds) Spinal Cord Monitoring. Basic principles, regeneration and pathophysiological aspects. Springer, Wien New York, pp 283–363

Correspondence: Dr. H. S. Sharma, Laboratory of Neuroanatomy, Department of Medical Cell Biology, Biomedical Centre, Post Box 571, Uppsala University, S-751 23 Uppsala, Sweden.

Section III

Spinal cord electrophysiology: methods and various techniques

Spinal Cord Monitoring
Eds: E. Stålberg, H. S. Sharma, Y. Olsson

Chapter 9

A cerebellar evoked potential for spinal cord monitoring

R. J. Hurlbert

The University of Calgary Spine Program, Division of Neurosurgery and Department of Clinical Neurosciences Foothills Hospital, Calgary, Alberta, Canada

Introduction

Evoked potential monitoring has become widely used throughout modern clinical practice as a means to evaluate the anatomical and functional integrity of the spinal cord. The recording of these responses is very important to the physician in circumstances where the patient either cannot or is unwilling to participate in the "gold standard" neurological exam. Most commonly this involves individuals undergoing operative procedures that place the spinal cord at risk of injury. Although clear evidence is lacking, it is generally believed that the detection of electrophysiological changes will alert the surgeon and allow timely intervention, thus preserving or perhaps even improving neurological function. However, despite a large background of basic science and clinical experience, the role of evoked potentials in defining nervous system function remains poorly established.

Not always is a change in evoked potentials reflected by a change in neurological status. *False positive* changes occur when significant alterations of amplitude or latency are encountered without evidence of change in clinical function. It is generally believed that there is a close correlation of evoked potential properties to the underlying anatomical substrates, namely populations of neurons and axons. However in contrast, clinical neurological function (being much more complex) is not related to anatomical integrity in a simple linear fashion. In the spinal cord, large populations of axons can be injured or killed before serious neurological sequelae result. Hence, it is possible for a relatively large number of neurons or axons to be lost resulting in significant amplitude and latency changes in evoked potentials, before neurological deficits become clinically apparent. True false positive changes must be distinguished from spurious shifts in amplitude and latency caused by alterations in anaesthesia, body temperature, pCO_2, electrode impedance, etc.

Perhaps more importantly, not always is a change in neurological status reflected by a change in evoked potentials. *False negative* monitoring is encountered when evoked potentials remain stable or unchanged only to have the patient awake with a new and sometimes devastating neurological deficit. Although the reasons for this phenomenon have not been clearly established, it is most likely due to selective injury in regions of the spinal cord distinct from the pathways responsible for conduction of routine evoked potentials. In the case of SSEP recordings, transmitted primarily through the dorsal columns, injury to more anterior motor pathways can theoretically result in a serious motor deficit (such as lower limb paralysis) without change in the SSEPs. Reports of false negatives are infrequent but the clinical implications for a patient are very serious (Ginsburg *et al.*, 1985; Lesser *et al.*, 1986; Levy, 1988; Takaki and Okumura, 1985; Zornow and Drummond, 1989).

It is clearly in the best interests of the surgeon and patient to develop techniques that can reliably indicate change and predict injury under all circumstances. A variety of methods have been described to overcome some of these limitations (see Chapters 10, 11, 13, 14 and 22). Alternatively, since sensory pathways are located throughout all compartments of the spinal cord, recording electrical activity from regions of the brain other than the somatosensory cortex might reflect sensory pathways in regions of the spinal cord other than the dorsal columns. This chapter explores the characteristics of signal-averaged sensory activity recorded in proximity to the cerebellum from rodents, non-human primates, and a series of neurosurgical patients to determine the contribution of this *cerebellar evoked potential* to monitoring of the spinal cord.

A novel sensory response recorded from the rat – a cerebellar evoked potential

In 1988, we began a series of experiments in rats to better assess non-pyramidal motor pathways as originally described by Levy *et al.*, 1986. Through epidural electrical stimulation over the rat cerebellum, we demonstrated non-selective activation of ascending and descending motor pathways whose progressive recruitment was determined by current intensity. An increase in stimulation intensity altered the anatomical substrate of evoked potential conduction by recruiting pathways more distal to the stimulating electrodes (Hurlbert *et al.*, 1992). It was clear to us that at least in the rat, these evoked responses would be difficult to interpret due to their sensitivity to amplitude of stimulation. Subsequently, we undertook to record pure ascending activity from the rodent cerebellum hypothesizing that its rich sensory supply would provide an index of spinal cord integrity in regions other than the dorsal columns (Hurlbert *et al.*, 1993).

Methods

All animals were anaesthetized with intraperitoneal injections of alpha-chloralose and urethane. Following tracheostomy, arterial and venous femoral lines were inserted. Muscle relaxation was achieved with pancuronium bromide and ventilation assisted mechanically. Body temperature and blood oxygenation were maintained within physiological limits.

Evoked potentials were elicited by direct sciatic stimulation using non-cutting paired needle electrodes (cathode and anode) inserted into the sciatic nerve. The cathode was located 1 mm distal to the anode; both needles were insulated with cork after skewering the nerve. Averaged responses were recorded from platinum disc electrodes placed in the epidural space directly over the cerebellar hemispheres and the sensory-motor cortex. All recording electrodes were referenced to a palatal electrode (Fig. 1). Square-wave pulses of 50 μsec in duration were

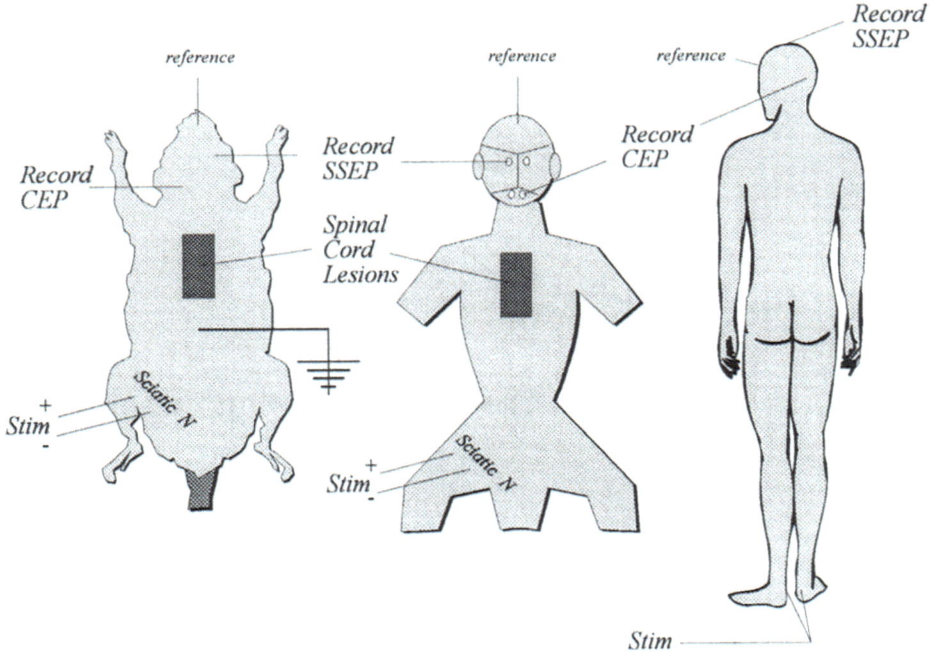

Fig. 1. Schematic representations of evoked potential recording techniques in rats, monkeys, and in man. In rats and monkeys the sciatic nerve was stimulated directly whereas in humans the posterior tibial nerves were stimulated by percutaneous needle electrodes inserted at the ankle. Recordings were made from electrodes placed in the epidural space over the somatosensory cortex and cerebellar hemispheres in animal experiments. In human patients, recordings were made from needle electrodes inserted into the scalp at Cz (*SSEP*) and 2–3 cm. lateral to the inion (*CEP*). Selective spinal cord lesions were created in the lower cervical region following laminectomies in rats and monkeys

delivered to the sciatic nerve at 5 mA through a constant current stimulator. Stimuli were produced at 8.1 Hz. One thousand and twenty-four responses were averaged over a 90 msec sweep, stored, and reproduced at least once (Model 8400, Cadwell Laboratories, Inc., Kennewick, Washington). Band-pass filters were set for 30 and 3000 Hz.

Laminectomies were completed from C5-T1 in eight animals to allow selective spinal cord lesions to be performed using microsurgical techniques. The extent of all lesions was verified histologically. In an additional 5 animals, craniectomies were performed exposing the rostral cerebellum so that ablation of the cerebellar peduncles might be carried out.

Results

Evoked responses were easy to record and reliably reproduced. Activity simultaneously recorded from over the cerebellum and somatosensory cortex was distinctly different. The cerebellar evoked potential (CEP) consisted of an early latency negative peak followed by a second, less well defined negative component (Fig. 2). In contrast, the SSEP consisted of an initial positive wave followed by a large negative deflection. Individual components were labeled according to their mean latencies.

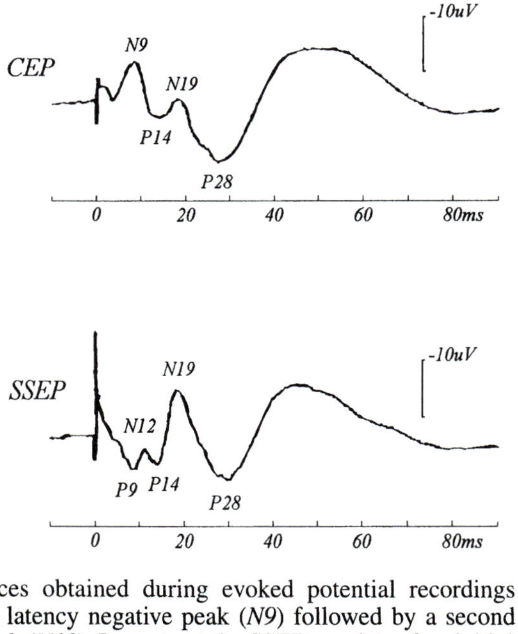

Fig. 2. Typical traces obtained during evoked potential recordings in rats. The CEP consists of an early latency negative peak (*N9*) followed by a second later latency, more variable negative peak (*N19*). In contrast the SSEP consists of an initial positive component (*P9*). The P14–N19 complex of the SSEP is the largest in amplitude. It corresponds roughly in latency to the N19 component of the CEP

Positioning of the reference electrode on the hindlimb opposite to the side of stimulation had no effect on overall waveform morphology, indicating the electrical activity to originate from the recording rather than the reference electrodes. Selective spinal cord lesions through the dorsal columns resulted in little effect on the CEP. In contrast the amplitude of the SSEP was significantly attenuated. Further sectioning of the dorsolateral funiculi essentially wiped out the SSEP with little or no effect on the CEP. Selective sectioning of the ventral spinal cord obliterated the CEP while the SSEP remained unchanged or increased in amplitude (Fig. 3).

Those animals subjected to lesions through the cerebellar peduncles demonstrated incomplete loss of the CEP recorded from the same side as the lesion. The CEP recorded from the opposite cerebellar hemisphere remained unchanged. When the peduncles on both sides were cut, the CEP was abolished, confirming the cerebellum to be the source of the evoked responses. SSEPs were maintained throughout, confirming their transmission through the brainstem and independence from the CEP.

Conclusions

From this experiment, it is clear that sciatic nerve stimulation in the rat produces an evoked response recorded from over the cerebellum, that ascends through the spinal cord in the ventral funiculus. The CEP is earlier in latency than the SSEP and shows less variation in amplitude and latency. The origin of the CEP is within the cerebellum as evidenced by progressive sectioning of the cerebellar peduncles. Based on these results it is likely that the CEP is transmitted through spino-olivo-cerebellar or spino-reticulo-cerebellar pathways. Although ventral spino-cerebellar afferents might also contribute to the CEP, selective spinal cord lesions demonstrate that the dorsal spino-cerebellar tract does not play a major role.

The primate cerebellar evoked potential – a different anatomical substrate?

Proposing a CEP in humans for routine monitoring in the operating room demands a large leap of faith from the experiments performed in rats. Besides the obvious size difference between rodents and humans, significant differences are also known to exist in the structural organization of the spinal cord (Brodal, 1981; Paxinos and Watson, 1982). Therefore, it seemed important to evaluate the CEP in a larger animal model more closely related to humans. In 1996 I had the opportunity to examine the CEP in a number of cynomologous monkeys being terminated as part of an unrelated experiment. Using mainly the same techniques as were successful in the rat, the purpose of this experiment was to confirm or disprove the presence of a cerebellar evoked response in primates, and to establish the region of the spinal cord through which it was transmitted.

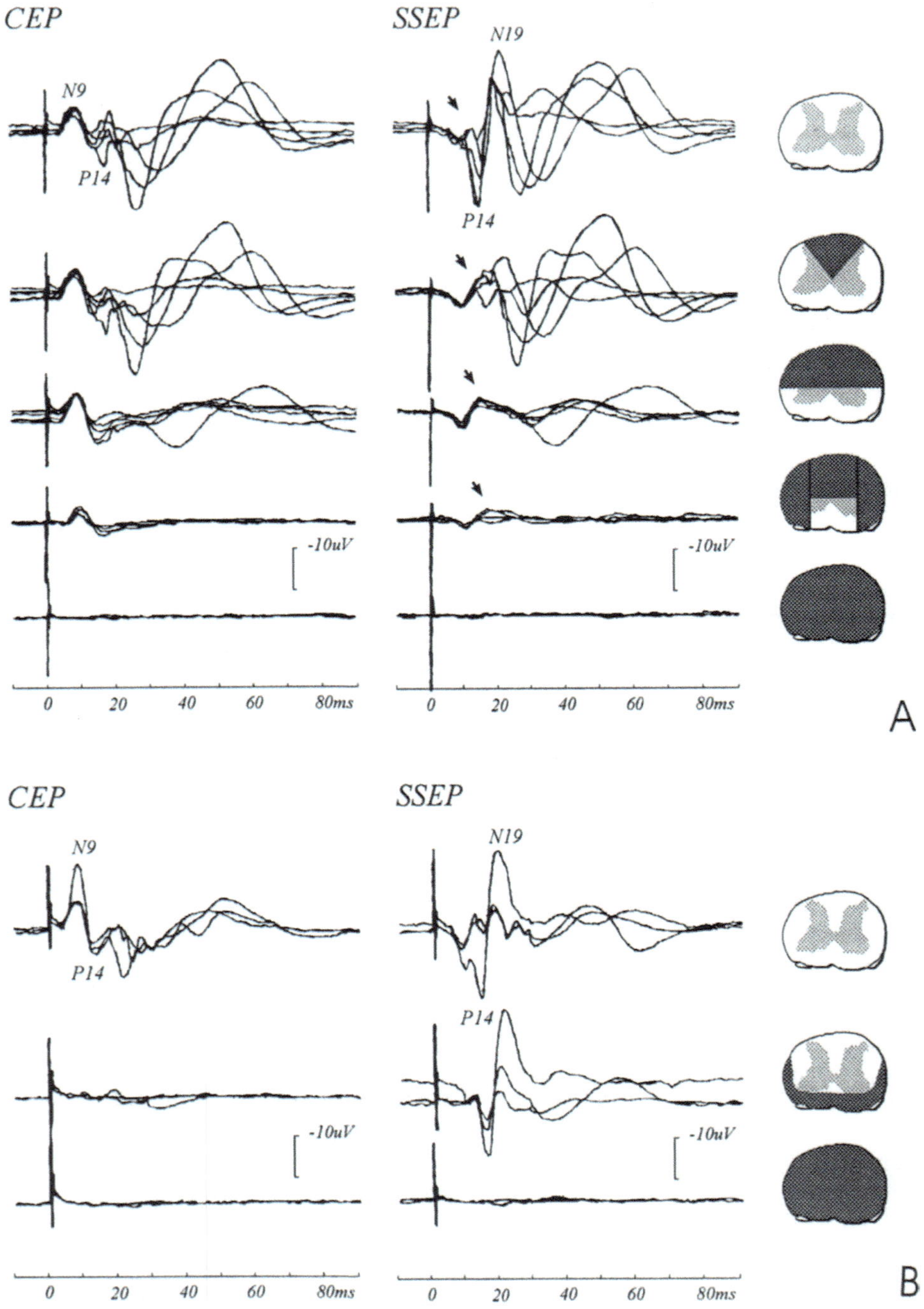
CEP
SSEP
N19
N9
P14
P14
-10uV
-10uV
0
20
40
60
80ms
0
20
40
60
80ms
A
CEP
SSEP
N9
N19
P14
P14
-10uV
-10uV
0
20
40
60
80ms
0
20
40
60
80ms
B

Methods

Nine adult male cynomologous monkeys were induced with intramuscular Ketamine and maintained with a general anaesthetic consisting of inhalational nitrous oxide and continuous Propofol infusion. Body temperature was maintained between 36.5 and 38.0 degrees Celsius by means of a warming blanket. Blood pressure and heart rate were monitored continuously. Arterial blood gases were sampled hourly and maintained within physiological ranges by alterations in mechanical ventilation.

Evoked potentials were elicited by constant current stimulation delivered directly to the surgically exposed sciatic nerves (Fig. 1). Current was determined by doubling the voltage required to elicit maximal muscle contraction prior to administration of an intravenous muscle relaxant. Pulse width was set to 250 μsec and stimuli delivered at 4.1 Hz. Optically coupled stimulus isolation units provided protection from grounding contamination. Evoked potentials were recorded with a PC driven MS-DOS based 32-channel signal averager (SynAmps, Neuro Scan Inc., Herndon, VA). One thousand responses were averaged for each epoch and then reproduced at least once. All results were stored to disc for subsequent off-line analyses. Recording electrodes were placed in an epidural location over the somatosensory cortex and cerebellum on both sides of midline.

Laminectomies at C7 and T1 were completed through a dorsal midline incision in the neck ($n = 6$). Using the operating microscope and microsurgical techniques, the dura was opened and selective spinal cord lesions performed sequentially on the exposed spinal cord in a rostral to caudal direction. In a further three animals posterior fossa craniectomies were followed by cerebellar or brainstem resection to determine relative contributions to the evoked potentials.

Results

Unilateral sciatic stimulation produced a bilateral response in the monkeys. Activity was larger in amplitude when recorded from epidural electrodes located over the contralateral somatosensory cortex, and when recorded from similar electrodes over the ipsilateral cerebellar hemisphere. Posterior

◄

Fig. 3. Summary of spinal cord lesion experiments performed in rats. (**A**) Superimposed recordings from 5 animals undergoing progressive dorsal spinal lesions from C5-T1 (as indicated by schematics on the right). Pre-lesion responses are shown at the top, prior to spinal cord injury. Dorsal column sectioning and dorsal hemisection have little effect on the CEP but essentially wipe out the P14–N19 component of the SSEP. Arrows indicate how the P9–N12 component of the SSEP is unmasked by these dorsal lesions. With bilateral lateral hemisection amplitude of the CEP is diminished. Complete cord transection obliterates all activity. (**B**) Superimposed recordings from 3 animals undergoing ventral spinal cord lesions. Pre-lesion responses show normal activity. Ventral sectioning erases the CEP in all three animals while SSEP amplitude is unchanged or increased. Complete cord transection produces electrical silence

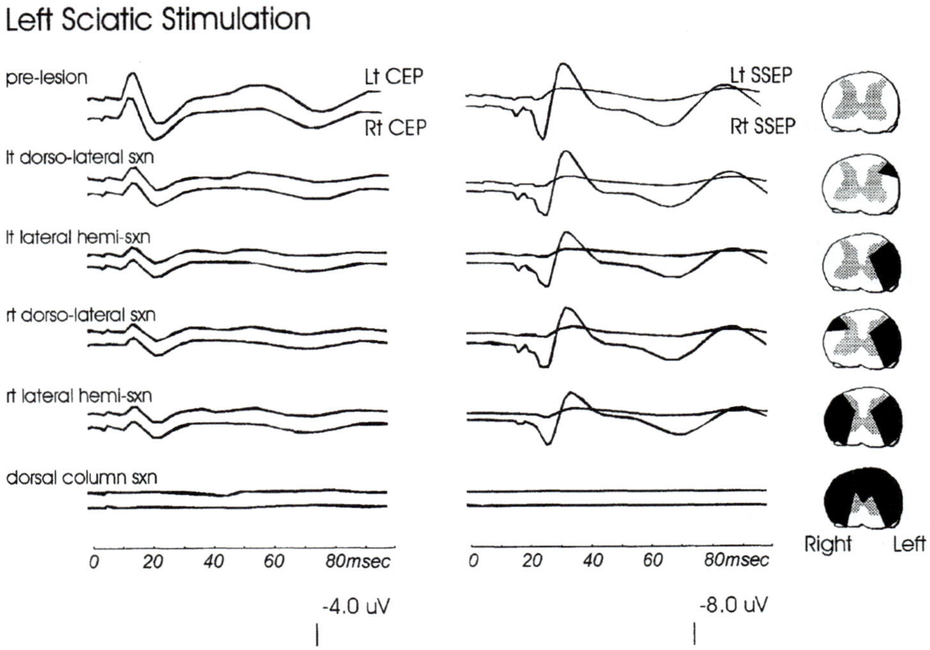

Fig. 4. Example of dorsal spinal cord lesions in a monkey. Pre-lesion (normal) responses from left sciatic nerve stimulation are shown at the top. Following left dorso-lateral sectioning (in the region of the dorsal spino-cerebellar tract) a significant amplitude decrement is observed in both the left and right CEP, whereas the SSEP amplitude is affected only slightly. Left lateral hemisection causes further attenuation of the CEP with a marginal effect on the SSEP. Mirror image lesions produced on the right side have little effect on either response, suggesting the spinal cord ipsilateral to the side of stimulation most important in carrying the CEP. Dorsal column sectioning has a pronounced effect on SSEP amplitude, also eradicating the remaining CEP

tibial nerve stimulation produced evoked responses similar in morphology but later in latency.

Selective spinal cord lesions were studied in six animals. Dorsal column sectioning produced significant attenuation in the somatosensory responses in each of three animals. Although the cerebellar evoked potentials were also diminished, the effect of dorsal column division was manifest much more as an amplitude decrease in SSEPs. In three separate animals, small incisions made through the dorso-lateral funiculi (in the region of the dorsal spino-cerebellar pathway) caused a large amplitude decrement in the CEP with much less effect on simultaneously recorded SSEPs (Fig. 4). Further attenuation of the CEP was produced by lateral hemisection, whereas SSEPs remained largely unaffected. When the dorsal columns were transected in these animals, the SSEPs were completely extinguished as were the residual small amplitude CEPs.

In two monkeys, microsurgical removal of the cerebellar hemispheres resulted in severe attenuation of the CEP. Although the SSEP also diminished in amplitude, the overall effect was much less. Conversely, in the third primate, destruction of the dorsal column nuclei caused a profound reduction in SSEP amplitude while the CEP exhibited only a small amplitude decrement.

Conclusions

This experiment demonstrates that a potential similar in morphology to that recorded in the rat can be recorded from over the primate cerebellum as a result of lower limb stimulation. However, in contrast to rodents, the results from selective spinal cord ablation in primates indicate the CEP to be transmitted through the spinal cord in a more dorsal and lateral location, possibly corresponding to the dorsal and ventral spino-cerebellar pathways. In addition it is clear that a minor yet significant proportion of the CEP is also transmitted through the dorsal columns. Similarly, a proportion of the SSEP is transmitted through the lateral and more ventral funiculi. Indeed, based on other animal studies, it has been suggested that the SSEP in humans ascends in part through the dorsal spino-cerebellar pathways when elicited by lower limb stimulation (Jones, 1982). Therefore, although the primate SSEP and CEP share common pathways to a certain degree, the two responses are transmitted through largely independent areas of the spinal cord: the CEP through the dorso-lateral funiculus and the SSEP through the dorsal columns. This independence is confirmed through the experiments involving resection of the cerebellar hemispheres or dorsal column nuclei.

Clinical experience with intraoperative monitoring of the CEP in humans

Recording of the CEP from rodents and primates obviously begs the question of whether or not such an evoked response can be recorded from humans. Although the primate studies described above suggest that should such activity be possible to record from patients, it would likely be transmitted posteriorly rather than through the ventral or anterior portion of the spinal cord. Nonetheless, independence of any type from the classical SSEP, even if through the dorso-lateral funiculi, would add an extra dimension to the assessment of overall cord integrity during intra-operative monitoring. Therefore the pursuit of recording a CEP in humans remains a worthwhile one. We have recently reported our initial experience with the CEP in humans undergoing a variety of neurosurgical procedures (Hurlbert *et al.*, 1995; Hurlbert *et al.*, 1995).

Methods

Intra-operative evoked potentials were recorded from twenty-five patients undergoing spinal or skull base surgery in a six month period at one institution. The median age of the patients was 48 years with a range of 24–78 years. Seventeen were men and eight were women. All patients were undergoing a spinal or skull base procedure and all were subjected to a general anaesthetic consisting of nitrous oxide and enflurane. In contrast to direct sciatic nerve stimulation performed on rodents and non-human primates, our patients underwent percutaneous needle stimulation of the right or left posterior tibial nerve. Similarly, responses were recorded percutaneously from needle electrodes inserted over the vertex of the skull (SSEP) and 2–3 cm lateral to the inion along the superior nuchal line (CEP). All recordings were reference to the forehead (Fpz). A grounding electrode was inserted into the earlobe or at T1. Bandpass filters were set for 30–1000 Hz. Between 800 and 1000 stimuli were delivered over 250 μsec at 4.72 Hz, digitized, averaged, and replicated at least once for each record (Cadwell 8400 or Spectrum 32 signal averagers; Cadwell Laboratories, Kennewick, WA).

Results

Activity similar to that recorded in both rodents and non-human primates could be recorded from the scalp overlying the cerebellum in all but one of the 25 patients (Fig. 5). In one instance of a patient with severe myelopathy secondary to a herniated thoracic disc, no reproducible responses could be elicited. Similar to activity recorded in animals, the human CEP consists of an initial negative peak of earlier latency than the SSEP. Through regression analyses we determined that the latency of the N33 peak of this evoked potential could be predicted based on the latency of the SSEP as described by the formula:

$$N33^{CEP} = (0.573 \times N45^{SSEP}) + 2.83$$

In five patients undergoing posterior fossa decompression, similar activity but larger in amplitude was recorded from electrodes placed directly on the surface of the cerebellum. Although selective spinal cord lesion studies could not be performed, we were able to demonstrate amplitude independence from the SSEP in isolated cases (Fig. 6) and statistically through all patients. The appearance of the human CEP was similar to the appearance of the C2 response recorded from over the second cervical vertebrae, but of slightly later latency.

Conclusions

Activity similar to the CEP recorded in rodents and non-human primates can be recorded with scalp and cerebellar surface electrodes in human patients under a general anaesthetic. These responses demonstrate independence in amplitude from the SSEP and can even be recorded in

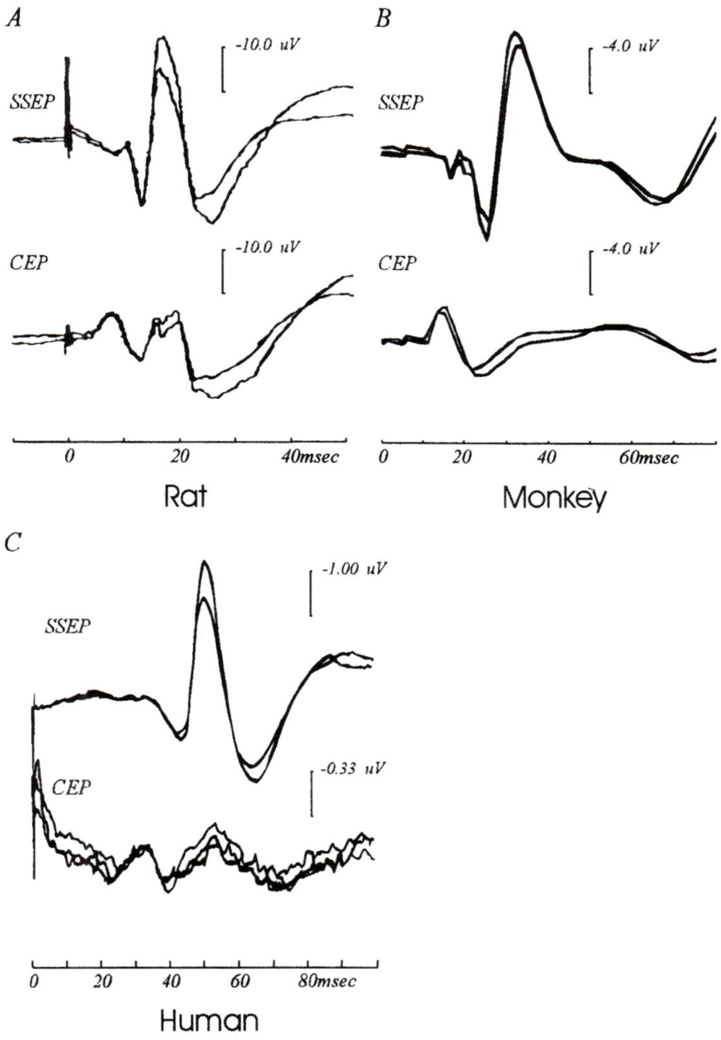

Fig. 5. Comparison of the morphology of somatosensory and cerebellar evoked potentials. Note the similarities between each of the study groups. SSEP activity is typified by a large positive-negative complex, which has been shown to be transmitted primarily through the dorsal columns in many animal species, including those presented here. Activity recorded from the vicinity of the cerebellar hemispheres (*CEP*) is also similar across the three study groups, consisting of an initial negative component with a later more variable negative wave similar in latency to the large negative component of the SSEP. However, although the CEP is conducted ventrally in the rat, the response is transmitted dorsally in primates. It appears that the dorsal spino-cerebellar pathway is most important in conducting the primate (and possibly the human) CEP. (**A**) Rat, (**B**) Primate, (**C**) Human

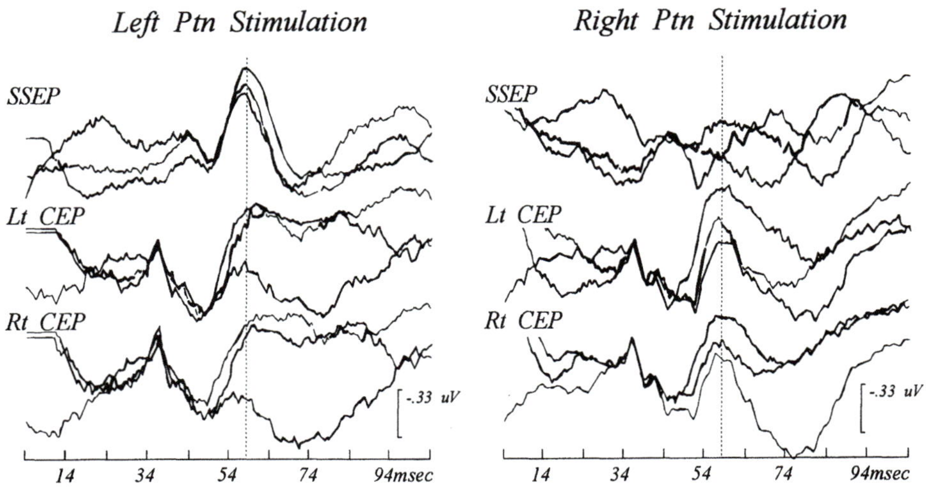

Fig. 6. Evidence of SSEP and CEP independence recording in the operating room. Left lower limb stimulation produced normal responses recorded from both the somatosensory cortex and from over the cerebellum. In the same patient, right lower limb stimulation produced CEPs with similar morphology and latency compared to those produced from left posterior tibial nerve (*ptn*) stimulation. However, despite recording from the same electrode at Cz, a somatosensory response could not be obtained. The dotted line indicates the observed latency of the SSEP from left *ptn* stimulation and the expected latency of the SSEP from right *ptn* stimulation

the absence of any recognizable somatosensory activity. Alternatively the SSEP can be recorded without demonstrable activity from over the cerebellar hemispheres. These results suggest that the equivalent of the rodent or primate CEP exists in humans. Similarities between the CEP and C2 spinal response are notable. It is possible that the C2 spinal response arises as a composite form of impulses transmitted through the dorsal columns and spino-cerebellar pathways. Further studies are necessary to more clearly elucidate the pathways involved in the conduction of the human CEP.

Summary

Through animal studies we have been able to demonstrate that a discrete evoked potential exists, separate from the SSEP, that can be recorded from over the cerebellar hemispheres. When elicited by sciatic nerve stimulation in rats, this evoked response travels in the anterior spinal cord. When elicited by sciatic nerve stimulation in monkeys, this evoked response travels primarily in the dorso-lateral spinal cord, although part of its amplitude is mediated through the dorsal columns. Selective brainstem or cerebellar ablation studies confirm that this activity arises largely from within the cerebellum. Studies in humans suggest that a similar evoked potential can be recorded from patients undergoing a general anaesthetic. Indeed, preliminary results confirm independent amplitude characteristics of the CEP compared to the SSEP.

It is likely that the human CEP, similar to the primate CEP, is conducted through the postero-lateral regions of the spinal cord. Therefore the more anterior regions of the spinal cord, those responsible for posture and locomotion, cannot be directly assessed through this type of monitoring. Motor evoked potentials may still provide the best means to assess electrophysiological integrity within the anterior spinal cord. Nonetheless, based on our animal and human studies, the addition of one or two extra recording electrodes in the region of the posterior fossa when recording lower limb SSEPs is likely to provide the clinician with novel information about the integrity of the dorso-lateral funiculi. When combined with the specificity of the SSEP in reflecting integrity of the dorsal columns, simultaneous recording of both evoked potentials is clearly advantageous in obtaining a more detailed picture of overall spinal cord function. We believe that recording of the CEP should be part of the routine efforts of those performing intra-operative evoked potential monitoring.

References

Brodal A (1981) Neurological anatomy in relation to clinical medicine, 3rd Edn. Oxford Press, New York

Ginsburg HH, Shetter AG, Raudzens P (1985) Post-operative paraplegia with preserved intraoperative somatosensory evoked potentials. A case report. J Neurosurg 63: 296–300

Hurlbert RJ, Fehlings MG, Moncada MS (1995) Cerebellar-evoked potentials to assess integrity of the ventral tracts of the spinal cord. Perspect Neurol Surg 6: 107–117

Hurlbert RJ, Fehlings MG, Moncada MS (1995) Use of sensory-evoked potentials recorded from the human occiput for intra-operative physiologic monitoring of the spinal cord. Spine 20: 2318–2327

Hurlbert RJ, Koyanagi I, Tator CH (1993) Sensory evoked potentials for selective monitoring of the ventral rat spinal cord: a cerebellar evoked potential for assessing ventral cord integrity. J Neurotrauma 10: 181–200

Hurlbert RJ, Tator CH, Fehlings MF, Niznik G, Linden RD (1992) Evoked potentials from direct cerebellar stimulation for monitoring of the rodent spinal cord. J Neurosurg 76: 280–291

Jones SJ, Edgar MA, Ransford AO (1982) Sensory nerve conduction in the human spinal cord: Epidural recordings made during scoliosis surgery. J Neurol Neurosurg Psychiatry 45: 446–451

Lesser RP, Raudzens P, Lueders H, *et al* (1986) Postoperative neurological deficits may occur despite unchanged intraoperative somatosensory evoked potentials. Ann Neurol 19: 22–25

Levy WJ, McCaffrey M, Goldman D, *et al* (1986) Nonpyramidal motor activation produced by stimulation of the cerebellum, direct or transcranial: a cerebellar evoked potential. Neurosurgery 19: 163–176

Levy WJ (1988) Use of motor evoked potential as a monitoring tool. In: Rossini PM, Marsdens CH (eds) Neurology and Neurobiology, Vol 41. Alan R Liss, New York, pp 275–296

Paxinos G, Watson C (1982) (eds.): The rat brain in stereotaxic coordinates. Academic Press, San Diego

Takaki O, Okumura F (1985) Application and limitations of somatosensory evoked potential monitoring during thoracic aortic aneurysm surgery: A case report. Anesthesiology 63: 700–703

Zornow MH, Drummond JC (1989) Intraoperative somatosensory evoked responses recorded during the onset of the anterior spinal artery syndrome. J Clin Monit 5: 243–245

Correspondence: Dr. J. Hurlbert, Division of Neurosurgery and Department of Clinical Neurosciences, Foothills Hospital, 1403, 29th Sd. N.W., Calgary, Alberta, Canada T2N 2T9.

Spinal Cord Monitoring
Eds: E. Stålberg, H. S. Sharma, Y. Olsson

Chapter 10

Generation of magnetic evoked spinal cord potentials

H. Kitagawa, Y. Kawaguchi, H. Nakamura, S. Nakatoh, and **H. Tsuji**

Department of Orthopaedic Surgery, Toyama Medical and Pharmaceutical University, Toyama City, Japan

Introduction

Patton and Amassian (1954) elucidated the physiological nature of multiple descending volleys through the pyramidal tract after a single shock of direct electrical stimulation of the motor cortex: the first volley stimulated at the initial segment or the first node of the pyramidal tract neurons was labelled the direct (D) wave, and later volleys activated trans-synaptically were labelled indirect (I) waves. Merton and Morton (1980) introduced the technique of corticospinal tract activation using high-voltage transcranial stimulation. Later, Barker *et al.* (1985) introduced painless activation of the motor tract using transcranial magnetic stimulation, which achieved widespread acceptance. These methods produce evoked responses that have different characteristics, as discussed in this chapter. Experimental studies of the conducting pathways following transcranial magnetic stimulation, and the clinical application of magnetic stimulation in intraoperative neuromonitoring is also discussed.

Magnetic evoked spinal cord potentials in humans

Transcranial magnetic stimulation of the motor cortex can induce repetitive discharges in descending corticospinal tracts by direct and indirect excitation of the corticospinal neuron in monkey (Amassian *et al.*, 1990; Edgley *et al.*, 1990; Kitagawa and Møller, 1995) and in human (Burke *et al.*, 1993; Fujiki *et al.*, 1996; Kitagawa *et al.*, 1995). The structures activated by transcranial magnetic stimulation were first analyzed by EMG responses from upper extremities (Day *et al.*, 1989). The latency of EMG responses at threshold intensity was 1–2 ms longer than that elicited by transcranial electrical stimulation. Day *et al.* (1989) suggested that electrical stimulation of the motor cortex was capable of activating nerve cells of the pyramidal tract directly, whereas magnetic stimulation activated nerve cells of the pyramidal tract transsynaptically. These investigators

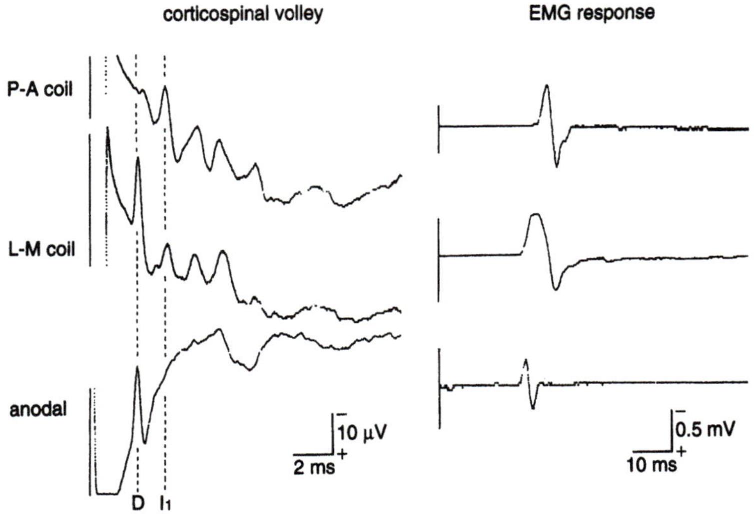

Fig. 1. Corticospinal volleys and EMG responses after transcranial magnetic and electrical stimulation. Responses were recorded from spinal cord (C3-4, left) and abductor pollicis brevis muscle (right) after stimulation of the hand motor area. P-A stimulation (upper traces) preferentially generated an I1 wave that was 1.3 ms longer in latency than was the electrical D wave in response to anodal stimulation (lower traces). In contrast, the L-M stimulation (middle traces) preferentially generated a magnetic D wave. [Nakamura H, Kitagawa H, Kawaguchi Y, Tsuji H (1996) Direct and indirect activation of human corticospinal neurons by transcranial magnetic and electrical stimulation. Neurosci Lett 210: 45–48]

reasoned that because magnetic stimulation produces a major current flow in gray matter parallel to the surface of the brain, it would be relatively ineffective in stimulating vertically oriented pyramidal cells.

Our results of simultaneously recording evoked spinal cord potentials (ESCP) and EMG responses (Nakamura *et al.*, 1996) support the deduction of Day *et al.* (1989). Nakamura *et al.* (1996) elicited evoked potentials using both magnetic transcranial stimulation (MAGSTIM Co., Model 200) and electrical transcranial stimulation (Digitimer Ltd, D180A) in awake humans (Fig. 1). Transcranial magnetic stimulation was delivered through a figure-eight coil (double 70mm coil). When the induced current in the neural tissue traveled in a posterior-to-anterior direction (P-A stimulation), the magnetic stimulation preferentially generated I waves. When the induced current in the neural tissue traveled in a lateral-to-medial direction (L-M stimulation), the magnetic stimulation preferentially generated D waves. There were, however, some I waves evoked by L-M stimulation that were smaller than those evoked by P-A stimulation, which was in agreement with results of recording muscle responses (Amassian *et al.*, 1989, 1990; Werhahn *et al.*, 1994) in the study of Nakamura *et al.* (1996). L-M stimulation evoked a large D wave. The latency of the EMG responses evoked by the L-M stimulation was identical to that evoked by anodal stimulation, and it was 3 ms shorter than that evoked by the P-A stimulation.

The mode of activation by transcranial magnetic stimulation alters, depending on the direction of the induced eddy current. The determinant of the difference in the mode of activation between L-M stimulation and P-A stimulation may be the directions of the trajectories of corticospinal neurons and corticocortical neurons, as demonstrated by modelling experiments (Amassian *et al.*, 1992). The two characteristics of magnetic P-A stimulation – the high sensitivity to excitability of the motor cortex and the long latency of the EMG responses – can be explained by the "D and I" hypothesis (Nakamura *et al.*, 1996).

Another theory was proposed basis of the difference in the site of activation (Burke *et al.*, 1990, 1993; Edgley *et al.*, 1990): magnetic stimulation excites corticospinal neurons at their initial segment or proximal nodes, whereas electrical stimulation excites these neurons deep in white matter, which conflicts with the "D and I" hypothesis. However, Nakamura *et al.* (1996) showed that a monophasic electrical D wave can result in an EMG response without the earlier component that was thought to be activated in deep white matter (Fig. 1). This implies that the deep D wave was not elicited by near motor-threshold electrical stimulation shown in Fig. 1, which is in agreement of Rothwell *et al.* (1994). Thus, our results support the "D and I" hypothesis (Nakamura *et al.*, 1996).

Conducting pathways of magnetic evoked spinal cord potentials in monkeys

Baseline recording

ESCP following transcranial magnetic stimulation were recorded and the distribution of field potentials of spinal cord signals in monkeys anesthetized with ketamine were analyzed (Kitagawa and Møller, 1994). Magnetic ESCP recorded from the epidural space had an initial negative wave (N1) followed by 6 or 7 negative waves (N2-7) (Fig. 2). At near threshold stimulus intensity, the latency of the N1 peak after magnetic stimulation was 3.2 ms, which was the same as that of wave N1 elicited by direct electrical stimulation (Fig. 2). However, the N1 wave of the magnetic ESCP recorded at threshold levels was difficult to evoke at near threshold stimulus. Near threshold the latency of the N1 wave of the response to transcranial electrical stimulation was the same as that of peak N1 of the magnetic ESCP at threshold, which decreased remarkably when stimulus intensity was increased.

The induced current from anterior to posterior by tangentially placed coil to the skull in this experiment was not the most optimal orientation for direct corticospinal activation that controls the muscles of the foot (Amassian *et al.*, 1992). In humans, a figure-eight coil placed over the vertex with the eddy current oriented in an anterior-to-posterior direction

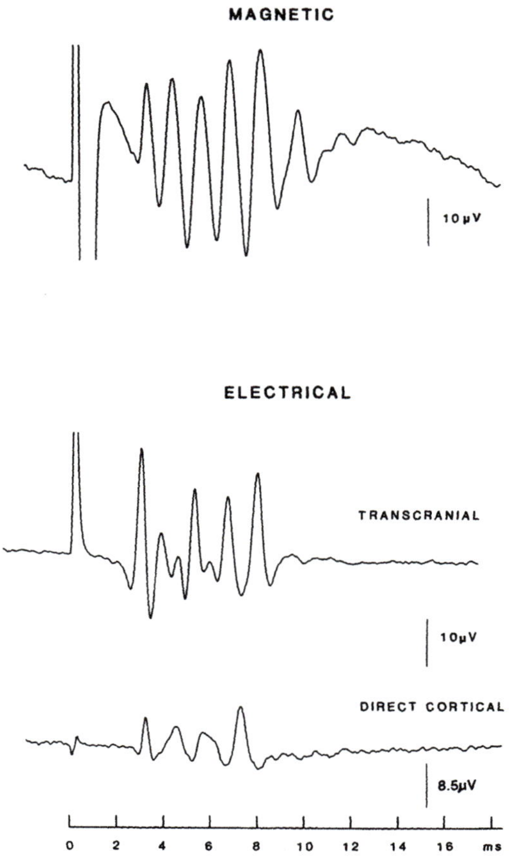

Fig. 2. Evoked spinal cord potentials elicited by magnetic and electrical motor cortex stimulation in monkey. ESCPs were recorded from T11 spinal cord. Magnetic ESCP (upper trace) were elicited by stimulation with a figure-eight coil (MAGSTIM, double 70) placed tangentially over the vertex with the orientation of the eddy current being anterior to posterior. Transcranial electrical ESCP (middle trace) were elicited by a high-voltage and low-output impedance electrical stimulator (Digitimer, D180), and they were delivered through pairs of surface electrodes, with the anode placed on the vertex and the cathode placed 3 cm anterior to the vertex. Direct electrical ESCP (lower trace) were elicited by a single stimulus of 0.1 ms duration through silver wire electrodes, with the anode placed on the primary motor cortex and the cathode placed 3 mm anterior to the anode. Stimulus intensity was 2 times the threshold. [Kitagawa H, Møller AR (1994) Conducting pathways and generators of magnetic evoked spinal cord potentials: a study in monkeys. Electroencephalogr Clin Neurophysiol 93: 57–67]

evokes D waves, but these waves are not as clear as those that are evoked when the eddy current is oriented in a medial-to-lateral orientation (unpublished data). The orientation of the radial fibers of the precentral sulcus in the infolded precentral gyrus and their tortuous complex trajectory (Amassian *et al.*, 1992) may explain the evocation of D waves as a result of the tangential coil orientation.

Field potential distribution of magnetic evoked spinal cord potential

The spinal cord was exposed at the T11 level in order to study the intraspinal field potential distributions of the magnetic ESCP by extracellular microelectrode recording (Kitagawa and Møller, 1994). Responses to 15–40 stimuli were averaged, and isopotential maps were then plotted using the topography system of a signal processor (NEC SANEI, Type #7T17). Six or seven negative deflections were seen within 15 ms after magnetic stimulation (Fig. 3). It is seen that the potentials that appear with the same latency as that of wave N1 have their maximal amplitude at a depth of 2.5 mm in track C (Fig. 3). The evoked potential of

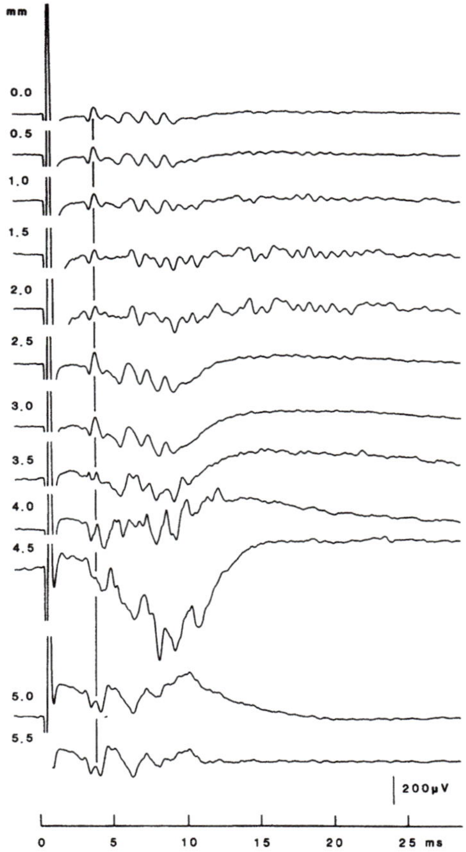

Fig. 3. Magnetic ESCP recorded within the left half of the T11 spinal cord by a microelectrode. Sample recordings are illustrated in Fig. 4, track C. The depth of the microelectrode is indicated on the left. Note that the N1 negative peak was inverted to the positive component between depths of 3.0 and 4.5 mm. Later, small waves appeared at depths of 1.5 and 2.0 mm in tracks C and D in Fig. 4. [Kitagawa H, Møller AR (1994) Conducting pathways and generators of magnetic evoked spinal cord potentials: a study in monkeys. Electroencephalogr Clin Neurophysiol 93: 57–67]

maximal amplitude in track C corresponds to the dorsolateral funiculus. The potentials that appear at the latency of negative peak N1 were inverted positively when they were recorded at a depth between 3.0 and 4.5 mm below the location where the microelectrode had been inserted. When the

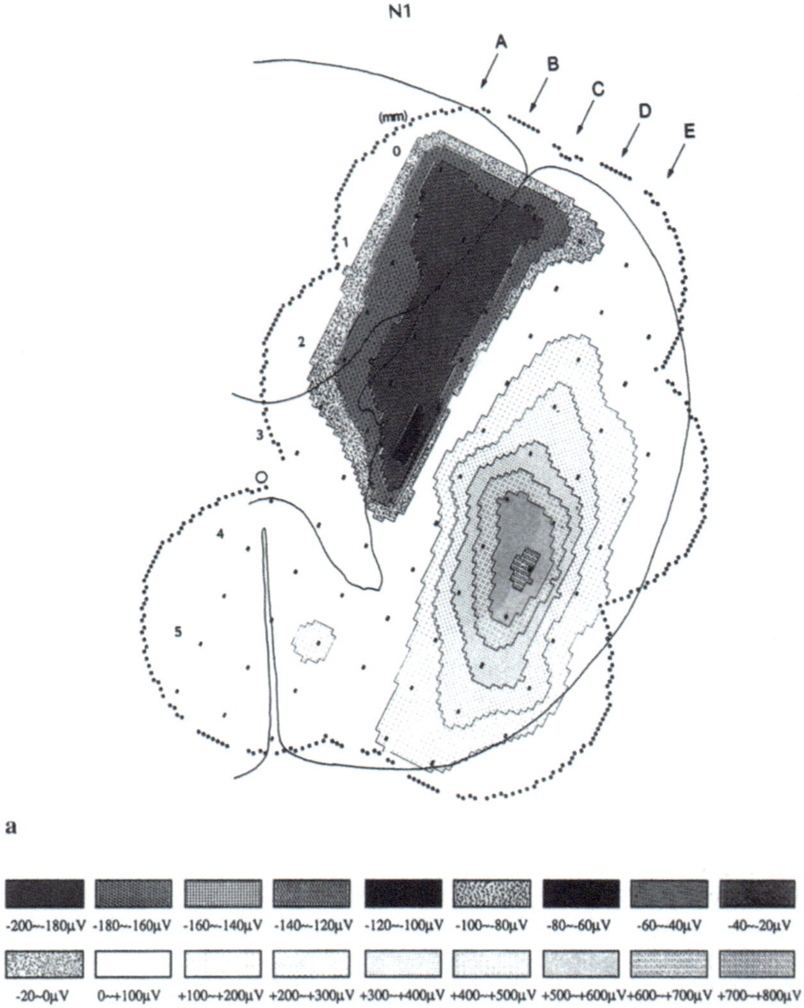

Fig. 4. Isopotential map of the N1 wave of the magnetic ESCP. The isopotential map after magnetic stimulation showed its maximum negativity at the medial dorsolateral funiculus, whereas the maximum positivity was shown to be at the ventrolateral funiculus. The black square shows the recording points. The depth of the microelectrode was shown on the left of the N1 isopotential map. Arrow C shows the sample track shown in Fig. 3. The absolute values of the amplitudes are expressed below (bottom). [Kitagawa H, Møller AR (1994) Conducting pathways and generators of magnetic evoked spinal cord potentials: a study in monkeys. Electroencephalogr Clin Neurophysiol 93: 57–67]

recording electrode was located in or near the dorsolateral funiculus, the responses were small deflections that appeared at longer latencies – up to 30 ms (Fig. 3), which is in agreement with recordings of later irregular discharges in pyramidal tract activity following electrical stimulation in the range of latencies between 15 and 40 ms (Kernell and Wu, 1967). Amassian and co-workers (1990) pointed out that phase cancellation might affect recordings made from the surface of the spinal cord, and this may have been the reason why we found it difficult to record these later components from the epidural surface.

The distribution of field potentials was sampled at the latency of each of the negative peaks of the epidural recording and were presented as isopotential maps superimposed on a cross-section of the resected spinal cord specimen (Fig. 4). Field potential distribution at the latency of wave N1 of the magnetic ESCP had its maximum negativity at the location of the medial dorsolateral funiculus, whereas the maximum positivity of the ESCP at the latency of the N1 peak was reached when the recording electrode was located in the ventrolateral funiculus. The maximal negative field potential distribution at the latency of the N1 wave, as seen in the isopotential maps in Fig. 4, corresponds to the location of the dorsal corticospinal tract distribution, which is suggested to be a current sink and assumed to be a generator of ESCP. The positive field potential is suggested to be a current source. The distribution of the negative field potential at the latencies of peaks N2, N3, N4, and N5 also corresponds to the anatomical distribution of the dorsal corticospinal tract. The ventral corticospinal tracts that are distributed in ventromedial funiculus may contribute to the negative field distribution of waves N2, N3, N4, N5, and N6. The reticulospinal tract may also contribute to the ventral negative field potential distribution in waves N2, N3, N4, N5, and N6 by the indirect cortico-reticulo-spinal connections (Alstermark *et al.*, 1983; Shapovalov, 1972).

The isopotential map of the earlier negative deflection of the magnetic ESCP (N1 wave) revealed the maximum negative field potentials corresponding to the dorsolateral corticospinal tracts. The N1 wave of the magnetic ESCP had the same latency as the D-wave of the electrical ESCP elicited by either direct cortical or transcranial electrical stimulation. We assumed that the N1 wave of the magnetic ESCP was generated by direct excitation of pyramidal axons. Later waves (I waves) of the magnetic ESCP were assumed to reflect not only the dorsal corticospinal tracts but also ventromedial spinal cord function.

Generation of magnetic evoked spinal cord potential in cats

Weissman *et al.* (1992) described the ineffectiveness of stimulating small brain by magnetic coil, because of the decrease in magnetic flux. Magnetic ESCP have been reported to contain corticospinal tract volleys in the N3 wave (Hiraizumi *et al.*, 1991) or in the N4 wave in a series of 4–5 multiple volleys (Fujiki *et al.*, 1990) in cats. However, the possibility of selective corticospinal activation in small animals has not yet been ascertained.

Authors (Nakatoh *et al.*, 1995) studied local magnetic stimulation of the motor cortex using the figure-eight coil and the magnetic field shield. ESCP and evoked compound muscle action potentials were recorded after magnetic stimulation in cats under ketamine anesthesia. The center of the figure-eight coil was placed over the motor cortex. Nakatoh *et al.*, 1995, showed that the optimal direction was a lateromedial direction of the induced current for effective activation of the forearm muscle in cats. Further localization of motor cortex stimulation was studied using an application of the magnetic field shield, Permalloy (NKK Co., Permalloy C-N thin board, 2 mm thick, maximum magnetic permeability 465,000 μm) that has a rhomboid window over the motor area. The thresholds of ESCP to transcranial magnetic stimulation decreased under a shield of Permalloy that had a 5×5 cm window (Fig. 5) (Nakatoh *et al.*, 1995).

Measurement of the magnetic field induced at the center of the figure-eight magnetic coil revealed that the size of the magnetic field using a Permalloy shield without a window was 62% smaller than that without shield (Nakatoh *et al.*, 1995). The variable magnetic field under the Permalloy shield was presumed to be smaller and less homogeneous than that without a Permalloy shield. These changes in size and homogeneity of the variable magnetic field may be the cause of the alteration of the induced

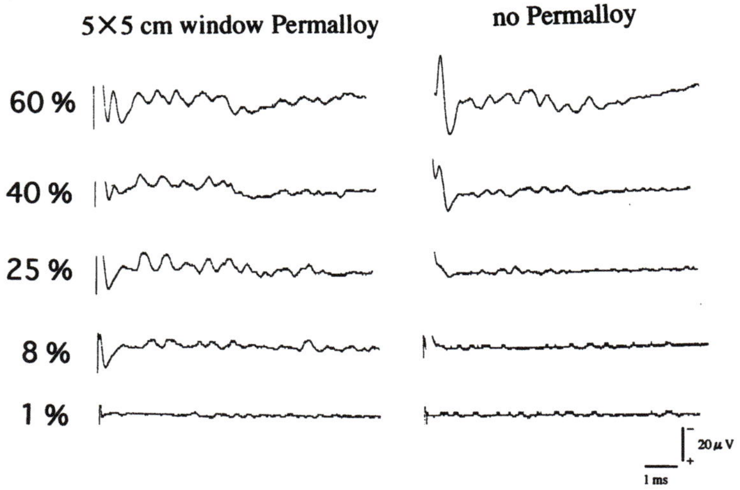

Fig. 5. Effects of Permalloy on the evoked spinal cord potential in cats. Note that the threshold of magnetic ESCP (right) decreased remarkably by covering the cranium, except in the motor area when using the magnetic shield Permalloy (left). The figure-eight coil (double 70-mm coil) was in contact tangentially with the cranium and the eddy current direction was lateromedial. [Nakatoh S, Kitagawa H, Kawaguchi Y, Nakamura H, Takano H, Tsuji H (1998) Effects of coil orientation and magnetic field shield on transcranial magnetic stimulation in cats. Muscle Nerve (in press)

current in the brain of the cat. Thus thresholds of muscle potentials and ESCP under Permalloy change. However, the conduction velocity of ESCP was 70–130 m/s, regardless of the intensity of the magnetic stimulation, which was too fast to attribute to the corticospinal tract. This suggests the involvement of the extrapyramidal pathway. Selective activation of the corticospinal tract was not achieved after transcranial magnetic stimulation of the cortical motor area when using the figure-eight coil and the magnetic shield.

Clinical application of transcranial magnetic stimulation in operating room

Forty-five patients who underwent surgery of the spine were monitored using magnetic-evoked compound muscle action potentials and magnetic ESCP (Kitagawa *et al.*, 1995). It has been shown that intraoperative monitoring of motor function by magnetic ESCP is as reliable (Fujiki *et al.*, 1996) as transcranial electrical ESCP (Kitagawa *et al.*, 1989). However, the clinical neurophysiologist has to be careful when interpreting the descending spinal cord signal after any kind of transcranial stimulation in operating room. Because electrical transcranial stimulation has a tendency to promote the spread of electrical current to anatomical structures deep within the brain near the craniocervical junction following bipolar stimulation (Rothwell *et al.*, 1994), and to recruit extrapyramidal activation by hard palate – vertex stimulation in humans (Hashimoto *et al.*, 1993) and cats (Kitagawa *et al.*, 1991). Magnetic transcranial stimulation can also possibly stimulate tissue deep within the brain (Kitagawa and Møller, 1995; Yunokuchi *et al.*, 1993). Stimulus intensity and the method of stimulation (electrode configuration and orientation) in transcranial stimulation are always crucial when determining which neural structure is to be activated.

The epidural recording technique has limitations in that it is less sensitive to neural injury. Recording evoked compound muscle action potentials via the synapse of the spinal motoneurons provides further information on the motor function of the spinal cord. Ketamine, fentanyl, nitrous oxide and oxygen mixture anesthesia enabled us to record muscle action potentials after magnetic stimulation, because this type of anesthesia has a less suppressive effect on magnetic ESCP than does inhalation anesthesia (Kitagawa *et al.*, 1995) (Fig. 6). Paired transcranial magnetic stimulation was delivered to the motor cortex. Activation of synaptic transmission in the motor cortex and temporal summation of the corticospinal volley at the spinal motoneurons gave rise to the facilitation of muscle signals after paired magnetic stimulation. The optimal interstimulus interval was 10 ms under ketamine-based anesthesia for facilitation of magnetic evoked muscle action potentials (Nakamura *et al.*, 1995a), whereas it was 25 ms in awake humans (Nakamura *et al.*, 1995b). Under propofol anesthesia, the optimal interstimulus interval decreased to 2 ms (unpublished data). The optimal interstimulus interval depends on the

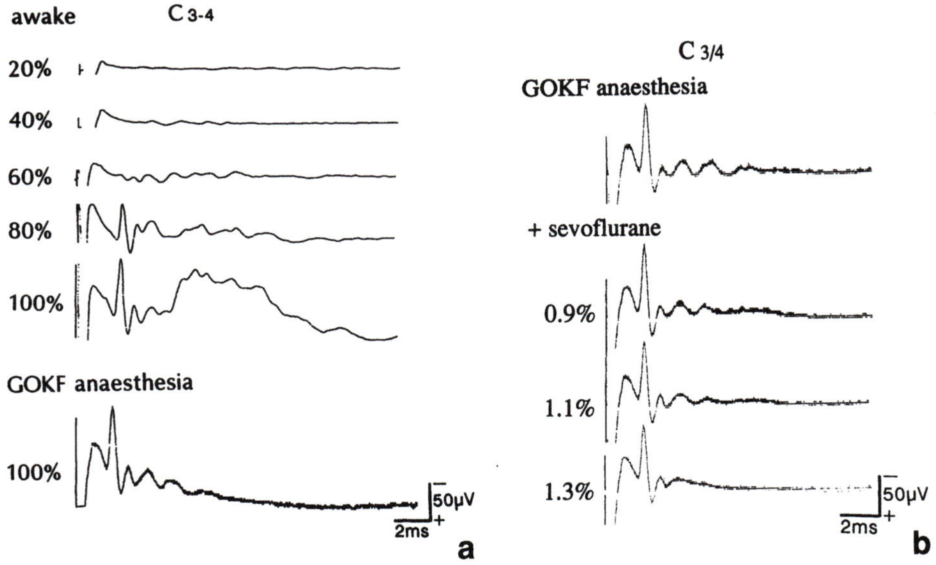

Fig. 6. Magnetic ESCP in awake and anesthetized human. Well preserved signals under ketamine-based anesthesia (**a**) were suppressed by sevoflurane (**b**). The negative peaks disappeared beginning with the later peaks (**b**). Direct wave showed a mild decrease in amplitude. The conduction velocity of the D wave was 57.6 m/s. [Kitagawa H, Nakamura H, Kawaguchi Y, Tsuji H, Satone T, Takano H, Nakatoh S (1995) Magnetic evoked compound muscle action potential neuromonitoring in spine surgery. Spine 20: 2233–2239]

degree of anesthetic suppression on the central nervous system; i.e., the deeper the anesthetic depression, the shorter the optimal interstimulus interval for the facilitation of magnetic evoked muscle action potentials due to the decreased number of I waves (Fig. 6) (Kitagawa *et al.*, 1995).

Neuromonitoring was reliable in 82% of the monitored cases of a recent series after reducing the dosage of fentanyl (Kitagawa *et al.*, 1995). Magnetic evoked compound muscle action potentials neuromonitoring is more selective monitoring method than that of ESCP, and is better to monitor the function of the entire motor pathway including the white and gray matter of the spinal cord than ESCP. Furthermore, the muscle signal of magnetic evoked compound muscle action potentials is more sensitive than that of ESCP. However, from the viewpoint of forecasting a postoperative neurological prognosis, epidural recording of ESCP is more reliable, either in response to transcranial electrical or magnetic or spinal cord stimulation. Kitagawa *et al.* (1989) proposed that the critical point in a spinal cord signal at which the patient can be expected to have a postoperative neurologic deficit corresponds to a 50% decrease in amplitude as compared to the baseline recording in normal spinal cord function. Kitagawa *et al.* (1995) questioned that the critical point of neural deficit in muscle signal as 50% amplitude decrease of the baseline, as same as that of spinal cord signals. In the experiments done in cats with spinal cord injury, a muscle

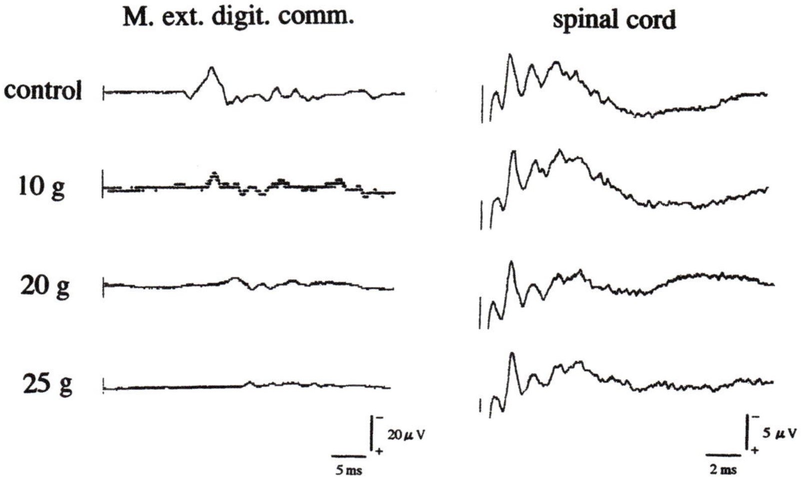

Fig. 7. Spinal and muscle responses following subacute spinal cord compression in cats. When the spinal cord was compressed at the C2 level, the amplitude of the muscle signal decreased (left) and the latency increased. However, spinal cord signals (right) elicited by direct electrical stimulation of the cortex were preserved. [Nakatoh S, Kitagawa H, Takano H, Nakamura H (1996) Changes of evoked compound muscle action potential and evoked spinal cord potential following motor cortex stimulation caused by acute compression of spinal cord. Presented at the 18th annual meeting of the electrodiagnosis of the spinal cord, Tokyo]

signal was far more sensitive than ESCP following direct motor cortex stimulation (Nakatoh *et al.*, 1996) (Fig. 7). The spinal cord signal in these experiments was well preserved when the muscle signal was abolished even after the muscle signal had been abolished. Although the evoked compound muscle action potential is very sensitive to neural injury, it must be emphasized that the absence of this potential in the recording does not indicate a postoperative neural deficit but rather the possible risk of a postoperative neural deficit. Hence, an insignificant manipulation to the neural tissue could cause an abolishment of the evoked compound muscle action potential that should not be taken as an indication of postoperative neural deficit. A return of an evoked compound muscle potential could be taken as an indication that neural injury has been averted. Thus, a change in the evoked compound muscle action potential should be considered an indication of neural manipulation that may or may not cause a postoperative neural injury but which should be brought to the attention of the surgeon to permit the choice to alter the manipulation.

For the clinical purpose of intraoperative neurophysiological monitoring, transcranial electrical stimulation has an advantage when the patient is under general anesthesia, because of it involves direct stimulation of pyramidal neurons, whereas magnetic stimulation does trans-synaptic activation predominantly. In 1995 the authors began using repetitive (3–5 times) transcranial high voltage electrical stimulation with an interstimulus

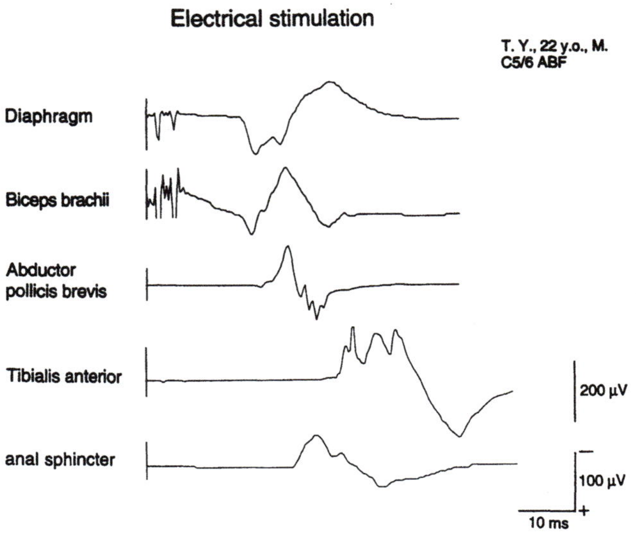

Fig. 8. Compound muscle action potentials after multiple transcranial electrical stimulation. The muscle signals were recorded with good reproducibility from axial and distal muscles with the patient under propofol anesthesia throughout surgery on the spine. [Kitagawa H, Kawaguchi Y, Nakamura H, Tsuji H, Takano H, Nakatoh S (1996) Intraoperative spinal cord monitoring after transcranial magnetic stimulation. Presented at the 25th annual meeting of the Japan spine research society, Nagoya]

interval of 2 ms (Digitimer, D185 Multipulse) while the patient was under propofol anesthesia, because when using this technique the muscle signal is more reproducible than the muscle signal that normally follows magnetic stimulation (Fig. 8). This technique also enables easy maintenance of anesthetics. Thus, transcranial neuromonitoring of electrically evoked compound muscle action potentials can be the chosen standard method for intraoperative neurophysiological monitoring.

References

Alstermark B, Pinter M, Sasaki S (1983) Brain stem relay of disynaptic pyramidal EPSPs to neck motoneurons in the cat. Brain Res 259: 147–150

Amassian VE, Cracco RQ, Maccabee PJ (1989) Focal stimulation of human cerebral cortex with the magnetic coil: a comparison with electrical stimulation. Electroencephalogr Clin Neurophysiol 74: 401–416

Amassian VE, Quirk GJ, Stewart M (1990) A comparison of corticospinal activation by magnetic coil and electrical stimulation of monkey motor cortex. Electroencephalogr Clin Neurophysiol 77: 390–401

Amassian VE, Eberle L, Maccabee PJ, Cracco RQ (1992) Modelling magnetic coil excitation of human cerebral cortex with a peripheral nerve immersed in a brain-shaped volume conductor: the significance of fiber bending in excitation. Electroencephalogr Clin Neurophysiol 85: 291–301

Barker AT, Jalinous R, Freestone IL (1985) Non-invasive magnetic stimulation of human motor cortex. Lancet 1: 1106–1107

Burke D, Hicks RG, Stephen JPH (1990) Corticospinal volleys evoked by anodal and cathodal stimulation of the human motor cortex. J Physiol (Lond) 425: 283–299

Burke D, Hicks R, Gandevia SC, Stephen J, Woodforth I, Crawford M (1993) Direct comparison of corticospinal volleys in human subjects to transcranial magnetic and electrical stimulation. J Physiol (Lond) 470: 383–393

Day BL, Dressler D, Maertens de Noordhout A, Marsden CD, Nakashima K, Rothwell JC, Thompson PD (1989) Electric and magnetic stimulation of human motor cortex: surface EMG and single motor unit responses. J Physiol (Lond) 412: 449–473

Edgley SA, Eyre JA, Lemon RN, Miller S (1990) Excitation of the corticospinal tract by electromagnetic and electrical stimulation of the scalp in the macaque monkey. J Physiol (Lond) 425: 301–320

Fujiki M, Isono M, Hori S (1990) Spinal and muscle motor evoked potentials following magnetic stimulation in cats. Neurol Med Chir (Tokyo) 30: 234–241

Fujiki M, Isono M, Hori S, Ueno S (1996) Corticospinal direct response to transcranial magnetic stimulation in humans. Electroencephalogr Clin Neurophysiol 101: 48–57

Hashimoto Y, Tamaki T, Matsuura S, Andou M, Nakamoto K, Yamada H, Keho H, Nishiura H (1993) Experimental and clinical study of spinal cord evoked potentials elicited by transcranial stimulation. J Electrodiagn Spinal Cord 15: 9–12

Hiraizumi Y, Fujimaki E, Transfeldt EE, Kawahara N (1994) Transcranial magnetic motor evoked potentials recorded from the spinal cord in cats: normal waves and spinal cord injuries. In: Jones SJ, Boyd S, Hetreed M, Smith NJ (eds) Handbook of Spinal Cord Monitoring. Kluwer, Lancester, pp 398–406

Kernell D, Wu CP (1967) Responses of the pyramidal tract to stimulation of the baboon's motor cortex. J Physiol 191: 653–672

Kitagawa H, Itoh T, Takano H, Takakuwa K, Yamamoto N, Yamada H, Tsuji H (1989) Motor evoked potential monitoring during upper cervical spine surgery. Spine 14: 1078–1083

Kitagawa H, Takano H, Takakuwa K, Yamamoto N, Kida H, Tsuji H (1991) Origins and conducting pathways of motor evoked potentials elicited by transcranial (vertex-hard palate) stimulation in cats. Neurosurgery 28: 358–363

Kitagawa H, Møller AR (1994) Conducting pathways and generators of magnetic evoked spinal cord potentials: a study in monkeys. Electroencephalogr Clin Neurophysiol 93: 57–67

Kitagawa H, Nakamura H, Kawaguchi Y, Tsuji H, Satone T, Takano H, Nakatoh S (1995) Magnetic evoked compound muscle action potential neuromonitoring in spine surgery. Spine 20: 2233–2239

Kitagawa H, Kawaguchi Y, Nakamura H, Tsuji H, Takano H, Nakatoh S (1996) Intraoperative spinal cord monitoring after transcranial magnetic stimulation. Presented at the 25th annual meeting of the Japan spine research society, Nagoya

Merton PA, Morton HB (1980) Stimulation of the cerebral cortex in the intact human subject. Nature 285, p 227

Nakamura H, Kitagawa H, Kawaguchi Y, Tsuji H, Takano H, Nakatoh S (1995) Intracortical facilitation and inhibition after paired magnetic stimulation in humans under anesthesia. Neurosci Lett 199: 155–157

Nakamura H, Kitagawa H, Kawaguchi Y, Tsuji H, Takano H, Nakatoh S (1997) Intracortical facilitation and inhibition after transcranial magnetic stimulation in conscious humans. J Physiol 498: 817–823

Nakamura H, Kitagawa H, Kawaguchi Y, Tsuji H (1996) Direct and indirect activation of human corticospinal neurons by transcranial magnetic and electrical stimulation. Neurosci Lett 210: 45–48

Nakatoh S, Kitagawa H, Takano H, Nakamura H (1996) Changes of evoked compound muscle action potential and evoked spinal cord potential following motor cortex stimulation caused by acute compression of spinal cord. Presented at the 18th annual meeting of the electrodiagnosis of the spinal cord, Tokyo

Nakatoh S, Kitagawa H, Kawaguchi Y, Nakamura H, Takano H, Tsuji H (1998) Effects of coil orientation and magnetic field shield on transcranial magnetic stimulation in cats. Muscle Nerve (in press)

Patton HD, Amassian VE (1954) Single and multiple unit analysis of cortical stage of pyramidal tract activation. J Physiol 17: 345–363

Rothwell JC, Burke D, Hicks R, Stephen J, Woodforth I, Crawford M (1994) Transcranial electrical stimulation of the motor cortex in man: further evidence for the site of activation. J Physiol (Lond) 481: 243–250

Shapovalov AI (1972) Extrapyramidal monosynaptic and disynaptic control of mammalian alpha-motoneurons. Brain Res 40: 105–115

Weissman JD, Epstein CM, Davey KR (1992) Magnetic brain stimulation and brain size: relevance to the animal studies. Electroencephalogr Clin Neurophysiol 85: 215–219

Werhahn KJ, Fong JKY, Meyer BU, Priori A, Rothwell JC, Day BL, Thompson PD (1994) The effect of magnetic coil orientation on the latency of surface EMG and single motor unit responses in the first dorsal interosseous muscle. Electroencephalogr Clin Neurophysiol 93: 138–146

Yunokuchi K, Kawakubo M, Tokimura H, Ueno S (1993) Measurement of induced current after transcranial magnetic stimulation in human brain model. Presented at the 23rd annual meting of Japanese Electroencephalography and Electromyography Society, Kagoshima

Correspondence: Dr. H. Kitagawa, Department of Orthopaedic Surgery, Faculty of Medicine, Toyama Medical and Pharmaceutical University, 2630 Sugitani, Toyama City, 930-01 Japan.

Spinal Cord Monitoring
Eds: E. Stålberg, H. S. Sharma, Y. Olsson

Chapter 11

Magnetic stimulation of spinal nerve roots and cortical somatosensory evoked potentials

S. Tsuji

Department of Neurology, University of Occupational and Environmental Health, School of Medicine, Fukuoka, Japan

Introduction

Somatosensory evoked potentials (SEPs) elicited by the electrical stimulation at the peripheral nerves have many clinical applications since the work of Dawson (1947). This electrical stimulation, however, is painful and difficult to apply to the spinal roots at the thoracic and lumbar levels and proximal nerve trunks. Magnetic stimulation is a new method which was developed for non-invasive stimulation of the human motor cortex by Barker *et al.*, in 1985. Since then, transcranial magnetic stimulation of the motor cortex has been used to evoke potentials (motor evoked potentials, MEPs) from the hand and foot muscles for the analysis of upper and lower motor neuron functions in patients with neurological diseases (Mills *et al.*, 1987; Uozumi *et al.*, 1991). This method is non-invasive, and less painful than the electrical stimulation of the human motor cortex through the scalp (Merton *et al.*, 1980). Recently, magnetic rather than electrical stimulation of the human motor cortex has been used in the analysis of neurological diseases.

Polson *et al.* (1982) reported that the magnetic stimulator is capable of stimulating the peripheral nerves. In another application of magnetic stimulation to the somatosensory pathway, Tsuji *et al.* (1988, 1991, 1993) and Kunesch *et al.* (1993) recorded the cortical SEPs following magnetic stimulation at the peripheral nerves, and lumbar and thoracic spines. Furthermore, Zhu *et al.* (1991) found that the cerebral potentials accompanying magnetic stimulation of muscles in the extremities appear to be due to activation by muscle afferent.

Cortical and short latency SEPs elicited by electrical stimulation at the lower extremity are not optimally useful in the precise evaluation of the posterior column function in patients with thoracic or lumbar cord lesions or those with polyneuropathy. In addition, patients with thoracic radiculopathy have normal SEPs after the electrical stimulation at the ankle because of the normal function of the posterior column. Non-invasive electrical stimulation of somatosensory pathways at the spinal roots and proximal sites of the peripheral nerves is technically difficult, although Goodridge *et al.* (1987) developed paraspinal electrical stimulation at the thoracic and lumbar levels to elicit SEPs. Their method allows a reasonable estimation of the level of a lesion within the spinal cord, but it is difficult to apply the electrical stimulation to the spinal roots at the lower thoracic and lumbar regions. Magnetic stimulation, however, can be easily applied to the spinal roots at the thoracic and lumbar spine levels, proximal nerves and plexus, resulting in clear cortical SEPs.

Methods of cortical SEPs following magnetic stimulation of the spines and peripheral nerves

Magnetic stimulation

The magnetic stimulator (Nihon Kohden) is composed of two types of flat helical coil with inside diameters of 12.0 cm and 2.2 cm and a twin coil (figure-eight coil designed by Ueno *et al.*, 1990) with an inside diameter of 5 cm and an outside diameter of 10 cm in each loop. Large currents were passed through the two types of round coil (7,500 A = 900 V and 8,400 A = 900 V at maximal output, respectively) and the twin coil (10,000 A = 900 V at maximal output) by the discharge of capacitors (1,500 μF). The magnetic field of the twin coil, which approached 1.0 tesla, had a peak of about 130 μsec. The maximum magnetic fields of the two types of round coil were 0.4 tesla with a peak at 166 μsec and 2.6 tesla with a peak at 136 μsec, respectively.

The center of the magnetic coil was placed on the skin overlying the spinous processes of the second (Th2), fifth (Th5), eighth (Th8), tenth (Th10), and twelfth (Th12) thoracic vertebrae, the third (L3) and fifth (L5) lumbar vertebrae, at the midpoint of the gluteus muscle, and at the medial malleolus (Fig. 1). Cervical vertebral regions were not stimulated because of the risk of cervical cord damage due to repetitive neck movement. The repetition rate of the stimulus was approximately 1 Hz. The strength and direction of currents were adjusted to produce a muscle twitch in the paraspinal muscles or leg muscles. Twitching of the paraspinal and proximal muscles of the legs was induced by the stimulation at Th10 and Th12, and twitching of foot muscles by stimulation at L5 and at the mid-gluteus muscle. Routine stimulus intensities were between 450 and 560 V. We speculate that the eddy current induced by the magnetic coil stimulates the spinal roots entering the spinal canal at thoracic and lumbar levels and the peripheral nerves at the mid-gluteus level. The small flat helical coil with an inside diameter of 2.2 cm was only used to stimulate the posterior tibial nerve at the medial malleolus. The twin coil has the advantage of being able to stimulate more focally than the round coil because the current concentration area of the twin coil is substantially smaller than that of the round coil (Ueno *et al.*, 1990).

Turning the magnetic coil 180 degrees for half of the sweeps in each trial effectively reversed the phase of the induced current flow allowing cancellation in the averaging process of the stimulating artifact. This was due to a tremendous stimulus artifact from the magnetic coil. The disadvantage of this method is restriction of the number of stimuli because the magnetic coil is heated by repetitive stimulation. Subjects did not feel any pain at the sites of magnetic stimulation but complained of slight shocks at the stimulation sites.

Cortical SEP recordings

The SEP recordings were made from silver-silver chloride disks. The following derivations and filter settings were used on all subjects (Fig. 1): Ez-Fz, and 8-1,500 and 32-300 Hz (-3dB). Ez was the midpoint between Cz and Pz (international 10-20 system). For scalp topography of cortical SEPs, sixteen recording electrodes were placed on Fp1, Fp2, C3, C4, P3, P4, O1, O2, F7, F8, T5, T6, Fz, Cz, Pz and F4 (international 10–20 system) connected to linked ear electrodes (A1A2). The electrodes were filled with conductive jelly, and electrode impedance was below 5,000 ohms. Sixty-four responses were averaged with a Sanei 7T17 averager. The analysis times were 80 to 120 msec. The recordings were obtained in a waking state and prone position with eyes closed. The responses were stored on flexible diskettes by means of a Sanei 7T17 storage unit and then plotted with a

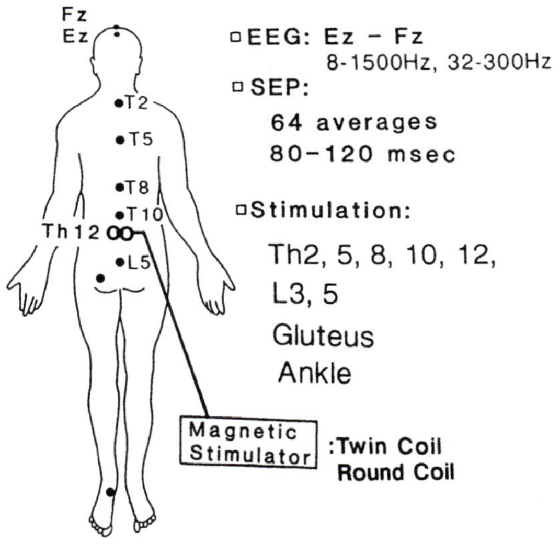

Fig. 1. Methods of cortical SEPs by magnetic stimulation

Sanei Thermal Printer. White noise was given to both ears to prevent the auditory artifact.

P2 and N2 were the initial cortical positive and negative potentials obtained by the electrical stimulation in the leg (Tsuji *et al.*, 1984, 1987).

Morphology of cortical SEPs by magnetic stimulation at the spines

Typical wave form in normal subjects

Figure 2 illustrates the principal cortical potentials evoked by twin coil magnetic stimulation at the Th2, Th5, Th12, L5, left mid-gluteus muscle, and left ankle levels in a normal subject. The cortical components P2, N2,

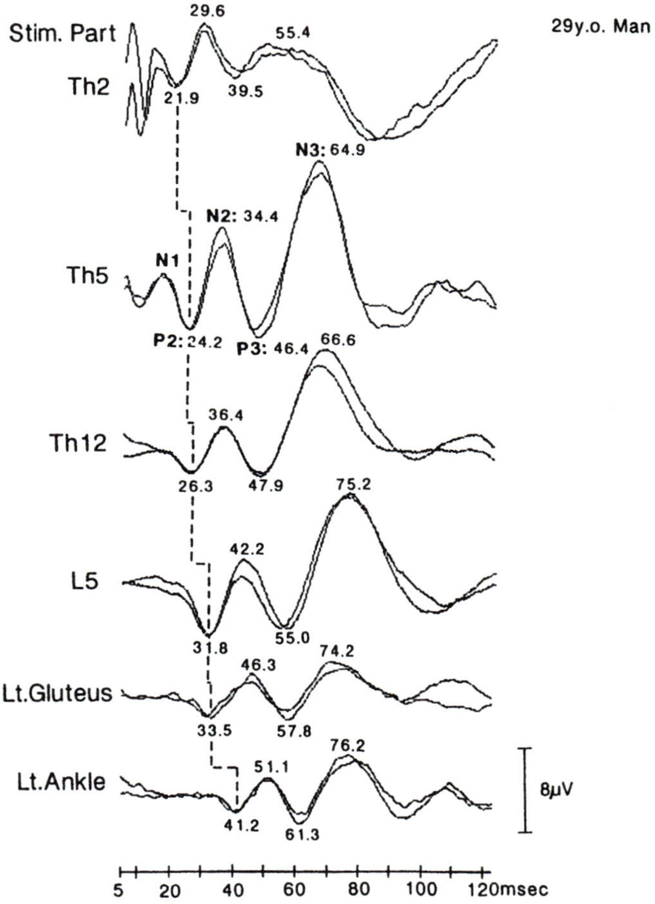

Fig. 2. Normal cortical SEPs by twin coil magnetic stimulation at each level. Derivation: Ez-Fz. Filter setting: 8–1,500 Hz. Number of stimuli = 64. (From Tsuji *et al.*, 1993)

Table 1. Mean peak latencies of cortical SEPs following twin coil stimulation in normal subjects

Stimulation levels	P2	N2	P3	N3
Th2	20.8±1.3 (15)	30.7±2.4 (15)	42.1±2.2 (15)	58.4±2.1 (14)
Th5	22.0±1.5 (15)	32.2±1.9 (15)	43.8±3.2 (15)	62.6±7.9 (14)
Th12	24.3±1.7 (15)	33.8±2.2 (15)	45.7±4.1 (15)	63.8±6.2 (15)
L5	29.2±2.4 (15)	38.3±2.9 (15)	50.0±4.3 (15)	69.1±5.9 (14)
Gluteus *	29.0±2.3 (40)	38.9±3.1 (40)	50.7±4.4 (40)	65.6±5.7 (39)
Ankle *	39.4±2.3 (39)	48.6±2.7 (39)	60.8±3.4 (37)	76.5±4.3 (31)

Latency (msec) mean ±S.D. (number of subjects); *filter setting* 8–1,500 Hz (−3 dB); * cortical SEPs following round coil stimulation.

P3 and N3 were clearly identified by magnetic stimulation at each level. N1 component was recorded inconstantly in normal subjects. Because the distance between the thoracic regions and the cerebral cortex is short, peak latencies of P2 and N2 evoked by the stimulation of the thoracic regions were faster than those evoked by the stimulation of the lumbar region and leg. P2 and N2 evoked by stimulation of the thoracic and lumbar regions had an amplitude approximately twice as great as that following stimulation of leg nerves. There were no substantial differences in the morphologies of P2 and N2 after either magnetic or electrical stimulation at the thoracic, lumbar and ankle levels (Tsuji *et al.*, 1988). Amplitudes of P2 and N2 were maximal after magnetic stimulation at the motor threshold.

Table 1 shows the mean and the standard deviation (S.D.) for peak latencies of P2, N2, P3 and N3 in 15 normal subjects for spinal magnetic stimulation, and in 40 normal subjects for lower extremity magnetic stimulation. P2 and N2 evoked by magnetic stimulation at each level could be identified in all subjects examined. However, P3 and N3 were unclear in a few cases, and the morphology of N3 was not constant in successive SEP studies. As would be expected, the S.D. for the peak latency of P2, N2 and P3 slightly increased with the more caudal stimuli, although the coefficients of variation (S.D./mean×100: 5.2–8.9%) for these components remained relatively constant. Relatively large coefficients of variation were found for N3 following Th5 and Th12 stimulation (9.7–12.6%). The mean latencies of P2 evoked by magnetic and electrical stimulation at the ankle were the same as 38.1±1.8 (20) and 38.07±3.47 (29) msec, respectively (Tsuji *et al.*, 1984, 1988). No substantial differences in morphologies and latencies of P2 and N2 were seen between the magnetic and electrical stimulation at the ankle, L5 and Th12 levels (Tsuji *et al.*, 1988). The interpeak latencies between P2 and N2 were 8.9±2.0 msec at Th10, 9.1±1.6 msec at Th12, 9.1±1.6 msec at L5, 9.8±2.0 msec at gluteus and

9.3 ± 1.7 msec at the ankle in 45 normal subjects (Tsuji *et al.*, 1991). These interpeak latencies of P2–N2 were substantially constant to stimulation at each level.

Effect of body height on peak latency

As would be expected, body height was significantly correlated with peak latencies of P2 evoked by stimulation at the lumbar, gluteus and ankle levels ($R = 0.355$–0.724, $P < 0.05$, R: regression coefficient, Fig. 3). Especially, P2 evoked by stimulation at the ankle showed the highest correlation factor ($R = 0.724$, $P < 0.01$). This is in good agreement with previous studies on cortical SEPs following electrical stimulation at the ankle (Tsuji *et al.*, 1984). However, the peak latencies of P2 evoked by stimulation at the thoracic regions (Th10, Th12) were independent of body height ($R = 0.223$–0.354, $P > 0.05$, Fig. 3). On the other hand, N2 latencies evoked by stimulation at each level (ankle, gluteus, L5, Th12 and Th10) were significantly correlated with body height ($R = 0.428$–0.556, $P < 0.01$). Interpeak latencies of P2–N2, however, were not correlated with body height ($R = -0.095$–0.326, $P > 0.05$). P3 latencies evoked by the stimulation at each level were also not correlated with body height ($R = 0.163$–0.294, $P > 0.05$).

Effect of age on peak latency

There were no significant correlations between age (15–67 years) and peak latencies of all components elicited by magnetic stimulation at Th10, Th12, L5 and mid-gluteus muscle ($R = -0.337$–0.222, $P > 0.05$, Tsuji *et al.*, 1991). P3 and N3 latencies following magnetic stimulation at the ankle were weakly correlated with age ($R = 0.358$ and 0.478, $P < 0.05$). This observation is in good agreement with previous reports on SEPs for lower extremity electrical stimulation (Lastimosa *et al.*, 1982, Tsuji *et al.*, 1984), but not in agreement with Eisen and Odusote (1980).

Effects of stimulus intensities

P2 and N2 latencies at Th12 were almost constant at the stimulus intensities of between 380 and 560 V, although these latencies were slightly delayed at the lower stimulus intensity of 340 V (Tsuji *et al.*, 1991). Various stimulus intensities ranging from 280 to 600 V were applied in this study. The N2 component showed double peaks at stimulus intensities of 380 and 440 V. In three of eight control subjects (38%), double peaks of N2 were noted at the stimulus intensities below 440 V. In contrast, there were no double peaks of N2 in 45 normal subjects who were stimulated by moderate to high intensities (above 450 V). The latencies of the second double peaks of N2 evoked at the intensities of 380 and 440 V were almost

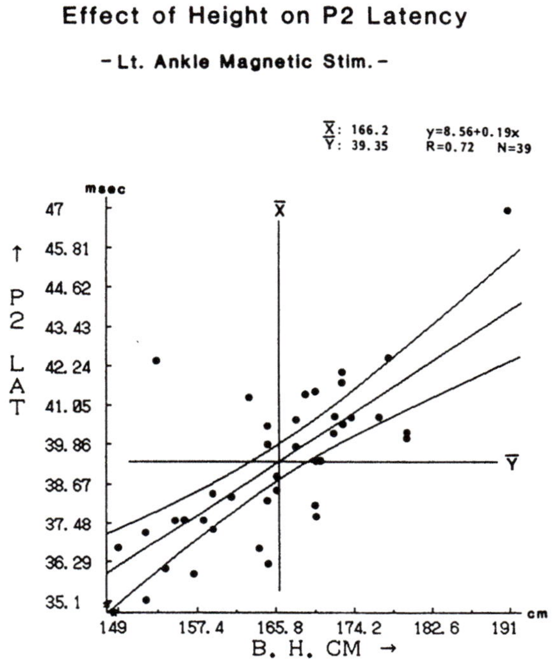

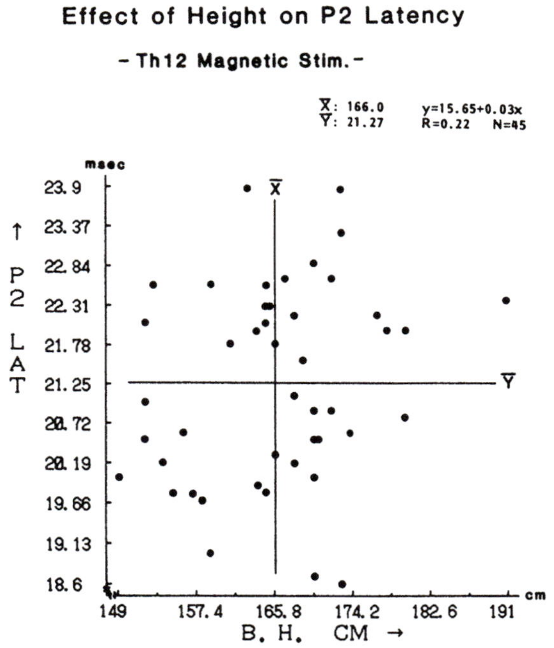

Fig. 3. Effect of body height on the latency of P2 evoked by magnetic stimulation at the left ankle and Th12 level in 45 normal subjects. The middle line represents the regression line. The upper and bottom lines represent the upper and lower 95% confidence limits for individual values. *BH* body height; *P2 LAT* P2 latency (From Tsuji *et al.*, 1991)

the same as that of N2 at the intensity of 560 V. No double peaks were seen on P2, P3 and N3 components at various stimulus intensities.

Topography of cortical SEPs on the scalp by magnetic stimulation at the spines

Amplitudes of cortical SEPs for scalp topography were measured as the vertical distance between two successive peaks. Figure 4 illustrates the typical distribution of cortical SEPs on the scalp elicited by magnetic stimulation at the L5 level. Cortical components P2, N2, P3 and N3 distributed over the centro-parietal areas bilaterally. On the other hand, the amplitudes of SEPs recorded from the frontal electrodes were small. Frontal responses showed reversed polarities compared with centro-parietal responses (P2, N2). Figure 5 shows typical topographies of components P2 and N2 by L5 magnetic stimulation, which were distributed widely over the posterior quadrants bilaterally. The maximal potential of P2 was at the mid-parietal areas, and that of N2 was at the midline of the parieto-central areas. P3 was distributed over the posterior quadrants bilaterally with a maximal potential at the vertex. N3 was distributed over the centro-parietal areas bilaterally and had a maximal amplitude at the vertex.

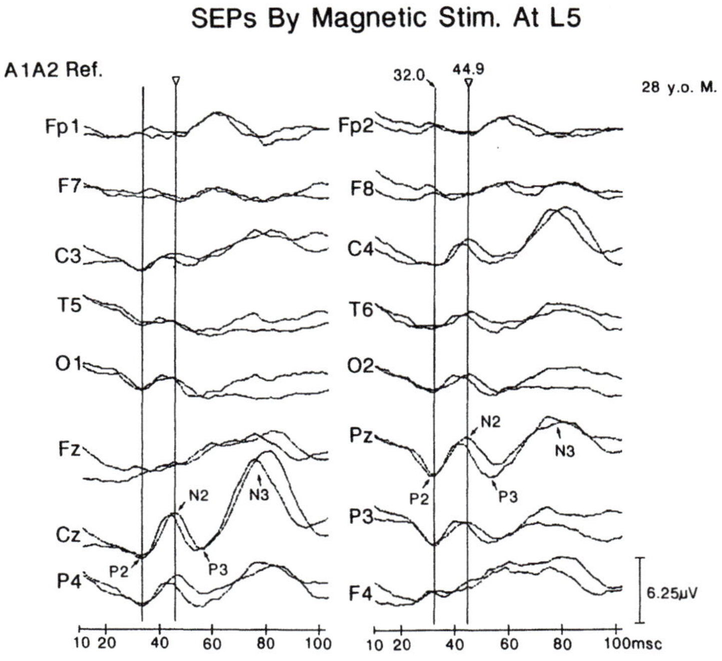

Fig. 4. Normal scalp distribution of cortical SEPs (P2, N2, P3 and N3) by magnetic stimulation at L5 level. Filter setting: 8–300 Hz (−3dB)

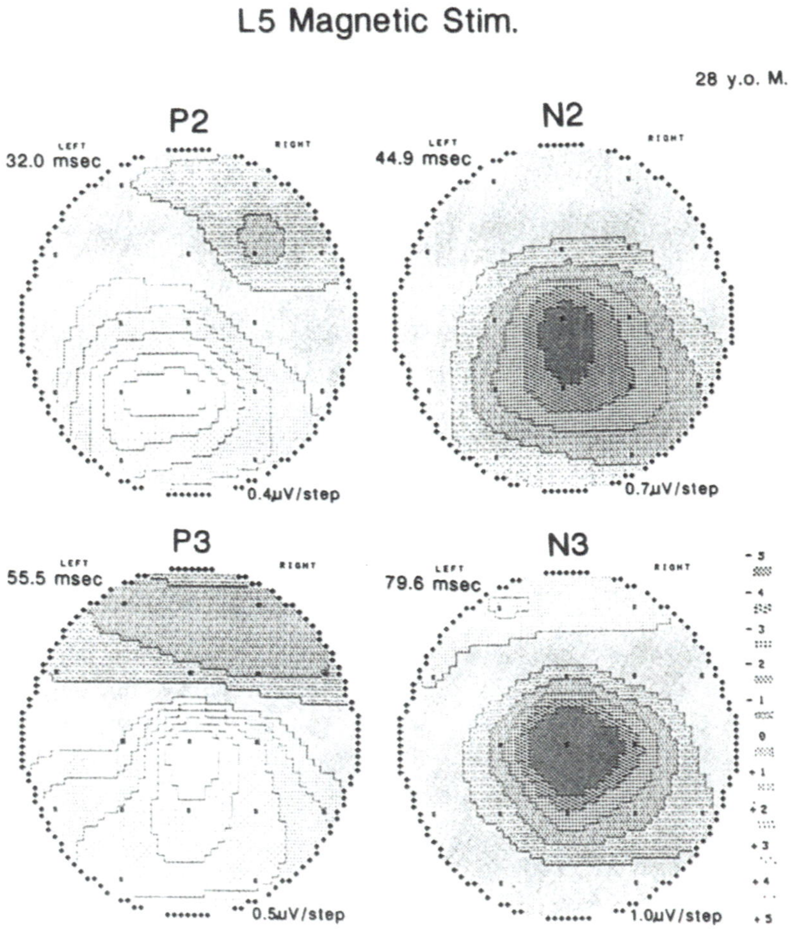

Fig. 5. Fixed time topographies of cortical SEPs (P2, N2, P3 and N3) by magnetic stimulation at L5 level. Scalp distribution of these components are shown in Fig. 4. The amplitude scales for positive (P2 and P3) and negative (N2 and N3) were divided into 5 classes: +1 to +5 and −1 to −5, respectively. 0 refers to the level of the baseline. The 16 black spots show the locations of the recording electrodes

Figure 6 shows the topographies of cortical SEPs elicited by magnetic stimulation at the Th5 level. There were no substantial differences in the topographies of P2, N2, P3 and N3 components evoked by Th5, Th12, L5 and mid-gluteus muscle stimulation.

Mid-parietal and ipsilateral parieto-central distributions of cortical SEPs were noted when the ankle was stimulated magnetically and electrically (Tsuji *et al.*, 1987).

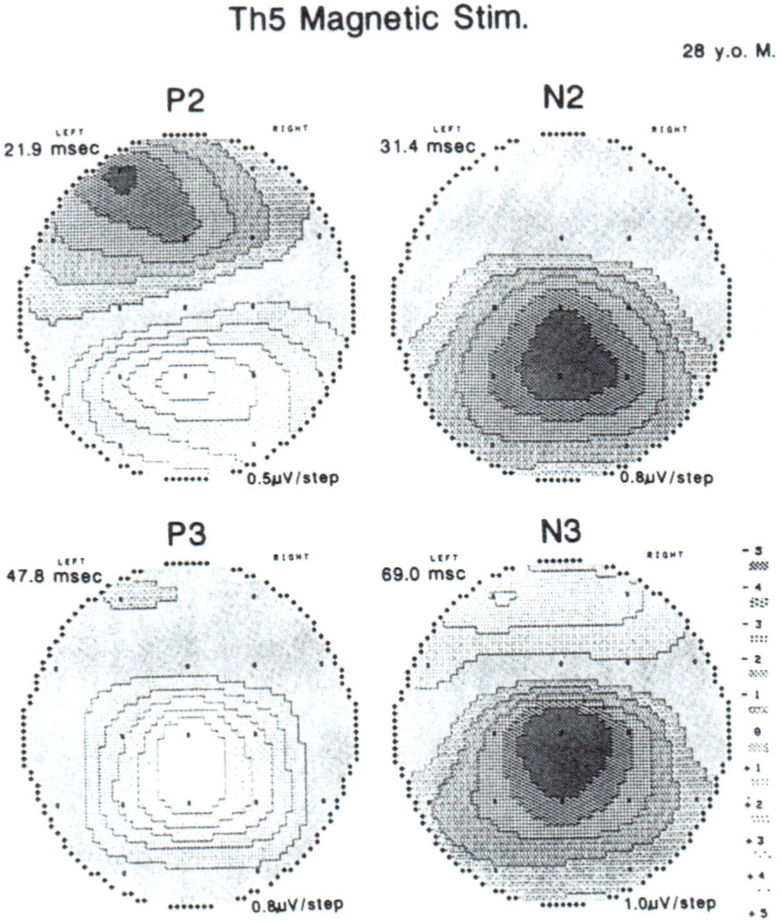

Fig. 6. Fixed time topographies of cortical SEPs (P2, N2, P3 and N3) by magnetic stimulation at Th5 level

Clinical applications in neurological patients

Figure 7 shows cortical SEPs by magnetic stimulation in a patient with a girdle sensation between Th3 and Th6 levels due to thoracic radiculopathy. The peak latencies of P2 were significantly delayed when Th2 and Th4 were stimulated. A further delay in the peak latency of P2 was noted when Th2 was stimulated. However, the peak latencies were normal when the levels below Th6 were stimulated. The peak latencies of P2, N2, P3 and N3 evoked by Th6, and Th10 stimulation were faster than those evoked by Th2 and Th4 stimulation. Short latencies and cortical SEPs elicited by electrical stimulation at the ankle were normal, and not optimally useful in this patient.

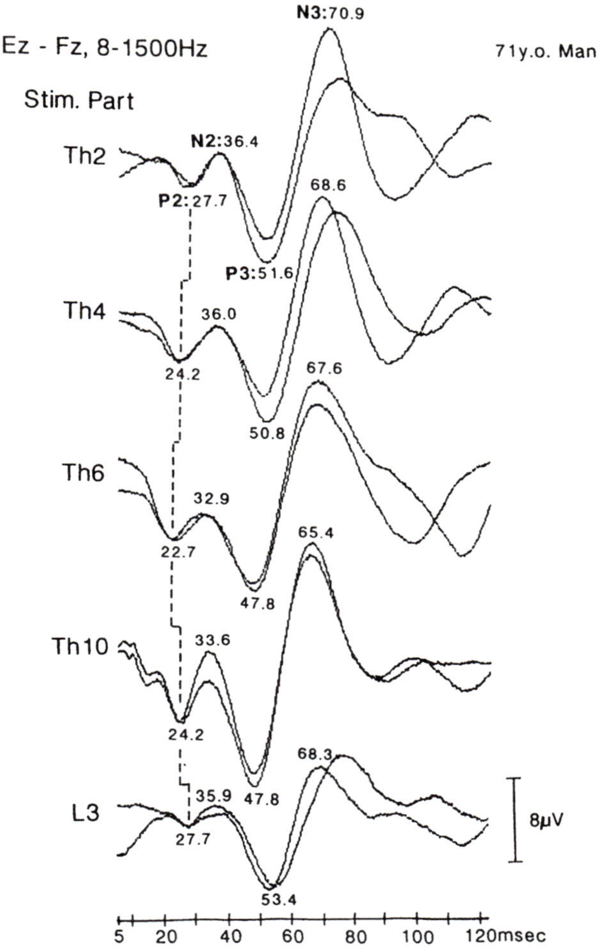

Fig. 7. Cortical SEPs by magnetic stimulation in a patient with a girdle sensation between Th3 and Th6 levels due to thoracic radiculopathy (From Tsuji *et al.*, 1993)

The patient with an acute disseminated encephalomyelitis (ADEM) between Th5 and Th10 showed normal cortical SEPs after the stimulation at Th2 and Th5 levels, and no response after the stimulation at Th10 and L3 levels (Fig. 8). Cortical SEPs elicited by electrical stimulation at the ankle were not induced in this patient.

In ten patients with polyneuropathy, P2 and N2 were normal in peak latencies when Th10, Th12, and the gluteus muscle were stimulated, but no responses were elicited by magnetic and electrical stimulation at the ankle. In eight patients with polyradiculoneuropathy, peak latencies and morphology of P2 and N2 evoked by stimulation at Th10 and Th12 levels were normal. The peak latencies of P2 and N2, however, were significantly delayed when L5 and ankle levels were stimulated magnetically.

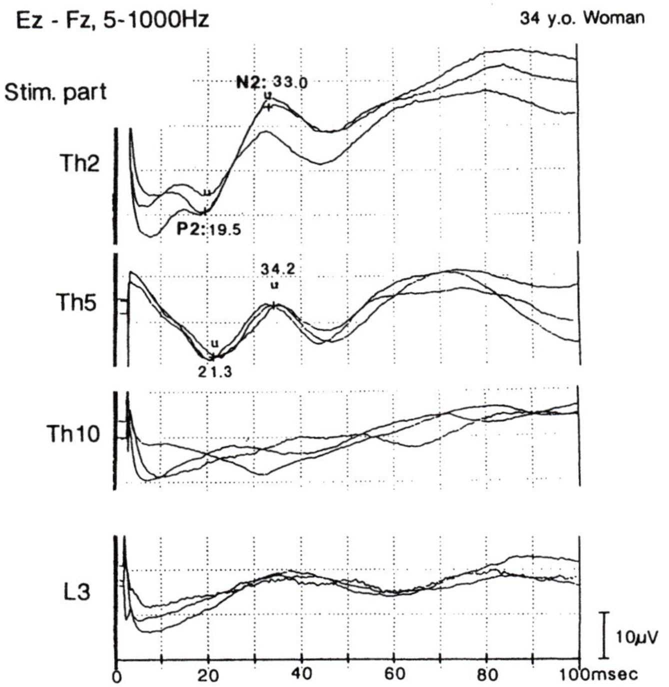

Fig. 8. Cortical SEPs by magnetic stimulation in a patient with an acute disseminated encephalomyelitis between Th5 and Th10 levels

In nine patients with thoracic myelopathy, peak latencies of P2 and N2 were significantly delayed when Th10 and Th12 were stimulated magnetically. There were no clear cortical responses by stimulation at the L5 and gluteus sites. Furthermore, no cortical SEPs were elicited by electrical stimulation at the ankle.

Serial SEPs following twin coil stimulation at each level were investigated in a patient with multiple sclerosis affecting the spinal cord between Th6 and Th8 levels. Before steroid therapy, abnormal SEPs were recorded when Th12 and L5 were stimulated. After the steroid therapy, SEPs elicited by twin coil stimulation at Th12 were normal in the morphology and peak latencies. In addition, there were clear responses elicited by the stimulation at L5 which showed a delayed peak latency of P2. SEPs by magnetic stimulation at the thoracic and lumbar root levels are useful for following their clinical course noninvasively.

In a patient with cervical spondylotic myelopathy at the C5 level, P2 latencies were significantly delayed when upper and lower thoracic levels were stimulated. Eight patients with cervical myelopathy also showed absent or delayed cortical SEPs by magnetic stimulation at the upper thoracic levels.

Discussion

SEPs elicited by electrical stimulation in the legs cannot be used to evaluate the function of the rostral somatosensory pathways when no subcortical nor cortical SEPs are evoked by stimulation of the leg nerves in the patients with polyneuropathy or lumbar radiculopathy. Lüders *et al.* (1982) developed a direct cauda equina stimulation technique that elicited SEPs more clearly. Their technique, however, is invasive because it requires lumbar puncture for the electrical stimulation at the cauda equina level. Goodridge *et al.* (1987) used two independent electrical stimulators for bilateral paraspinal stimulation which induced clear cortical SEPs. However, their peak latencies of cortical SEPs were slower than ours. These differences in peak latencies between magnetic and electrical stimulation at the thoracic and lumbar levels could be related to the difference in stimulation sites because the paraspinal electrical stimulation is probably applied to the trunk of the cutaneous branch of the dorsal primary ramus of the spinal nerves. On the other hand, we believe that the magnetic coil stimulates the spinal roots entering the spinal canal at the thoracic and lumbar levels. But, it is very difficult to determine the exact site of magnetic stimulation because the conductors of the human body are not homogeneous.

We applied magnetic stimulation to the spinal roots, and proximal and distal parts of the peripheral nerves, which is noninvasive and less painful than the electrical stimulation (Tsuji *et al.*, 1988). Furthermore, magnetic stimulation can be easily applied to any sections of the body between the upper thoracic vertebrae and foot. In normal subjects, cortical SEPs (P2, N2) which originate from the sensory cortex (Tsuji *et al.*, 1984, 1987), are easily recorded when the spinal roots or peripheral nerves are stimulated magnetically (Fig. 2).

Peak latencies of P2 evoked by ankle, mid-gluteus and L5 stimulation were correlated with body height, although P2 latencies evoked by Th10 and Th12 stimulation were independent of body height. This is in good agreement with our studies on SEPs following electrical stimulation at the ankle (Tsuji *et al.*, 1984). It was surprising to realize that all N2 latencies evoked by magnetic stimulation at Th10, Th12, L5, mid-gluteus and ankle were significantly correlated with body height. This suggests that P2 and N2 are induced by different neural networks in the cortex, or that N2 is induced by the slow afferent fibers. In the intensity study, N2 showed double peaks when low to moderate stimulus intensities were applied to the Th12 vertebra. Furthermore, in two patients with mild somatosensory dysfunction of lumbar roots or cord, N2 evoked by L5 stimulation revealed double peaks with prolonged interpeak latencies between the first peak of N2 and P3, which might be related to the temporal dispersion of the afferent fibers (Tsuji *et al.*, 1991). These data suggest that the N2 component is induced by both fast and slow afferent fibers, and slow afferent fibers of the somatosensory pathways might be correlated with body height.

The other possibility is that a spindle afferent discharge, related to the sudden paraspinal muscle contraction induced by magnetic stimulation, may initiate a delayed afferent volley which contributes to the N2 component.

There was no correlation between P3 or N3 latency and body height. In the topographical study, P3 and N3 components were distributed over the centro-parietal areas bilaterally and were at maximal amplitude at the vertex (Figs. 5, 6). These findings are compatible with that the origins of P3, and N3 components are a reflection of multiple generators in the postrolandic sensory cortex.

No peak latencies of cortical components elicited by magnetic stimulation at each level were affected by age within the age group studied (15–67 years). This observation is in good agreement with previous reports for lower extremity electrical stimulation (Cracco *et al.*, 1979, Dorfman 1977, Tsuji *et al.*, 1984, Chu, 1986).

Topographical studies of cortical SEPs by magnetic stimulation support the fact that the generators of early cortical SEPs (P2, N2, P3 and N3) following magnetic stimulation at the thoracic and lumbar levels are located in the sensory cortex of the interhemispheric fissure, and not in the frontal cortex.

Cortical SEPs by magnetic stimulation at the thoracic and lumbar root levels are useful in detecting lesions of the spinal cord or roots and for following their clinical course in patients with neurological diseases noninvasively. There were normal cortical SEPs with stimulation of spinal roots above the levels of lesions, and abnormal cortical SEPs with stimulation below the levels of lesions (Figs. 7, 8). However, the magnetic stimulation technique may mask unilateral spinal cord lesions because of the bilateral stimulation.

In other applications of magnetic stimulation, sympathetic skin responses and sudomotor potentials are recorded easily by magnetic stimulation of the neck, and are noninvasive and useful methods for evaluating the postganglionic sympathetic function clinically (Uozumi *et al.*, 1993; Matsunaga *et al.*, 1995).

References

Barker AT, Jalinous R, Freeston IL (1985) Non-invasive magnetic stimulation of human motor cortex. Lancet 2: 1106–1107

Chu N-S (1986) Somatosensory evoked potentials: correlations with height. Electroencephalogr Clin Neurophysiol 65: 169–176

Cracco JB, Cracco RQ, Stolove R (1979) Spinal evoked potential in man: a maturational study. Electroencephalogr Clin Neurophysiol 46: 58–64

Dawson GD (1947) Cerebral responses to electrical stimulation of peripheral nerve in man. J Neurol Neurosurg Psychiatry 10: 134–140

Dorfman LJ (1977) Indirect estimation of spinal cord conduction velocity in man. Electroencephalogr Clin Neurophysiol 42: 26–34

Eisen A, Odusote K (1980) Central and peripheral conduction times in multiple sclerosis. Electroencephalogr Clin Neurophysiol 48: 253–265

Goodridge A, Eisen A, Hoirch M (1987) Paraspinal stimulation to elicit somatosensory evoked potentials: an approach to physiological localizations of spinal lesions. Electroencephalogr Clin Neurophysiol 68: 268–276

Kunesch E, Knecht S, Classen J, Roick H, Tyercha C, Benecke R (1993) Somatosensory evoked potentials (SEPs) elicited by magnetic nerve stimulation. Electroencephalogr Clin Neurophysiol 88: 459–467

Lastimosa ACB, Bass NH, Stanback K, Norvell EE (1982) Lumbar spinal cord and early cortical evoked potentials after tibial nerve stimulation: effects of stature on normative data. Electroencephalogr Clin Neurophysiol 54: 499–507

Lüders H, Hahn J, Gurd A, Tsuji S, Dinner D, Lesser R, Klem G (1982) Surgical monitoring of spinal cord function: Cauda equina stimulation technique. Neurosurgery 11: 482–485

Matsunaga K, Uozumi T, Tsuji S, Murai Y (1995) Sympathetic skin responses evoked by magnetic stimulation of the neck. J Neurol Sci 128: 188–194

Merton PA, Morton HB (1980) Stimulation of the cerebral cortex in the intact human subject. Nature 285: 227–230

Mills KR, Murray NMF, Hess CW (1987) Magnetic and electrical transcranial brain stimulation: physiological mechanisms and clinical applications. Neurosurgery 20: 164–168

Polson MJR, Barker AT, Freeston IL (1982) Stimulation of nerve trunks with time varying magnetic fields. Med Biol Eng Comput 20: 243–244

Tsuji S, Lüders S, Lesser RP, Dinner DS, Klem G (1984) Subcortical and cortical somatosensory potentials evoked by posterior tibial nerve stimulation: normative values. Electroencephalogr Clin Neurophysiol 59: 214–228

Tsuji S, Murai Y (1987) Variability of initial cortical sensory evoked potentials to posterior tibial nerve stimulation. J UOEH 9: 287–298

Tsuji S, Murai Y, Yarita M (1988) Somatosensory potentials evoked by magnetic stimulation of lumbar roots, cauda equina, and leg nerves. Ann Neurol 24: 568–573

Tsuji S, Murai Y (1991) Cortical somatosensory potentials evoked by magnetic stimulation: effect of body height, age, and stimulus intensity. Electroencephalogr Clin Neurophysiol 80: 32–38

Tsuji S, Murai Y, Yarita M (1993) Cortical somatosensory potentials evoked by magnetic stimulation of thoracic and lumbar roots. Neurology 43: 391–396

Ueno S, Matsuda T, Hiwaki O (1990) Localized stimulation of the human brain and spinal cord by a pair of opposing pulsed magnetic fields. J Appl Physiol 67: 5838–5840

Uozumi T, Tsuji S, Murai Y (1991) Motor potentials evoked by magnetic stimulation of the motor cortex in normal subjects and patients with motor disorders. Electroencephalogr Clin Neurophysiol 81: 251–256

Uozumi T, Nakano S, Matsunaga K, Tsuji S, Murai Y (1993) Sudomotor potential evoked by magnetic stimulation of neck. Neurology 43: 1397–1400

Zhu Y, Starr A (1991) Magnetic stimulation of muscle evokes cerebral potentials. Muscle Nerve 14: 721–732

Correspondence: Dr. S. Tsuji, Department of Neurology, University of Occupational and Environmental Health, School of Medicine, Yahatanishi-ku, Kitakyushu, Fukuoka 807-8555, Japan.

Spinal Cord Monitoring
Eds: E. Stålberg, H. S. Sharma, Y. Olsson

Chapter 12

A combination of microdialysis, electrophysiology and histology for exploring secondary damage upon spinal cord injury

D. Liu

Departments of Neurology and Human Biological Chemistry and Genetics, University of Texas Medical Branch, Galveston

Secondary injury

Injury to the spinal cord is a major health care issue, causing lifelong disability to the victims. The damage worsens for hours after the initial trauma by secondary destructive processes that include release of agents that kill neurons, post injury ischemia, edema, inflammation, and oxidative damage. Agents suspected of causing secondary damage include excitatory neurotransmitters, free fatty acids, neuropeptides and free radicals. Oxidative damage to proteins, DNA and membrane lipids could be the final common pathway by which secondary injury substances cause neuronal degeneration. Reduction of secondary damage is the chief hope in the near future for reducing long term disability from spinal cord injury. A better understanding of the mechanisms behind secondary damage that accompany central nervous system trauma will pave the way to therapies that reduce damage following spinal cord injury.

Glutamate mediated neurotoxicity and oxidative damage

The excitatory neurotransmitter glutamate (Glu) kills neurons at high concentrations (Choi, 1985, 1987; Choi *et al.*, 1987; Olney *et al.*, 1971; Rothman and Olney, 1986). The release of excitatory amino acids (EAAs) (Demediuk *et al.*, 1989; Faden *et al.*, 1989; Panter *et al.*, 1990, Liu *et al.*, 1991; Liu and McAdoo, 1993a) is probably the first event following acute traumatic injury to the spinal cord. The excitation of neurons opens calcium channels increasing intracellular calcium levels ($[Ca^{2+}]_i$) (Young *et al.*, 1982; Young and Koreh, 1986; Young, 1992). This in turn activates phospholipases liberating free fatty acids (FFA), such as arachidonic acid from membrane phospholipids. Arachidonate is rapidly converted to prostaglandins (PGs) by cyclooxygenase, leukotrienes (LXs) and thromboxanes (TXs) by lipoxygenase, with the superoxide anion ($O_2^{\cdot -}$) produced as a by-product (see Fig. 1, Chan and Fishman, 1980; Chan *et al.*, 1988; Kuehl and Egan, 1980; Kontos *et al.*, 1984, 1985). Elevated $[Ca^{2+}]_i$ also activates calpain (Kinuta *et al.*, 1989; McCord, 1985; Traystman *et al.*, 1991) which converts xanthine dehydrogenase to xanthine oxidase. The latter enzyme produces $O_2^{\cdot -}$. There are two distinct forms of glutamate toxicity. The acute form – which is dependent on neuronal excitation – is characterized by neuronal swelling that results from the massive influx of water through Glu-gated cation channels. The delayed form – which is distinct from neuronal excitation – appears to lead to neuronal degeneration through pathways which involve oxidative damage (Coyle and Puttfarcken, 1993). NMDA significantly worsens trauma to the spinal cord (Faden and Simon, 1988). Administration of MK-801, an open channel blocker of the NMDA-type Glu receptor, reduces NMDA neurotoxicity (Olney *et al.*, 1987), and significantly improves neurological outcome following trauma to the spinal cord (Faden *et al.*, 1988). Lafon-Cazal and co-workers (1993) demonstrated that $O_2^{\cdot -}$ was produced following NMDA receptor stimulation in cultured cerebellar granule cells. MK-801, the competitive NMDA receptor antagonist APV, superoxide dismutase (SOD), the phospholipase A_2 inhibitor mepacrine, and elimination of external Ca^{2+} all reduced $O_2^{\cdot -}$ production. This suggests that NMDA induces a calcium- and arachidonate-dependent $O_2^{\cdot -}$ formation. Nitric oxide ($^{\cdot}NO$), which is involved in the neurotoxicity of glutamate under certain conditions (Dawson *et al.*, 1991, 1992), was also induced following NMDA receptor activation (Garthwaite *et al.*, 1989). Furthermore, pretreatment with either MK-801 or the antioxidant U74006F significantly reduced the $FeCl_2$-induced increase in capillary permeability (Zuccarello and Anderson, 1993), suggesting that free radicals may also enhance glutamate release, thereby providing a feedback amplification mechanism. In our previous studies, we found that generation of both the hydroxyl radical ($^{\cdot}OH$) (Liu *et al.*, 1994) and $O_2^{\cdot -}$ (Liu *et al.*, 1995) in rat spinal cord elevated extracellular levels of glutamate as well as other amino acids. Methylprednisolone (MP), a drug which is believed to be an antioxidant that improves long term recovery after spinal cord injury in humans, significantly reduces excitatory amino acid (EAA) release upon impact injury (Liu and McAdoo, 1993). These findings support the hypothesized feedback mechanism and correlate glutamate toxicity with oxidative damage.

Free radicals and oxidative damage

It has been hypothesized that free radical attack on membrane phospholipids, protein and DNA triggers oxidative damage to these molecules following CNS injury, and so destroy neurons. Free radical attack on polyunsaturated fatty acids (PUFA) of cell membranes triggers chain reactions that destroy membrane phospholipids (Ginsberg *et al.*, 1988; Watson and Ginsberg, 1989; Ikeda and Long, 1990; Jesberger and Richardson, 1991; Traystman *et al.*, 1991; Siesjo, 1993). Extensive experimental evidence suggests that peroxidative degradation of neuronal membrane phospholipids follows CNS injury (Demopoulos, 1980; Balentine 1985; Anderson *et al.*, 1985a,b; Demediuk *et al.*, 1985a,b, 1988; Hall and Braughler, 1986; Faden *et al.*, 1987; Cohadon et al., 1989; Saunders *et al.*, 1987). Free radical attack on proteins modifies amino acids, fragments polypeptide chains and generates cross links. There is compelling evidence demonstrating that oxidative damage to proteins occurs during brain injury caused by an ischemia/reperfusion insult (Oliver *et al.*, 1990). Oxidative damage rapidly inactivates enzymes (Oliver *et al.*, 1990; Stadtman, 1990). The aromatic amino acids tryptophan, tyrosine and phenylalanine, the sulfur-containing amino acids cysteine and cystine (Jesberger and Richardson, 1991), and lysine, proline, histidine, and arginine (Stadtman, 1990, Stadtman and Oliver, 1991) are all sensitive to oxidative damage. Exposed organic components of DNA are susceptible to $^{\cdot}OH$ attack, and $O_2^{\cdot -}$ has been postulated to break DNA strands (Jesberger and Richardson, 1991). Early events in mammalian cells subjected to oxidative stress seem to involve DNA damage and consequent activation of poly (ADP-ribose) synthetase (Halliwell, 1987, 1989). The predominant lesion is strand breakage (Birnboim, 1988), but DNA base modification products are also formed (Floyd and Carney, 1992). Oxidative damage to DNA by iron or copper ions produces mutagenic lesions (Loeb *et al.*, 1988; Tkeskelashvili *et al.*, 1991) and modifies bases such as thymine glycol, 5-hydroxymethyl-uracil and 8-hydroxyguanine (Floyd and Carney, 1992). Oxidative damage caused by free radical attack on DNA (Aruoma, *et al.*, 1989; Floyd *et al.*, 1986a,b, 1988; Imlay and Linn, 1988; Ito *et al.*, 1992; Mullar and Gurster, 1993) and on protein (Amici *et al.*, 1989; Cross *et al.*, 1992; Oliver *et al.*, 1990; Linseman *et al.*, 1993) has been demonstrated in in vitro experiments. Although a variety of free radicals may be involved in oxidative damage, $^{\cdot}OH$ is the most destructive one. Two major pathways for generating $^{\cdot}OH$ in vivo have been proposed. One is the iron-catalyzed Haber-Weiss reaction. The other is $^{\cdot}NO$- and $O_2^{\cdot -}$-mediated $^{\cdot}OH$ generation.

Oxygen radicals and Haber Weiss/Fenton reaction. Highly reactive oxygen-containing species that form upon CNS injury include $O_2^{\cdot -}$, $^{\cdot}OH$, hydroperoxides and hydrogen peroxide (H_2O_2). Superoxide is produced during normal metabolism through several pathways in aerobic cells (Ikeda and Long, 1990; Watson and Ginsberg, 1989). The major cellular sources are the electron transport chain of mitochondria and the endoplasmic reticulum. Other possible sources of $O_2^{\cdot -}$ include purine metabolism, PG synthesis, catecholamine metabolism and activated neutrophils. SOD converts $O_2^{\cdot -}$ into H_2O_2, and H_2O_2 is reduced to H_2O by catalase and selenium-dependent glutathione peroxidase. A dynamic equilibrium exists between the oxidative damage potential and the antioxidant defense capacity at all times (Floyd and Carney, 1992). Disruption of this balance can produce excess H_2O_2 and $O_2^{\cdot -}$. Excess $O_2^{\cdot -}$ production can damage tissue (Chan and Fishman, 1980, 1982; Chan *et al.*, 1988). This is supported by the observation that SOD has apparent therapeutic efficacy when administered by itself (Fridovich, 1986). However, the damage produced is slight in the absence of the ferrous ion, a strong catalyst for generating $^{\cdot}OH$ from H_2O_2 (Halliwell, 1989). Therefore, it has been hypothesized that the damaging species in vivo is $^{\cdot}OH$ produced by the iron catalyzed Haber-Weiss reaction:

$$2O_2^{\cdot -} + 2H^+ \rightarrow H_2O_2 + O_2 \quad (1)$$

$$Fe^{3+} + O_2^{\cdot -} \rightarrow Fe^{2+} + O_2 \quad (2)$$

$$Fe^{2+} + H_2O_2 \rightarrow Fe^{3+} + {}^{\cdot}OH + OH^- \quad (3)$$

Step (3) in the Haber-Weiss reaction is also named the Fenton reaction.

The hydroxyl radical reacts very rapidly with almost every molecule found in the cell, and there is no enzyme system to remove it (Jesberger and Richardson, 1991). The brain and the rest of the nervous system are especially prone to radical damage because they are highly enriched in easily peroxidizable PUFA side chains and iron. Cell injury by almost any mechanism has the potential to accelerate free radical reactions, partly because injured and lysed cells release their intracellular iron pools, thereby presumably generating $^{\cdot}OH$. Furthermore, the cerebrospinal fluid (CSF) that surrounds neurons has little or no iron-binding capacity and contains few antioxidants (Halliwell, 1989; Halliwell and Gutteridge, 1990). It has been demonstrated that brain injury increases $O_2^{\cdot -}$ (Kontos and Wei, 1986; Ikeda *et al.*, 1989a,b,c), $^{\cdot}OH$ (Cao *et al.*, 1988; Hall *et al.*, 1993) and low-molecular-weight bound iron (Cerchiari *et al.*, 1987). Injection of iron salts into rat brain causes transient focal epileptiform discharge, neuron death and lipid peroxidation (Willmore, 1980, 1986). Therefore, iron-catalyzed $^{\cdot}OH$ generation from $O_2^{\cdot -}$ probably is an important factor of secondary injury following trauma.

$^{\cdot}NO$- and $O_2^{\cdot -}$-mediated $^{\cdot}OH$ generation. Nitric oxide, a free radical, is synthesized from L-arginine (Knowles *et al.*, 1989; Palmer *et al.*, 1988; Schmidt *et al.*, 1988) by nitric oxide synthase (NOS). In the nervous system, $^{\cdot}NO$ acts as a neurotransmitter and mediates a variety of neuronal functions (Garthwaite, 1991; Snyder and Bredt, 1991; Snyder, 1992). At high concentrations, $^{\cdot}NO$ serves as a neurotoxin, mediating the neurotoxicity of glutamate (Dawson *et al.*, 1991, 1992). It plays a role in neuronal death after cerebral ischemia (Nowicki *et al.*, 1991). The degree of neuroprotection achieved with N^G-nitro-L-arginine, a NOS inhibitor, is greater than that achieved by MK-801. Superoxide dismutase and N^G-nitro-L-arginine both protect neurons from hypoxia/reoxygenation-induced injury (Cazevieille *et al.*, 1993). Beckman et al. (1990) proposed that

$^{\cdot}NO$ reacts with $O_2^{\cdot -}$ to form peroxynitrite anion ($ONOO^-$) in high yield. Peroxynitrite decomposes to $^{\cdot}OH$ and $NO_2^{\cdot -}$ when protonated.

$$O_2^{\cdot -} + {}^{\cdot}NO \rightarrow ONOO^- \xrightarrow{H^+} ONOOH \rightarrow {}^{\cdot}OH + NO_2^{\cdot -} \rightarrow NO^{3-} + H^+$$

This hypothesis is widely accepted (Kirsch *et al.*, 1992; Stamler *et al.*, 1992). Application of 3-morpholinosydnonimine (to generate $^{\cdot}NO$ and $O_2^{\cdot -}$) or S-nitrosocysteine (to generate $^{\cdot}NO$) to cerebrocortical cultures induced neuronal death in a dose-dependent manner. Neuronal death caused by these $^{\cdot}NO$ generators was completely blocked by SOD. In contrast, SOD/catalase did not prevent the lethal action of peroxynitrite on neurons (Lipton *et al.*, 1993). Peroxynitrite itself is neurotoxic (Ischiropoulos *et al.*, 1992). The $O_2^{\cdot -}$- and $^{\cdot}NO$-mediated neurotoxicity has been strongly supported by evidence from in vitro experiments (Hogg *et al.*, 1992; Radi *et al.*, 1991; Lipton *et al.*, 1993).

An *in vivo* experimental model for studying secondary injury

Our strategy for studying secondary injury includes two major procedures. The first is to measure the substances released following impact injury to the rat spinal cord using microdialysis sampling. A weight drop injury model originally devised by Allen (1911) and slightly modified by us is used to induce injury. The second is to administer those agents identified in post-trauma microdialysis to the spinal cord and to characterize the resulting damage. We established an experimental model to evaluate the damage caused by the administered agent in rat spinal cord (Liu and McAdoo, 1993b; Liu, 1994). In this model, microdialysis is used to administer suspected damaging agents and to sample the release of substances caused by the administered agent. To characterize the effects of the administered agent on neuronal damage, microdialysates are analyzed by HPLC to evaluate the biochemical changes caused by candidate administration, evoked potentials are monitored during sampling to evaluate blockage of axonal conduction caused by the administered agent, and neuronal death is assessed histologically. This model enables us to correlate electrophysiological, histological, and neurochemical parameters from the same experiment to assess a substance as an agent of secondary damage and to study the mechanism of the damage. It also can be used to study the efficacies of drugs in reducing secondary damage. Microdialysis is a powerful approach for exploring biochemical events during secondary damage in in vivo experiments.

Excitotoxicity in spinal cord injury studied using the *in vivo* model

To study excitotoxicity in spinal cord injury, neurotransmitter levels such as biogenic amines (Liu *et al.*, 1990) and excitatory amino acids (Liu *et al.*,

1991; Liu and McAdoo, 1993a) were measured from microdialysates following impact injury to the rat spinal cord. We found that extracellular levels of these neurotransmitters dramatically increased. We then tested whether potassium (Liu and McAdoo, 1993b) and glutamate (Liu *et al.*, submitted) could serve as agents of secondary damage. We also assessed the effects of glutamate receptor agonists in spinal cord injury (Liu, 1994). Microdialysis was used to administer glutamate or its agonists into the spinal cord and to sample the resulting release of amino acids. Damage caused by the administered substances was assessed by postmortem histological examination. Glutamate destroyed neurons for several hundred μm around the microdialysis fiber and caused the release of other amino acids. Estimated thresholds for destruction of cell bodies imply that glutamate damages spinal cord neurons *in vivo* at concentrations approaching those released upon spinal cord injury. Administering NMDA + kainate caused blockade of postsynaptic potentials while stimulating the sciatic nerve, release of other amino acids, and neuronal death (Liu, 1994). The neurotoxicity of NMDA + kainate is lower than the same concentration of glutamate suggesting that other glutamate receptor subtypes are involved.

Free radicals and peroxidative damage to membrane lipids

Figure 1 proposed mechanisms of membrane phospholipid degradation. Trauma causes Ca^{2+} influx which initiates the membrane lipid hydrolysis processes. Trauma also causes free radical formation which triggers membrane lipid peroxidation. There is a feedback mechanism between these two process. Our studies have strongly supported these pathways and their interactions.

Free radical formation following impact injury to the rat spinal cord

Microdialysis has limitations in the study of free radicals. The free radicals are typically too short-lived to cross the dialysis membrane, and the enzymes used to generate or remove free radicals are too high in molecular weight to cross the membrane in the dialysis zone. Therefore, it is desirable to retain the advantages of microdialysis while overcoming the inherent difficulties of this technique. To do this, we developed a perfusing cannula that can be inserted into the grey matter of the cord to collect the substances released after trauma and to administer substances into the cord. Two pairs of holes are made within a 2 mm zone through the wall of the cannula. The section of cannula containing the holes is placed in the gray matter of the cord. Since there is no membrane between tissue and the perfusing solution, molecules of any size with any chemical property may pass through the holes. In addition, short-lived free radicals can be reacted with trapping agents to produce stable substances that can be sampled by microdialysis. For example, $^{\cdot}OH$ will react with salicylate administered

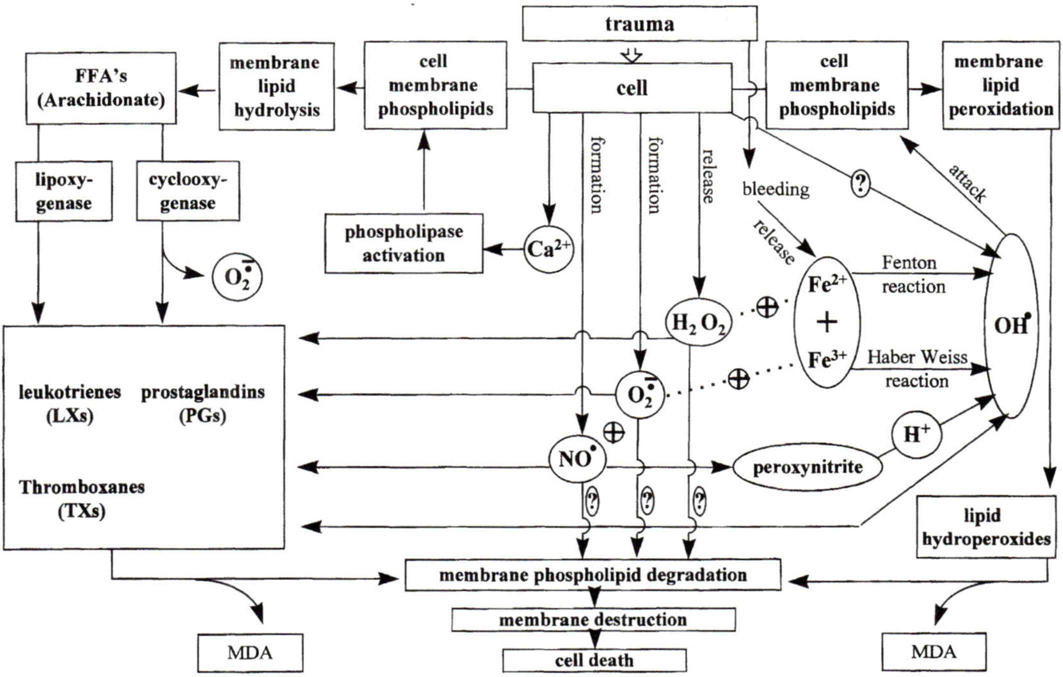

Fig. 1. Diagram of proposed pathways for membrane lipid degradation. Traumatic injury to neurons causes Ca^{2+} influx. The elevated $[Ca^{2+}]_i$ activates phospholipases which hydrolyze lipids to produce PGs, LXs and TXs and $O_2^{\cdot-}$. Trauma causes H_2O_2, $O_2^{\cdot-}$ and $^{\cdot}NO$ formation, and releases iron ions from binding protein which in turn catalyses $^{\cdot}OH$ formation by the Haber Weiss or Fenton reaction. Elevated $O_2^{\cdot-}$ and $^{\cdot}NO$ can react to produce $^{\cdot}OH$ through peroxynitrite. Free radicals trigger peroxidation of the membrane lipids to produce MDA. Free radicals can also attack the double bonds in FFAs to release MDA. Therefore the hydrolysis and peroxidation of membrane phospholipids cause membrane destruction and ultimately destroy neurons. This diagram also illustrates the interactions between these two pathways

through the dialysis fiber and the products are measured. Alteratively, the byproducts of the synthesis of such radicals can be analyzed.

The time courses of formation of $O_2^{\cdot-}$, $^{\cdot}OH$, and $^{\cdot}NO$ following impact injury to the rat spinal cord were measured using these techniques. The results reported below demonstrate that these free radicals are indeed generated in the extracellular space following trauma, supporting the hypothesis that these free radicals play a role in secondary CNS injury.

Nitric oxide formation. Analysis of $^{\cdot}NO$ in biological models has been reviewed by Archer (1993). The most powerful method to measure $^{\cdot}NO$ levels employs a NO-sensitive microelectrode (Takemura *et al.*, 1995; Mitsuhata *et al.*, 1995; Nagafuji *et al.*, 1995). It can measure changes in $^{\cdot}NO$ levels in second intervals, far exceeding the time limitation of other assays, and can be used directly in vivo. We used a nitric oxide-sensitive electrode to monitor the real time changes in nitric oxide levels following

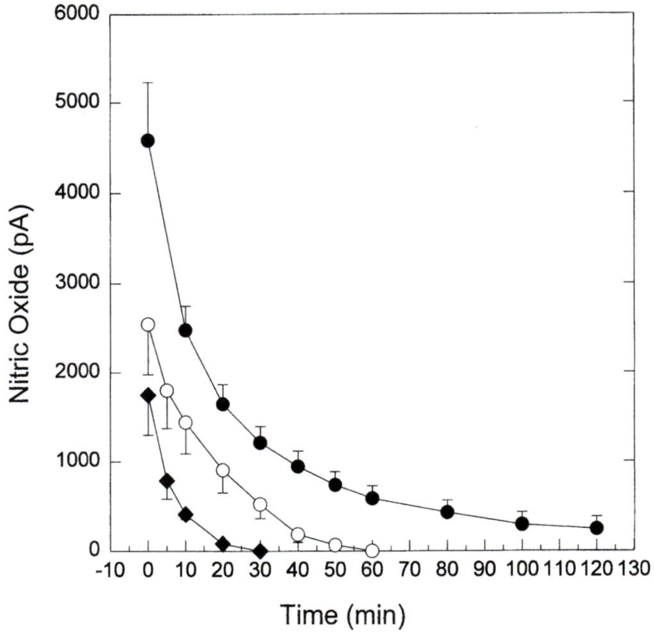

Fig. 2. The time course of ˙NO formation measured by the NO-selective electrode following spinal cord injury. The working electrode was inserted into the spinal cord, and the reference electrode was set close to the working electrode. After the initial rise of the current caused by electrode insertion reduced to a stable basal level, the instrument was adjusted to a current reading of zero, then a weight (5 g) was dropped 10 cm down to the cord (time 0). The recordings were continued until the current returned to the basal level. Nitro-L-arginine (1 and 10 mg/kg body weight), a NOS inhibitor, was injected (ip) 15 min prior to injury. Solid circle, impact injury without treatment ($N = 6$, $\pm$ SEM); open circle, 1 mg/kg nitro-L-arginine treated 15 min before impact injury ($N = 5$, $\pm$SEM); square, 10 mg/kg nitro-L-arginine treated before injury ($N = 3$, $\pm$SEM). The dose related inhibition of ˙NO formation by nitro-L-arginine demonstrates that the current changes detected by the electrode are due to ˙NO production

traumatic injury to the rat spinal cord (Liu *et al.*, manuscript in preparation). Figure 2 illustrates the time course of ˙NO formation following impact injury. The concentration of ˙NO dramatically increased to approximately 4500 pA immediately following the impact injury and gradually reduced to close to the basal level within 2 h. Nitro-L-arginine (1 and 10 mg/kg respectively, ip injection), significantly reduced the ˙NO level reached immediately following injury and shortened the time required to return to the basal level. The results demonstrate that traumatic injury indeed causes ˙NO formation.

Superoxide formation. As reviewed by Weber (1990), current methods for measuring $O_2^{\cdot -}$ levels are not specific. We have recently developed a unique *in vivo* detection method to measure $O_2^{\cdot -}$(Liu *et al.*, 1998). In this method, cytochrome *c* in artificial cerebrospinal fluid (ACSF) was administered through a perfusing cannula inserted laterally through the

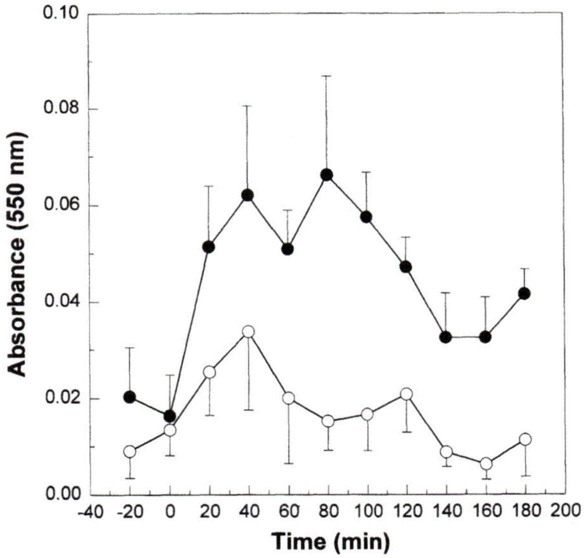

Fig. 3. The time course of $O_2^{\cdot -}$ formation following impact injury to the rat spinal cord. Cytochrome *c* (50 µM in ACSF) was pumped through a cannula inserted laterally through the cord. Perfusate collection was begun 75 min after insertion of the cannula into the cord. This allows the increase in $O_2^{\cdot -}$ formation due to the insertion of the cannula to subside to a stable baseline. Two control samples were collected at 20 min intervals to obtain the basal level, a weight drop injury (75 g.cm) was performed at time 0, and sample collection was continued for 3 h (closed circles, N = 5, ± SEM). In another series of experiments, SOD (60 U/ml) was included in the cytochrome *c* solution. The cytochrome *c* and SOD mixture was pumped through the cannula, and the cord was similarly injured (open circles, N = 5, ± SEM). The perfusates were measured spectrophotometrically at 550 nm. When SOD was present the absorbance was significantly reduced ($p < 0.01$) indicating that the reduction of cytochrome *c* was due to increased $O_2^{\cdot -}$ formation

spinal cord. Cytochrome *c* was reduced by $O_2^{\cdot -}$ formed following injury and reduced cytochrome *c* was determined spectrophotometrically. To validate the specificity of this method, SOD was added in the perfusion solution in another group of experiments. Figure 3 illustrates the time course of $O_2^{\cdot -}$ formation following impact injury to the rat spinal cord. The levels of $O_2^{\cdot -}$ increased about 3.5 fold following injury and remained at that level for 3 h. SOD significantly reduced this increase ($P < 0.01$), indicating that cytochrome *c* was reduced by $O_2^{\cdot -}$. These results demonstrate that trauma induces $O_2^{\cdot -}$ formation and that the combination of cannula sampling and cytochrome *c* perfusion can be used to measure $O_2^{\cdot -}$ in the spinal cord (Liu *et al.*, 1998).

Hydroxyl radical formation. Halliwell *et al.* have reviewed the methods for detecting $^{\cdot}OH$, including salicylate hydroxylation (Halliwell and Grootveld, 1987; 1988). The products of salicylate hydroxylation, 2,3- and 2,5-dihydroxybenzoic acid (2,3- and 2,5-DHBA) produced by $^{\cdot}OH$ attack on salicylate are separated by HPLC and detected by electrochemical detection (Floyd *et al.*, 1984, 1986c). In our studies, salicylate was administered through a microdialysis fiber into the rat spinal cord and 2,3- and 2,5-DHBA were measured by HPLC from microdialysates as a sign of ·OH formation following impact injury. We demonstrated that impact injury induced both 2,5- and 2,3-DHBA production (Fig. 4).

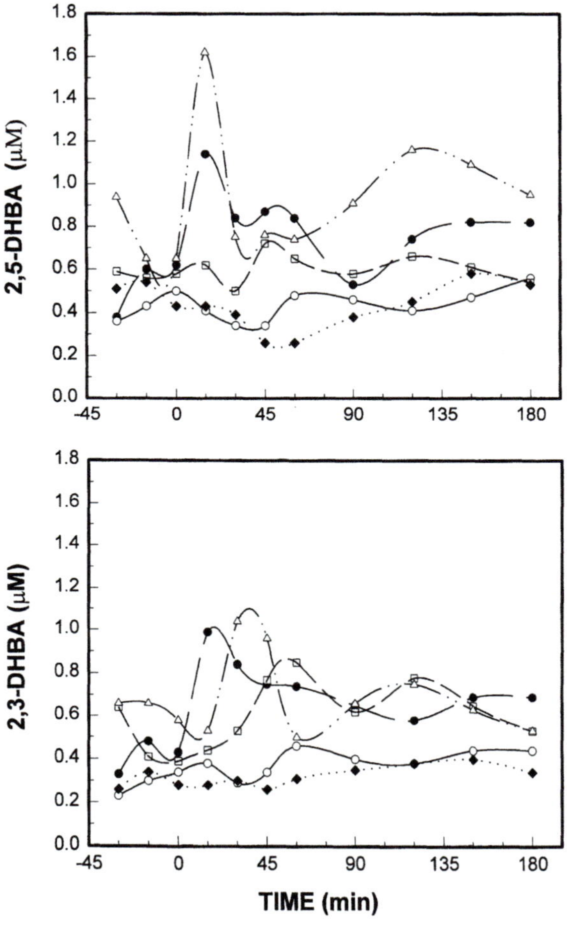

Fig. 4. The time course of hydroxyl radical formation following impact injury to the rat spinal cord. Following exposure of the cord, a microdialysis fiber was inserted laterally through the cord and the dialysis zone located in the cord. The ACSF was pumped through the fiber for 75 min to allow ·OH levels formed by fiber insertion to return to the basal level, then salicylate (5 mM in ACSF, pH adjusted to 7.2) was administered through the fiber as a trapping agent. After three control samples were collected, the cord was injured by the weight drop method (50 g.cm) and sampling was continued for another 3 h. The products of ·OH attack on salicylate, 2,3- and 2,5-DHBA were separated by HPLC with a C18 column (3 μm particle, 15 cm × 4.6 mm) and detected by electrochemical detection at 0.65 V. The mobile phase was 0.03 M potassium citrate and 0.03 M sodium acetate, and the flow rate was 0.7 ml/min. The data are presented for the individual experiments by different symbols (N = 5)

Neurotoxicity of free radicals characterized by the in vivo model

In an attempt to obtain in vivo evidence that free radicals kill neurons, we first generated free radicals in vivo, then characterized the damage caused by the radicals using our model. Since $^{\cdot}OH$ is extremely short-lived, with a half life of nanoseconds, it is very difficult to study *in vivo*. We generated $^{\cdot}OH$ in the rat spinal cord by administering Fenton's reagent through two parallel microdialysis fibers. Hydrogen peroxide and ferrous chloride were mixed in the gray matter of the cord around the fibers. Co-administration of phenylalanine and subsequent analysis of the dialysates indicated that large amounts of hydroxyphenylalanine isomers were formed, demonstrating that considerable levels of $^{\cdot}OH$ were generated (Liu, 1993). This unique method of generating $^{\cdot}OH$ provides a way to overcome the difficulty of studying the effects of free radicals *in vivo*. It also provides an animal model for studying mechanisms of radical damage and can be used to study the effects of radical scavengers and other drugs on radical damage. The damage caused by generated $^{\cdot}OH$ has been characterized using our model (Liu *et al.*, 1994). This provided direct *in vivo* evidence that $^{\cdot}OH$ causes neuron death, EAA release, $PGF_{2\alpha}$-like molecule release (Liu and Li, 1995), and axonal conduction blockage. Administration of either $FeCl_2$/EDTA or H_2O_2 alone had much smaller effects on the parameters measured (Liu *et al.*, 1994).

We also generated $O_2^{\cdot -}$ in rat spinal cord by administering paraquat – a $O_2^{\cdot -}$ generator – and characterized the resulting damage it caused (Liu *et al.*, 1995). The toxicity of paraquat is predominantly mediated by production of $O_2^{\cdot -}$ (Bus *et al.*, 1974; 1977). Neuronal damage was characterized using our model. We found that paraquat caused a great deal of neuronal death, increased amino acid concentrations and blocked electrical conduction (Liu *et al.*, 1995).

Degradation of membrane phospholipids

To study membrane lipid degradation after trauma, we measured the products of membrane lipid hydrolysis and peroxidation following impact injury to the rat spinal cord.

Extracellular levels of prostaglandin $F_{2\alpha}$-like (PGF_2-like) molecules rise following impact injury and MP reduces the increase. Prostaglandins are among the products of membrane lipid hydrolysis. We found that $PGF_{2\alpha}$-like molecule levels in the extracellular space, initially undetectable, increased to 304 nM in microdialysates following impact injury to the rat spinal cord. The $PGF_{2\alpha}$-like molecule was analyzed as its 3-bromomethy-6,7-dimethoxy-1-methyl-2(1H)-quinoxalinone (Br-DMEQ) derivative (Yamaguchi *et al.*, 1985, 1986) using fluorescence detection following HPLC separation. Methylprednisolone (30 mg/kg) reduced levels of $PGF_{2\alpha}$-like molecules (Liu *et al.*, manuscript in preparation). This result demonstrates that membrane lipid hydrolysis occurs and that MP blocks this degradation following traumatic injury.

Extracellular levels of malondialdehyde rise upon impact injury. To test the hypotheses in in vivo experiments that in CNS injury free radical attack on PUFA in cell membrane lipids triggers peroxidative damage processes, we measured malondialdehyde (MDA) – an end product of membrane lipid peroxidation – as a marker of peroxidative damage from microdialysates upon impact injury to the rat spinal cord.

MDA can be measured colorimetrically as its thiobarbituric acid (TBA) derivative (Uchiyama and Mihara, 1978; Ohkawa *et al.*, 1979). To obtain higher sensitivity, fluorometric detection of the MDA-TBA derivative was used (Yagi, 1976). To assure specificity, HPLC was used to separate MDA from other TBA positive substances prior to fluorescence detection (Tatum *et al.*, 1990; Li and Chow, 1994). We have recently improved this method (Qian and Liu, 1997). This is the most specific and sensitive method to measure MDA. Figure 5 illustrates experimental data obtained using this method following impact injury to the rat spinal cord. The extracellular levels of MDA gradually increased, peaking 5 hours post-trauma to 4.5-fold that of the basal level and then gradually reduced to about 2-fold of the basal level over the next 9 h. This time course of MDA production demonstrates that membrane lipid peroxidation occurs following trauma (Qian and Liu, 1997).

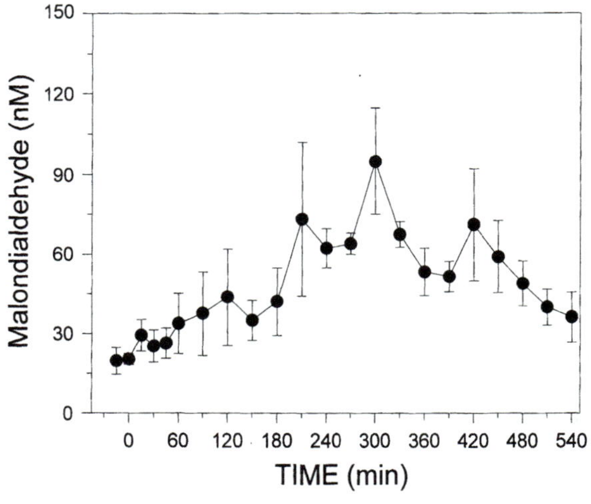

Fig. 5. The time course of MDA formation following impact injury to the rat spinal cord. The microdialysis fiber insertion and the impact injury method are the same as described in Fig. 4. Two 15 min samples were collected to obtain basal levels, and a weight drop injury (75 g.cm) was performed. The sampling continued for 9 h following injury. MDA in microdialysates was quantified as its TBA-derivative by the method developed in our lab. The microdialysates are directly reacted with TBA reagent at 95°C for 1 hour. The products of the TBA reaction are separated by HPLC with a C18 column (3 μm particle, 4.6 mm × 15 cm) at a flow rate of 1 ml/min and detected using a fluorescence detector at an excitation wavelength of 515 nm and an emission wavelength of 565 nm. The mobile phase was 33.3% methanol in water, containing 0.01 M NaAc (pH 3.8). MDA levels did not completely return to the basal level after 9 h (N = 5, ± SEM)

The correlation between free radical triggered peroxidation and hydrolysis of membrane lipids

We have investigated the mechanisms of membrane lipid peroxidation and studied the correlation between peroxidation and hydrolysis. We have also explored the mechanisms of action of MP in reducing secondary damage. We found that MP reduced the release of excitatory amino acids (Liu and McAdoo, 1993a), blocked ˙OH generation and MDA formation, and reduced $PGF_{2\alpha}$-like molecule production.

Evidence that free radicals trigger membrane lipid peroxidation. The most straightforward way to correlate oxygen radical production with membrane phospholipid peroxidation is to generate the radicals in vivo, measure the products of peroxidation and examine the resulting damage. We administered Fenton's reagent (estimated concentrations outside of fiber were 0.5 mM H_2O_2 and 0.02 mM $FeCl_2$/0.05 mM EDTA) to generate ˙OH in the rat spinal cord and measured MDA from microdialysates. We found that ˙OH generation induced MDA production in the rat spinal cord (Fig. 6). This provided direct in vivo evidence correlating free radical generation to

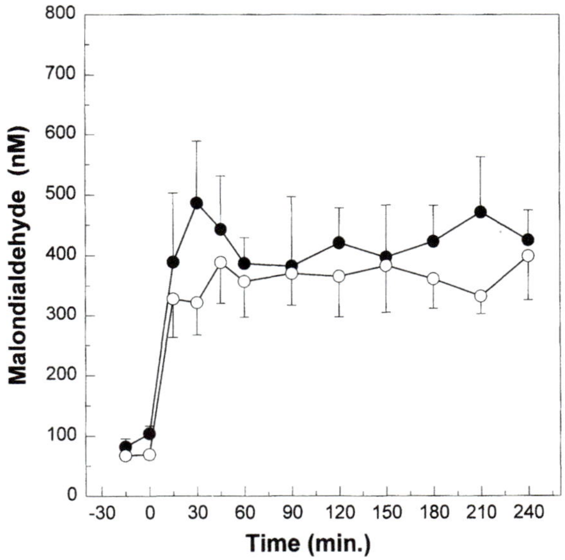

Fig. 6. The time course of MDA production in response to ˙OH generation and the effect of MP in reducing MDA production. Two 15 min samples were collected as controls during which only ACSF was administered through microdialysis fibers. At time 0, Fenton's reagent was administered, and sample collection continued. MP (60 mg/kg) was given in the MP-treated group (open circles, N = 9, ± SEM). The vehicle group received only the solution to dissolve MP (closed circles, N = 6, ± SEM). The bolus dose of MP was started at the same time as was the administration of Fenton's reagent and was completed in 15 min. The maintenance dose of MP was started 30 min after completion of the bolus dose and continued until the end of the experiment. MP did not significantly reduce MDA release ($p = 0.24$)

peroxidation of membrane lipids. MP (60 mg/kg) did not significantly reduce MDA release (p = 0.24, Liu *et al.*, manuscript in preparation) suggesting that the major mechanism by which MP reduces injury is by blocking the cascade of secondary membrane degradation, rather than by scavenging ·OH.

Evidence that free radicals induce membrane hydrolysis. To obtain direct evidence that ·OH can induce membrane hydrolysis, Fenton's reagents were administered (estimated concentrations outside of the fiber were 5 mM H_2O_2 and 0.5 mM $FeCl_2$/0.82 mM EDTA) to generate ·OH in the rat spinal cord. The time course of $PGF_{2\alpha}$-like molecule release in response to ·OH generation was determined from microdialysates collected from the fiber in which $FeCl_2$/EDTA was administered. We found that ·OH induced PGF_2-like molecule production. Administration of either 5 mM H_2O_2 or 0.5 mM $FeCl_2$/0.82 mM EDTA alone did not cause detectable release (Liu and Li, 1995).

Prostaglandin $F_{2\alpha}$ induces and MP blocks membrane lipid peroxidation. Administration of prostaglandin $F_{2\alpha}$ ($PGF_{2\alpha}$) into the rat spinal cord through a microdialysis fiber increased the concentration of MDA about 5 fold. MP (60 mg/kg) almost completely blocked MDA production (Fig. 7,

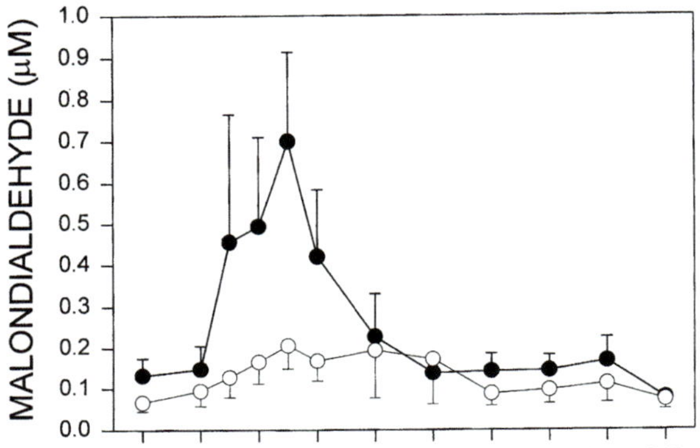

Fig. 7. The time course of MDA production in response to $PGF_{2\alpha}$ administration and the effect of MP in reducing MDA production. To determine the basal level of MDA production, two 15 min samples were collected during which only ACSF was administered through the microdialysis fiber into the gray matter of the rat spinal cord. At time 0, $PGF_{2\alpha}$ (1 mM) was administered through the fiber for one hour, then a maintenance concentration of $PGF_{2\alpha}$ (40 nM) was continued for another 3 h. This gives estimated concentrations outside the fiber wall of 100 μM and 4 nM. The latter is the concentration measured in rat CSF, while the former concentration is used to cause tissue damage. The vehicle group received only the solution used to dissolve MP (Solid circles, N = 5, ± SEM). In the MP-treated group, MP (60 mg/kg) was given (open circles, N = 6, ± SEM). The bolus doses started 30 min before $PGF_{2\alpha}$ administration and finished in 15 min. The maintenance dose of MP started 30 min after the bolus dose finished and continued until the end of the experiment. At 60 mg/kg MP almost completely prevented MDA production

Liu *et al.*, in preparation). This result demonstrates that the lipid hydrolysis product $PGF_{2\alpha}$ causes further peroxidative damage to membrane lipids, and that this can be blocked by MP.

Prostaglandin $F_{2\alpha}$ induces and MP blocks ˙OH generation. We also demonstrated that the lipid hydrolysis product $PGF_{2\alpha}$ causes free radical generation (Liu *et al.*, manuscript in preparation). Salicylate and $PGF_{2\alpha}$ were co-infused through a microdialysis fiber. The generation of ˙OH in response to $PGF_{2\alpha}$ administration was detected by measuring the products of ˙OH attack on salicylate, 2,3- and 2,5-DHBA. We found that 2,3- and 2,5-DHBA increased up to 3.6- and 2.6- fold, resp., and that MP (30 mg/kg) completely blocked ˙OH generation (Fig. 8). Thus, the product of membrane hydrolysis correlates with generation of ˙OH in in vivo experiments. A possible conclusion is that $PGF_{2\alpha}$ administration generates ˙OH, and ˙OH triggers the peroxidation of membrane lipids to produce MDA.

In summary, ˙OH generation induces membrane lipid degradation producing $PGF_{2\alpha}$-like molecules (Liu and Li, 1995) and MDA (Fig. 6), and PGF_2-like molecules in turn causes ˙OH (Fig. 8) and MDA (Fig. 7) production. These results demonstrate in in vivo experiments that (1) there is a feedback mechanism between free radicals and $PGF_{2\alpha}$; and (2) one of the mechanisms by which MP serves as an antioxidant is to block the pathway from $PGF_{2\alpha}$ to ˙OH, rather than to directly scavenge free radicals as previously presumed by others.

Oxidative damage to protein

Protein carbonyl content rises following impact injury. As an indicator of oxidative damage to protein, the carbonyl content of oxidized proteins is most popularly measured (Carney *et al.*, 1991; Levine *et al.*, 1990; Oliver *et al.*, 1987; 1990). We have measured the carbonyl content in protein using the method reported by Oliver *et al.*, (1990, 1987). The protein carbonyl content was determined spectrophotometrically using the 2,4-dinitrophenylhydrazine (DNPH)-labeling procedure (Oliver *et al.*, 1987). Injury was performed using the weight drop technique (75 g.cm). The spinal cord was first frozen by liquid nitrogen, then tissue was taken at 3 or 9 h after either impact injury or surgical control groups. The tissue was divided into three sections, with section 1 centered at the injury site, section 2 centered 7.5 mm away from the injury site, and section 3 centered 15 mm away. The protein carbonyl content of the control samples was approximately 3 nmol/mg. We found that there was a significant increase in the protein carbonyl content for each of the post-trauma sections ($p < 0.05$), that there was no significant difference for the protein carbonyl content between sections 1 and 2, and that section 3 had significantly higher level of oxidation at 3 h and lower level of oxidation at 9 h following injury than did sections 1 and 2. These results suggest that oxidative damage to proteins plays a role in tissue damage following traumatic spinal cord injury.

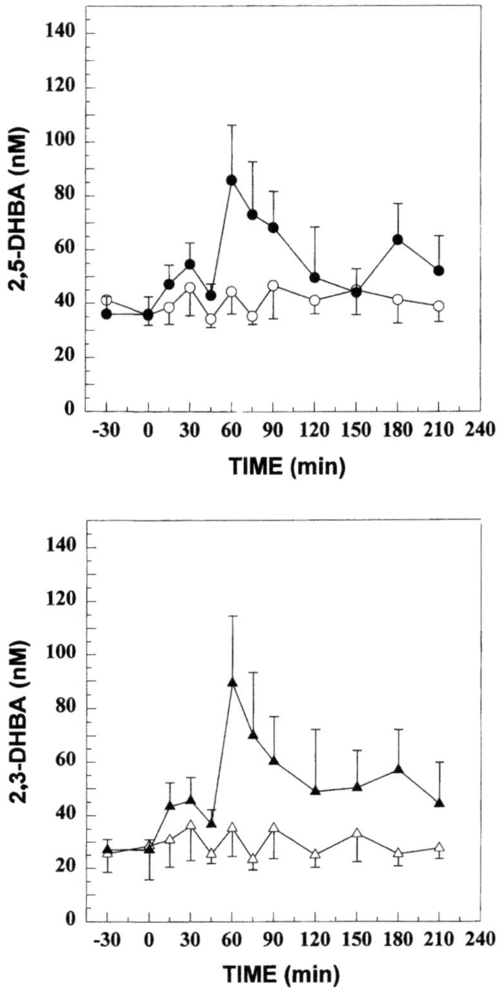

Fig. 8. The time course of ˙OH generation in response to $PGF_{2\alpha}$ administration and the effect of MP in blocking ˙OH generation. Salicylate was administered throughout the experiment to detect ˙OH as described in Fig. 4. To determine the basal levels of ˙OH, two 15 min samples were collected during which only ACSF and salicylate (5 mM in ACSF) were administered through a microdialysis fiber into the gray matter of the rat spinal cord. At time 0, $PGF_{2\alpha}$ (2 mM in ACSF) was administered through the fiber for 1.5 hours to cause damage, then a maintenance concentration of $PGF_{2\alpha}$ (40 nM) was continued for another 2.0 hours and the sample collection continued. This gives estimated concentrations outside the fiber wall of 200 μM and 4 nM of $PGF_{2\alpha}$. MP (30 mg/kg) was given in the MP-treated group. The vehicle group received only the solution used to dissolve MP. The bolus dose of MP was administered at the same time as $PGF_{2\alpha}$ and completed in 15 min. The maintenance dose of MP started 30 min after the bolus dose finished and continued until the end of the experiment. Solid symbols, vehicle group (N = 5, ± SEM); open symbols, MP-treated group (N = 6, ± SEM). At 30 mg/kg MP completely blocked ˙OH generation

The population of damaged protein determined by western blot. We also performed SDS-PAGE and immunoblot analysis on samples for each section from surgical control and injured tissue (Fig. 9). The Coomassie-stained gel of surgical control samples (no impact injury) for sections 1–3

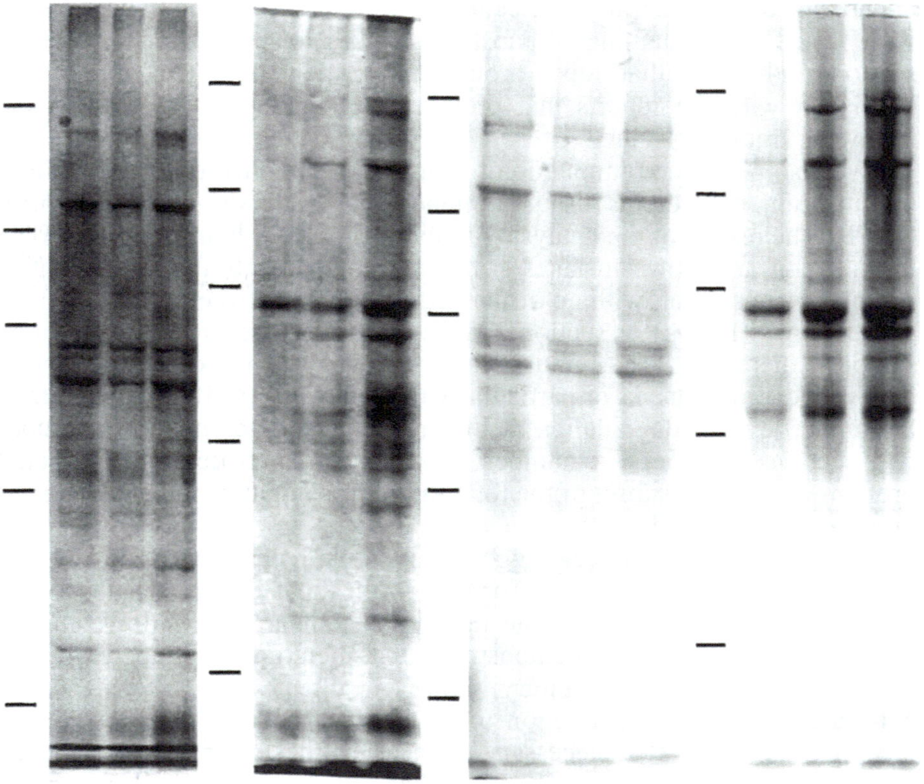

Fig. 9. SDS-PAGE and immunoblot analysis following impact injury. The spinal cord was first frozen by liquid nitrogen, then 3 sections of tissue were taken at 3 h after impact injury as described in the text. The total protein of these sections was analyzed by SDS-PAGE and immunoblot. Coomassie-stained SDS-PAGE gels of DNPH-labeled proteins are shown in lanes a–f. Lanes a–c show surgical control samples from tissue section 1–3, resp. Lanes d–f show the separated proteins of section 1–3, resp., from tissue 3 h post-injury. Substantial proteolysis occurred in section 1 and 2 following injury, which accounts for the lower overall staining and absence of some bands. The immunoblots of aliquots of the same samples run in lanes a–f are shown in lanes g–l. These blots were probed with an antibody directed against DNPH. Lanes g–i show surgical control samples from section 1–3, resp., lanes j–l show the samples from 3 h post-injury (section 1–3 resp.). A lower level of labeling is evident in section 1 than in section 2 and 3, probably due to proteolysis that occurred in section 1. The position of molecular weight markers are indicated by the bars, which are from the top 203 kDa, 118 kDa, 86 kDa, 51.6 kDa, and 31.4 kDa. Gels containing 8% (lanes a–c and lanes g–i) or 10% acrylamide (lanes d–f and lanes j–l) were run in 4 M urea; 25 µg protein was applied to each lane

are shown in lanes a–c, resp., while lanes d–f show samples from tissue 3 hours post-trauma. Samples from section 1 consistently showed a large amount of degradation in the 3 hour post-trauma sample (compare lane a with lane d). Samples from section 2 were more variable, but generally showed similar levels of degradation in the 3 hour post-trauma sample, (compare lane b with lane e), while samples from section 3 showed no apparently degradation in the 3 hour post-trauma sample (compare lane c with lane f). The staining of most bands was lower in the post-trauma section, indicating that most proteins were degraded. Our results are consistent with the reports that proteins are degraded at the injury site following traumatic injury.

We also performed immunoblot analysis on samples following DNPH-modification of proteins using a monoclonal antibody directed against DNPH. The results of the analysis for aliquots of the 3 hour samples shown in lanes a–f are shown in lanes g–l. The immunoreactivity of the control samples were similar, with no differences observed between section 1 (lane g), section 2 (lane h) or section 3 (lane i). In the case of the post-trauma samples, there was less immunoreactivity observed in section 1 (lane j) than in section 2 (lane k) or particularly in section 3 (lane l) probably due to the higher level of degradation that occurred in section 1 than in the other two sections. Although a number of proteins are oxidized, the labeled proteins represent a subset of the total sample, as not all Coomassie-stained proteins were labeled.

In summary, free radical-triggered oxidative damage to protein, DNA and membrane phospholipids probably is the final common pathway by which various secondary damage factors cause neuronal death. Extensive experimental evidence has demonstrated that membrane lipid peroxidation increases following CNS injury; however, direct *in vivo* evidence to correlate free radical production to membrane lipid peroxidation is still lacking. Therefore, it is necessary to measure the levels of free radicals following injury and to generate free radicals *in vivo* at the levels measured and assess the resulting damage to finally confirm that free radical-induced membrane lipid peroxidation is an important factor that contributes to neuronal death. Furthermore, oxidative damage to protein and DNA following spinal cord injury is almost unexplored and little information exists about oxidation of these macromolecules following any CNS trauma. To confirm that free radicals play an important role in CNS injury by causing oxidative damage to protein and DNA, the products of protein and DNA damage should be measured upon free radical generation. To address whether the delayed toxicity of excitatory neurotransmitters is related with free radical-triggered oxidation, the levels of free radicals formed and the products of oxidation to protein, DNA and membrane lipids should be measured following administration of these neurotransmitters at the levels induced by trauma.

Acknowledgments

The author gratefully thanks Michael L. Leski for his enthusiastic help in preparing this manuscript, the Paralyzed Veterans of America Spinal Cord Research Foundation (#1025), the Texas Higher Education Coordinating Board's Advanced Research Program (#4952) and NIH (NS34048) for financial support.

References

Allen AR (1911) Surgery of experimental spinal cord injury equivalent to crush injury of fracture dislocation of spinal column. JAMA 57: 878–880

Amici A, Levine RL, Tsai L, Stadtman ER (1989) Conversion of amino acid residues in proteins and amino acid homopolymers to carbonyl derivates by metal-catalyzed oxidation reactions. J Biol Chem 264: 3341–3346

Anderson DK, Demediuk P, Saunders RD, Dugan LL, Means ED, Horrocks LA (1985a) Spinal cord injury and protection. Ann Emer Med 14: 816–821

Anderson DK, Saunders RD, Demediuk P, Dugan LL, Braughler JM, Hall ED, Means ED, Horrocks LA (1985b) Lipid hydrolysis and peroxidation in injured spinal cord: partial protection with methylprednisolone or vitamin E and selenium. J Neurotrauma 2: 257–267

Archer S (1993) Measurement of nitric oxide in biological models. FASEB 7: 349–360

Aruonma OI, Halliwell B, Dizdaroglu M (1989) Iron ion-dependent modification of bases in DNA by the superoxide radical-generating system hypoxanthine/xanthine oxidase. J Biol Chem 264: 13024–13028

Balentine JD (1985) Hypotheses in spinal cord trauma research. In: Becker DP, Povlishock JT (eds) Central nervous system trauma status report. NIH, Bethesda, pp 455–461

Beckman JS, Beckman TW, Chen J, Marshall PA, Freeman BA (1990) Apparent hydroxyl radical production by peroxynitrite: implications for endothelial injury from nitric oxide and superoxide. Proc Natl Acad Sci USA 87: 1620–1624

Birnboim HC (1988) Superoxide anion may trigger DNA strand breaks in human granulocytes by acting at a membrane target. Ann NY Acad Sci 551: 82–94

Brien JF, McLaughlin BE, Nakatsu K, Marks GS (1991) Quantitation of nitric oxide formation from nitrovasodilator drugs by chemiluminescence analysis of headspace gas. J Pharmacol Meth 25: 19–27

Bus JS, Aust SD, Gibson JE (1977) Lipid peroxidation as a proposed mechanism for paraquat toxicity. In: Autor AP (ed) Biochemical mechanisms of paraquat toxicity. Academic Press, New York, pp 157–174

Bus JS, Aust SD, Gibson JE (1974) Superoxide- and singlet oxygen-catalyzed lipid peroxidaton a possible mechanism for paraquat (methyl viologen) toxicity. Biochem Biophys Res Commun 58: 749–755

Cao W, Carney JM, Duchon A, Floyd RA, Chevion M (1988) Oxygen free radical involvement in ischemia and reperfusion injury to brain. Neurosci Lett 88: 233–238

Carney JM, Starke-Reed PE, Oliver CN, Landum RW, Cheng MS, Wu JF, Floyd RA (1991) Reversal of age-related increase in brain protein oxidation, decrease in enzyme activity, and loss in temporal and spatial memory by chronic administration of the spin-trapping compound *N*-tert-butyl-α-phenylnitrone. Proc Natl Acad Sci USA 88: 3633–3636

Cazevieille C, Muller A, Meynier F, Bonne C (1993) Superoxide and nitric oxide cooperation in hypoxia/reoxygenation-induced neuron injury. Free Rad Biol Med 14: 389–395

Cerchiari EL, Hoel TM, Safar P, Sclabassi RJ (1987) Protective effects of combined superoxide dismutase and deferotamine on recovery of cerebral blood flow and functions after cardiac arrest in dogs. Stroke 18: 869–878

Chan PH, Chen SF, Yu ACH (1988) Induction of intracellular superoxide radical formation by arachidonic acid and by polyunsaturated fatty acids in primary astrocytic cultures. J Neurochem 50: 1185–1193

Chan PH, Fishman RA (1980) Transient formation of superoxide radicals in polyunsaturated fatty acid-induced brain swelling. J Neurochem 35: 1004–1007

Chan PH, Fishman RA (1982) Alterations of membrane integrity and cellular constituents by arachidonic acid in neuroblastoma and glioma cells. Brain Res 248: 151–157
Choi DW (1985) Glutamate neurotoxicity in cortical cell culture is calcium dependent. Neurosci Lett 58: 293–297
Choi DW (1987) Ionic dependence of glutamate neurotoxicity. J Neurosci 7: 369–379
Choi DW, Maulucci-Gedde M, Kriegstein AR (1987) Glutamate neurotoxicity in cortical cell culture. J Neurosci 7: 357–368
Cohadon F, Rigoulet M, Averet N, Arrigoni E (1989) Membrane damage in acute brain trauma. Ital J Neurol Sci 10: 147–155
Coyle JT, Puttfarcken P (1993) Oxidative stress, glutamate, and neurodegenerative disorders. Science 262: 689–695
Cross CE, Reznick AZ, Packer L, Davis PA, Suzuki YJ, Halliwell B (1992) Oxidative damage to human plasma proteins by ozone. Free Rad Res Commun 15: 347–352
Dawson VL, Dawson TM, London ED, Bredt DS, Snyder SH (1991) Nitric oxide mediates glutamate neurotoxicity in primary cortical cultures. Proc Natl Acad Sci USA 88: 6368–6371
Dawson TM, Dawson VL, Snyder SH (1992) A novel neuronal messenger molecule in brain: the free radical, nitric oxide. Ann Neurol 32: 297–311
Demediuk P, Daly MP, Faden AI (1989) Effect of impact trauma on neurotransmitter and nonneurotransmitter amino acids in rat spinal cord. J Neurochem 52: 1529–1536
Demediuk P, Faden AI, Romhanyi R, Vink R, McIntosh TK (1988) Traumatic brain injury in the rat: effects of lipid metabolism, tissue magnesium, and water content. J Neurotrauma 5: 105–119
Demediuk P, Saunders RD, Anderson DK, Means ED, Horrocks LA (1985a) Membrane lipid changes in laminectomized and traumatized cat spinal cord. Proc Natl Acad Sci USA 82: 7071–7075
Demediuk P, Saunders RD, Clendenon NR, Means ED, Anderson DK, Horrocks LA (1985b) Changes in lipid metabolism in traumatized spinal cord. Prog Brain Res 63: 211–226
Demopoulos HB, Flamm ES, Pietronigro DD, Seligman ML (1980) The free radical pathology and the microcirculation in the major central nervous system disorders. Acta Physiol Scand 492: 91–119
Faden AI, Chan PH, Longar S (1987) Alterations in lipid metabolism, Na^+, K^+-ATPase activity, and tissue water content of spinal cord following experimental traumatic injury. J Neurochem 48: 1809–1816
Faden AI, Demediuk P, Panter SS, Vink R (1989) The role of excitatory amino acids and NMDA receptors in traumatic brain injury. Science 244: 798–800
Faden AI, Lemke M, Simon RP, Noble LJ (1988) *N*-methyl-D-aspartate antagonist MK801 improves outcome following traumatic spinal cord injury in rats: behavioral, anatomic, and neurochemical studies. J Neurotrauma 5: 33–45
Faden AI, Simon RP (1988) A potential role for excitotoxins in the pathophysiology of spinal cord injury. Ann Neurol 23: 623–626
Floyd RA, Carney JM (1992) Free radical damage to protein and DNA: mechanisms involved and relevant observations on brain undergoing oxidative stress. Ann Neurol 32: S22–S27
Floyd RA, Henderson R, Watson JJ, Wong PK (1986c) Use of salicylate with high pressure liquid chromatography and electrochemical detection (LCED) as a sensitive measure of hydroxyl free radicals in driamycin treated rats. J Free Rad Biol Med 2: 13–18
Floyd RA, Watson JJ, Harris J, West M, Wong PK (1986a) Formation of 8-hydroxydeoxyguanosin, hydroxyl free radical adduct of DNA in graulocytes exposed to the tumor promotor, tetradeconylphorbolacetate. Biochem Biophys Res Communn 137: 841–846
Floyd RA, Watson JJ, Wong PK, Altmiller DH, Rickard RC (1986b) Hydroxyl free radical adduct of deoxyguanosine: sensitive detection and mechanisms of formation. Free Rad Res Comms 1: 163–172

Floyd RA, Watson JJ, Wong PK (1984) Sensitive assay of hydroxyl free radical formation utilizing high pressure liquid chromatography with electrochemical detection of phenol and salicylate hydroxylation products. J Biochem Biophys Meth 10: 221–235
Floyd RA, West MS, Eneff KL, Hogsett WE, Tingey DT (1988) Hydroxyl free radical mediated formation of 8-hydroxyguanine in isolated DNA. Arch Biochem Biophys 262: 266–272
Fridovich I (1986) Biological effects of the superoxide radical. Arch Biochem Biophys 247: 1–11
Garthwaite J (1991) Glutamate, nitric oxide and cell-cell signaling in the nervous system. Trends Neurosci 14: 60–67
Garthwaite J, Garthwaite G, Palmer RMJ, Moncada S (1989) NMDA receptor activation induces nitric oxide synthesis from arginine in rat brain slices. Eur J Pharmacol – Mol Pharmacol Sect 172: 413–416
Ginsberg MD, Watson BD, Busto R, Yoshida S, Prado R, Nakayama H, Ikeda M, Dietrich WD, Globus MY-T (1988) Peroxidative damage to cell membranes following cerebral ischemia. Neurochem Path 9: 171–193
Hall ED, Braughler JM (1986) Role of lipid peroxidation in post-traumatic spinal cord degeneration: A review. J Neurotrauma 3: 281–294
Hall ED, Andrus PK, Althaus JS, VonVoigtlander PF (1993) Hydroxyl radical production and lipid peroxidation parallels selective post-ischemic vulnerability in gerbil brain. J Neurosci Res 34: 107–112
Halliwell B (1987) Oxidants and human disease: some new concepts. FASEB J 1: 358–364
Halliwell B (1989) Oxidants and the central nervous system: some fundamental questions. Is oxidant damage relevant to Parkinson's disease, Alzheimer's disease, traumatic injury or stroke? Acta Neurol Scand 126: 23–33
Halliwell B, Grootveld M (1988) Methods for the measurement of hydroxyl radicals in biochemical systems: Deoxyribose degradation and aromatic hydroxylation. Meth Biochem Anal 33: 59–90
Halliwell B, Grootveld M (1987) The measurement of free radical reactions in humans. FEBS Lett 213: 9–14
Halliwell B, Gutteridge JMC (1990) The antioxidants of human extracellular fluids. Arch Biochem Biophys 280: 1–8
Hogg N, Darley-Usmar M, Wilson MT, Moncada S (1992) Production of hydroxyl radicals from the simultaneous generation of superoxide and nitric oxide. Biochem J 281: 419–424
Ikeda Y, Anderson JH, Long DM (1989a) Oxygen free radicals in the genesis of traumatic and peritumoral brain edema. Neurosurgery 24: 679–685
Ikeda Y, Brelsford KL, Ikeda K, Long DM (1989b) Oxygen-free radicals in traumatic brain edema. Neurol Res 11: 213–216
Ikeda Y, Ikeda K, Long DM (1989c) Comparative study of different iron-chelating agents in cold-induced brain edema. Neurosurgery 24: 820–824
Ikeda Y, Long DM (1990) The molecular basis of brain injury and brain edema: the role of oxygen free radicals. Neurosurgery 27: 1–11
Imlay JA, Linn S (1988) DNA damage and oxygen radical toxicity. Science 240: 1302–1309
Ischiropoulos H, Zhu L, Chen J, Tsai M, Martin JC, Smith CD, Beckman JS (1992) Peroxynitrite-mediated tyrosine nitration catalyzed by superoxide dismutase. Arch Biochem Biophys 298: 431–437
Ito K, Yamamoto K, Kawanishi S (1992) Manganese-mediated oxidative damage of cellular and isolated DNA by isoniazid and related hydrazines: non-Fenton-type hydroxyl radical formation. Biochemistry 31: 11606–11613
Jesberger JA, Richardson JS (1991) Oxygen free radicals and brain dysfunction. Intern J Neurosci 57: 1–17
Kinuta Y, Kimura M, Itokawa Y, Ishikawa M, Kikuchi H (1989) Changes in xanthine oxidase in ischemic rat brain. J Neurosurg 71: 417–420

Knowles RG, Palacios M, Palmer RMJ, Moncada S (1989) Formation of nitric oxide from L-arginine in the central nervous system: a transduction mechanism for stimulation of the soluble guanylate cyclase. Proc Natl Acad Sci USA 86: 5159–5162

Kontos HA, Wei EP (1986) Superoxide production in experimental brain injury. J Neurosurg 64: 803–807

Kontos HA, Wei EP, Ellis EF, Jenkins LW, Povlishock JT, Rowe GT, Hess ML (1985) Appearance of superoxide anion radical in cerebral extracellular space during increased prostaglandin synthesis in cats. Circ Res 57: 142–151

Kontos HA, Wei EP, Povlishock JT, Christman CW (1984) Oxygen radicals mediate the cerebral arteriolar dilation from arachidonate and bradykinin in cats. Circ Res 55: 295–303

Kuehl FA Jr, Egan RW (1980) Prostaglandins, arachidonic acid, and inflammation. Science 210: 978–984

Lafon-Cazal M, Pietri S, Culcasi M, Bockaert J (1993) NMDA-dependent superoxide production and neurotoxicity. Nature 364: 535–537

Levine RL, Garland D, Oliver CN, Amici A, Climent I, Lenz A-G, Ahn B-W, Shaltiel S, Stadtman ER (1990) Determination of carbonyl content in oxidatively modified proteins. Meth Enzymol 186: 464–478

Li XY, Chow CK (1994) An improved method for the measurement of malondialdehyde in biological samples. Lipids 29: 73–75

Linsemann KL, Larson P, Braughler JM, McCall JM (1993) Iron-initiated tissue oxidation: lipid peroxidation, vitamin E destruction and protein thiol oxidation. Inhibition by a novel antioxidant. Biochem Pharmacol 45: 1477–1482

Lipton SA, Chol Y-B, Pan Z-H, Lel SZ, Chen H-SV, Sucher NJ, Loscalzo J, Singel DJ, Stamler JS (1993) A redox-based mechanism for the neuroprotective and neurodestructive effects of nitric oxide and related nitrosocompounds. Nature 364: 626–632

Liu D (1993) Generation and detection of hydroxyl radical in vivo in rat spinal cord by microdialysis administration of Fenton's reagents and microdialysis sampling. J Biochem Biophys Meth 27: 281–291

Liu D (1994) An experimental model combining microdialysis with electrophysiology, histology, and neurochemistry for studying excitotoxicity in spinal cord injury: effect of NMDA and kainate. Mol Chem Neuropath 23: 77–92

Liu D, Li L (1995) Prostaglandin $F_{2\alpha}$ rises in response to hydroxyl radical generated *in vivo*. Free Rad Biol Med 18: 571–576

Liu D, McAdoo DJ (1993a) Methylprednisolone reduces excitatory amino acid release following experimental spinal cord injury. Brain Res 609: 293–297

Liu D, McAdoo DJ (1993b) An experimental model combining microdialysis with electrophysiology, histology, and neurochemistry for exploring mechanisms of secondary damage in spinal cord injury: the effect of potassium. J Neurotrauma 10: 349–362

Liu D, Sybert TE, Qian H, Liu J (1998) Superoxide production after spinal cord injury detected by microperfusion of cytochrome C. Free Rod Biol Med 25: 298–304

Liu D, Thangnipon W, McAdoo DJ (1991) Excitatory amino acid rise to toxic levels upon impact injury to rat spinal cord. Brain Res 547: 344–348

Liu D, Valadez V, Sorkin LS, McAdoo DJ (1990) Norepinephrine and serotonin release upon impact injury to rat spinal cord. J Neurotrauma 7: 219–227

Liu D, Xu GY, Pan E, McAdoo DJ (1998) Glutamate neurotoxicity in the spinal cord: A histological and neurochemical study. Neuroscience (submitted)

Liu D, Yang J, Li L, McAdoo DJ (1995) Paraquat – a superoxide generator – kills neurons in the rat spinal cord. Free Rad Biol Med 18: 861–867

Liu D, Yang R, Yan X, McAdoo DJ (1994) Hydroxyl radicals generated in vivo kill neurons in spinal cord: electrophysiological, histological, and neurochemical results. J Neurochem 62: 37–44

Loeb LA, James EA, Waltersdorph AM (1988) Mutagenesis by the autoxidation of iron with isolated DNA. Proc Natl Acad Sci USA 85: 3918–3922

McCord JM (1985) Oxygen-derived free radicals in postischemic tissue injury. N Engl J Med 312: 159–163

Mitsuhata H, Shimizu R, Yokoyama MM (1995) Role of nitric oxide in anaphylactic shock, J Clin Immun 15: 277–283
Muller K, Gurster D (1993) Hydroxyl radical damage to DNA sugar and model membranes induced by anthralin (dithranol) Biochem Pharmacol 17: 1695–1704
Nagafuji T, Sugiyama M, Matsui T, Muto A, Naito S (1995) Nitric oxide synthase in cerebral ischemia. Possible contribution of nitric oxide synthase activation in brain microvessels to cerebral ischemic injury. Mol Cell Neuropath 26: 107–157
Nihei H, Kanemitsu H, Tamura A, Oka H, Sano K (1989) Cerebral uric acid, xanthine, and hypoxanthine after ischemia: the effect of allopurinol. Neurosurgery 25: 613–617
Nowicki JP, Duval D, Poignet H, Scatton B (1991) Nitric oxide mediates neuronal death after focal cerebral ischemia in the mouse. Eur J Pharmacol 204: 339–340
Ohkawa H, Ohishi N, Yagi K (1979) Assay for lipid peroxides in animal tissues by thiobarbituric acid reaction. Anal Biochem 95: 351–358
Oliver CN, Levine RL, Stadtman ER (1987) A role of mixed-function oxidation reactions in the accumulation of altered enzyme forms during aging. JAGS 35: 947–956
Oliver CN, Starke-Reed PE, Stadtman ER, Liu GJ, Carney JM, Floyd RA (1990) Oxidative damage to brain proteins, loss of glutamine synthetase activity, and production of free radicals during ischemia/reperfusion-Induced injury to gerbil brain. Proc Natl Acad Sci USA 87: 5144–5147
Olney JW, Ho OL, Rhee V (1971) Cytotoxic effects of acidic and sulphur containing amino acids on the infant mouse central nervous system. Exp Brain Res 14: 61–76
Olney J, Price M, Salles KS, Labruyere J, Frierdich G (1987) MK-801 powerfully protects against *N*-methyl aspartate neurotoxicity. Eur J Pharmacol 141: 357–361
Palmer RMJ, Ashton DS, Moncada S (1988) Vascular endothelial cells synthesize nitric oxide from L-arginine. Nature 333: 664–666
Panter SC, Yum SW, Faden AI (1990) Alteration in extracellular amino acids after traumatic spinal cord injury. Ann Neurol 27: 96–99
Qian H, Liu D (1997) The time course of malondialdehyde production following impact injury to rat spinal cord as measured by microdialysis and high pressure liquid chromatography. Neurochem Res 22: 1231–1236
Radi R, Beckman JS, Bush KM, Freeman BA (1991a) Peroxynitrite oxidation of sulfhydryls: the cytotoxic potential of superoxide and nitric oxide. J Biol Chem 266: 4244–4250
Radi R, Beckman JS, Bush KM, Freeman BA (1991b) Peroxynitrite-induced membrane lipid peroxidation: the cytotoxic potential of superoxide and nitric oxide. Arch Biochem Biophys 288: 481–487
Rothman SM, Olney JW (1986) Glutamate and the pathophysiology of hypoxic-ischemic brain damage. Ann Neurol 19: 105–111
Saunders RD, Dugan LL, Demediuk P, Means ED, Horrocks LA, Anderson DK (1987) Effects of methylprednisolone and the combination of α-tocopherol and selenium on arachidonic acid metabolism and lipid peroxidation in traumatized spinal cord tissue. J Neurochem 49: 24–31
Schmidt HHHW, Nau H, Wittfoht W, Gerlach J, Prescher K-E, Klein MM, Niroomand F, Bohme E (1988) Arginine is a physiological precursor of endothelium-derived nitric oxide. Eur J Pharmacol 154: 213–216
Siesjo BK (1993) Basic mechanisms of traumatic brain damage. Ann Emerg Med 22: 959–969
Snyder SH (1992) Nitric oxide: first in a new class of neurotransmitters? Science 257: 494–496
Snyder SH, Bredt DS (1991) Nitric oxide as a neuronal messenger. Trends Pharmacol Sci 12: 125–128
Stadtman ER (1990) Metal ion catalyzed oxidatives of proteins: biochemical mechanism and biological consequences. J Free Rad Biol Med 9: 315–325
Stadtman ER, Oliver CN (1991) Metal-catalyzed oxidatives of proteins. J Biol Chem 266: 2005–2008

Stamler JS, Singel DJ, Loscalzo J (1992) Biochemistry of nitric oxide and its redox-activated forms. Science 258: 1898–1902
Takemura H, Tamaoki J, Tagaya E, Chiyotani A, Konno K (1995) Isoproterenol increases Cl diffusion potential difference of rabbit trachea through nitric oxide generation. J Pharmacol Exp Ther 273: 584–588
Tatum VL, Changchit C, Chow CK (1990) Measurement of malondialdehyde by high performance liquid chromatography with fluorescence detection. Lipids 25: 226–229
Tkeshelashvili LK, McBride T, Spence K (1991) Mutative spectrum of copper-induced DNA damage. J Biol Chem 266: 6401–6406
Traystman RJ, Kirsch JR, Koehler RC (1991) Oxygen radical mechanisms of brain injury following ischemia and reperfusion. J Appl Physiol 71: 1185–1195
Uchiyama M, Mihara M (1978) Determination of malondialdehyde precursor in tissues by thiobarbituric acid test. Anal Biochem 86: 271–278
Watson BD, Ginsberg MD (1989) Ischemic injury in the brain. Role of oxygen radical-mediated processes. Ann NY Acad Sci 559: 269–281
Weber GF (1990) The measurement of oxygen-derived free radicals and related substances in medicine. J Clin Chem Clin Biochem 28: 569–603
Willmore LJ, Triggs WJ, Gray JD (1986) The role of iron-induced hippocampal peroxidation in acute epileptogenesis. Brain Res 382: 422–426
Willmore LJ, Ballinger WE Jr, Boggs W, Sypert GW, Rubin JJ (1980) Dendritic alterations in rat isocortex within an iron-induced chronic epileptic focus. Neurosurgery 7: 142–146
Yagi K (1976) A simple fluorometric assay for lipoperoxide in blood plasma. Biochem Med 15: 212–216
Yamaguchi M, Fukuda K, Hara S, Nakamura M (1986) Fluorimetric high-performance liquid-chromatography of prostaglandins and its application to their determination in human seminal fluid. J Chromatogr 380: 257–265
Young W (1992) Role of calcium in central nervous system injuries. J Neurotrauma 9: 1 S9–S25
Young W, Yen V, Blight A (1982) Extracellular calcium ionic activity in experimental spinal cord contusion. Brain Res 253: 105–113
Young W, Koreh I (1986) Potassium and calcium changes in injured spinal cords. Brain Res 365: 42–53
Zuccarello M, Anderson DK (1993) Interaction between free radicals and excitatory amino acids in the blood-brain barrier disruption after iron injury in the rat. J Neurotrauma 10: 397–403

Correspondence: Dr. D. Liu, Gail Borden Bldg. 0653, Department of Neurology, University of Texas Medical Branch, Galveston, TX 77555-0653, USA.

Section IV

Spinal cord evoked potentials: pathophysiology and pharmacological aspects

Spinal Cord Monitoring
Eds: E. Stålberg, H. S. Sharma, Y. Olsson

Chapter 13

Spinal cord bioelectrical activity, edema and cell injury following a focal trauma to the rat spinal cord. An experimental study using pharmacological and morphological approaches

T. Winkler[1], **H. S. Sharma**[2], **E. Stålberg**[1], and **J. Westman**[2]

[1] Department of Clinical Neurophysiology, University Hospital,
[2] Laboratory of Neuroanatomy, Department of Medical Cell Biology, Biomedical Centre,
Uppsala University, Uppsala, Sweden

Introduction

Trauma to the spinal cord is associated with problems of sensory and motor disturbances along with abnormalities of reflex activities (Faden, 1987; Tator and Fehlings, 1991; Faden and Salzman, 1992; Faden, 1993a; Schwab and Bartholdi, 1996). The prevalence of spinal cord injury is about 30 to 50 cases per million population per year in the United States of America which is quite comparable to Europe and other continents (Tator and Edmonds, 1979; Schwab and Bartholdi, 1996) (Table 1). The victims of cord trauma are mainly young men in their early 20 to 30 years of age and only 20 to 30% of spinal cord injury cases involves women. The major cause of spinal cord injury is motor vehicle accidents followed by fall, penetrating injuries like gun shot, knife wounds or sports injuries (Tator and Fehlings, 1991; Schwab, and Bartholdi, 1996). The cervical spinal cord and thoracolumbar junctions are mostly affected following such injuries (Kakulas, 1984; Kakulas and Taylor, 1991). About half of the patients with cord trauma have complete injuries without any signs of voluntary motor or sensory perception below the level of the lesion (Schwab and Bartholdi, 1996). The symptoms of spinal cord injury includes quadruplegia that involves paralysis of the upper and lower extremities followed by paraplegia involving only the lower extremity. The other leading causes of paralysis involving spinal cord are multiple sclerosis, ischemia and tumours (Kakulas, 1984; Ducker, 1976).

Although spinal cord injury is not so common in human population it still requires great attention because the victims depending on the severity of injury can suffer a permanent disability (Schwab and Bartholdi, 1996).

Table 1. Epidemiology of spinal cord injury

Statistics
USA: 30–50 cases per million population per year, about 8–10,000 new cases per year
UK: 20–30 cases per million population per year, about 750 new cases each year
Sweden: 8–12 per million population per year, about 80 cases per year
Traumatic cases of SCI
motor vehicle accidents (<50%)
falls
penetrating injuries
gun shot
knife wound
sports accidents
Non-traumatic SCI and paralysis
cancer
multiple sclerosis
Location of SCI
cervical spinal cord
thoracolumbar junctions
Functional deficits
quadriplegia (about 54%)
paraplegia (about 46%)
complete injury (<50%)
Gender
young men (<61%)
16–30 years of age
women (<20–30%)

Modified after Schwab and Bartholdi (1996); compiled from various sources.

Still today, no suitable therapy for victims of spinal cord trauma is available. Thus in order to improve the quality of life of trauma patients development of new therapeutic strategies and possibility of regeneration of lesioned spinal cord axons are needed in near future (see Schwab and Bartholdi, 1996; Hagg and Oudega; Sharma *et al.*, in this volume).

To understand the spinal cord function in both clinical and in experimental situation, bioelectrical activity of the cord has been used since last several decades after the fundamental description of "intermediary cord potentials" by Gasser and Graham (1933), (for reviews see Gasser and Graham, 1933; Ducker, 1976; Willis Jr, 1980; 1984). However, its use in assessing spinal cord dysfunction following trauma in relation to the neurological outcome still remains controversial.

Recent developments in the techniques of recording spinal cord evoked potentials (SCEP) using electrodes placed in different places such as bone, muscles and in epidural space after stimulation of either peripheral nerves, parts of cerebral structures or even spinal cord has increased our knowledge in this field tremendously (Homma and Tamaki, 1984; Schramm and Jones,

1985; Dimitrijevic *et al.*, 1992). New knowledge from these studies suggest that spinal cord bioelectrical activity can be used to assess spinal cord functions which seems to be a reliable indicator of spinal cord pathology (Sharma *et al.*, 1991; Winkler *et al.*, 1993; Winkler, 1994; Winkler *et al.*, 1994a,b; 1995a,b; 1997; 1998).

This review is focused on the use of spinal cord bioelectrical activity in assessing spinal cord pathology based on our own investigations in a rat model of spinal trauma. In addition, the basic aspects of spinal cord bioelectrical activity and pathophysiology of spinal cord injury is reviewed. Finally, pharmacological influence on SCEP following trauma in relation with edema and cell injury is also discussed.

Spinal cord bioelectrical activity

Various extracellular potentials from the spinal cord can be recorded following stimulation of primary afferent fibers, motor axons and long spinal tracts (Willis Jr, 1980; 1984). These spinal cord potentials are generated as a result of current flow in the extracellular space which mainly comprise of (a) compound action potentials in nerve fibres or spinal neurons, (b) synaptic potentials generated in spinal cord interneurons, and/or (c) depolarisation of nerve terminals in primary afferent fibers. The shape, amplitude and latency of these bioelectrical potentials can be used as signals for assessment of the spinal cord function. Bioelectrical signals can be recorded either from the dorsal nerve roots (Gasser and Graham, 1933), spinal segments (Eccles *et al.*, 1954), higher segments on the dura (Shimoji *et al.*, 1972), vertebral elements (Nemeth *et al.*, 1978) and/or from the cerebral cortex (Dawson, 1947; Perot, 1973; Ducker and Brown, 1989).

Spinal cord potentials are quite comparable to the cerebral evoked potentials produced by afferent volleys or by antidromic volleys in corticofugal neurons (for details see Purpura, 1959). The neural elements are mainly responsible for these bioelectrical activity. It may be possible that glial cells which outnumber the neurons in the CNS (Duffy, 1984) also contribute to these bioelectrical potentials. However, this is not established yet and require additional investigation.

Spinal cord electrical potentials are first demonstrated more than 100 years ago by Gotch and Horsley (1891). The subject, however, did not get much attention before the description of "intermediary cord potentials" often called as cord dorsum potentials by Gasser and Graham (1933). The intermediary cord potential represents a series of electrical events which can be recorded from the dorsal surface of the spinal cord following stimulation of primary afferent fibers. These potentials exhibit a compound action potential followed by a series of negative waves and a long-lasting positive wave. The negative waves reflect excitatory events in the interneurons of the dorsal horn (Bernhard, 1953) whereas the positive wave is associated with the inhibitory activity (Hughes and Gasser, 1934).

Electrical activity generated in the spinal cord results in development of a potential gradient in the extracellular space along the dorsal and ventral root filaments. Potentials recorded from the dorsal root is known as "dorsal root potentials" (Barron and Matthews, 1938) whereas the most common potentials from ventral root is recorded as reflex discharges (Renshaw, 1940; Araki *et al.*, 1960) and slow potentials reflecting synaptic activity in motoneurons (Brooks and Fuortes, 1952; Araki *et al.*, 1960).

Generation of spinal cord potentials

Several mechanisms are involved in the generation of spinal cord potentials (see Willis Jr, 1980; Willis *et al.*, 1980; Willis Jr, 1984) which include compound action potentials in nerve fibers or in spinal neurons, excitatory and inhibitory postsynaptic potentials in interneurons or motoneurons, and depolarisation of primary afferent nerve fibres.

The compound action potential recorded extracellularly in a volume conductor such as spinal cord (Rall, 1977) basically has a biphasic (positive-negative) or a triphasic (positive-negative-positive) wave form which depends on the site of recording electrode (see Lorente de Nó, 1947). On the other hand, synaptic potentials are monophasic spinal cord potentials which arise due to asymmetric changes in membrane potential (cf Klee and Rall, 1977; Nicholson, 1973). Since synapses are most common on dendrites of interneurons and motoneurons, and also found on the nerve cell body, activation of soma-dendritic synapse is mainly responsible for synaptic potential (Eccles, 1964). In case of excitatory post-synaptic potential, the permeability of the membrane to sodium and potassium is increased and will result in a net influx of positively charged ions into the excited neurons (Eccles, 1957). In this case the potential in the extracellular space is positive outside the axon and negative outside the somatic-dendritic area. On the other hand, inhibitory post-synaptic potentials will result in positive extracellular fields around the nerve cell bodies and dendrites and negative fields along the axons (Coombs *et al.*, 1956).

In addition to these potentials, electrotonic conduction potentials can be recorded from axons in the dorsal and ventral roots which are primarily synaptic potentials due to the activity of axo-axonal synapses, a feature not common in the spinal cord (Schmidt, 1971; Burke and Rudomin, 1977). Depolarization of primary afferent fibers may also contribute to these potentials because of the leakage of potassium into the extracellular environment (Barron and Matthews, 1938; Kríz *et al.*, 1974). The process of "primary afferent depolarisation" is common event in many spinal cord potentials (Wall, 1958; Eccles *et al.*, 1962).

Apart from origin of these potentials in the extracellular space, recording of electrical activity also depends on the geometric arrays of neuronal population. Some neuronal arrangements are aligned to produce potentials which are added together, whereas an opposite alignment of neurons may produce potential fields that will cancel the electrical field potentials. Thus placement of the recording electrode is an important aspect in recording these bioelectrical activity from the spinal cord (Willis Jr, 1984).

Analysis of spinal cord potentials

The spinal cord electrical activity is generated by neurons located within the spinal cord which can be recorded either from the dorsal surface of the spinal cord or from dorsal root filaments. Several studies in the past have recorded spinal cord electrical activity within the spinal cord near the source of generation of active electrical field (Campbell, 1945; Austin and McCouch, 1955; Coombs *et al.*, 1956). These investigations suggest that if a microelectrode is inserted into the spinal cord at various lengths and the potentials are recorded following stimulation of peripheral afferents, the field potentials near the dorsal surface of the cord is negative followed by the positive wave. However if the microelectrode will be inserted further deep, the size and signs of the different component of these potentials are altered. The negative wave reverses to positive in sign when the tip of the microelectrode reaches near the central canal. In this situation, the positive wave reverses to become negative within the dorsal horn (see Willis Jr, 1984).

These field mapping studies suggest that the lateral half of the dorsal horn is negative, whereas most of the ventral horn is electrically positive. The exact distributions of the field potentials will vary with the particular nerve stimulated and with the different segmental level where the recordings are made.

Factors influencing spinal cord evoked potentials

The shape, latency and amplitude of spinal cord evoked potentials are influenced by several factors (see Reger *et al.*, 1987).

Noise and stimulation

Evoked potentials are recorded using differential amplifier using a gain of 4×10^5, analogue filtered between 0.1 to 2200 Hz and averaged between 20 to 600 individual responses (Reger *et al.*, 1980). These parameters enhance signal to noise ratio to an acceptable level. In such a set-up, the combined electronic, environmental and biological noise, excluding large artefact noise such as electrocardiogram signals, exhibits normal distribution with an average root mean square amplitude of 1.7 microvolts. This level of noise is independent of the evoked response and, the signal to noise ratio following stimulation of peripheral nerves ranged between 0.4 to 2.6 when the recordings were made from the various levels of the spinal cord (Reger *et al.*, 1980).

Changes in stimulus parameters also considerably influence the shape of evoked potentials. The peak of individual evoked potential amplitudes appears to be smoothened with increase in stimulus pulse frequency without any marked alterations in latency. With an increase in stimulus pulse amplitude, the peak to peak amplitude of the evoked potentials increased in size without any appreciable changes in the latency. Increased stimulus pulse duration from 50 to 200 microseconds did not influence either amplitude or latency of the evoked potentials (Reger *et al.*, 1987).

Method of recording

The shape of evoked potential amplitude is dependent on the position of recording electrodes. The main negative wave is always present which can be used for analysis of spinal cord function. Recordings with electrodes placed in spinal column over thoracic, lumbar and cervical regions in cats suggest that the shape of evoked potentials after stimulation of peripheral sural nerves at the ankle depends on the positions of the recording electrodes on the spinal cord and their depth within the spinal column (for details see Reger *et al.*, 1987). Increase in latency is evident with movement of recording electrodes towards cranium. A decrease in depth of recording electrodes within the spinal column shows a reduction in signal complexity, flattening of the peaks and broadening of the signals. Placement of the recording electrodes over the surface of dura without in contact with the underlying tissue resulted in further shortening of amplitude by about 50% and decreased the quality of evoked signals. These observations point out the importance of the placement of recording electrodes over spinal cord in assessing the nature of evoked potentials and their relationship with the cord function.

Effect of anaesthesia

The depth of anaesthesia significantly influences the shape of evoked spinal cord potentials (see Salzman *et al.*, 1990; Chapter 21, this volume). With increasing dose of pentobarbital the level of anaesthesia was increased which is associated with a decrease in background noise activity. This is followed by a slight decrease in amplitude of the evoked potential response without any significant effect on latency for the first 30 min of recording in cat experiments (Reger *et al.*, 1987). With further increase in depth of anaesthesia, the amplitude is further reduced and a slight but significant increase in latency was observed in rats (Winkler *et al.*, unpublished observations).

Recordings made under urethane in a dose which achieve stage IV of anaesthesia showed most reliable recordings of spinal cord evoked potentials throughout a period of 5 to 6 h (Sharma *et al.*, unpublished observation). These observations suggest that anaesthetics can influence remarkably the shape of evoked spinal cord potentials (Salzman *et al.*, 1990).

Effect of spinal cord injury

In both clinical and in experimental conditions spinal evoked potentials are used to assess the magnitude and complexity of the cord injury. Mechanical, chemical, ischemic or anoxic lesions significantly influence the spinal cord conduction. Mechanical trauma caused by contusion of the dura results in the rapid loss of the spinal cord evoked signals. Impact injury to the cord produced by 400 g cm at T12 level in cats results in an immediate disappearance of evoked potentials (Reger *et al.*, 1980). A broad positive injury potential appeared 15 min after trauma in this experiment which reflects severe contusion of the dura and the spinal cord (Reger *et al.*, 1987).

These observations show that spinal evoked potentials are sensitive to cord injury and reflect the severity of trauma. Both neural and vascular mechanisms contribute to the alterations in spinal cord conduction. Haemorrhages, edema and anoxia also seem to participate in the disturbances of spinal cord evoked signals, however, this aspect is not well examined in details.

Studies carried out in the last 25 years suggest that in severe spinal cord injuries, electrical conduction of the injured cord ceases immediately which later spreads to the entire length of the spinal cord by mechanisms of neuronal depression and/or neuronal shock (Ducker *et al.*, 1971; Perot, 1973; Ducker, 1976). A complete loss of the cortical evoked potential occurs if a direct impact of 500 g cm force on the lower level of the dorsal spinal cord in monkeys results in paraplegia (Singer *et al.*, 1970; Ducker *et al.*, 1971; Ducker, 1976). On the other hand, some recovery of evoked potentials can be seen after trauma, if animals do not develop paraplegia. In few cases, cord conduction was good at the time of injury but worsened later. These observations suggest that evoked potentials reflect cord pathology which is progressive with time.

On the other hand, there are reports which suggest that spinal cord evoked potentials are not related to the progressive cord injury or spinal cord function. The spinal cord conditions of animals or patients improve with time, or remain static while the pathology of the spinal cord shows a progressive increase (Ducker *et al.*, 1971; Ducker, 1976). Although the pathological, biochemical and ischemic changes within the cord were progressing, the magnitude of depression of evoked potentials did not differ between 5 min and 1 h after the initial impact. In many cases, spinal cord conduction returned to normal after trauma and the animal showed considerable signs of improvement. In some paraparetic animals, spinal cord conduction appeared normal, whereas in some cases, the animal behaved like normal but spinal cord conduction was significantly depressed. These observations on evoked potentials are misleading. Thus, the use of spinal cord bioelectrical activity as an indicator of cord pathology is still controversial.

Evoked potentials are related earlier with the neurological outcome in experimental and in clinical situations (Ducker, 1976; Schramm and Jones, 1985). No attempts were made to modify the evoked potentials with drug-treatments before injury and to follow the consequences of trauma in relation to the cord pathology (Homma and Tamaki, 1984). Thus, use of pharmacotherapeutic agents to influence SCEP following trauma in relation to cord pathology is still a new subject which require further investigation.

Studies carried out in our laboratory show that the neuroprotective compounds which reduce the cell injury and edema seem to have beneficial effects on the SCEP changes seen immediately after injury (Sharma *et al.*, 1991; Winkler, 1994). This observation opened a new vista of research and suggest that the cord neurochemistry before injury is important in determining the outcome of the electrical events and/or pathological disturbances in the spinal cord.

Basic mechanisms of spinal cord injury

In the following section a brief survey over the currently established mechanisms responsible for trauma induced cord pathology is described.

The concept of primary and secondary injury mechanisms

There are two basic mechanisms of spinal cord injury induced tissue damage (de la Torre, 1981). The primary injury mechanisms represent the direct physical injury to the spinal cord. This is followed by secondary injury mechanisms which comprise the release of one or several factors following primary injury leading to tissue destruction (Sandler and Tator, 1976a,b; Tator, 1991; Tator and Fehlings, 1991). The concept of secondary injury mechanisms was first described by Allen (1911) on the basis of his meticulous observation in dogs that myelotomy and removal of post-traumatic hematomyelia resulted in improvement of neurological function. Allen (1914) suggested that the noxious agent present in haemorrhagic necrotic material is responsible for tissue damage and the injurious substance is a "biochemical factor". This was the first experimental evidence of a biochemical theory of post-traumatic "autodestruction" of spinal cord. Since then, research in many laboratories resulted in discovery of numerous biochemical substances that are now established as major causes of cell and tissue injury in the cord following trauma (for review see Tator and Fehlings, 1991; Faden and Salzman, 1992; Faden, 1993a; Sharma *et al.*, 1995a,b,c; Schwab and Bartholdi, 1996).

Factors involved in spinal cord tissue damage are also responsible for cell injury or cell death in brain injury, ischemia, infarction, stroke, hypoxia, anoxia and in neurodegeneration (Faden, 1987; Lipton, 1993; Schwab and Bartholdi, 1996). This indicates that basic mechanisms of cell injury in the CNS are quite similar in nature irrespective of the primary nature of the insult. No single chemical compound or factor is thus responsible for all the pathological events seen in a particular situation. A wide spectrum of chemicals/factors are responsible for cell injury or death following different kinds of CNS insult.

Secondary injury mechanisms

A focal trauma to the spinal cord leads to various degree of functional disturbances resulting in permanent disability due to cell and tissue destruction within the cord. The degree of tissue injury is influenced by several factors such as extent and type of the primary physical injury, secondary or "perifocal" changes around the site of the initial impact and the efficiency of medical treatment and rehabilitation (Balentine, 1978; 1988; Bracken and Holford, 1993).

There is an immediate loss of electrical activity, increase in extracellular potassium, decrease in sodium-potassium ATPase after spinal cord injury indicating membrane dysfunction and failure of ionic pump mechanisms (Clendendon *et al.*, 1978). The secondary injury processes include ischemia and impairment of autoregulation of blood flow (Lipton, 1993). These changes altogether make the spinal cord vulnerable to reductions in arterial pressure and oxygen tension, both of which are frequent after cord injury (Sandler and Tator, 1976b) (Table 2).

The degree of the initial impact is the major factor influencing magnitude and severity of the various circulatory disturbances which occur around the primary injury following cord trauma. Rheological disturbances occur very early (Young and Flamm, 1982; Faden, 1993a). Alteration of microvascular permeability in the perifocal segments of the spinal cord is evident within minutes after injury (Griffiths and Miller, 1974; Hsu *et al.*, 1985; Olsson *et al.*, 1990). Edema of the cord is present as soon as 30 min after trauma (Griffiths and Miller, 1974; Lewin *et al.*,1974; Wagner and Stewart, 1981; Sharma and Olsson, 1990; Sharma *et al.*,1990a,b,c; Olsson *et al.*, 1992; 1995). The extracellular edema spreads in both longitudinal and transverse direction which is progressive with time (Griffiths and Miller, 1974).

Proposed secondary injury factors include products of phospholipid hydrolysis such as polyunsaturated fatty acids, eicosanoids, free radicals, neuropeptides, monoamines and changes of cations and amino acids (Tator and Fehlings, 1991; Faden and Salzman, 1992; Schwab and Bartholdi, 1996). The list is further expanding with the new knowledge derived from recent observations. Nitric oxide (NO) and carbon monoxide (CO) which are newly discovered gases involved in neuronal communication also play important roles in the pathophysiology of cell injury following spinal trauma (Sharma *et al.*, 1996a; 1997a,c; 1998a,b).

Trauma to the spinal cord is thus associated with numerous reactions in the cells and the tracts of the cord, changes in their surrounding extra-

Table 2. Pathophysiology of acute spinal cord injury

1. Bioelectrical disturbances
alterations in amplitude and latency of
spinal cord evoked potentials
motor evoked potentials
cerebellar evoked potentials
somatosensory evoked potentials
electrospinogram
2. Vascular disturbances
local ischemia[a]
edema formation[a]
necrosis
inflammation
3. Biochemical disturbances
release of chemicals
products of membrane breakdown
arachidonic acid
leukotrienes
thromboxane
cations
excitatory amino acids
monoamines[a]
neuropeptides[a]
4. Pathological disturbances
damage to microvasculature of the gray matter[a]
multifocal petaecheal microhaemorrhages[a]
microthrombi formation
vascular congestion
extravasation of blood into the parenchyma[a]
breakdown of the microvascular permeability[a]
widening of tight junctions
increased transendothelial transport[a]
vacuolation[a]
swelling[a]
haemorrhagic necrosis
glial reaction
proliferation
scar formation
astrocytic reaction[a]
activation of microglia
neuronal reaction
dark and distorted nerve cells[a]
swollen and shrunken cells[a]
chromatolysis[a]
loss of Nissl substance[a]
myelin damage
vesiculation[a]
swelling[a]
lamellae formation[a]

Compiled from various sources; [a]authors own investigation (for details see text).

cellular fluid and in the microcirculation (Faden and Simon, 1988; Tator and Fehlings, 1991; Faden, 1993a). The outcome after injury is influenced by several mechanisms: (1) the severity of cell and microfluid disturbances (edema, haemorrhages) at the site of the primary physical injury, (2) the degree of disruption and block of descending and ascending neuronal pathways including axonal and myelin alterations, (3) additional lesions in the segments cranial and caudal to the primary injury caused by edema, circulatory disturbances and other factors, so-called perifocal or secondary injuries, and (4) repair mechanisms. In order to understand the pathogenesis and effects of therapy of experimental spinal cord injury all these changes must be taken into account.

Animal models of spinal cord injury

Current information regarding the pathophysiology of traumatic spinal cord injury is based on research in experimental animals. Although there are great variations in the results obtained from experimental spinal cord injury research, the data somehow point to a general mechanism of tissue destruction following trauma.

Experimental models of spinal cord injury comprise weight drop technique, controlled contusion and maintained compression (de la Torre, 1981, Fehlings and Tator, 1988; Tator and Fehlings, 1991) (Table 3). Studies on spinal cord injury and repair mechanisms are addressed in models involving partial to complete transection or hemisection of the spinal cord (Schwab and Bartholdi, 1996). The incision model allows better control of lesion to the axons and fibre tracts. Thus behavioural, biochemical, physiological, vascular and pathological changes can be well reproduced in these models (see Winkler, 1994; Sharma, *et al.*, 1995a–c). Another advantage of the incision model is the possibility to study the details of morphological changes in a precise way. Thus, morphological changes can be examined in the incision model in both rostro-caudal and/or ipsi- and contra-lateral sides in order to understand the progression of pathological changes around a focal lesion.

All the existing animal models have essentially some advantages and drawbacks (Table 3). Human spinal cord injury cannot be simulated in anyway by animal experiments. Basic differences between the existing animal models and human spinal cord injury are summarised in Table 3. Human spinal cord injury results from four principal vectors of forces such as flexion, extension, rotation and compression which are working simultaneously or in combination (Ikata *et al.*, 1989; Sypert, 1990). Fracture of vertebral bodies or luxation of vertebrae will lead to concussion, contusion or laceration of the spinal cord as a result of these multiple sheer forces (Table 4). Thus human spinal cord injury represents several kinds of physical and mechanical damage altogether that will finally determine the nature and extent of the cord injury and vary in each individual case. Human spinal cord injury usually represents an anterior

Table 3. Experimental models of spinal cord injury

Weight drop technique	2nd Century AD
(on exposed spinal cord)	revised by Allen (1911)
widely used today in many species	
advantages	*shortcomings*
can be graded	
reproducible	results are variable
compression injury	
closely simulates some of the biomechanics of human SCI	differs from clinical conditions because of posterior injury
Compression injury	
histopathology mimics human SCI	variable results
haemorrhage, necrosis, cyst formation	
revised with electromechanical impact device	(Brosnan *et al.*, 1987)
Balloon inflation in the spinal extradural space	
Photochemical infarction of vascular endothelium	
no laminectomy required	
Incision[a]	
reproducible	differs from clinical conditions
can be graded	
reproducible results	
transverse or longitudinal	does not mimic clinical injury
white matter mainly intact	(Sharma and Dey, 1982;
minimal and focal injury	Winkler *et al.*, 1992)
changes can be seen ipsilateral/contralateral	
spread of cell injury can be examined in both rostro-caudal and ipsi-/contra-lateral directions	
a good model for studying secondary injury mechanisms	involves simple "cut" factor
	haemorrhage due to direct blood vessel damage

Compiled from various sources; [a]authors own investigation.

Table 4. A comparison of facts between clinical and experimental spinal cord injury

Clinical	*Experimental*
Multifactorial	simple
Flexion, extension, rotation, compression	weight-drop, incision[a]
Concussion, contusion, laceration	compression, transaction hemisection, electrolytic lesion
Complex forces are active at the time of injury	only one injury factor is present
Closed vertebral system injury	open vertebral system injury after laminectomy
Usually anterior compression	mainly posterior compression
Mainly occurs during conscious state	produced under anaesthesia
Unlimited or uncontrolled time interval between injury and first hand hospital examination	carefully controlled time schedule

Compiled from various sources (for review see Schwab and Bartholdi, 1996); for details see text; [a]authors own investigation.

cord compression or damage in conscious state and in the closed vertebral system (Table 4).

The animal models of spinal cord injury involves one force vector which is applied to the cord from posterior approach after laminectomy under anaesthesia (Table 4). Thus, at the moment no experimental design of spinal cord injury is close to mimic the human cases.

Pathophysiology of spinal cord injury

Our knowledge on the basic aspect of pathological reactions in the spinal cord following trauma has widely expanded in the last 20 years. Pathophysiology of a focal spinal trauma comprise a mixture of biochemical changes, disturbances in bioelectrical activity, alterations in ionic and cellular energy metabolism, vascular reactions, ischemia, edema and cell changes. A brief overview of key events responsible for spinal cord dysfunction following trauma is described below.

Pathological changes

Trauma-induced morphological changes constitute one of the major aspect of spinal cord dysfunction (Tator and Fehlings, 1991; Schwab and Bartholdi, 1996). Pathology of the spinal cord can be classified as acute, sub-acute and late phases. Most observations are made 2–3 days to weeks or months after the initial injury. Only a few studies examined the spinal cord pathology within 8 h after the primary insult. However, use of immunohistochemistry or other molecular markers in early phase of spinal trauma are still lacking.

Early phase after trauma is very important in elucidating alterations in the cord fluid micro-environment which seems to be instrumental in causing adverse neuronal, glial and myelin reactions. If these changes will continue for sometime, an irreversible cell damage will ensue leading to the life time disability.

Our laboratory focused great attention on early cell reactions occurring within the cord during a 5 h period after trauma using modern technology. Thus, specific markers for nerve cell, glial cell, myelin basic proteins, stress proteins, enzymes responsible for synthesis of neurotransmitters, and/or immediate early genes were used to generate new information which appears to be highly relevant for our understanding of early pathological reactions of the cord following trauma (Sharma *et al.*, 1990a–d; 1991; 1992a,c; 1993a,c,d; 1994a–d; 1995a–c, e–h; 1996a,b; 1997a,c,d; 1998b). A brief description of these early pathological events are described below.

Nerve cell reactions

Using Nissl stain to identify nerve cell reaction within the cord following 5 h spinal cord injury we detected profound dark and distorted nerve cells in the perifocal segments (Fig. 19). The Nissl substance was significantly degraded and chromatolysis is quite clear in many nerve cells. These nerve cells were swollen or shrunken and most of them were devoid of a clear nucleus (Figs. 18 and 19). A general sponginess and edema are frequent (Sharma *et al.*, 1995a).

The nerve cell reactions in the edematous zone represent disturbances in the micro-fluid environment. This is apparent from the findings that drugs inhibiting serotonin or prostaglandin synthesis or blocking the receptors of opioids are able to attenuate nerve cell reaction following spinal injury (Sharma *et al.*, 1990a–c, 1993a–d; Olsson *et al.*, 1992; 1995; Sharma *et al.*, 1997a–d).

Glial cell reactions

The role of glial cells in traumatic injuries to the spinal cord is not yet well characterised (Sharma *et al.*, 1993c). One of the important cell types of glia are astrocytes (Arenander and Vellis, 1984; Cervós-Navarro *et al.*, 1998) which play important role in the regulation of fluid microenvironment (Arenander and Vellis, 1984). Astrocytes respond to noxious insults by upregulation of glial fibrillary acidic protein (GFAP), a marker of astrocytic activation (Bignami *et al.*, 1980; Duffy, 1984; Fedoroff and Vernadakis, 1986). We found marked upregulation of GFAP in the perifocal segments indicating that astrocytes are sensitive to trauma induced alterations in the microfluid environment of the cord (Fig. 21) (Sharma *et al.*, 1993c; Cervós-Navarro *et al.*, 1998). Trauma induced activation of astrocytes is thus regulated by several neurochemicals involved in secondary injury processes. This idea is well supported by the fact that pretreatment of rats with p-CPA, indomethacin or naloxone prevented the upregulation of GFAP response (Sharma *et al.*, 1993c; Olsson *et al.*, 1995; Sharma and Westman, 1997). A direct inhibitory effect of neurochemicals on astrocyte activation is unlikely because treatment with these drugs alone did not influence GFAP immunoreactivity within the cord.

Axonal damage

Myelin is an important component for nerve conduction (Hogan and Banik, 1985) and damage to myelin following trauma leads to disturbances in the spinal cord conduction (Balentine and Spector, 1977; Balentine, 1978; Hogan and Banik, 1985; Fehlings *et al.*, 1989). Morphological studies show that myelin vesiculation and damage to myelinated axons are present even 5 min after trauma (Balentine, 1988) which reaches its peak at about 6–8 h after injury. This indicates that myelin damage is quite rapid and involved in slowing of the cord conduction.

The mechanisms underlying myelin damage following trauma is not clear. Alterations in micro-fluid environment and release of neurochemicals and ions, particularly calcium is involved in vesiculation of myelin. Using myelin basic protein (MBP) immunohistochemistry in our laboratory we found a rapid degradation of myelin following trauma (Fig. 22) which corroborates well with the axonal disruption seen at ultrastructural level (Fig. 24) (Sharma and Olsson, 1990; Olsson *et al.*, 1992; Sharma *et al.*, 1993a; Olsson *et al.*, 1995). Degradation of MBP was considerably reduced by drug treatments influencing serotonin, prostaglandins and opioid neurotransmission within the spinal cord. This indicates that loss of myelin is influenced with neurochemicals involved in secondary injury processes (Balentine, 1988; Olsson *et al.*, 1992; Sharma *et al.*, 1993a).

Molecular mechanisms of spinal cord injury

It has been shown that trauma to the spinal cord is capable to influence various cellular proteins or enzymes which may serve a key role in cell injury (Faden, 1993a; Schwab and Bartholdi, 1996).

Heat shock protein response

Trauma to the cord is a severe stress at cellular and molecular levels (Faden, 1993a; Sharma *et al.*, 1995c; Sharma and Westman, 1997) which results in an upregulation of intracellular stress proteins, known as heat shock proteins (HSP) (Lindquist, 1986; Kaufmann, 1992; Sharma and Westman, 1997; Westman and Sharma, 1998). HSP response is protective in nature, however, this issue is controversial (Westman and Sharma, 1998).

Experiments carried out in our laboratory show that trauma to the spinal cord was associated with a marked upregulation of HSP (72 kD) response in many neurons located within the adjacent perifocal segments (Sharma *et al.*, 1995c; Sharma and Westman, 1997). The HSP reaction product was present in the cytoplasm and in the nucleus (Fig. 20). Ultrastructural studies revealed the reaction product in the cytoplasm of dendrites and neurons attached to the endoplasmic reticulum (Sharma *et al.*, 1992d; 1995c). This indicates that stress induced by trauma is sufficient to upregulate expression of HSP within the cord.

The secondary injury induced release of neurochemicals play important role in inducing stress response. Pretreatment with drugs influencing serotonin (Sharma *et al.*, 1995c), prostaglandin (Sharma and Westman, 1997) and opioid transmission (Olsson *et al.*, 1995) markedly reduced the HSP response after cord injury. This suggests that neurochemicals are capable of reducing the stress protein response following injury. A reduction in HSP response following drug treatment represents a reduction in trauma induced stress response, rather than a direct inhibitory influence on HSP expression (Sharma *et al.*, 1995c; Sharma and Westman, 1997; Westman and Sharma, 1998).

c-fos immunoreactivity

Cellular fos (*c-fos*) is a component of cell nucleus which can be activated following several signals reaching to the cell directly by nerve stimulation or indirectly by local alterations in cell metabolism (Hughes and Dragunow, 1995). Upregulation of *c-fos* occurs following cellular and metabolic stress (Hughes and Dragunow, 1995; Nyberg *et al.*, 1995). This indicates that expression of *c-fos* represents an intracellular activation or excitation mechanisms.

Alterations in spinal cord injury induced *c*-fos is not well documented (Faden, 1993a; Schwab and Bartholdi, 1996). Since trauma induces upregulation of HSP response, it seems quite likely that some nerve cells will also enhance *c-fos* expression. Examination of *c-fos* immunoreactivity in our model of spinal trauma exhibited a marked upregulation in both dorsal and ventral gray matter 5 h after injury (Sharma *et al.*, 1996b). This upregulation of *c-fos* was prevented in rats pretreated with serotonin synthesis inhibitor, p-CPA (H. S. Sharma, unpublished observation). This indicates that expression of this immediate early gene is also influenced by trauma induced release of secondary injury factors and/or disturbances in the microfluid environment of the cord.

Alterations in cellular enzymes

Spinal cord trauma induces either degradation or over-expression of several key enzymes involved in synthesis of various neuroactive compounds. NO and CO the new class of neurotransmitters, are synthesised by enzymes known as nitric oxide synthase (NOS) and heme oxygenase (HO) respectively. Different isoforms of these enzymes are present in nerve cells, glial cells and in endothelial cells (see Dawson and Dawson, 1996; Kimura and Steinbusch, 1996; Sharma *et al.*, 1998a). Upregulation of these enzymes represent increased NO and CO activity.

A focal trauma to the spinal cord significantly enhances the expression of NOS and HO in the perifocal segments (Sharma *et al.*, 1996a; 1997a; 1998b) indicating that NO and CO are involved in the pathophysiology of cord injury (Sharma *et al.*, 1997a,c; Sharma, 1998; Sharma *et al.*, 1998b).

Disturbances in microfluid environment and release of neurochemicals following trauma induced secondary injury mechanisms play some role in activation of NOS and HO. Drugs inhibiting serotonin and prostaglandin synthesis or capable of blocking opioid receptors attenuate upregulation of these enzymes (see Sharma *et al.*, 1998a).

Vascular changes

The most important instant event following spinal cord injury is the alteration in spinal cord blood flow (SCBF). Disturbances in SCBF ranging from mild to ischemic episodes is well correlated with altered bioelectrical potentials and the morphological damage of the cord (see Chapters by Li and Tator; Marsala, in this volume).

Trauma induces mechanical damage to the cord microvasculature. Release of several vasoactive secondary injury factors lead to ischemia which is progressive in nature (Dohrmann *et al.*, 1971; Ducker *et al.*, 1971; Griffith and Mc Culloh, 1983). Severe reduction of SCBF in certain areas leads to tissue injury and loss of structure and function of the cord tissue. However, the critical blood flow necessary to maintain cord function is not well characterised.

Using autoradiographic technique, Landau (1955) for the first time measured feline SCBF which was 63 ml/100 g tissue/min in the gray matter and only about 14 ml/100 g tissue/min in the white matter. This observation is well supported later by studies using several other techniques (see Marsala, in this volume). A significant reduction in SCBF occurs within the first few hours after trauma which corresponds well with the degree of initial injury (Tator and Fehlings, 1991; Schwab and Bartholdi, 1996). The white matter is more resistant to blood flow changes compared to gray matter (Kobrine *et al.*, 1975). However, discrepancies still exist with regard to blood flow changes in white matter compared to the gray matter (Kobrine *et al.*, 1975). It is not certain whether this reduction in blood flow is due to trauma induced loss of autoregulation or to hypotension and/or changes in local energy metabolism (see Marsala, in this volume).

Our laboratory showed a progressive reduction in the SCBF following 1 to 5 h after trauma which is well correlated with elevation in plasma and spinal serotonin levels (Fig. 14) (Sharma *et al.*, 1990a; 1993b). These observations strongly indicate that trauma induced release of vasoactive neurochemicals will influence SCBF alterations (see below).

Microvascular permeability disturbances

The microvessels of the spinal cord are equipped with a blood-spinal cord barrier (BSCB) (Olsson *et al.*, 1990; 1992; 1995; Sharma *et al.*, 1995b). The BSCB comprise tight junctions between the adjacent endothelial cells and a lack of vesicular transport (see Sharma *et al.*, 1998c). The permeability properties of the BSCB is thus similar to that of an extended plasma membrane (Sharma *et al.*, 1990d; 1995d; 1998c). The passage of tracers like horseradish peroxidase, lanthanum, or radiochemical when injected into the blood stream is stopped at the tight junctions which constitute the physical barrier.

Trauma induces disruption of the BSCB due to a direct physical injury of the microvessels (Tator and Fehlings, 1991). Secondary injury factors and release of neurochemicals also influence the permeability of the BSCB probably via receptor mediated mechanisms (Figs. 14 and 25). Spinal cord microvessels contain receptors to several vasoactive neurochemical agents.

Our laboratory showed that serotonin, prostaglandins, histamine, opioid peptides are involved in the disruption of the BSCB following trauma to the cord (Sharma *et al.*, 1990a; Olsson *et al.*, 1992; Sharma *et al.*, 1993b; Winkler *et al.*, 1995b). Drugs inhibiting the synthesis or receptors of these neurochemicals significantly attenuated the trauma induced breakdown of the BSCB permeability. In these experiments, it appears that the trans-cellular permeability of endothelial cell is increased to tracers rather than intercellular widening of the tight junctions (Olsson *et al.*, 1990; 1992; 1995; Sharma *et al.*, 1995a,b). Ultrastructural studies revealed that the lanthanum was stopped at tight junctions following trauma but the endothelial cells are infiltrated with the tracer (Fig. 25) (Olsson *et al.*, 1990; 1992; 1995; Sharma *et al.*, 1995b).

Edema formation of the spinal cord

Edema is a serious complication of various neurological diseases following traumatic, ischemic or hypoxic injuries to the brain and spinal cord (Cervós-Navarro and Ferszt, 1980; James *et al.*, 1997, for details see Sharma *et al.*, 1998c).

Much attention has been paid to the development of brain edema following traumatic or ischemic injuries (Mohanty *et al.*, 1985; 1989; Dey and Sharma, 1983; 1984; Sharma and Olsson, 1990; Sharma *et al.*, 1992a; Sharma *et al.*, 1995a,b; 1996a; James *et al.*, 1997). However, edema formation following spinal cord trauma is not considered well in the literature. Research from our laboratory during the last one and half decade suggests that spinal trauma is associated with profound edema development which seems to be a key factor in cord pathology (Sharma and Olsson, 1990; Sharma *et al.*, 1990a–c; 1991, 1993a–d; 1994a–d; 1995a–c; 1996a; 1997a–d; 1998a–c).

Chemical mediators influence spinal cord edema formation probably via increasing the BSCB permeability of many solutes, ions and proteins (Sharma *et al.*, 1990a–c; 1995a,b). Pretreatment with serotonin synthesis inhibitor p-chlorophenylalanine, the prostaglandin synthesis inhibitor indomethacin or the opioid receptor antagonist naloxone significantly reduced the extravasation of tracers and markedly attenuated edema formation of following cord trauma (Fig. 14) (Sharma *et al.*, 1990a–c; 1993a–d; 1995a–c; 1996b; 1997a–c).

Biochemical mechanisms of spinal cord dysfunction

Apart from alterations in neurochemicals disturbances in ionic and other molecular metabolism play important roles in cell injury following cord trauma.

Calcium and potassium metabolism

Even small variations in the concentrations of some ions and molecules greatly influence the neural excitability and synaptic transmission of the cord (Faden and Salzman, 1992; Faden, 1993a). Calcium plays a crucial role in the regulation of sodium and potassium conductance and is involved in neural excitability, regulation of many critical cellular enzymes as well as storage and release of many neurotransmitters at synaptic levels (Rasmussen, 1986). Balentine and Spector (1977) were the first to demonstrate intra-axonal accumulation of calcium 30 min after spinal injury which is responsible for myelin vesiculation. Extracellular calcium store is decreased rapidly following trauma to the cord (Young and Flamm, 1982). Total calcium level increased 45 min following trauma and reached the peak level at 8 h. This increase in the cord calcium level persisted even 1 week after injury (Happel *et al.*, 1981; Balentine, 1988; Tator and Fehlings, 1991).

Mechanisms behind increased tissue calcium level and a decrease in extracellular calcium levels are not well understood. Voltage-dependent calcium channels and N-methyl-D-aspartate (NMDA) receptor channel are involved in the mechanisms of increased calcium level (Balentine, 1988). Another possibility includes failure of calcium^{2+}-ATPase-mediated calcium extrusion and release from intracellular organelles (Faden, 1993). A decrease in extracellular calcium may be due to its sequestration with organic phosphates (Young, 1985).

Increased levels of calcium induce many adverse effects on cellular functions (Farber, 1982; Siesjö, 1988). One possible cause of cell death is activation of calcium-dependent phospholipase C and A_2 which induces a rapid breakdown of cellular membranes and production of arachidonic acids or its metabolites thromboxane, leukotrienes and free radicals which may directly or through inflammatory processes contribute to cell and membrane injury (Demeduik and Faden, 1988; Sharma and Cervós-Navarro, 1990; Tator and Fehlings, 1991; Sharma and Olsson, 1994a,b; Winkler, 1994; Winkler *et al.*, unpublished observations).

Spinal cord trauma is also associated with a marked elevation of tissue potassium levels (Stokes *et al.*, 1985). Increased potassium levels together with calcium induces depolarisation of neuronal population in both gray and white matter which is mainly responsible for conduction block in the damaged spinal cord segments (Eidelberg *et al.*, 1975). Chronic loss of calcium and potassium following trauma is related to disturbances in cell membrane function leading to cell injury or cell death (Anderson *et al.*, 1976; Lewin *et al.*, 1974).

Phospholipids and free radicals formation

Activation of phospholipase A_2 increases the levels of platelet-activating factor (PAF) which seems to be related with the post-traumatic disturbances in the SCBF and the BSCB (Faden and Halt, 1992; Frerichs and Feuerstein, 1990). Phosphorylasis is also responsible for generation of free radicals, the molecules which damage cell membranes due to a very high random reactivity in presence of unpaired electrons (see Chieuh *et al.*, 1994; Liu in this volume). The source of free radical generation after spinal cord trauma is not well known. Haemorrhage and extravasation of blood components, such as haemoglobin within the cord is one potent source of iron to catalyse oxygen radicals and lipid peroxidation (Tator and Fehlings, 1991). Another factors include alterations in arachidonic acid metabolism and modification in cyclooxygenase pathways, invading neutrophils undergoing oxidative bursts or auto oxidation of catecholamines (McCord, 1987; Misra and Freidovich, 1972).

Recently it has been shown that formation of free radicals is associated with production of NO and CO (Dawson and Dawson, 1996; Sharma, 1998; Sharma *et al.*, 1998a; Winkler *et al.*, 1998). This is supported by our observations that pretreatment with antioxidant compound H-290/51 significantly attenuated the NO or CO production by inhibiting their converting enzymes NOS or HO-2 respectively within the spinal cord (Sharma, unpublished observation; Sharma *et al.*, 1997c; Sharma *et al.*, in this volume).

Excitatory and inhibitory neurotransmitters

Release of excitatory amino acids such as glutamate and aspartate occurs within the very first minutes after primary insult to the spinal cord (Panter *et al.*, 1990; Liu *et al.*, 1991). The detailed mechanisms of excitotoxins induced cell death in spinal cord injury is not known. In cell culture, glutamate and aspartate induce cell injury via stimulation of NMDA or AMPA receptors because blockade of these receptors attenuated cell death. However further studies are needed to characterise the involvement of excitotoxins in the pathophysiology of spinal cord injury (Schwab and Bartholdi, 1996).

Involvement of the inhibitory amino acids, such as glycine and GABA, in spinal trauma is not well characterised. Preliminary investigations from our laboratory revealed a marked down regulation of these inhibitory amino acids following trauma to spinal cord (Sharma *et al.*, 1995e). Pretreatment with p-chlorophenylalanine, which is found to be neuroprotective in this model significantly restores the GABA and glycine levels in the spinal cord following trauma (Sharma *et al.*, 1995e,f; Sharma, unpublished observations). These observations strongly point out that a balance between excitatory and inhibitory amino acids in the spinal cord following trauma is crucial for normal spinal cord function.

Metabolic changes in the spinal cord after trauma

One of the earliest changes in the spinal cord is a linear decrease in tissue oxygen tension at the lesion site which continue for several hours (Hedeman and Sil, 1974; Hayashi *et al.*, 1980). Decreased tissue oxygen is responsible for local tissue hypoxia which leads to local ischemia.

Local ischemia and hypoxia will alter the normal tissue response for energy utilisation leading to overproduction of lactic acid (de la Torre,

1981). Increased lactic acidosis further reduces glucose and oxygen consumption and induces production of circulating prostaglandins along with stimulation of ADP (Siesjö, 1988). Increased production of prostaglandins initiate a cascade of cell injury by production of free radicals (Matsumura *et al*, in this volume). A shift from aerobic glucose to anaerobic glucose utilisation is harmful and reduce the high-energy metabolism of the traumatised tissue (Walker *et al.*, 1977). During anaerobic glycolysis, release of lysosomal enzymes occurs which causes cell autolysis (Ashford *et al.*, 1966).

Pharmacotherapy of spinal cord injury

Treatment approach for spinal cord injury is divided into two major goals (Table 5). The first approach is directed towards limiting the process of secondary injury cascade thereby reducing the tissue damage. This approach also influences sensory-motor function. The second approach is based on the rehabilitation concept. The idea is to stabilise the current status of sensory-motor dysfunction from further deterioration to achieve an optimal living condition by training of reflex activities (Schwab and Bartholdi, 1996).

Another emerging aspect of treatment approach is to establish axonal regeneration with trophic support or foetal tissue implants in the injured spinal cord segments (Table 5). This approach is currently limited to experimental stage (see Hagg and Oudega, in this volume). Application of neurotrophins on the traumatised cord, or antibody raised against regeneration inhibiting factors help in axonal regeneration (see Schwab and Bartholdi, 1996; Hagg and Oudega, in this volume; Sharma *et al.*, in this volume). Preliminary research from our laboratory show that application of neurotrophins on the traumatised cord restores spinal cord conduction and

Table 5. Treatment strategies in spinal cord injury

1. *Neuroprotective approach*
directed against interrupting the cascade of
secondary injury processes[a]
limiting tissue damage[a]
arrest or reverse sensory/motor function impairment[a]
2. *Rehabilitating approach*
directed against consequences of spinal cord injury
stabilisation of current status with trauma
training of reflexes and residual circuits
for optimal living conditions
3. *Regenerative approach*?
directed towards enhancement of axonal regeneration
purely experimental at this stage[a]
no experience in human spinal cord injury

Compiled from various sources; [a]authors own investigations.

attenuates cell injury (Winkler *et al.*, unpublished observations; Sharma *et al.*, in this volume).

Neuroprotective drugs in basic and clinical studies

Drugs currently used to achieve neuroprotection in spinal cord trauma are summarised in Table 6. In clinical strategy, methylprednisolone is the only

Table 6. Current therapeutic strategy of spinal cord injury

Clinical spinal cord injury	
methylprednisolone	multicentre large clinical trials in humans
advantages	*shortcomings*
widely used in clinical practice	not based on drug's pharmacokinetics
beneficial on behavioural parameters	not correlated with histological outcome
naloxone	not different from placebo
Experimental spinal cord injury	
methylprednisolone	not comparable to each other
	behavioural parameters are main outcome measures that is difficult to interpret
	no histological evaluation with behavioural studies
	different survival periods
	different models
neuroprotective/beneficial effects observed	
α-methyl-p-tyrosine	
clonidine	beneficial in chronic spinal cord injury
21-aminosteroids	
MK-801	
naloxone[a]	
U-50488 H (κ-agonist)	
nor-BNI (κ-antagonist)	
TRH	
GM-1 ganglioside	
NOS inhibitors?	
p-CPA[a]	
diazepam[a]	
clonazepam	
baclofen	
indomethacin[a]	
ibuprofen[a]	
cimetidine[a]	
nimodipine[a]	
dynorphin antiserum[a]	
NOS antiserum[a]	
serotonin antiserum[a]	
no significant neuroprotection/beneficial effects can be seen or conflicting reports exist	
propranolol[a]	
phenoxybenzamine[a]	
dexamethasone[a]	
mepyramine[a]	

Compiled from various sources; [a]authors own investigations.

drug used to improve the outcome following spinal trauma. However, problems regarding dosage and side effects are quite common (see Schwab and Bartholdi, 1996; Faden and Salzman, 1992).

Majority of investigations using pharmacotherapy are limited to the evaluation of motor function, however, it is not certain whether beneficial effects on motor function also reflect spinal cord tissue protection. Most drugs can induce neuroprotection if given as a pretreatment. However only a few drugs exhibit beneficial effects if given as a post-trauma treatment (Faden and Salzman, 1992; Faden, 1993a). Further research is thus highly needed in this area because clinical conditions require drugs to be given as a post-trauma treatment only.

Methylprednisolone and 21-amino steroids

Use of methylprednisolone in spinal cord trauma is based on its ability to stabilise lysosomal enzymes, inhibit lipid peroxidation and edema formation (Faden, 1993a). Steroids also reduce the microvascular permeability disturbances and cellular stress response (Sharma, 1982; Winkler, 1994). Neuroprotection is achieved by very high doses (30 to 60 mg/kg) of steroids which is often associated with adverse side effects (Faden and Salzman, 1992). High dose of steroids inhibit lipid peroxidation, improve SCBF and attenuate rise of tissue calcium following trauma in experimental studies (Faden, 1987; Braughler and Hall, 1982).

A new class of synthetic 21-aminosteroids is used to avoid the side effects of steroids (Faden, 1993a). These compounds lack glucocorticoid and mineralcorticoid activity but are able to reduce lipid peroxidation more efficiently than methylprednisolone (Faden, 1987; Faden and Salzman, 1992). Scavengering of free radicals with antilipolytic actions (Faden and Salzman, 1992), inhibition of arachidonic acid release following membrane disruptions are important factors for neuroprotective action of these compounds.

Clinical evidence suggests a therapeutic window for the treatment with steroids which is limited to 8 h. Beneficial effects can be seen if steroid treatment is initiated within 8 h following injury. Treatment schedule after this period have deleterious effects (see Schwab and Bartholdi, 1996).

Drugs modifying arachidonic acid metabolism and prostaglandin synthesis

Pharmacological compounds which either block the synthesis or receptors of arachidonic acid or its metabolites thromboxane, leukotrienes, hydroxyeicosatetraenoic acids as well as free radicals can induce beneficial effects by reducing the cascade of secondary injury mechanisms (Faden and Salzman, 1992). Several cyclooxygenase inhibitors such as ibuprofen or meclofenamic acid, indomethacin and other salicylates have been used to reduce the consequences of spinal cord injury. The results seems to be promising because most of the pretreatment studies induced neuroprotection (for details see Matsumura *et al.*, in this volume). A combination of treatment using thromboxane synthetase inhibitor (U63447A) and a stable PGI_2 analogue (ciprostene, but not a lipoxygenase inhibitor, piriprost) significantly reduced white matter ischemia in cats following spinal trauma. Pretreatment with indomethacin, a mixed potent cyclooxygenase inhibitor markedly reduces, edema and BSCB permeability (Hsu *et al.*, 1985; Sharma *et al.*, 1993b). Several oligomeric prostaglandin derivatives from PGE_1 also appear to be beneficial (Faden, 1987; Faden and Salzman, 1992). Further research using these groups of compounds is necessary on other parameters of spinal trauma apart from behavioural recovery in order to explore their therapeutic potentials.

Compounds influencing lipid-peroxidation and free radical production

Endogenous antioxidants such as α-tocopherol (vitamin E), ubiquinones (coenzyme Q), retinoic acid (vitamin A) and ascorbic acid (vitamin C) levels are decreased following CNS trauma (Faden, 1993a). Exogenous supplement of these chemicals following trauma will thus enhance scavengering capacity of free radicals and inhibition of lipid peroxidation process (Faden and Salzman, 1992).

Several other anti-oxidant compounds are now available commercially which provide various degree of neuroprotection following CNS trauma. A new anti-oxidant compound H-290/51 which inhibit the generation of free radicals significantly attenuated the BSCB permeability, edema formation and cell changes in spinal cord injury (Mustafa *et al.*, 1995; Sharma *et al.*, 1995g; Thörnwall *et al.*, 1995; 1997). These observations provide a promising approach for the use of anti-oxidants as suitable compounds for therapy in future.

Gangliosides, a new class of potential therapeutic compounds

Exogenous gangliosides as a neuroprotective compound is the subject of many investigations in CNS trauma (for details see Faden and Salzman, 1992; Schwab and Bartholdi, 1996). Gangliosides are sialic acid-containing glycolipids and are present in abundant quantities in the cell membranes. Gangliosides play important roles in neurite outgrowth and sprouting (Gorio *et al.*, 1984). In addition, they reduce NMDA receptor mediated release of amino acids and influence the regulation of protein kinase C induced signal transduction mechanisms (Nicoletti *et al.*, 1989).

Gangliosides exerts beneficial effects on behavioural recovery even administered 72 h after trauma in humans (Geisler *et al.*, 1991). This has encouraged clinical use of the compound in patients of CNS injury. Gangliosides GM-1 in one clinical study showed a beneficial effect in muscle function of the lower extremities involving 16 cases of spinal cord injury. However, the results are complicated because these patients have received an earlier dose of methylprednisolone before GM-1 administration (Geisler *et al.*, 1991). In rats when GM-1 and methylprednisolone were given simultaneously, the neuroprotective effects of methylprednisolone were blocked by GM-1 (Constantini and Young, 1994; see Schwab and Bartholdi, 1996).

Detail mechanisms of gangliosides induced neuroprotection are not known. Modulation of NO activity by these compounds appears to play major role in experimental models of brain or spinal cord injury (see Dawson and Dawson, 1996; Sharma *et al.*, in this volume).

Drugs influencing monoamines neurotransmission

A focal trauma to the cord influences biogenic amines concentration in the spinal cord (Sharma and Olsson, 1990). Thus, drugs modifying monoamines transmission are used to influence the outcome of spinal cord injury.

Involvement of monoamines in spinal cord injury was first postulated by Osterholm and Mathews (1972) which was based on the accumulation of norepinephrine in the traumatised cord. However, inhibition of catecholamines synthesis with α-methyltyrosine, or blockade of α-receptor with clonidine yielded controversial results (see Osterholm *et al.*, 1987; Faden and Salzman, 1992; Schwab and Bartholdi, 1996). Barbeau *et al.* (1993) suggested recently that noradrenergic agonist may be useful in later phase of chronic spinal cord injury.

Activation of cholinergic activity is somehow related to behavioural abnormalities following trauma (Faden, 1993a). Acetylcholine receptor blockers, scopolamine and atropine exerted some beneficial effect in cerebral concussion and human head injury (Faden and Salzman, 1992; Schwab and Bartholdi, 1996), however their use in spinal cord injury is not well documented yet.

Modification of serotonin transmission received new interest in spinal cord trauma research recently (Sharma and Olsson, 1990; Salzman, 1991). Blockade of serotonin synthesis or receptors attenuate trauma induced behavioural and histopathological outcome indicating an involvement of the amine in spinal injury (see Sharma *et al.*, 1990a–c, 1991; 1993a–c, 1995c). Results from our laboratory show that inhibition of serotonin synthesis with p-chlorophenylalanine reduces trauma induced stress response, BSCB permeability, edema and cell injury (Sharma *et al.*, 1990a–c). Mianserin and (s)-emopamil, mixed serotonin 1 and 2 receptor antagonists, significantly improves motor dysfunction in animal models of spinal injury (Salzman, 1991). These observations support a serotonin mediated neurotoxicity in spinal cord trauma.

Drugs influencing opioid neurotransmission

Opioid and non-opioid peptides influence synaptic transmission within the cord by receptor mediated mechanisms (Hökfelt *et al.*, 1978; 1987; Faden,

1993b; Nyberg *et al.*, 1995). Faden and his group were the first to show a beneficial effect of an opioid receptor antagonist naloxone in traumatic injuries to the CNS (for details see Faden, 1990; 1993b; Winkler *et al.*, 1994a,b; Olsson *et al.*, 1995). Naloxone in high doses improves behavioural outcome and reduces cell damage in spinal cord injury (see Chapter 19 in this volume).

Apart from its classic effects on μ-opioid receptors, naloxone in high doses will inhibit κ-opioid receptors which seems to be related with its beneficial effects in spinal trauma (Sharma *et al.*, 1995a). This is evident from the studies using nor-BNI, a specific κ-opioid receptor antagonist, which exerted a significant improvement on behaviour recovery (Faden, 1993b). These findings are, however, complicated by the beneficial results obtained with one selective κ-receptor agonist U-50488H (see Faden, 1993b).

Experiments carried out in our laboratory show that naloxone in high doses significantly improved spinal cord conduction (Winkler *et al.*, 1994a) and attenuated microvascular permeability disturbances, edema and cell injury (Olsson *et al.*, 1995). On the other hand low doses of the compound were ineffective (Winkler *et al.*, 1994a).

Opioid receptor antagonists inhibit release of excitotoxins, induction of phospholipases as well as increased release of Ca^{2+} and Mg^{2+} levels and reduce post-traumatic ischemia. These compounds thus improve cellular bioenergetic states which is crucial for reduction in cell damage (Faden, 1993b; Olsson *et al.*, 1995). However, clinical trials failed to show any beneficial effect of naloxone compared to placebo (Schwab and Bartholdi, 1996) suggesting further studies to understand the role of opioid receptor blockers in spinal trauma.

Thyrotropin releasing hormone and its analogues

Thyrotropin releasing hormone (TRH) and its analogues antagonise some physiological actions of endogenous opioids in the CNS without influencing their analgesic properties. These compounds exhibit neuroprotection by decrease in edema formation and necrosis in some animal models of brain and spinal trauma (see Faden *et al.*, 1989; Faden and Salzman, 1992; Schwab and Bartholdi, 1996). When administered in high dose, far above than their endocrine actions, significant improvement in neurological behaviours is seen in cat and rat models of spinal cord injury. These beneficial effects persist even if the compounds are administered 24 h after the primary insult. This property of TRH and related compounds is quite suitable for clinical studies.

Stable TRH analogues are beneficial on motor function recovery in animal models of CNS trauma (Faden and Salzman, 1992). Pyroglutamyl substituted compounds such as CG3509, CG3703 and YM14673 as well as several imidazole (histidyl)-substituted compounds exhibited beneficial effects in several animal models of spinal cord injury. The C-terminus of the tripeptide appears critical for its neuroprotective effects (Faden and Salzman, 1992). Reduction in tissue acidosis, phospholipid degradation, edema formation and antagonism of some physiological actions of endogenous opioids seems to play important roles in neuroprotection by these compounds (see Faden *et al.*, 1989; Faden and Salzman, 1992).

Blockers of NMDA and AMPA receptors

Increased levels of excitatory amino acids are involved in the mechanism of cell death in brain or spinal cord following injury. Substances which block the receptors of excitatory amino acids such as glutamate or aspartate reduces the consequences of cell injury (Choi *et al.*, 1991; Choi, 1992; Liu, 1994; Okumura *et al.*, 1992). Blockade of selective and non-selective NMDA receptor antagonists attenuate pathophysiology of cell injury in ischemia, brain injury and in some models of spinal trauma (see Schwab and Bartholdi, 1996). Preliminary observations suggest a positive role of AMPA receptors blockers in attenuating cell death in some models of cell injury as well (Faden and Salzman, 1992: Schwab and Bartholdi, 1996). However, serious side effects of these compounds make them incapable to use in clinical studies.

Blockers of voltage-gated ion-channels

Ion channels blockers can improve conduction deficits because trauma induces redistribution of ion fluxes across cell membrane which constitute the basis of altered spinal cord conduction and cell injury (Tator and Fehlings, 1991; Schwab and Bartholdi, 1996).

One potent blocker of voltage dependent potassium channels is 4-aminopyridine (4-AP) which exerts beneficial effects by influencing calcium influx and phosphorylation of membrane proteins (Lundh and Thesleff, 1977; Molgo *et al.*, 1980). The drug is able to restore conduction in patients following chronic spinal cord injury (Hayes *et al.*, 1994) indicating its possible therapeutic use in future.

Intracellular calcium influx is the final common pathways of cell death in the CNS (Siesjö, 1988). Thus, use of calcium channel antagonists appears to be beneficial in spinal cord injury if the associated hypotension is controlled by other supporting drugs. Tator's laboratory was the first to show a beneficial effect of calcium channel blocker nimodipine in spinal cord injury which reduces post-traumatic ischemia (see Tator and Fehlings, 1991). In this investigation only 25% reduction in blood pressure was seen with a dose of 0.05 mg/kg.

Results from our laboratory showed that constant infusion of low dose of nimodipine improved spinal cord conduction and exerted marked reduction in BSCB permeability, edema formation and cell injury without inducing profound hypotension (Winkler, 1994; Sharma and Olsson, 1994b; Sharma, unpublished investigation). These observations suggest that calcium channel antagonists may be useful in spinal injury (see Tator and Fehlings, 1991; Faden and Salzman, 1992; Faden, 1993a; Schwab and Bartholdi, 1996).

Platelet activating factor antagonists

Platelet activating factor (PAF) is an alkyl-phospholipid which is produced by platelets, white blood cells and endothelial cells (Lindsberg *et al.*, 1991; Kochanek *et al.*, 1991; see Faden and Salzman, 1992; Faden, 1993a). PAF reduces SCBF, disrupt the BSCB permeability and is toxic to nerve cell membranes. Thus antagonists of PAF seem to be beneficial in spinal cord injury. Few studies indicate beneficial effects of PAF antagonists on neurological recovery following brain trauma, ischemia and hypoxia. However, detailed investigation of PAF antagonists in spinal trauma is still lacking.

NOS and HO antagonists

A focal trauma to the brain or spinal cord upregulates the expression of different isoforms of NOS and HO in the CNS (Fukuda *et al.*, 1996a,b; Sharma *et al.*, 1996a; 1997a,c; 1998a,b). Blockade of these enzymes by compounds capable of inhibiting their synthesis appears to be neuroprotective in CNS trauma (see Dawson and Dawson, 1996; Sharma *et al.*, 1998a). Thus NOS and HO inhibitors offer neuroprotection in some models of brain and spinal trauma. The issue is however controversial due to lack of dose response, pharmacokinetics and selective inhibition of these compounds in *in vivo* studies (see Kimura and Steinbusch, 1996; Dawson and Dawson, 1996; Sharma *et al.*, 1998a).

Antisera as possible therapeutic agents

Antisera generated against specific compounds or molecules involved in neurotoxicity or cell injury are emerging as a new concept in pharmacotherapeutic strategy (Faden, 1990; Sharma *et al.*, 1995a; 1997d). Several antisera, interferon, oligo-peptides and gene therapy is proposed for the treatment of different CNS diseases. However, their use in spinal cord injury is still limited (Schwab and Bartholdi, 1996).

Faden (1990) was the first to apply antiserum of dynorphin A intrathecally in spinal cord injury which induced beneficial effects on motor functions. Sharma and his group employed antisera to several key components involved in cell injury to achieve neuroprotection in rat spinal cord injury. Topical application of highly specific antiserum directed against dynorphin A (Sharma *et al.*, 1994c; 1995a), serotonin (Sharma *et al.*, 1993e; 1997d) or constitutive isoforms of neuronal NOS (Sharma *et al.*, 1994d; 1995a; 1996a; 1998b; Sharma *et al.*, in this volume) significantly attenuated BSCB permeability, edema formation and cell injury following spinal trauma. Using immunohistochemistry Sharma and co-workers further show that topical application of antisera are able to penetrate the spinal cord in order to neutralise the tissue antigens *in vivo* (Sharma *et al.*, 1996a; 1997d; 1998b). These observations strongly support a possible therapeutic value of antisera in future studies.

Neurotrophins, growth factors and growth hormone

Trauma to the cord induces lesion of axons or other cellular elements (Balentine, 1988). Thus neurotrophins involved in regeneration and neurite outgrowth in the CNS are useful as therapy in spinal injury (Schwab and Bartholdi, 1996). Brain derived neurotrophic factor (BDNF), nerve growth factor (NGF), fibroblast growth factor (FGF), ciliary neurotrophic factor (CNTF) and insulin like growth factor (IGF) offer neuroprotection in brain trauma and ischemia (for review see Schwab and Bartholdi, 1996; Sharma *et al.*, 1998d).

Alterations in neurotrophin receptors occur in spinal trauma, however use of neurotrophic factors as a therapy in spinal cord injury is not well explored (see Schwab and Bartholdi, 1996; Sharma *et al.*, 1997a; 1998a,b; Sharma *et al.*, in this volume). Results from our laboratory show that topical application of BDNF and IGF-1 on the traumatised cord reduces BSCB permeability, edema and cell changes (Sharma *et al.*, 1997a; 1998b) indicating neuroprotective effects of neurotrophins in spinal cord injury (Haag and Oudega, in this volume).

Trauma to the spinal cord decreases growth hormone levels in the plasma (Mustafa *et al.*, 1995). Thus, exogenous application of growth hormone in trauma may have some beneficial effect. Topical application of growth hormone on the traumatised spinal cord in high concentrations is neuroprotective (Sharma HS, unpublished observation). These preliminary results are encouraging and require further investigation.

Other compounds and miscellaneous therapy

Several other compounds and treatment strategy in the past yielded some positive results in spinal trauma. Dimethyl sulfoxide, thyroid hormones, hypertonic agents, phenytoin, aminophylline, isoproterenol exert beneficial effects. Hypothermia, surgical intervention such as myelotomy or treatment with hyperbaric oxygenation is also beneficial in some studies (see de la Torre, 1981; Faden, 1987; 1993a; Schwab and Bartholdi, 1996). However, these strategies usually results in controversial data and often complicated with many adverse side effects.

Pharmacotherapy of spinal cord injury is thus a complex subject. A combine treatment after post-trauma regimen with several beneficial pharmacological substances may have some synergistic good or deleterious bad effects. This possibility, however, is the subject of future investigation.

Clinical aspects of spinal cord injury

The characteristics of human spinal cord injury is quite similar to that observed in several animal models of spinal cord injury (Dohrmann *et al.*, 1971). Microhaemorrhages, vacuolation and swelling of the capillary endothelium are seen within five minutes after the primary insult (Ducker *et al.*, 1971). Ischemia and hypoxia predominates during the first 4 hours after injury, followed by a progressive gray matter necrosis (Nemecek, 1978). Pathological changes in the white matter are prominent after 3 or 4 hours after the injury. Edema, distorted perfusion and ischemia followed by necrosis of white matter continue over a period of several hours and days (Balentine, 1978). The magnitude of edema is usually related to the severity of the primary traumatic insult (Wagner and Stewart, 1981).

Humans do not show apparent neurological deficit after initial loss of segmental gray matter following trauma, however, loss of white matter will render them para- or tetraplegic (Ducker, 1976). Thus treatment after the primary insult must be initiated as early as possible in order to minimise the loss of white matter and to attenuate the consequences of secondary injury mechanisms.

Injury to the spine and the spinal cord

Trauma to the spinal cord is caused by mechanical injury to the spine. The initial mechanical trauma causes compression and contusion of the spinal cord. In most severe cases, laceration of the cord due to displacement of bone and disk fragments into the cord is quite common (Schwab and Bartholdi, 1996). About half of the spinal injuries are located to the cervical spine C2-C3 to C6-C7 and are as common as vertebral fractures (Kakulas and Taylor, 1991). Injury or fractures of upper thoracic or lower lumbar spine is uncommon.

Four different vectors of forces i.e., rotation, compression, flexion, and extension are acting simultaneously to induce a multiple injury pattern in humans (Table 4). Hyperextension results in rupture of ventral longitudinal ligament and transverse fracture in the junction of the disc. Hyperflexion causes rupture of the dorsal longitudinal ligament. On the other hand, rotational injuries are responsible for damage to both dorsal and ventral ligaments. Damage to bony fragments or penetrating injuries results in spinal cord compression leading to ischemia (Schwab and Bartholdi, 1996). Regions with high mobility, such as cervical and lumbar spine are more vulnerable compared to the regions with low mobility. In case of injury to the thoracolumbar spine, vertebral fractures and penetrating injuries mainly dominate the clinical picture (see Kakulas and Taylor, 1991; Kakulas *et al.*, in this volume).

Classification of spinal cord injury

Human spinal cord trauma represents concussion, contusion and laceration injuries. Analysis of cervical spinal cord injury cases revealed 46% flexion injury followed by compression (28%) and hyperextension (26%) injury (Kiewerski, 1991). Spinal cord concussion results in a temporary loss of function which rapidly returns to normal if pathological changes are absent following the initial impact. Contusion of the cord is often associated with petechial haemorrhages even the spinal cord is physically intact. Spinal cord laceration is a severe injury in which disruption of the cord tissue takes place. Occasionally the spinal cord is completely transected. However, this is largely uncommon and occurs only in the most heavily injured cases where other kinds of injuries cause an immediate death.

According to the level of the lesion

The intensity of functional dysfunction vary according to the level of injury which further depends on the injury severity. A concussed segment quickly regains sensory-motor function, whereas in case of severe trauma the functional deficit will continue for long time without any apparent restoration of the function (Homma and Tamaki, 1984). The neurological deficit after trauma is further complicated due to number of severed axons and traumatised nerve cells in and around the lesion site (Bennet, 1984).

Complete vs. incomplete injuries

Complete and piercing injuries of the spinal cord are rare (Schwab and Bartholdi, 1996). Irrespective of the complete or incomplete injury, loss of sensory and motor function caudal to the lesion is always prominent. In an incomplete lesion remaining motor and sensory fibers are responsible for functional and reflex activity for sometimes after trauma (Dimitrijevic *et al.*, 1992).

In 225 traumatic lesions of the spinal cord, necrosis is pronounced in central gray and white matter around the lesion site (Kakulas, 1984). However, subdural and extradural haemorrhages were absent (Kakulas, 1984). Continuity of the most spinal microvessels were preserved in the lateral, ventral and dorsal white matter and segmental feeding vessels, arterial arcades, and anterior and posterior spinal arteries were intact. Only in a few cases with massive injuries, complete disruption of spine or spinal cord was observed (Kakulas, 1984; Tator, 1991).

Clinical neurophysiology as a tool to diagnose severity of spinal cord injury

In the last decades, several reliable methods for neurophysiological assessment of spinal cord have been developed to assess the cord function in clinical situations after trauma (D'Anglo *et al.*, 1973; Salzman, 1990). Currently, the following methods are used in patients suffering from spinal cord dysfunction due to diverse origin.

Somatosensory evoked potentials

Somatosensory evoked potentials (SEP) were recorded for the first time about 50 years ago (Dawson, 1947) but their implications in clinical practise were not recognised until the early 1970s. When a sensory system is stimulated, SEP can be recorded in a localised region of the spinal cord or brain (Perot, 1973) (for details see Chapters 9–11, 14 and 22 in this volume). Cortically recorded SEP following peripheral nerve stimulation reflect impulse conduction in the dorsal column, leminiscal pathways, thalamus and in the sensory cortex (Ducker, 1976). Changes in SEP activity is related with ischemia due to cord compression (for details see Li and Tator, in this volume). However, the relationship between SEP latency/amplitude and spinal cord function is not well understood.

Several modifying factors from different neural centres are present between spinal cord and the cerebral cortex all of which contribute to the SEP changes (Reger and Henry, 1979). Patients which remain paraplegic after a lumbar spinal cord injury, often exhibit a complete loss of cortical SEP response if stimulation is performed below the lesion. Normal SEP and sensory functions can be seen in spite of damage to motor pathways resulting in para- or tetraplegia (Fehlings *et al.*, 1989). Thus, recording of cortical SEP with stimulation of intercostal nerves can be useful in evaluation of thoracic and lower cervical cord injuries.

Spinal cord evoked potentials

Spinal cord evoked potentials (SCEP) are compound action potentials recorded from the cord with the electrodes placed either in the epidural space, in the posterior bony elements of the lamina, in the spinous processes, interspinous ligaments, or in the cauda equina (Homma and Tamaki, 1984). SCEP is recorded either following stimulation of peripheral nerves, different levels of the spinal cord, and/or stimulation of the cortical motor area (Salzman, 1990). The amplitude of the negative peak of SCEP reflects the number of conducting axons in the investigated spinal tracts (Willis, 1984). The conduction velocity of the axons in the spinal tracts is reflected by the latency of the different peaks of SCEP (Reger *et al.*, 1987). SCEP is a direct method for assessment of spinal cord function, with a minimal interference from other neural centres (Reger *et al.*, 1980). Recording of SCEP from the proximal and distal as well as ipsi- and contra-lateral side of the spinal cord lesion is also possible (Fehlings *et al.*, 1989; see Chapter by Li and Tator, in this volume; Winkler *et al.*, unpublished observations).

Motor evoked potentials

Electrical or magnetic stimulation of cerebral cortex, cerebral motor area, motor tracts, and in distal motor axons in peripheral roots or nerves generate action potentials in prepyramidal neurons or pyramidal tracts (Hurlbert *et al.*, 1993). These action potentials can be recorded directly from the pyramidal tract in the spinal cord, from peripheral nerves, or from distal muscles. Injuries to motor tracts, motor neurons or peripheral motor axons will result in increased latency and decreased amplitude at the recording site. Motor evoked potentials (MEP) reveal injuries involving lateral and ventral corticospinal tracts and is therefore a good complement to SEP investigations which is mediated by dorsal sensory tracts (for details see chapters 16–18, in this volume). MEP is more sensitive to spinal cord injuries than SEP and is a technique of choice in situations where preservation and restoration of motor function is a major goal following spinal injury (for further details see Chapters 10, 16, 17, in this volume).

Combination of MEP and SCEP

With epidural electrodes, it is possible to record motor and sensory SCEP simultaneously following stimulation of motor cortical and peripheral sensory pathways together. The technique is valuable during surgery involving thoraco-lumbar spine or spinal cord when electrodes are advanced in epidural space from the laminectomised region in both rostral and caudal direction (Homma and Tamaki, 1984). Motor and sensory SCEP are recorded separately with different latencies from each epidural electrode. The technique implicits monitoring of both motor and sensory tracts cranially and caudally to the surgical site and thereby increases accuracy and safety in the monitoring of the spinal cord (see other chapters, in this volume).

Electromyography

Using electromyography (EMG) comprising polyphasic motor units with increased amplitudes and duration and/or the appearance of spontaneous activity in a distal muscle, it is possible to demonstrate loss of motor axons in the motor nerve or loss of motor cells in the cord segment supporting the muscle (Stålberg and Grimby, 1995; Stålberg, 1996).

It is possible to distinguish all injured segments in the injured spinal cord with EMG. Two–three weeks after the trauma all injured segments with involvement of the ventral horns demonstrate spontaneous activity in distal muscles. The loss of active motor units may be due to both loss of pyramidal tract axons or loss of motor cell bodies in the ventral horns. Intercostal and paravertebral muscles can be used to study thoracic lesions. The amount of viable motor neurons in segments without any voluntary activity can be examined by measurement of muscle response amplitude after peripheral nerve stimulation (Stålberg and Grimby, 1995; Stålberg, 1996).

In chronic cases, the size of reinnervating motor units gives a measurement of the total loss of motor cell bodies. With repeated stimulation of motor nerves simultaneously with the EMG it is possible to make this measurement also in completely paretic segments. EMG sometimes show signs of reinnervation in one or several segments cranial to the lesion. In these patients, there is often a higher level of damage initially than indicated clinically. Thus using repeated EMG recordings, or Macro-EMG recordings, any change in motor unit size can be easily demonstrated (Grimby *et al.*, 1994).

Nerve conduction studies

It is possible to gain insight regarding spinal cord function using nerve conduction studies in clinical situation. Distal motor amplitudes obtained by stimulation of motor nerves coming from injured spinal cord segments give a measurement of remaining motor neurons which can be seen in segments even without any voluntary function. In lesions proximal to the sensory ganglion, i.e., cauda lesions, root avulsions and spinal cord lesions, sensory amplitudes are usually preserved whereas motor responses are either low or absent. In these cases the central SEP responses are often absent (Ducker and Brown, 1989). Latencies and number of recordable F-responses, where action potentials migrate in motor axons in retrograde direction to the motor cell and then back in anterograde direction, reflect the number of functional motor cells in injured segments of the spinal cord (Ducker, 1976; Homma and Tamaki, 1984).

Another possibility to study nerve conduction is to use H-reflexes which are monosynaptic, single-segmental spinal cord reflexes (Ducker and Brown, 1989). H-reflexes are normally present in a few spinal cord segment (C7 and S1) in healthy adults (Ducker, 1976). In spinal cord injury, H-reflexes can be seen in segments that normally do not give this reflex. An increased latency or absence of reflex activity is a common feature if this reflex is associated with the injured segment. A long latency of H-reflexes demonstrate the function of intersegmental tracts which represent activity of extrapyramidal tracts (rubrospinal and reticulospinal tracts) when there is no preservation of function in the pyramidal tracts (Dimitrijevic *et al.*, 1984; see Dimitrijevic, in this volume).

Electrospinogram

Electrospinogram (ESG) is used to understand spinal cord function in some studies (Homma and Tamaki, 1984). Spontaneous ESG activity represents electrical activity of segmental neurons in the spinal cord which can be well reflected by measuring the amplitude and frequency of the neuronal discharge in normal and spinal-traumatised subjects. Amplitudes of ESG following cord injury show a marked decrease along with either an increased or decreased frequency (Homma and Tamaki, 1984). Decreased ESG amplitude represent spinal cord ischemia. However, this technique is often complicated by artefacts such as electrocardiogram or from respiratory muscles. Recording of ESG as routine measure in spinal cord injury is thus quite limited. However, this technique is useful in studying the arrangement and conduction velocities of various fibre bundles in the spinal cord (see Tamaki, in this volume).

Neurophysiological assessment in experimental spinal cord injury

In animal studies, several investigators recorded SCEP as a reliable indicator of spinal cord dysfunction following trauma (Sharma *et al.*, 1991; Tator and Fehlings, 1991). In addition, SEP response from cortical surface or epidural recordings, MEP recording from peripheral muscles, and SCEP elicited with cerebral motor or cerebellar stimulation (Hurlbert *et al.*, 1993) are also used in several models of experimental spinal cord injury.

Our own investigations in spinal cord injury

We used SCEP recordings obtained from epidural electrodes placed cranial and caudal to a focal spinal lesion following stimulation of peripheral nerves to study the cord function. This section describes the results obtained from our own investigations which clearly points out that SCEP are good indicator of early cord pathology following trauma.

Background and hypotheses

Trauma to the spinal cord includes mechanical distortion of the spinal cord tissue followed by secondary injury cascades such as disturbances in the microcirculation followed by ischemia, breakdown of the BSCB, vasogenic edema and cell injury. These early traumatic reactions are further complicated by release of various neurochemicals which may influence the spinal cord conduction of impulses. It appears that pathophysiological events are influenced by neurochemicals probably via receptor mediated mechanisms.

We have initiated a series of investigations to characterise the early SCEP response following a unilateral longitudinal incision of the spinal cord in relation with later development of edema and cell injury. A new approach in our study is to evaluate the influence of several pharmacological compounds modifying early neurochemical events following trauma on SCEP response and spinal cord pathology.

Methodological aspect

SCEP is recorded in a well characterised new model of spinal cord injury (see Sharma and Dey, 1982; Sharma and Olsson, 1990; Sharma *et al.*, 1991;

Winkler *et al.*, 1993; Winkler, 1994). The salient features of the model is described below.

The new model of rat spinal cord injury

This model consists of a longitudinal incision to the right dorsal horn of the T10–11 segments. The deepest part of the lesion was limited to the Rexed's laminae VIII (Sharma and Dey, 1982; Sharma and Olsson, 1990). The model offers new possibilities for SCEP studies because the primary injury, i.e., the knife wound, is limited to the gray matter leaving white matter mainly intact (Sharma *et al.*, 1991). In other injury models involving either impact, compression or transverse incision, damage to white matter due to a direct physical injury of the ascending and descending tracts significantly contribute to the disturbances in spinal cord conduction. In our model, it is possible to study the mechanisms of neurochemicals released from the gray matter following trauma on the SCEP activity in relation with edema formation and tissue damage. The model is quite reproducible in our hands (n = 20) because in formalin fixed injured segments (n = 20) the extent of the lesion was found within a very narrow range (4 ± 1 mm long, 2 ± 1 mm deep) (Sharma *et al.,* 1991; Sharma and Olsson, 1990).

Recording of spinal cord evoked potentials

The SCEP was recorded from epidural electrodes placed over the adjacent rostral (T9) and caudal (T12) segments around the lesioned (T10–11) site following stimulation of right tibial and sural nerves (see Sharma *et al.*, 1991). The recording electrodes are made from a 10 cm long copper wire (denuded outer diameter 0.1 mm), inserted into the lumen of a 7 cm polythene tubing (PE 10). The denuded 3 cm end of the wire is taken out

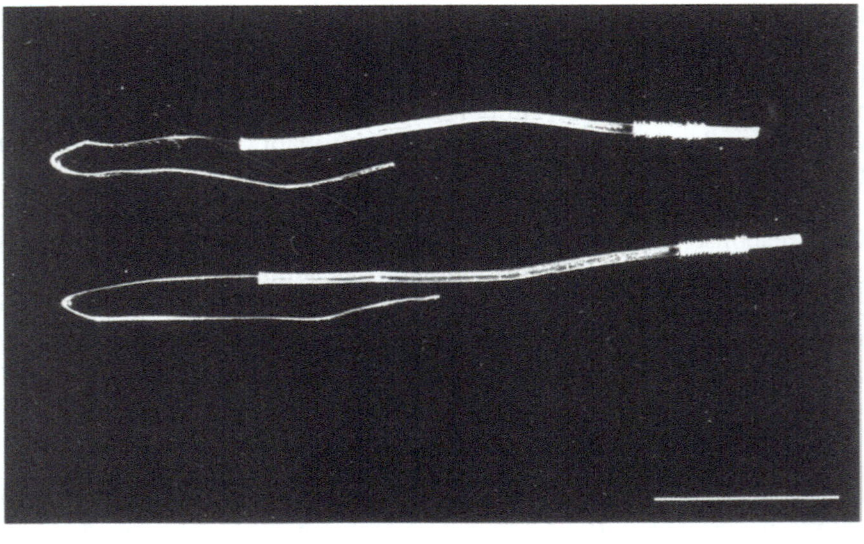

Fig. 1. Epidural electrodes used for SCEP recordings (bar = 5 mm)

through a hole made in the wall of the tubing (5 mm below the tip) and wound 12–15 turns around the tubing (covering 2.5 to 3 mm of the tubing (Fig. 1). Openings are sealed with Araldite. The electrical resistance of these unipolar electrodes is less than 0.11 Ohm (Sharma *et al.*, 1991; Winkler *et al.*, 1993; Winkler, 1994).

Under urethane anaesthesia (1.5 g/Kg, i.p.), unipolar electrodes are inserted from the laminectomized area into the spinal epidural space towards the right side of the cord and advanced rostrally and caudally to the levels of T9 and T12 respectively. The recording area of the active electrodes is located 2–3 mm away from the primary lesion. The reference needle electrodes are placed in the paravertebral muscles 5–10 mm apart from the corresponding levels of the explorative electrodes (Fig. 2).

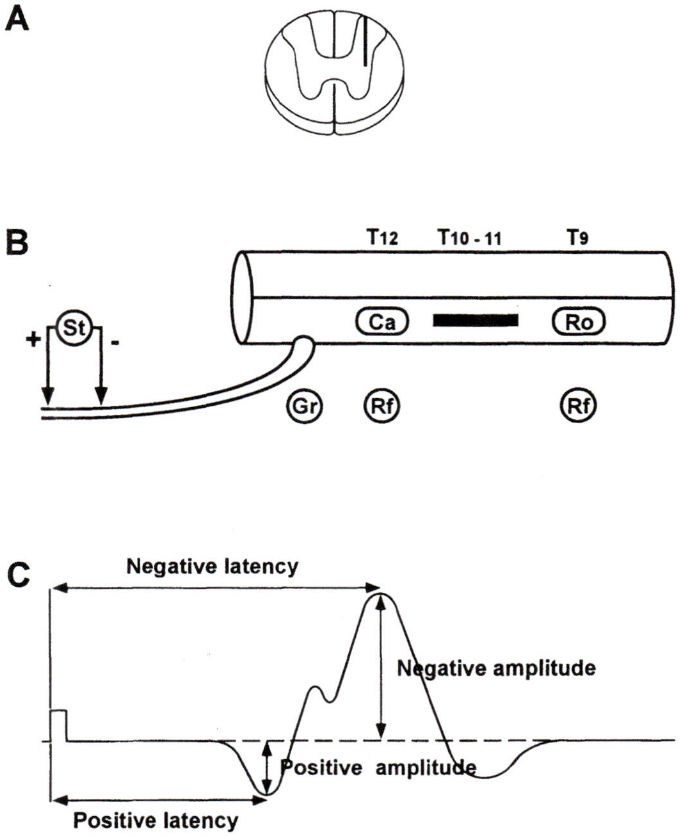

Fig. 2. (**A**) Cross section of the lesion site. The extent of injury is represented by a vertical black bar. (**B**) electrode set up for SCEP recording over the spinal cord around the lesion site represented by a longitudinal black bar. Epidural electrodes are placed rostral (*Ro*) and caudal (*Ca*) to the site of incision (black bar). The reference electrodes (*Rf*) are placed in paravertebral muscles and the ground electrode (*Gr*) is inserted into the hip. Stimulation of peripheral nerves of the right hind limb is indicated by (*St*). (**C**) Measurements of SCEP amplitudes (vertical arrowhead) and latencies (horizontal arrow heads) of SCEP recordings obtained using above set-up

The tibial and sural nerves in the right distal hind limb are stimulated simultaneously and supramaximally at a frequency of 7 Hz with needles (0.6 × 25 mm) placed behind the Achilles tendon (cathode) and below the plantar surface (anode), with a stimulus duration of 0.1 msec. Two hundred responses are collected and averaged by an EMG equipment (Medelec MS 20 "Mystro", or Dantec "Keypoint", filter 3 Hz – 5 kHz, sweep duration 20 msec). Rostral and caudal SCEP are recorded 30 min before, immediately before, immediately after and at 2, 4, and 10 min after injury. Then recordings are made at 30 min and at every hour after injury until the animal is sacrificed 5 h after injury.

The SCEP contains a small positive peak followed by a broad and high, multiphasic negative peak. The amplitudes and latencies of the maximal positive peak (MPP) and the maximal negative peak (MNP) are measured (Fig. 2). In calculations, the latencies and amplitudes measured 30 min before injury are used as references (100.0%). A complete loss of amplitude is denoted 0% (Sharma *et al.*, 1991; Winkler, 1994).

Pharmacological manipulation of SCEP

In order to understand the influence of neurochemicals on SCEP changes after trauma, several drugs were used in this investigation. A detailed description on the types of drugs, dosage and schedule of treatment is summarised in Table 7.

Characteristics of SCEP response

The rat SCEP consists of a small positive peak followed by a distinct negative peak with high amplitude (Fig. 3). SCEPs are stable in repeated recordings and occasionally represents very high voltage (100 μV to 500 μV). Normally it is possible to distinguish the SCEP from background activity without averaging (Sharma *et al.*, 1991; Winkler, 1994).Our SCEP recordings in the rat are very similar to those described by others in cats and monkeys (for details see Homma and Tamaki, 1984).

Influence of spinal cord injury

Repeated recordings of SCEP showed that the amplitude and latencies are quite stable before injury on both rostral and caudal electrodes (Sharma *et al.*, 1991) (Figs. 3 and 6). However, after injury a reduction in MNP was quite pronounced in rostral and in some cases in caudal recordings (Figs. 3, 6 and 7). In general, most of the traumatised animals exhibited loss of MNP amplitude on rostral recordings within 4 min after incision (Fig. 3). The injured animals often showed appearance of a broad MPP amplitude in rostral recordings (Fig. 3) which increased successively with time (Figs. 3 and 8). The increase in caudal MPP amplitude was also apparent in some animals, however, this increase was not statistically significant (Fig. 11).

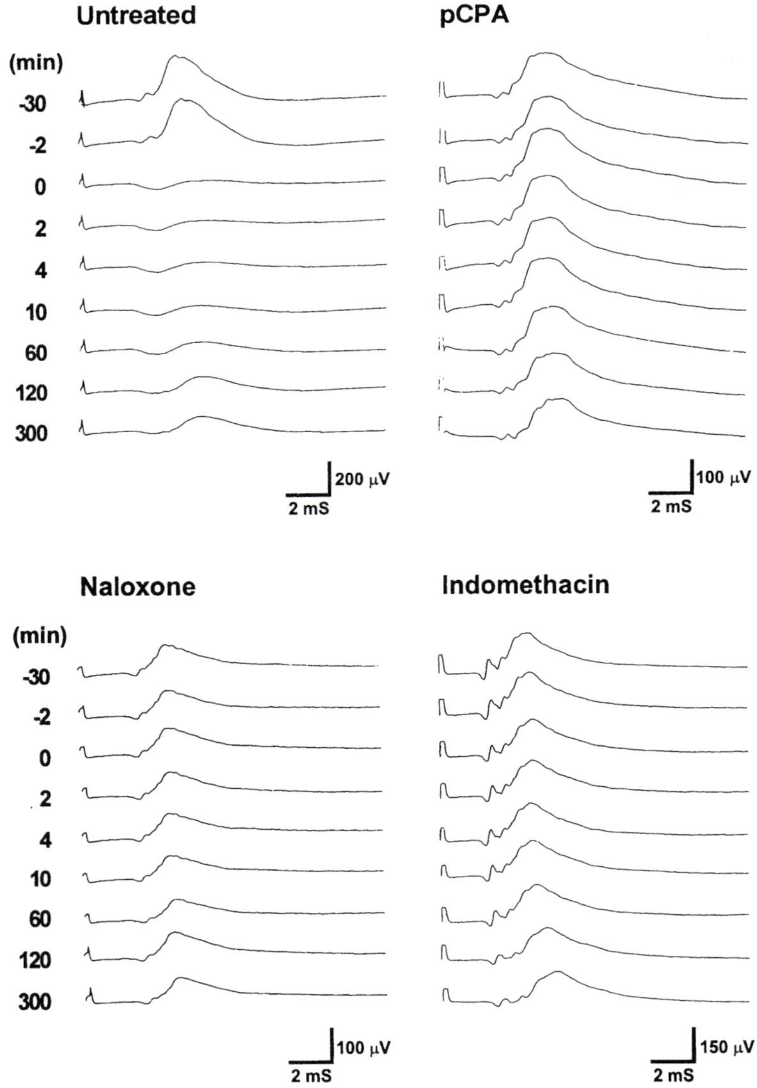

Fig. 3. Rostral SCEP from one untreated, one p-CPA pretreated, one naloxone pretreated, and one indomethacin pretreated rats. Untreated rat show a complete loss of MNP amplitude and an increase in MPP amplitude and latency after spinal cord injury (immediately 0 min). These changes in SCEP are not evident in drug-treated rats after injury (Data modified after Sharma *et al.*, 1991; Winkler *et al.*, 1993; Winkler *et al.*, 1994a)

bility exists that early changes in spinal cord neurochemistry can influence SCEP response and may be responsible for the later outcome of spinal cord pathology as well (Sharma *et al.*, 1998a,c). Therefore, we pretreated animals with several drugs (Table 7) and examined trauma induced changes in SCEP response and spinal cord pathology.

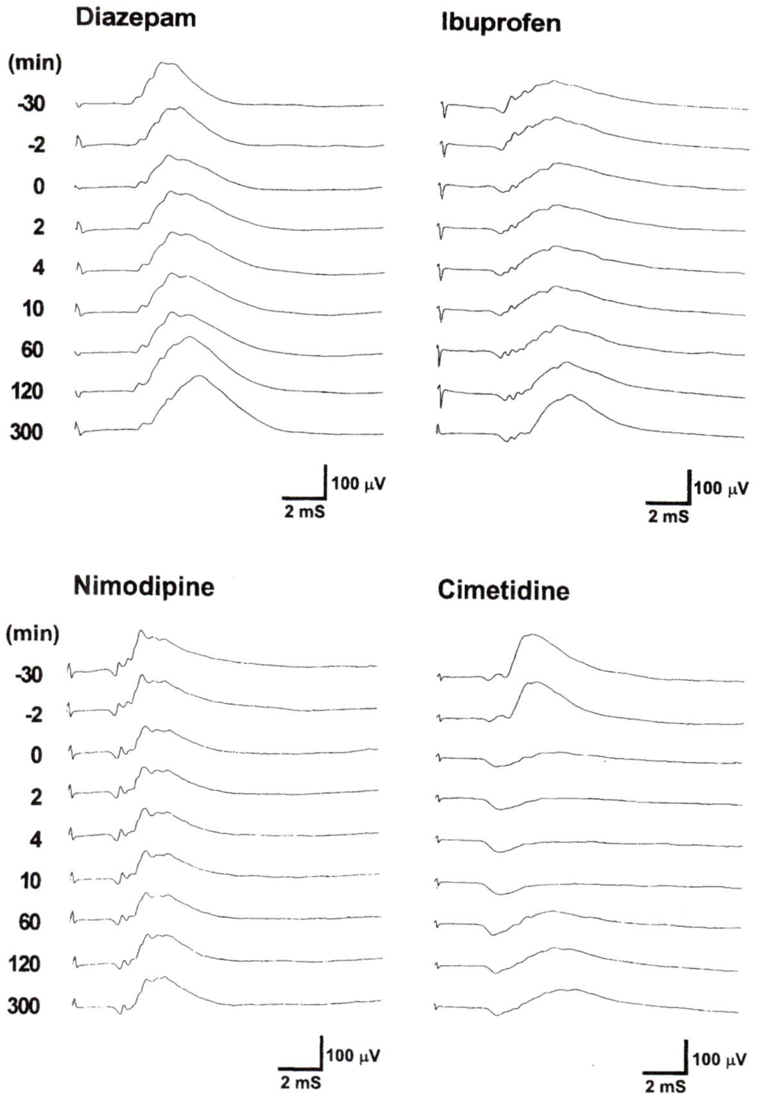

Fig. 4. Rostral SCEP from one diazepam pretreated, one ibuprofen pretreated, one nimodipine pretreated, and one cimetidine pretreated rat. Only cimetidine pretreated rat showed a complete loss of MNP amplitude and an increase in MPP amplitude and latency after spinal cord injury (immediately after 0 min). SCEP changes are not evident in diazepam, ibuprofen and nimodipine-treated rats after injury (Data modified after Winkler 1994; Winkler *et al.*, 1995a, 1997)

Representative examples of trauma induced changes in SCEP tracings of rostral MNP amplitude following drug treatments are displayed in Figures 3 to 5. It is quite clear from these tracings that p-CPA, naloxone, indomethacin (Fig. 3), diazepam, ibuprofen and nimodipine (Fig. 4) significantly attenuated the decrease of rostral SCEP amplitude (Sharma

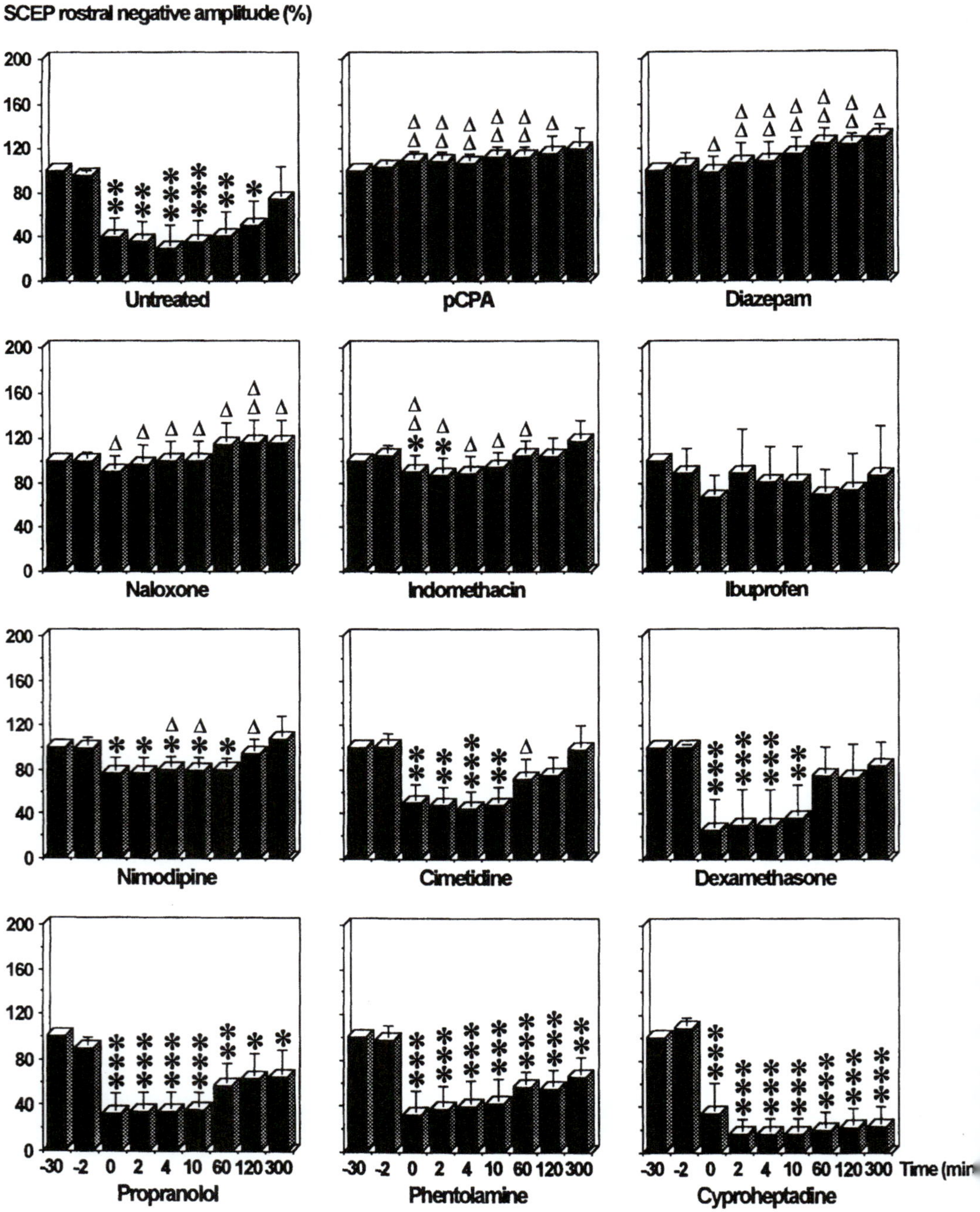

Fig. 6. Mean MNP amplitude on rostral recording following spinal cord injury and its influence by drugs. Data are Mean ± *SE*; *$P < 0.05$; **$P < 0.01$; ***$P < 0.001$; compared from pre injury level; Δ $P < 0.05$; ΔΔ $P < 0.01$; compared from untreated injured group (Student's unpaired *t* test)

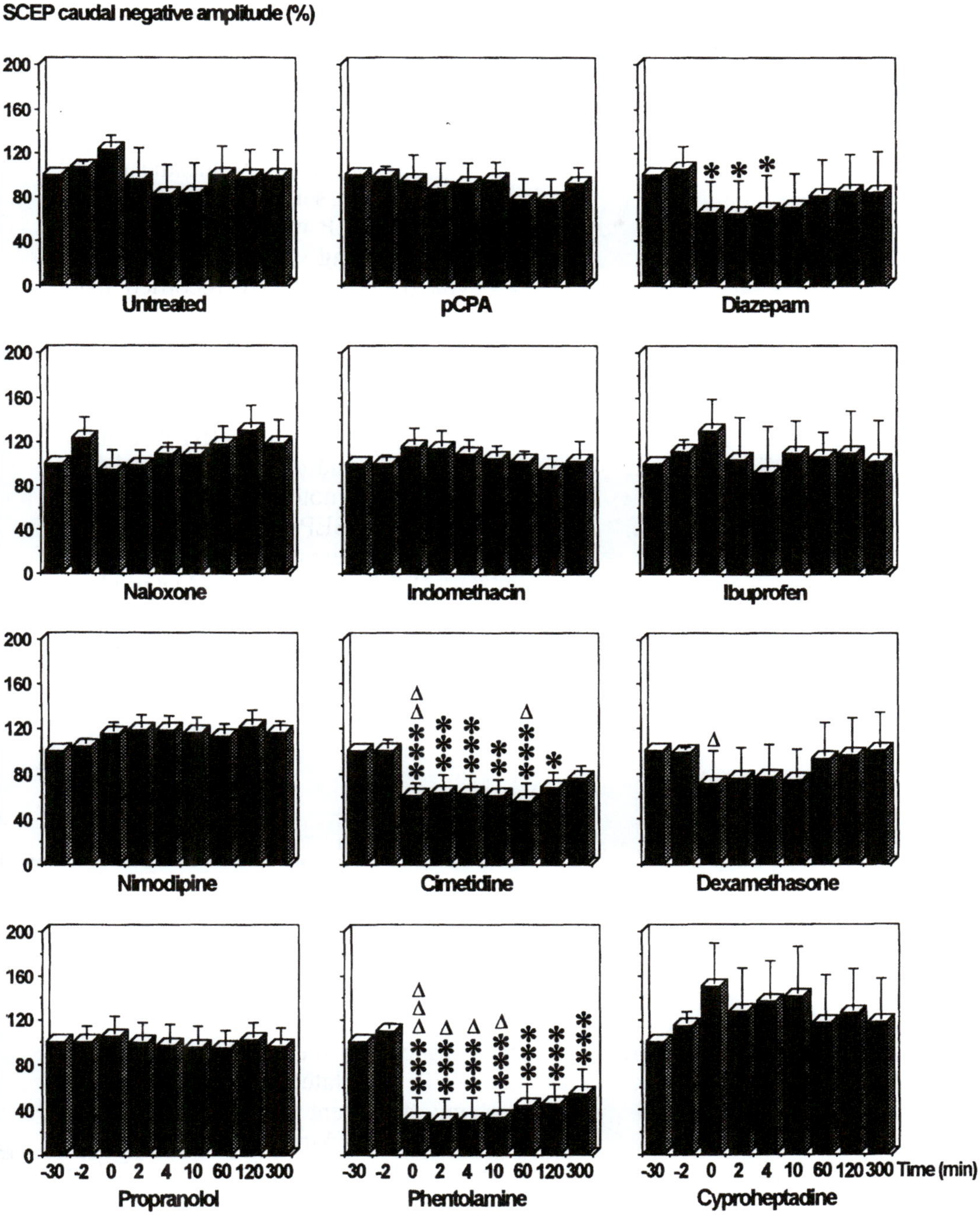

Fig. 7. Mean MNP amplitude on caudal recording following spinal cord injury and its influence by drugs. Data are Mean ± *SE*; *P < 0.05; **P < 0.01; ***P < 0.001; compared from pre injury level; Δ P < 0.05; ΔΔ P < 0.01; ΔΔΔ P < 0.001; compared from untreated injured group (Student's unpaired *t* test)

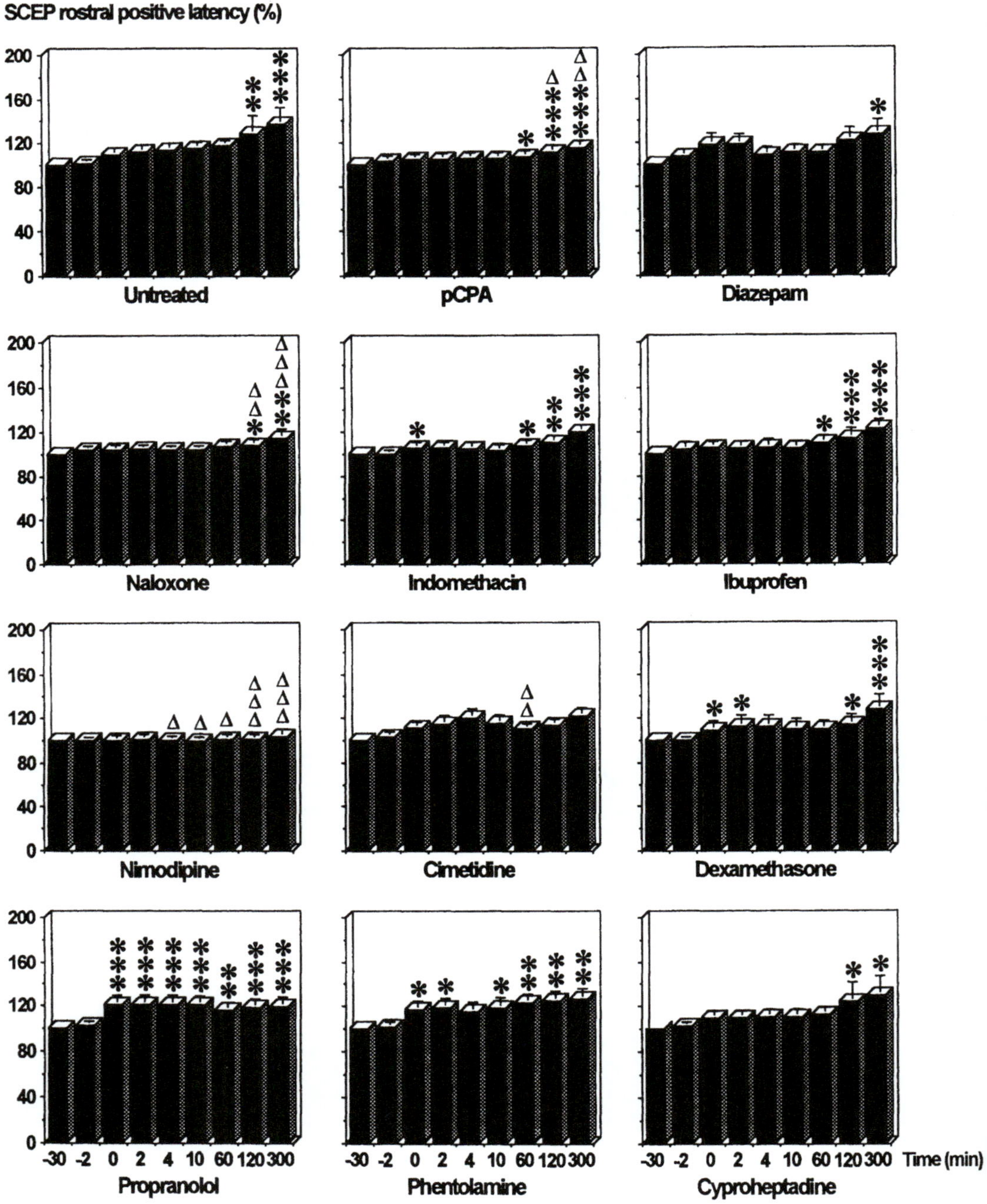

Fig. 10. Mean MPP latency on rostral recording following spinal cord injury and its influence by drugs. Data are Mean ± *SE*; *P < 0.05; **P < 0.01; **P < 0.001; compared from pre injury level; ** P < 0.05; ΔΔ P < 0.01; ΔΔΔ P < 0.001; compared from untreated injured group (Student's unpaired *t* test)

the SCEP disturbances on both amplitude and latency parameters (results not shown) (for details see Winkler *et al.*, 1994a,b).

Influence of indomethacin

Pretreatment with indomethacin significantly attenuated the decrease of rostral MNP amplitude (Figs. 3 and 6) after trauma (Winkler *et al.*, 1993). The compound was also able to block the effect of trauma on MPP amplitudes on rostral recordings (Fig. 8). However, no significant changes in caudal MNP or MPP amplitude (Figs. 7 and 11) or latencies of rostral MNP or MPP (Figs. 9 and 10) as well as caudal MNP and MPP (Figs. 12 and 13) were observed.

Influence of ibuprofen

Our results with ibuprofen attenuated the decrease of rostral MNP amplitude (Figs. 4 and 6) following trauma (Winkler, 1994). However, this drug did not influence caudal MNP or MPP amplitude (Figs. 7 and 11) changes. The increase in latencies of rostral and caudal MNP (Figs. 9 and 12) or MPP (Figs. 10 and 13) recordings were unaffected by ibuprofen compared to the untreated traumatised rats.

Influence of diazepam

Pretreatment with diazepam markedly attenuated SCEP disturbances on rostral MNP amplitude (Figs. 4 and 6) following trauma compared to untreated traumatised rats (Winkler *et al.*, 1997). Thus diazepam pretreatment significantly attenuated the decrease of rostral MNP amplitude (Fig. 6), whereas a late increase in MPP amplitude was evident (Fig. 8). On the other hand, a significant reduction in caudal MNP amplitude (Fig. 7) was evident with this drug immediately after trauma which persisted up to 4 min after injury and gradually recovered. However, this drug treatment did not influence MNP and MPP latency increase in either rostral (Figs. 9 and 10) or caudal (Figs. 12 and 13) recordings.

Influence of nimodipine

Pretreatment with nimodipine (2 μg/kg/min continued infusion) inhibited the decrease of rostral MNP amplitude (Figs. 4 and 6) after trauma (Sharma and Olsson, 1994b; Winkler, 1994) compared to untreated rats. On the other hand increase of MPP amplitude was also attenuated (Fig. 8). Whereas, caudal MNP (Fig. 7) and MPP (Fig. 11) amplitudes were not significantly affected from the untreated traumatised group. Nimodipine was found to be very effective in reducing latency changes on both rostral and caudal MNP (Figs. 9 and 12) or MPP (Figs. 10 and 13) recordings. Thus apparently no significant increase in latency was noted in this

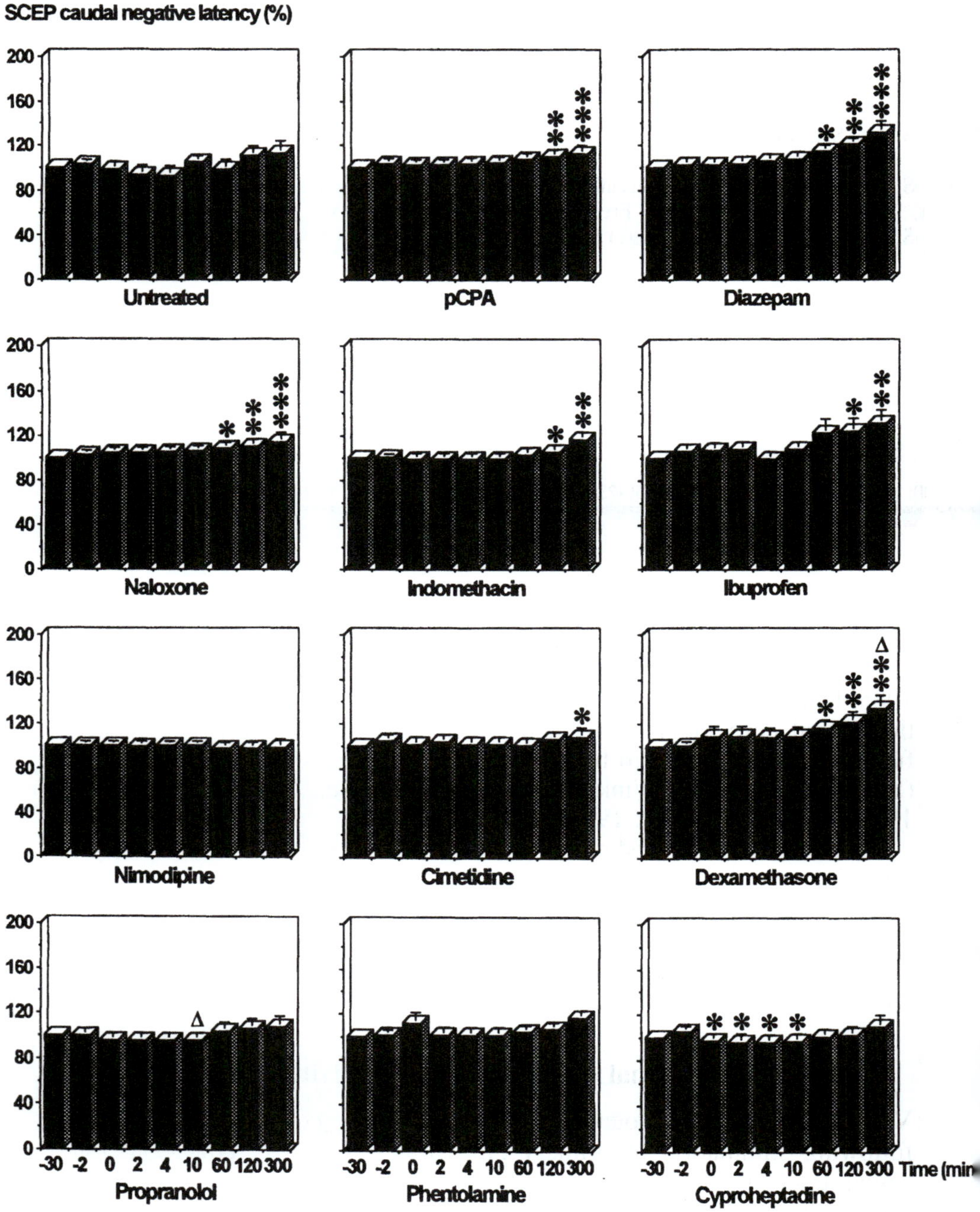

Fig. 12. Mean MNP latency on caudal recording following spinal cord injury and its influence by drugs. Data are Mean ± *SE*; *$P < 0.05$; **$P < 0.01$; ***$P < 0.001$; compared from pre injury level. Δ $P < 0.05$; compared from untreated injured group (Student's unpaired *t* test)

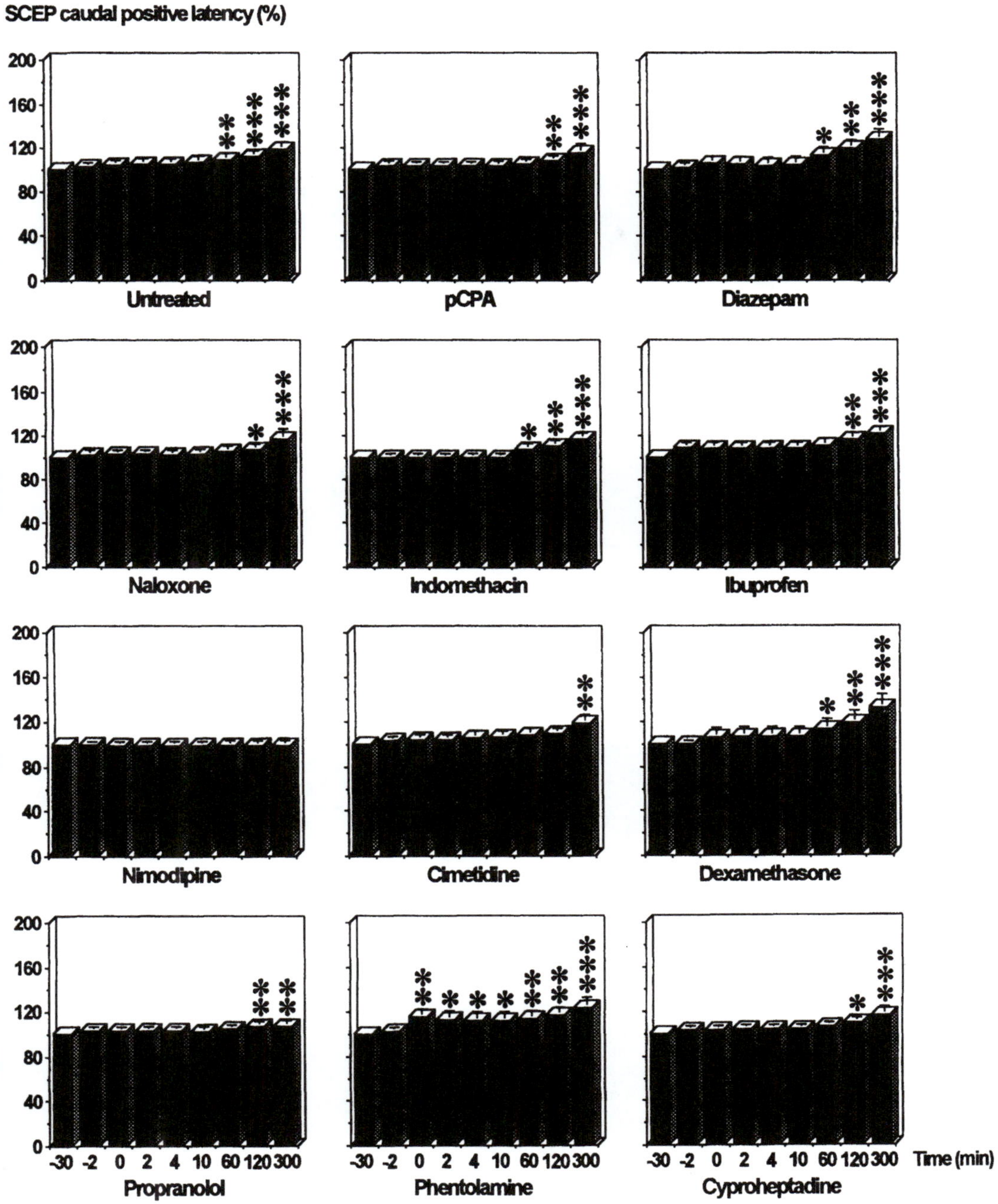

Fig. 13. Mean MPP latency on caudal recording following spinal cord injury and its influence by drugs. Data are Mean ± *SE*; *P < 0.05; **P < 0.01; ***P < 0.001; compared from pre injury level (Student's unpaired *t* test)

Spinal cord edema formation

Spinal cord water content was measured in the traumatised segment (Sharma and Olsson, 1990) to determine edema formation. The cord samples were removed at 5 h, weighed immediately and placed at 90 °C for 72 h in order to determine the dry weight of the tissue. The water content was determined from the differences in the wet and dry weight of the tissue sample (see Dey and Sharma, 1983; 1984; Sharma and Olsson, 1990; Sharma *et al.*, 1991; Sharma *et al.*, 1998c).

About 4% increase in spinal cord water content was observed in the traumatised rats compared to controls (Sharma *et al.*, 1991; Sharma and Olsson, 1990). A linear increase in water content of the spinal cord following 30 min, 1 h, 2 h and 5 h after injury in separate group of rats (Fig. 14) suggests that edema formation in the spinal cord is progressive in nature (Sharma and Olsson, 1990). This increase in water content is reliable because morphological examination also revealed edematous swelling and general sponginess of the traumatised spinal cord (Figs. 17, 18) (see Sharma and Olsson, 1990; Sharma *et al.*, 1990a).

A good correlation between early reduction in rostral MNP amplitude (within 0–4 min following trauma) and later development of edema formation (at 5 h after primary insult) was observed (Fig. 15) (Sharma *et al.*,

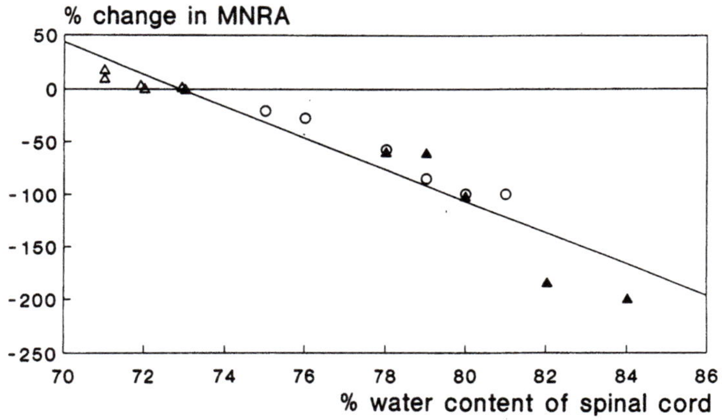

Fig. 15. Correlation between decrease in rostral mean MNP amplitude following spinal cord injury at 4 min and increase in water content of the injured spinal cord at 5 h in spinal cord injured (open circle), p-CPA pretreated (open triangle) and cyproheptadine pretreated (filled triangle) rats. ($r = 0.90$; $P < 0.001$; Data from Sharma *et al.*, 1991, with permission)

Fig. 14. Changes in blood-spinal cord barrier (*BSCB*) permeability, spinal cord water content, plasma and spinal cord serotonin (*5-HT*) levels and spinal cord blood flow (*SCBF*) following 1 h, 2 h, 5 h spinal cord injury in T9 (*rostral*), T10–11 (*traumatised*) and T12 (*caudal*) segments and their modification with p-CPA pretreatment. $^{*}P < 0.01$; $^{**}P < 0.001$; compared from control group. $\Delta P < 0.05$; compared from 5 h injured group. *A* control; *B* 1 h injury; *C* 2 h injury; *D* 5 h injury; *E* p-CPA + 5 h injury (Data modified after Sharma and Olsson, 1990; Olsson *et al.*, 1990; Sharma *et al.*, 1990a,b)

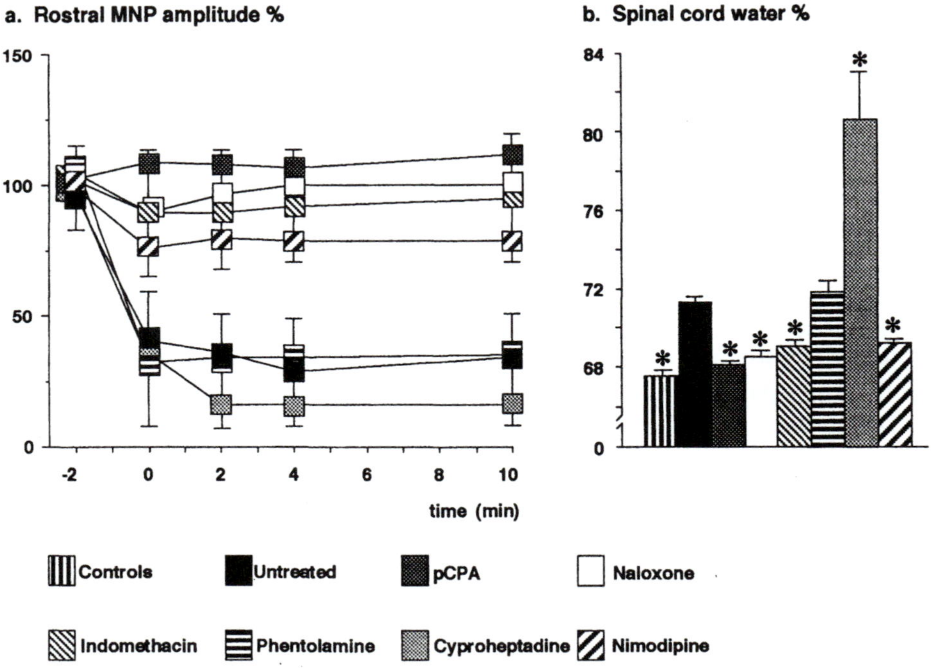

Fig. 16. (**a**) Decrease in rostral mean MNP amplitude following spinal cord injury and its modification with drugs. (**b**) Water content of the injured spinal cord and its modification with drugs. *P<0.01 compared from untreated injured group (Student's unpaired t-test). The drugs which attenuated the decrease of MNP amplitude within 4 min after injury were able to reduce edema formation of the injured spinal cord at 5 h (Data from Winkler, 1994, with permission)

1991; Winkler, 1994; Winkler *et al.*, 1993; 1994a,b; 1997; 1998). Pharmacological compounds which attenuated SCEP changes were able to reduce edema formation (Fig. 16). On the other hand, drugs which did not inhibit the loss of SCEP amplitude were unable to attenuate edema formation (Fig. 16). Thus pretreatment with p-CPA, indomethacin, ibuprofen, diazepam and nimodipine significantly attenuated edema formation of the spinal cord, whereas, pretreatment with cyproheptadine, phenoxybenzamine, propranolol, dexamethasone were unable to reduce the water content of the spinal cord (Fig. 16) (Winkler *et al.*, 1994; T Winkler, unpublished observations).

Pathophysiology of spinal cord injury

In addition to edema formation of the cord, early SCEP changes were compared with several other important pathophysiological events like disturbances in BSCB permeability, SCBF, plasma and spinal serotonin levels 5 h after trauma.

Spinal cord blood flow changes

Spinal cord blood flow (SCBF) was measured using radiolabelled carbonised microspheres labelled with [125]Iodine (for details see Sharma *et al.*, 1990a). In brief, about 10^6 microspheres were injected into the left cardiac ventricle thorough an indwelling polythene catheter implanted into the left common carotid artery for 45 sec (Sharma and Dey, 1986; Sharma, 1987; Sharma and Dey, 1987a). The timed arterial blood samples from right femoral artery was collected every 30 sec before administration of the microspheres at the rate of 0.8 ml/min. After administration of microspheres, the animals were killed immediately and the spinal cord samples from T9, T10, T11 and T12 segments were dissected out and the radioactivity counted in a 3-in well type gamma counter (see Sharma and Dey, 1986). Using this method, about 30% decrease in SCBF was noted after 5 h spinal cord injury (Fig. 14) (Sharma *et al.*, 1990a). The SCBF was restored near normal values by pretreatment with p-CPA (Fig. 14) indomethacin, naloxone, diazepam and nimodipine which significantly attenuated SCEP changes after trauma (Sharma *et al.*, 1990a; Sharma, unpublished observations). On the other hand, no improvement in SCBF was noted with drugs such as dexamethasone, phenoxybenzamine or propranolol which were unable to reduce trauma induced SCEP changes (Sharma, unpublished observations).

Blood-spinal cord barrier permeability

We measured BSCB permeability using Evans blue, ^{131}I-sodium, or ^{125}I-labelled plasma proteins or growth hormone in spinal cord segments ranging from T4 to T12 (Olsson *et al.*, 1990; Sharma *et al.*, 1990a; 1995a, b; Mustafa *et al.*, 1995; Sharma *et al.*, 1996a; 1997a,b). A focal trauma to the spinal cord induces widespread disruption of the BSCB to hormones and plasma proteins in the perifocal as well as far remote segments of the cord within a period of 5 h (Fig. 14). Increased BSCB permeability to rat growth hormone was seen as early as 30 min after trauma. This increase in BSCB permeability continued until 2 h and then declined at 5 h after trauma (Mustafa *et al.*, 1995).

Leakage of plasma proteins are responsible for vasogenic edema formation (Sharma *et al.*, 1991; 1997a–c,e; 1998a,b). Breakdown of the BSCB permeability to small (growth hormone, iodine) or large molecules (plasma proteins, Evans blue albumin) suggests widespread alteration in the fluid microenvironment of the cord which is crucial for edema formation and cell injury. Our results show that the drugs such as p-CPA (Fig. 14), indomethacin, naloxone, diazepam and nimodipine which are able to reduce the SCEP changes following trauma were capable to attenuate BSCB permeability. On the other hand, drugs like dexamethasone, phenoxybenzamine and propranolol which did not reduce trauma induced SCEP changes were unable to reduce the leakage of BSCB permeability to various tracers (Sharma *et al.*, 1995a,b; 1996a; 1997a–c; 1998b; Sharma, unpublished observations).

Plasma and spinal cord serotonin content

Using a very specific and sensitive spectrophotofluorometric method (Sharma, 1982; Sharma and Dey, 1986; 1987a,b; Sharma *et al.*, 1998a) we measured plasma and spinal cord serotonin content following trauma to the spinal cord (Sharma and Dey, 1982; Sharma and Olsson, 1990; Sharma *et al.*, 1990a; 1997a–c). A marked increase in plasma serotonin occurred even 30 min after trauma which showed a further linear increase at 1 h, 2 h and 5 h after injury (Fig. 14) (Sharma *et al.*, 1990a; Sharma, unpublished observation). The spinal cord serotonin also showed several fold increase in the traumatised and adjacent perifocal segments 5 h after trauma (Fig. 14) (Sharma *et al.*, 1990a; Sharma HS, unpublished observation).

In order to find out a role of serotonin in cell injury (Sharma *et al.*, 1989), we used several drugs as a pretreatment on trauma induced alterations in plasma and cord serotonin level. Our results suggest that pretreatment with p-CPA (Fig. 14), indomethacin and naloxone significantly attenuated increase in plasma and cord serotonin level, whereas pretreatment with phenoxybenzamine and cyproheptadine were without any effect. Pretreatment with cyproheptadine in fact augmented the plasma and spinal cord serotonin levels following trauma (Sharma *et al.*, 1991).

Increased circulating serotonin levels represents the magnitude of traumatic stress (Chaouloff *et al.*, 1993; Sharma and Dey, 1986; 1987b; 1988) which can influence cell injury directly or indirectly through manipulation of the BSCB permeability and vasogenic edema formation (Sharma *et al.*, 1989; 1990b,d; 1995a,c; 1998a,c).

Structural changes of the spinal cord

The neuronal, glia and axonal reactions using histochemical and morphological approaches at light and electron microscopy were examined.

Trauma to the spinal cord is associated with profound structural changes in the adjacent spinal cord T9 (Fig. 17) and T12 segments after 5 h injury (Sharma *et al.*, 1995c). These morphological changes at 5 h in the spinal cord correlates well with the early SCEP disturbances (Sharma and Cervós-Navarro, 1990; Sharma and Olsson, 1990; Sharma *et al.*, 1995a–c; 1996a; 1998a–c).

Fine structure of the spinal cord after trauma

The perifocal T9 and T12 segments of the spinal cord at 5 h showed profound swelling which was not observed in rats treated with p-CPA (Fig. 17), indomethacin, naloxone, diazepam or nimodipine (results not shown) which prevented the early SCEP changes. Whereas, pharmacological

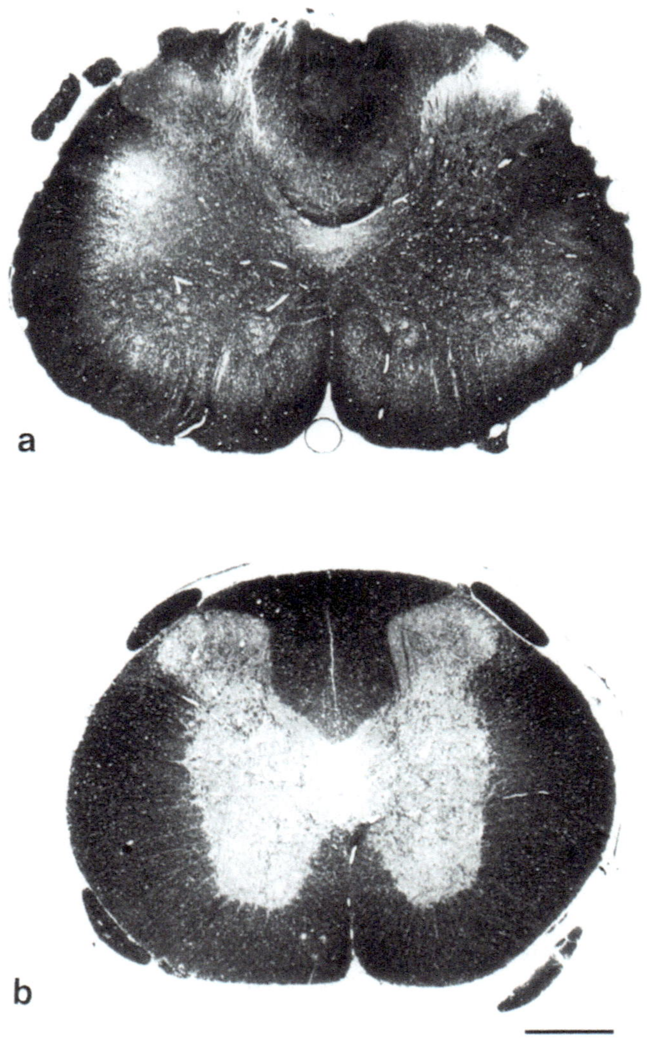

Fig. 17. Low power light micrograph showing cross section of the spinal cord T9 segment from one *untreated* 5 h injured rat (**a**) and one *p-CPA pretreated* traumatised rat (**b**). In drug treated injured rat, spinal cord gray matter and white matter appears to be quite preserved (Data from Sharma and Olsson, 1990, with permission, bar 2 mm)

agents such as dexamethasone, phenoxybenzamine or propranolol which failed to reduce early SCEP changes did not reduce the edematous expansion and visual swelling of the spinal cord.

The fine structure of the spinal cord after trauma was examined using high resolution light microscopy on 1 µm thick sections stained with toluidine blue staining (Sharma and Olsson, 1990). A general expansion of the spinal cord, widespread microhameorrhages and a clear distinction

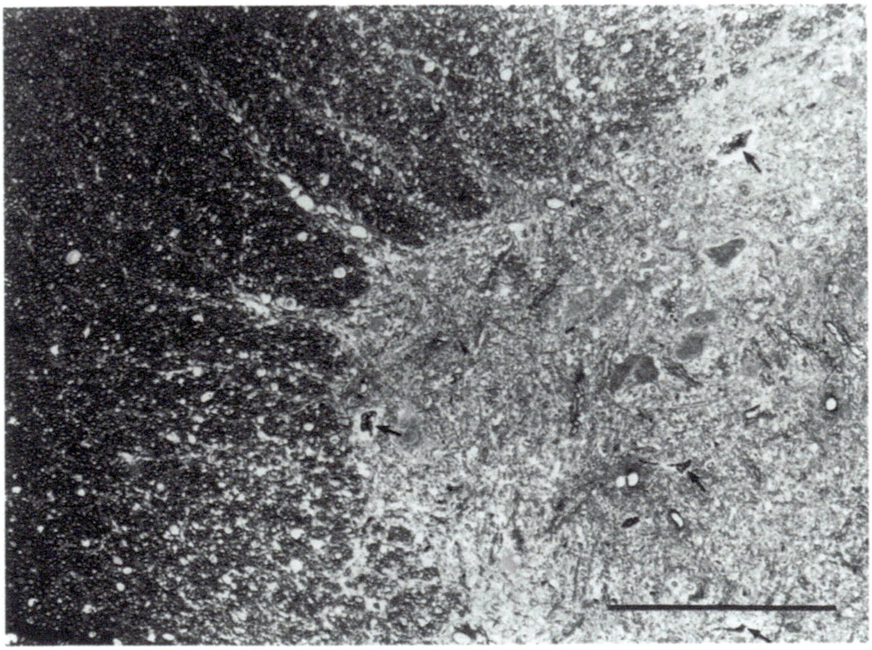

Fig. 18. High power light micrograph from the left ventral horn of the T9 segment from one 5 h injured rat. Many distorted nerve cells (arrows) are present. The neuropil shows profound sponginess and edema. Damage to white matter is also apparent (Data from Winkler *et al.*, 1997, with permission, bar 50 μm)

between gray and white matter were absent at 5 h after trauma (Fig. 18). Nissle staining showed many distorted nerve cells with condensed cytoplasm and a degenerated nucleus. Some nerve cells were swollen and some are shrunken (Fig. 19). Vacuolation, edema and distortion of other cellular structures are quite prominent in the spinal cord of traumatised animals. These nerve cell changes were most prominent in the ipsilateral side of the cord compared to the contralateral gray matter.

Pretreatment with drugs which prevented the early SCEP changes markedly attenuated these cell changes. On the other hand drugs that did not attenuate the early SCEP changes were also unable to reduce these adverse cell reactions.

Heat shock protein (HSP 72 kD) immunoreactivity

Heat shock protein (HSP) is a marker of stress reaction in the CNS (Westman and Sharma, 1998). A significant increase of HSP in motoneurons was seen at 5 h following trauma (Sharma *et al.*, 1995c). Pre-

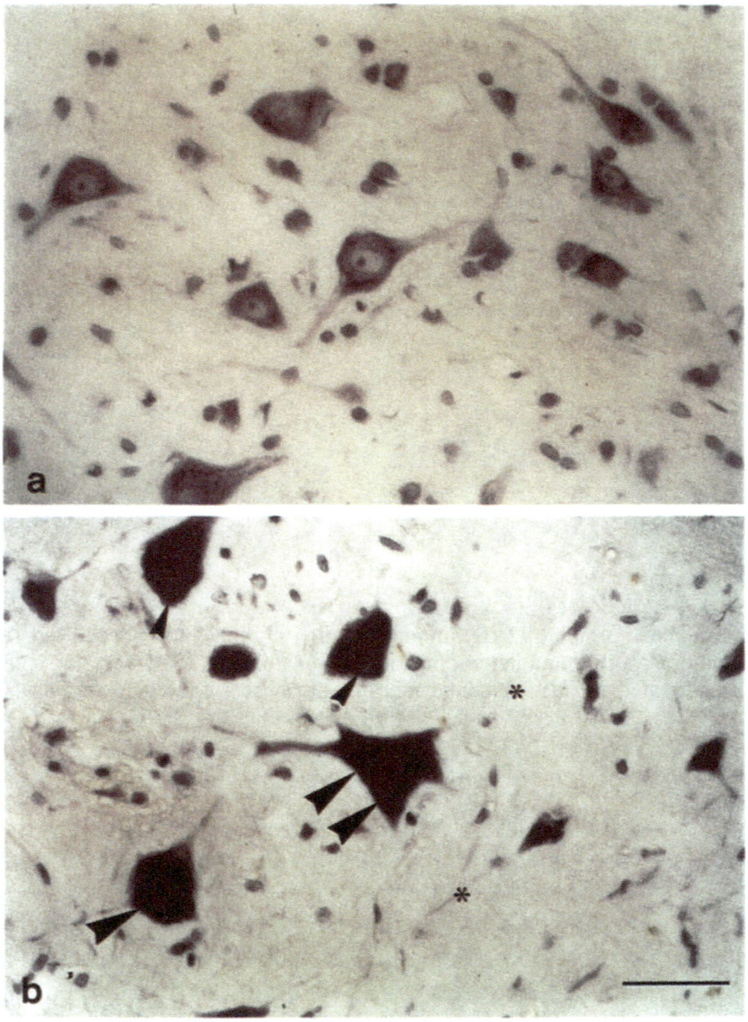

Fig. 19. Nissl stained nerve cells in the ventral horn of T9 segment from one control (**a**) and one 5 h injured rat (**b**). Many dark and distorted nerve cells (arrows heads) are present in the traumatised cord. A general sponginess and edema is quite clear in the injured rat (Data modified from Olsson *et al.*, 1992, bar 25 μm)

treatment with several drugs like naloxone (Fig. 20), p-CPA, indomethacin, diazepam or nimodipine which reduced the early SCEP changes were able to markedly attenuated the HSP response in the spinal cord at 5 h. On the other hand drugs which did not reduce SCEP changes were unable to attenuate HSP response. This indicates that a reduction in stress response is associated with less disturbances in SCEP changes following trauma.

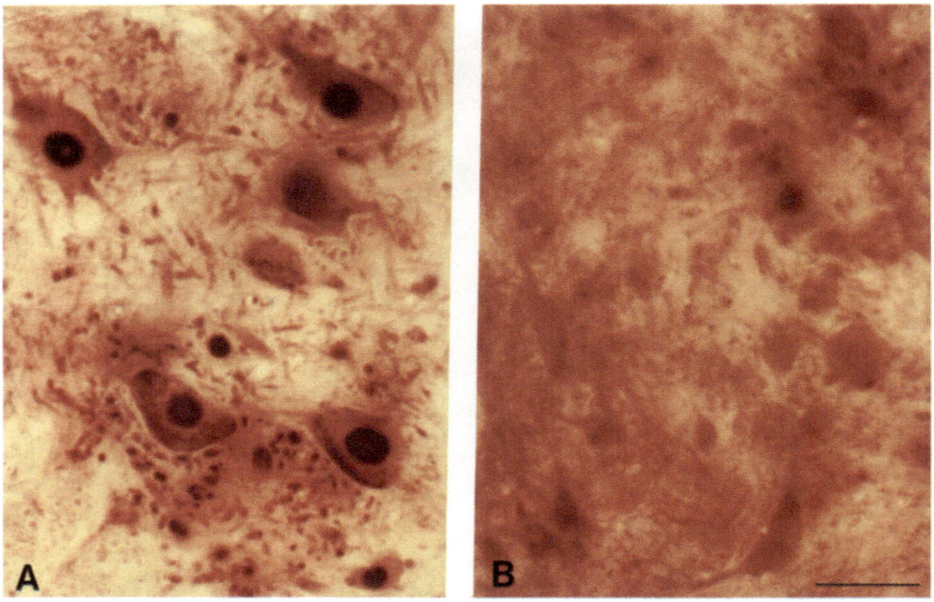

Fig. 20. Heat shock protein (72 kD) immunostaining in the ventral horn of the T9 segment of one 5 h injured rat (**A**) and its modification with naloxone (10 mg / kg, i.p.) pretreatment (**B**). In drug treated rat HSP labelled neurons are absent following trauma (Data after Olsson *et al.*, 1995, with permission, bar 25 μm)

Astrocytic response of the cord following trauma

Astrocytes participate in the traumatic response of the spinal cord (Sharma *et al.*, 1993c). We used glial fibrillary acidic protein (GFAP) immunoreactivity as a specific marker for astrocytic reaction (Sharma *et al.*, 1993c). Our observations showed a profound upregulation of GFAP in the perifocal segments of the spinal cord (Fig. 21). This activation of astrocytic response was markedly reduced by drugs which inhibited early SCEP disturbances following trauma (results not shown). Disturbances in microenvironment of the cord following trauma is thus a key factor in triggering astrocytic response which can be influenced by drug treatments.

Myelin reactions of the spinal cord

We used immunostaining of myelin basic protein (MBP) to study the myelin damage in the cord following trauma. A focal trauma to the cord induces widespread degeneration of immunostained nerve fibres and loss of MBP immunostaining (Fig. 22). Drugs which reduced the SCEP changes were able to attenuate myelin degeneration and loss of MBP immunostaining. This observation further support the hypothesis (Sharma *et al.*, 1993a) that damage of myelin sheaths are primarily responsible for disturbances in the spinal cord conduction.

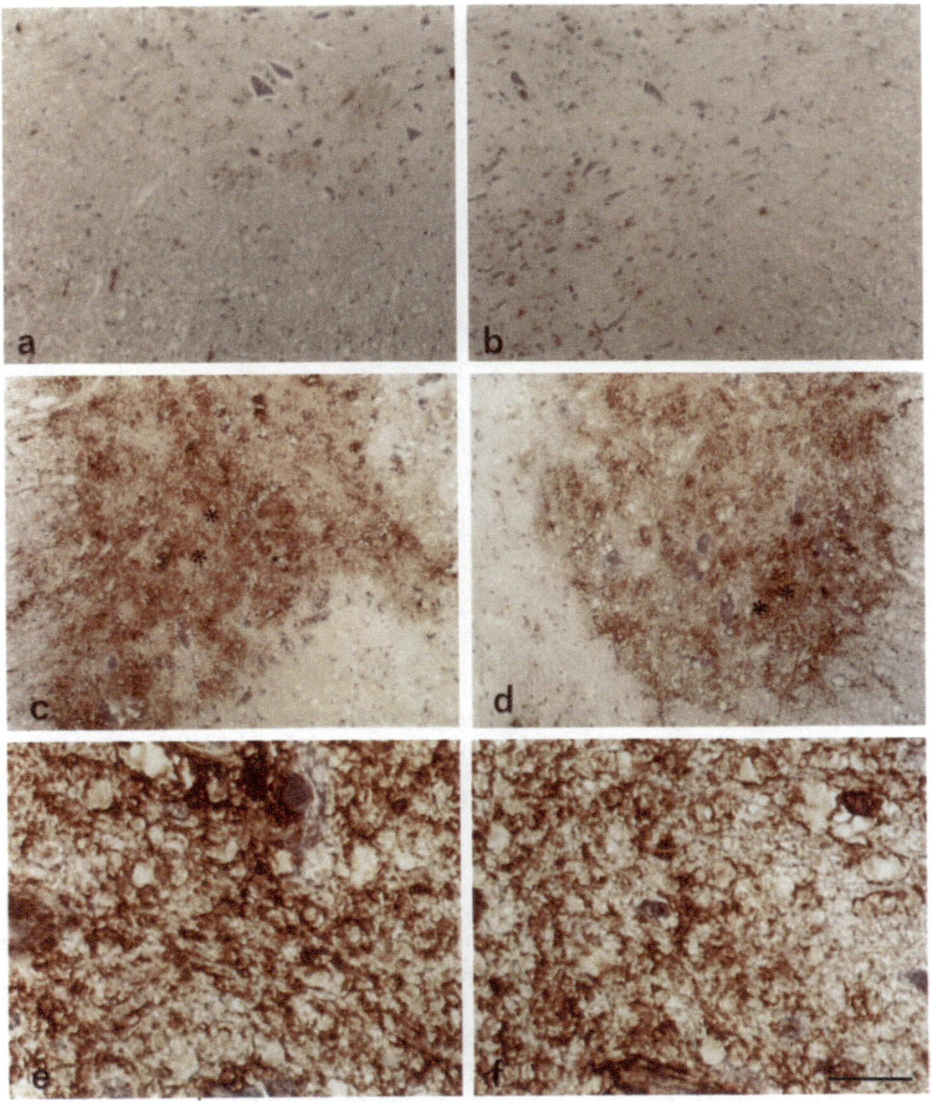

Fig. 21. Glial fibrillary acidic protein (*GFAP*) immunostaining in ventral horns of the T9 (**a**, **c**, **e**) and T12 (**b**, **d**, **f**) segments in one control (**a**, **b**) and in one 5 h traumatised rat (**c**–**f**). Marked increase in GFAP immunostaining is apparent in traumatised rat (**c**, **d**) compared to controls (**a**, **b**). GFAP immunostaining is present in damaged and distorted regions of the ventral horn (**e**, **f**) (Data modified after Olsson *et al.*, 1992; Sharma *et al.*, 1993c, bar: a,b 100 µm; c,d 120 µm; e,f 50 µm)

Ultrastructural changes of the spinal cord

At ultrastructural level dark and distorted nerve cells with eccentric nucleolus, degenerated synapses, membrane disruption, perivascular edema and swollen astrocytes are frequent findings in the perifocal segments of

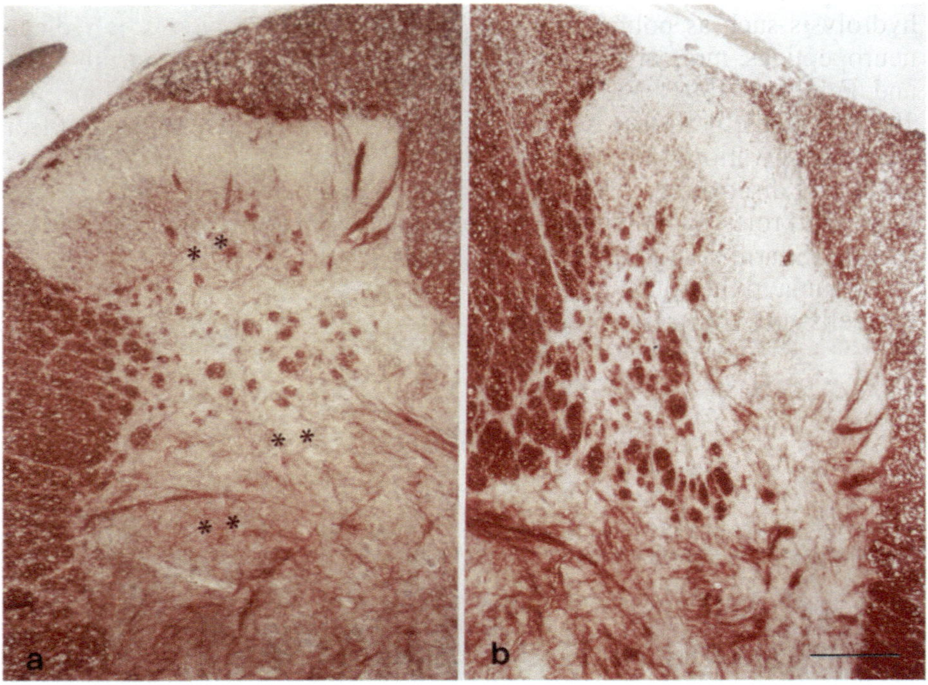

Fig. 22. Myelin basic protein (*MBP*) immunostaining in the left dorsal horn of the T9 segment in one control (**b**) and in one 5 h traumatised rat (**a**). Loss of MBP immunostained fibres (**) and a less intense red staining in the traumatised rat (**a**) is clearly visible compared to the control rat (**b**) (Data modified after Sharma *et al.*, 1993c, bar 80 μm)

the injured spinal cord. Vesiculation of myelin and damage to the myelinated and unmyelinated axons is quite common (Fig. 23).

Pretreatment with drugs which prevented the early SCEP changes showed less distortion of the spinal cord and changes in the myelinated and non-myelinated axons were considerably reduced. The neuropil appears to be quite compact and the signs of vacuolation, edema and membrane damage were significantly reduced (Fig. 24). Drugs which failed to inhibit SCEP changes did not attenuate these cell changes at ultrastructural level.

Passage of lanthanum across BSCB

At ultrastructural level Lanthanum seen as dark black particles is mainly confined within the lumen of the microvessels in the normal spinal cord (Olsson *et al.*, 1990; 1992; 1995; Sharma *et al.*, 1995b). Trauma to the cord resulted in a widespread extravasation of the lanthanum within the endothelial cell membranes and occasionally into the basal lamina (Fig. 25). Vesicular profiles containing lanthanum within the endothelium is common in untreated traumatised rats.

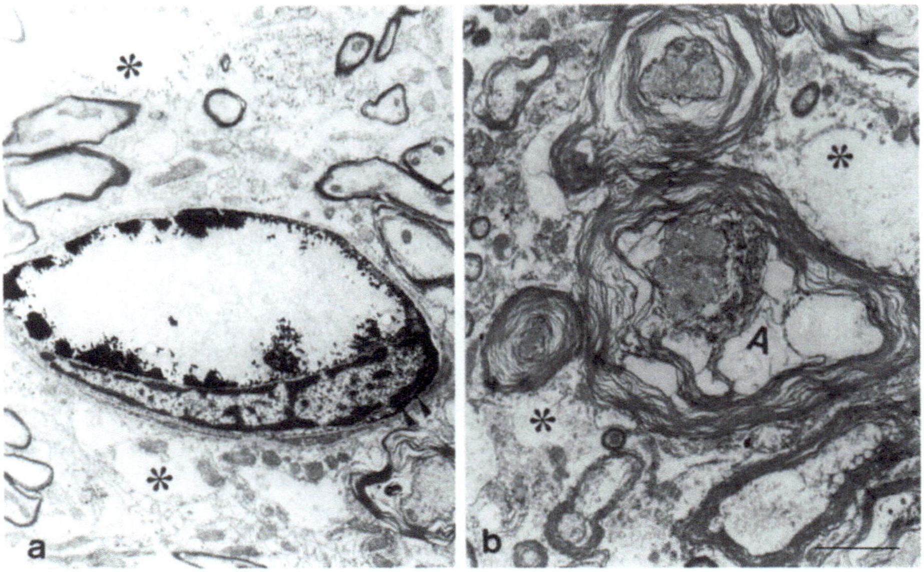

Fig. 23. Ultrastructural changes in the neuropil of the left ventral horn of the T9 segment in one 5 h traumatised rat. (**a**) Perivascular edema (*) and extravasation of lanthanum in the basal lamina (arrow head) is quite distinct. (**b**) Damage to myelinated axon (*A*); vesiculation of myelin, membrane damage and edema (*) is apparent (Data after Olsson *et al.*, 1992, with permission, bar 30 μm)

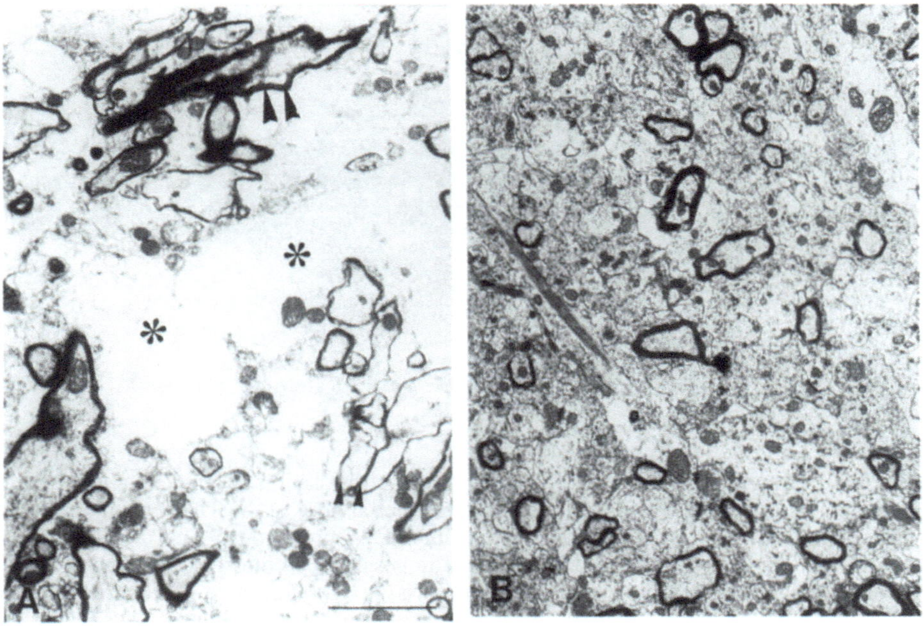

Fig. 24. Ultrastructural changes in the neuropil of the left ventral horn of the T9 segment in one untreated (left) and one p-CPA pretreated (right) and 5 h injured rat. The neuropil appeared quite compact and signs of edema, damage to myelinated axon, vesiculation of myelin and membrane disruption are minimal in drug treated rat (Data modified after Sharma and Olsson, 1990, bar 2 μm)

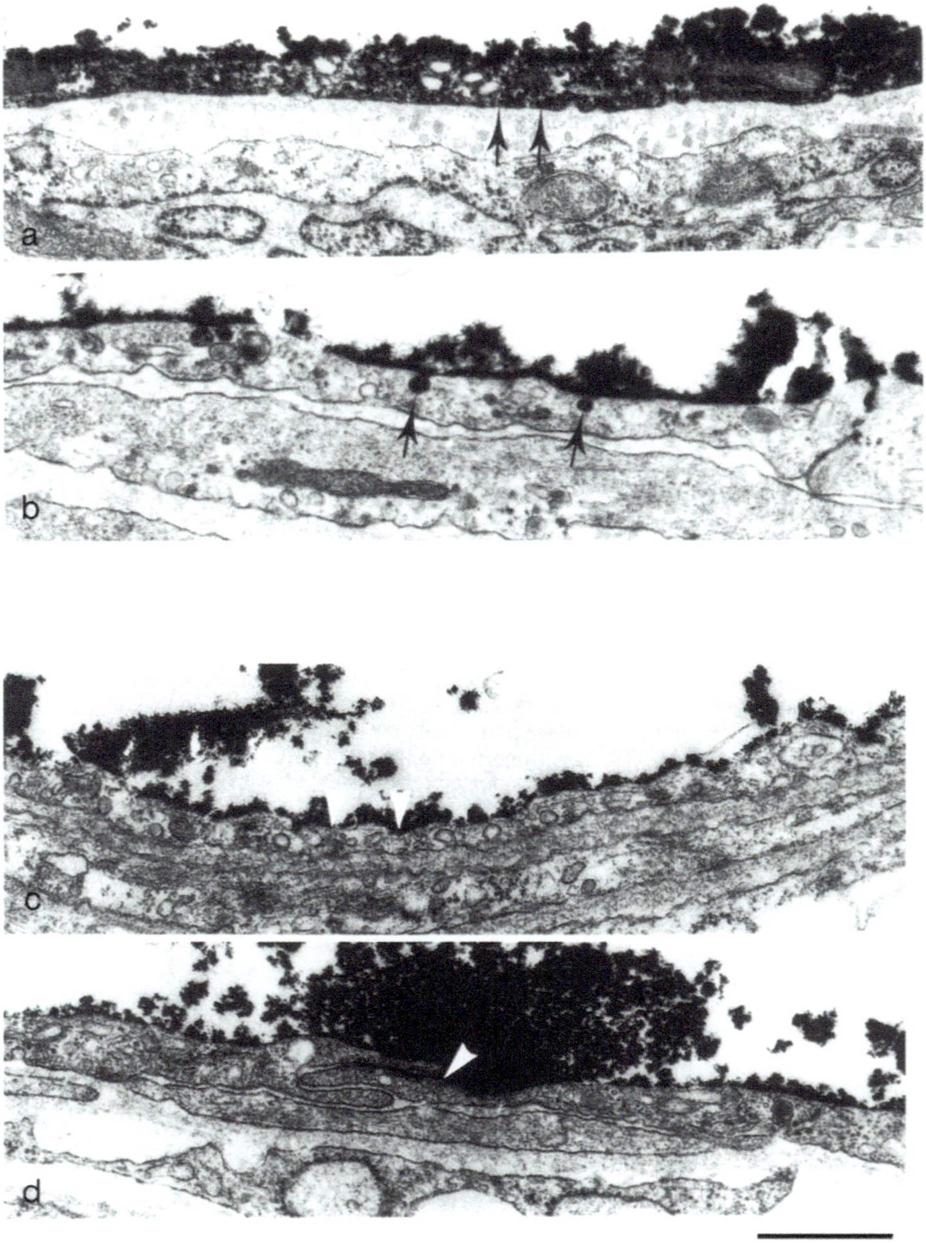

Fig. 25. High power electron micrograph showing extravasation of lanthanum (dark black particles) across the endothelial cells in the left ventral horn of the T9 segment in one 5 h traumatised rat (**a**,**b**) and its modification with p-CPA pretreatment (**c**,**d**). Infiltration of lanthanum within the endothelial cell (**a**) and in vesicular profiles (**b**) are clearly distinct (arrows) in untreated traumatised rats. The passage of lanthanum is stopped by tight junction in the traumatised rat (**b**). Pretreatment with p-CPA significantly attenuated the infiltration of lanthanum within the endothelial cell (white arrow heads, **c**). Appearance of lanthanum in vesicular profiles are also absent in drug treated rat. The passage of lanthanum is stopped at tight junctions (white arrow head, **d**) (Data from Olsson *et al.*, 1990, with permission, bar 200 nm)

Pretreatment with drugs which prevented the early SCEP changes significantly reduced the extravasation of lanthanum within the endothelium and in general no exudation of lanthanum can be seen in the basal lamina of these drug treated traumatised rats (Fig. 25). Our results are in line with the hypothesis that breakdown of the microvascular permeability is one of the key component in vasogenic edema formation of the spinal cord (Sharma *et al.*, 1995a,b,c; Sharma *et al.*, 1996a,b; 1997a).

Mechanisms of SCEP alterations in spinal cord injury

The mechanisms of SCEP changes following spinal cord injury are not well understood. Our observations indicate that SCEP can be influenced by neurochemical metabolism within the spinal cord (Sharma *et al.*, 1991; Winkler, 1994). Thus bioelectrical disturbances in the spinal cord closely reflect the underlying neurochemical metabolisms and seems to represent a reliable indicator of the cord dysfunction.

Field mapping studies within the spinal cord show a wide area of negative field potentials located within the dorsal horn up to lamina VII and adjacent dorsal columns (Foreman *et al.*, 1979; Bennett, 1984; Willis Jr, 1984). Thus, a lesion limited to this region will influence the negative component of the SCEP activity (Willis Jr, 1984; Schramm *et al.*, 1984; Sharma *et al.*, 1991). The other possibility of a loss of negative component of SCEP is a loss of synaptic excitation by direct physical injury to the cord. A balance between synaptic excitation and inhibition is responsible for normal spinal cord conductivity and SCEP response which is greatly perturbed by a focal lesion of the cord (Schramm *et al.*, 1984; Dimetrijevic *et al.*, 1992). An increase in MNP in caudal SCEP recording following lesion, represents a loss of synaptic inhibition (see Homma *et al.*, 1984; Tamaki *et al.*, 1984). Obviously, inhibitory events are often reflected as an increase in amplitude of one or several components of SCEP response (Willis Jr, 1984).

We observed disappearance of all negative peaks in rostral recordings after trauma and appearance of a positive broad peak at similar latency (Sharma *et al.*, 1991). The appearance of a broad positive peak appears to be similar to that of injury potentials or "killed end potentials" seen by various other workers in experimental spinal cord injuries (see Goodman *et al.*, 1985; Schramm and Jones, 1985; Ducker and Brown, 1989; Fehlings *et al.*, 1989). According to classical volume conduction theory the positive potentials represent a large positive 'killed end' signal which is measured by an electrode placed distal to it (Ducker and Brown, 1989). The local ionic changes within the extra- or intracellular spaces at the site of the lesion may also contribute to this potential (Goodman *et al.*, 1985). This injury potential suggests a major cord injury (Malmstadt *et al.*, 1981; Schramm and Jones, 1985; Fehlings *et al.*, 1989; Ducker and Brown, 1989; Sharma *et al.*, 1991).

An increase in latency of MNP and MPP following injury is in agreement with many other studies (see Homma and Tamaki, 1984; Schramm and Jones, 1985; Salzman, 1990; Sharma *et al.*, 1991; Winkler *et al.*, 1993; Winkler, 1994). A delay in SCEP latency is one of the most sensitive indicators of cord injury indicating slowing of axonal conduction in motor tracts across the lesion site (Bennett, 1984; Levy *et al.*, 1986). The cause of latency increase could be demyelination of axons which is in agreement with our findings showing a loss of myelin basic protein immunostaining (Tator, 1991; Fehlings *et al.*, 1988; Sharma *et al.*, 1993a). Amplitude and latency changes represent a combination of axonal injury and demyelination respectively and constitute important parameters in the assessment of severity of the spinal lesion (Sharma *et al.*, 1991; Winkler, 1994; Li and Tator, in this volume).

Our pharmacological studies added a new dimension in the understanding of SCEP response following trauma. Our observations demonstrate that neurochemical mediators markedly influence the SCEP activity after spinal injury. It seems likely that drugs influencing neurochemistry of the spinal cord influence the trauma induced stress response. Obviously a decrease in traumatic stress response will lead to less intensive changes in the spinal cord fluid microenvironment after injury. This hypothesis is in line with the fact that pretreatment with drugs which reduced the SCEP disturbances after trauma has also reduced the HSP response and *c*-fos immunoreactivity in the cord (Sharma *et al.*, 1995c; 1996b).

Trauma to the spinal cord can inhibit neuronal activation or synchronisation by mechanisms of "neuronal shock" which represents traumatic stress phenomena. Traumatic stress can temporarily arrest the conduction properties in many nerve cells for sometime. An immediate loss of MNP amplitude following trauma represents a good example of neuronal shock (Sharma *et al.*, 1991). A slow but significant recovery of MNP amplitude following 60 min and onwards are in line with this hypothesis. Pretreatment with drugs which can attenuate this neuronal stress response following trauma can also attenuate the neuronal shock and the spinal cord maintains its normal conduction properties. Drugs which fail to block the stress response are unable to attenuate neuronal shock induced response. A preservation of MNP amplitude following trauma by pretreatment with several drugs support the above hypothesis.

Drugs which prevented the synthesis or reduced the action of certain neurochemicals such as serotonin, prostaglandins and opioids are able to attenuate neuronal shock thereby improving the spinal cord conduction following trauma. These neurochemicals are well known stress hormones (Chrousos *et al.*, 1995). These observations are further supported by the fact that anxiolytic drug diazepam also attenuates the SCEP changes Winkler *et al.*, 1997). Obviously a decrease in traumatic stress will set a stage which will reduce the secondary mechanisms of cell injury leading to a significant reduction in the edema formation and spinal cord tissue damage. Since increased calcium is the common denominator of the action of many neurochemicals, blockade of calcium channel by nimodipine may be neuroprotective in spinal trauma.

A failure of dexamethasone in attenuating SCEP changes and cord pathology following trauma may be dose related. A failure of histamine (Winkler *et al.*, 1995a) and catecholamine receptor blockade (Winkler *et al.*, 1992) suggest that these neurochemicals may not be primarily responsible for traumatic stress response (see Chrousos *et al.*, 1995). Alternatively, blockade of α- and β-catecholaminergic receptors or histaminergic-2 receptors may not have a beneficial effects on stress reaction (Sharma, 1982). It may be that inhibition of synthesis or blockade of other catecholaminergic or histaminergic receptor subtypes are necessary to influence neuronal shock after trauma.

SCEP changes are not directly related with alterations in mean arterial blood pressure (MABP), arterial pH and/or blood gases measured as physiological variables in these group of animals. Although, the MABP was reduced significantly following 5 h spinal cord injury, the changes in blood gases were not significantly different from normal control group (Sharma and Olsson, 1990; Sharma *et al.*, 1990a–c; 1991; 1992a–c; 1993a–d; 1995a–c). Pretreatment with drugs irrespective of their effects on SCEP changes were unable to influence the physiological variables significantly (Winkler, 1994).

Conclusion

Our studies on early spinal cord bioelectrical activity and pathology of the spinal cord following trauma using a pharmacological approach strongly suggest that the SCEP changes are one of the earliest indicators of spinal cord dysfunction. A close correlation between early decrease in mean negative SCEP amplitude and increase in the water content of the cord is quite apparent (Figs. 14 and 15). Thus, the negative amplitude of SCEP recordings cranial to the lesion site is important in predicting cord pathology. These electrical events and tissue damage are modified by pretreatment with drugs known to influence either the synthesis or receptors of various endogenous neurochemicals. This indicates that neurochemistry of the cord is primarily responsible for the altered bioelectrical activity and pathological disturbances of the cord following trauma.

These observations open a new vista of research concerning modification of early bioelectrical activity of the cord by pharmacological agents. It remains to be seen whether modification of one or several neurochemicals by drugs, antisera, neurotrophins or gene therapy alone or in combination after trauma may have some beneficial effects on spinal cord bioelectrical activity or edema formation. This aspect should be examined in great details in order to find a suitable therapy for clinical purposes.

Future perspectives

Further studies are definitely needed to understand the relationship between early bioelectrical disturbances and later pathological outcome of the cord. Currently, we are using this approach to study the relationship between early SCEP changes and pathology of the cord by increasing the survival period of animals from 5 h to 12 h. In addition several drugs, hormones, antisera and neurotrophic factors are applied at different post-injury intervals to find out a therapeutic window where we can still get the beneficial effects on SCEP activity in relation with tissue damage. This approach will certainly provide the basis of evaluation of new drugs for therapy and form a sound platform for our understanding of the basic structure and function of the normal and traumatised spinal cord.

Acknowledgements

Due to space restrictions, only key references are given below. The original references can be found in the listed reviews for further study. This investigation is supported by Grants from Swedish Medical Research Council Project nr 135 (ES, TW), 2710 (JW, HSS); Göran Gustafsson Foundation (HSS), Stockholm; Astra Hässle (HSS), Mölndal, Pharmacia & Upjohn, Sweden; Merck Sharp and Dohme, Sweden; Alexander von Humboldt Foundation (HSS), Bonn, Bayer AB, Germany; University Grants Commission (HSS), New Delhi, Indian Council of Medical Research (HSS), New Delhi, India. Authors (HSS) are grateful to Agneta Lukinius for expert help and assistance during initial part of Electron Microscopy. The skilful technical assistance of Ingmarie Olsson, Kärstin Flink; Kerstin Rystedt, Madeleine Jarild, Gunilla Tibling (Sweden); Katja Deparade; Franziska Drum (Germany); Secretarial assistance of Mrs Angela Ludwig, Aruna Sharma, GunBritt Lind; photographic assistance of Frank Bittkowski; and computer assistance of William Schannong are highly appreciated.

References

Allen AR (1911) Surgery of experimental lesion of spinal cord equivalent to crush injury of fracture dislocation of spinal column. Preliminary report. JAMA 57: 878–880

Allen AR (1914) Remarks on the histopathological changes in the spinal cord due to impact. An experimental study. J Nerv Ment Dis 41: 141–147

Anderson DK, Prockop LD, Means ED, Hartley LE (1976) Cerebrospinal fluid lactate and electrolyte levels following experimental spinal cord injury. J Neurosurg 44: 715–722

Araki T, Eccles JC, Ito M (1960) Correlation of the inhibitory postsynaptic potential of motoneurones with the latency and time course of inhibition of monosynaptic reflexes. J Physiol 154: 354–377

Arenander AT, Vellis JD (1984) Frontiers of glial physiology. In: Rosenberg RN, Willis Jr WD (eds) The Clinical Neurosciences, Vol V. Neurobiology. Raven Press, New York, pp 53–91

Ashford T, Palmerio C, Fine J (1966) Structural analogue in vascular muscle to the functional disorder in refractory traumatic shock and reversal by corticosteroid: Electron microscopic evaluation. Ann Surg 164, p 575

Austin GM, McCouch GP (1955) Presynaptic component of intermediary cord potential. J Neurophysiol 18: 441–451

Balentine JD (1978) Pathology of experimental spinal cord trauma II. Ultrastructure of axons and myelin. Lab Invest 39: 254–266

Balentine JD (1988) Spinal cord trauma: in search of the meaning of granular axoplasm and vesicular myelin. J Neuropath Exp Neurol 47: 77–92

Balentine JD, Spector M (1977) Calcification of axons in experimental spinal cord trauma. Ann Neurol 2: 520–523
Barbeau H, Chau C, Rossignol S (1993) Noradrenergic agonists and locomotor training recovery after cord transection in adult cats. Brain Res Bull 30: 387–393
Barron DH, Matthews BHC (1938) The interpretation of potential changes in the spinal cord. J Physiol 92: 276–321
Bennet HM (1984) Correlations of ascending and descending conduction block following ischemic and compressive insult to the spinal cord of cat. In: Homma S, Tamaki T (eds) Fundamentals and clinical application of spinal cord monitoring. Saikon, Tokyo, pp 125–134
Bernhard CG (1953) The spinal cord potentials in leads from the cord dorsum in relation to peripheral source of afferent stimulation. Acta Physiol Scand 106 [Suppl] 29: 1–29
Bignami AD, Dahl D, Rueger DC (1980) Glial fibrillary acidic protein (GFAP) in normal neural cells and in pathological conditions. Adv Cell Neurobiol 1: 285–319
Bracken MB, Holford TR (1993) Effects of timing of methylprednisolone or naloxone administration on recovery of segmental and long-tract neurological function in NASCIS 2. J Neurosurg 79: 500–507
Braughler JM, Hall ED (1982) Correlation of methylprednisolone levels in cat with its effects on (Na^+ K^+)-ATPase, lipid peroxidation, and alpha motor neuron function. J Neurosurg 56: 838–844
Brooks C McC, Fuortes MGF (1952) The relation of dorsal and ventral root potentials to reflex activity in mammals. J Physiol 116: 380–394
Burke RE, Rudomin P (1977) Spinal neurons and synapses. In: Kandel E (ed) Handbook of Physiology: section I, The Nervous System. American Physiology Society, Bethesda, pp 877–944
Campbell B (1945) The distribution of potential fields within the spinal cord. Anat Rec 91: 77–88
Cervós-Navarro J, Ferszt R (1980) Brain Edema: Pathology, Diagnosis and Therapy. Adv Neurol 20: 1–450
Cervós-Navarro J, Sharma HS, Westman J, Bongcum-Rudloff E (1998) Glial cell reactions in the central nervous system following heat stress. In: Sharma HS, Westman J (eds) Brain Function in Hot Environment, Progress in Brain Research. Elsevier, Amsterdam 115: 241–274
Chaouloff F (1993) Physiopharmacological interactions between stress hormones and central serotonergic systems. Brain Res Rev 18: 1–32
Chieuh CC, Gilbert DL, Colton CA (1994) The neurobiology of NO˙ and OH˙. Ann NY Acad Sci 738: 1–471
Choi DW (1992) Excitotoxic cell death. J Neurobiol 23: 1261–1276
Choi DW, Barde Y-A, Kosic KS, Muller HW, Schwartz R, Schwartz M, Shooter EM, Siesjö BK, Unsicker K (1991) Group report: neuronal death and survival. In: Price D L, Thoenen H, Aguayo AJ (eds) Neurodegenerative Disorders: Mechanisms and Prospects for Therapy. Wiley, West Sussex, pp 233–248
Chrousos GP, McCarty R, Pacak KG, Sternberg E, Gold PW, Kvetnansky R (1995) Stress: Mechanisms and clinical implications. Ann NY Acad Sci 771: 1–471
Clasen RA, Cooke PM, Pandolfi S, Carnecki G (1963) The effects of focal freezing on the central nervous system. Presbyterian St Luke's Med Bull 2: 36–46
Clendendon NR, Allen N, Ito T, Gordon WA, Yashon D (1978) Response of lysosomal hydrolases of dog spinal cord and cerebrospinal fluid to experimental trauma. Neurology 28: 78–84
Constantini S, Young W (1994) The effects of methylprednisolone and the ganglioside GM1 on acut e spinal cord injury in rats. J Neurosurg 80: 97–111
Coombs JS, Curtis DR, Landgren S (1956) Spinal cord potentials generated by impulses in muscle and cutaneous afferent fibers. J Neurophysiol 19: 452–467
Dawson GD (1947) Cerebral responses to electrical stimulation of peripheral nerve in man. J Neurol Neurosurg Psychiatry 10: 137–139

Dawson VL, Dawson TM (1996) Nitric oxide neurotoxicity. J Chem Neuroanat 10: 179–190
D'Anglo CM, Gilder JCV, Taub AT (1973) Evoked cortical potentials in experimental spinal cord trauma. J Neurosurg 38: 322–336
de la Torre JC (1981) Spinal cord injury. Review of basic and applied research. Spine 6: 315–335
Demediuk P, Faden AI (1988) Traumatic spinal cord injury in rats causes increases in tissue thromboxane but not peptidoleukotrienes. J Neurosci Res 20: 115–121
Dey PK, Sharma HS (1983) Ambient temperature and development of traumatic brain edema in anaesthetized animals. Indian J Med Res 77: 554–563
Dey PK, Sharma HS (1984) Influence of ambient temperature and drug treatments on brain edema induced by impact injury on skull in rat. Indian J Physiol Pharmacol 28: 177–186
Dimitrijevic MR, Dimitrijevic MM, Faganel J, Sherwood AM (1984) Suprasegmentally induced motor unit activity in paralyzed muscles of patients with established spinal cord injury. Ann Neurol 16: 216–221
Dimitrijevic MR, Hsu CY, McKay WB (1992) Neurophysiological assessment of spinal cord and head injury. J Neurotrauma 9 [Suppl] 1: 293–300
Dohrmann GJ, Jr FCW, Bucy PC (1971) The microvasculature in transitory traumatic paraplegia. An electron microscopic study in the monkey. J Neurosurg 35: 263–271
Ducker TB (1976) Experimental injury of the spinal cord. In: Vinken PJ, Bruyn GW (eds) Handbook of Clinical Neurology, Vol. 9. Elsevier, New York, pp 26–68
Ducker TB, Brown RH (1989) Neurophysiology and Standards of Spinal Cord Monitoring. Elsevier, Amsterdam
Ducker TB, Kindt GW, Kempe LG (1971) Pathological findings in acute experimental spinal cord trauma. J Neurosurg 35: 700–708
Duffy PE (1984) Astrocytes: normal, reactive and neoplastic. Raven Press, New York
Eccles JC (1957) The physiology of nerve cells. Johns Hopkins, Baltimore
Eccles JC (1964) The physiology of synapses. Springer, Wien New York
Eccles JC, Fatt P, Landgren S, Winsbury GJ (1954) Spinal cord potentials generated by volly in the large muscle afferents. J Physiol 125: 590–606
Eccles JC, Kostyuk PG, Schmidt RF (1962) Central pathways responsible for depolarization of primary afferent fibres. J Physiol 161: 237–257
Eidelberg E, Sullivan J, Brigham A (1975) Immediate consequences of spinal cord injury: Possible role of potassium in axonal conduction block. Surg Neurol 3: 317–324
Faden AI (1987) Pharmacotherapy in spinal cord injury: a critical review of recent developments. Clin Neuropharmacol 10: 193–204
Faden AI (1990) Opioid and non opioid mechanisms may contribute to dynorphin's pathophysiological actions in spinal cord injury. Ann Neurol 27: 67–74
Faden AI (1993a) Experimental neurobiology of central nervous system trauma. Crit Rev Neurobiol 7: 175–186
Faden AI (1993b) Role of endogenous opioids and opioid receptors in central nervous system. Handbook Exp Pharmacol 104 (I): 325–341
Faden AI, Simon RP (1988) A potential role for excitotoxins in the pathophysiology of spinal cord injury. Ann Neurol 23: 623–626
Faden AI, Vink R, McIntosh TK (1989) Thyrotropin releasing hormone and central nervous system trauma. In: Metcalf G, Jackson IMD, (eds) Thyrotropin Releasing Hormone. Biomedical Significance. Ann NY Acad Sci, pp 380–384
Faden AI, Halt P (1992) Platelet-activating factor reduces spinal cord blood flow and causes behavioral deficits after intrathecal administration in rats through a specific receptor mechanism. J Pharmacol Exp Ther 261: 1064–1070
Faden AI, Salzman S (1992) Pharmacological strategies in CNS trauma. TiPS 13: 29–35
Farber JL (1982) Biology of disease: membrane injury and calcium homeostasis in the pathogenesis of coagulative necrosis. Lab Invest 47: 114–123
Fedoroff S, Vernadakis A (1986) Astrocytes, Vols 1–3. Academic Press, Orlando

Fehlings MG, Tator CH (1988) A review of experimental models of acute spinal cord injury. In: Illis L (ed) Spinal Cord Dysfunction: Assessment. Oxford University Press, Oxford, pp 3–33
Fehlings MG, Tator CH, Linden RD (1989) The relationships among the severity of spinal cord injury, motor and somatosensory evoked potentials and spinal cord blood flow. Electroenceph Clin Neurophysiol 74: 241–259
Foreman RD, Kenshalo DR Jr, Schmidt RF, Willis WD (1979) Field potentials and excitation of primate spinothalamic neurones in response to volleys in muscle afferents. J Physiol (Lond) 286: 197–213
Frerichs KU, Feuerstein GZ (1990) Platelet-activating factor; key mediator in neuroinjury? Cerebrovasc Brain Metab Rev 2: 148–160
Fukuda K, Panter SS, Sharp FR, Noble LJ (1996a) Induction of heme oxygenase-1 (HO-1) after traumatic brain injury in the rat. Neurosci Lett 199: 127–130
Fukuda K, Richmon JD, Sato M, Sharp FR, Panter SS, Noble LJ (1996b) Induction of heme oxygenase-1 (HO-1) in glia after traumatic brain injury. Brain Res 736: 68–95
Gasser HS, Graham HT (1933) Potentials produced in the spinal cord by stimulation of the dorsal roots. Am J Physiol 103: 303–320
Geisler FH, Dorsey FC, Coleman WP (1991) Recovery of motor function after spinal-cord injury – a randomized, placebo-controlled trial with GM-1 ganglioside. N Engl J Med 324
Goodman RM, Wachs K, Keller S, Black P (1985) Spontaneous spinal cord 'injury potential' in the rat. Neurosurgery 17: 757–759
Gorio A, Ferrari G, Fusco M, Janigro D, Zanoni R, Jonsson G (1984) Gangliosides and their effects on rearranging peripheral and central neural pathways. Cent Nerv Syst Trauma 1: 29–37
Gotch F, Horsley V (1891) On the mammalian nervous system, its functions, and their localization determined by an electrical method. Phil Trans B 182: 267–526
Griffiths IR, Miller R (1974) Vascular permeability to protein and vasogenic oedema in experimental concussive injuries to the canine spinal cord. J Neurol Sci 22: 291–304
Griffiths IR, McCulloch MC (1983) Nerve fibers in spinal cord impact injuries. J Neurol Sci 58: 335–349
Grimby G, Hedberg M, Henning G-B (1994) Changes in muscle morphology, strength and enzymes in a four-five year follow-up of post-polio subjects. Scand J Rehab Med 27: 121–130
Happel RD, Smith KP, Banik NL, Powers JM, Hogan EL, Balentine JD (1981) Ca^{2+}-accumulation in experimental spinal cord trauma. Brain Res 211: 476–479
Hayashi N, Torre de la JC, Green BA (1980) Regional spinal cord blood flow and tissue oxygen content after spinal cord trauma. Surg Forum 31: 461–467
Hayes KC, Potter PJ, Wolfe DL, Hsieh JTC, Delaney GA, Blight AR (1994) 4-aminopyridine sensitive neurological deficits in patients with spinal cord injury. J Neurotrauma 11: 433–436
Hedeman LS, Sil R (1974) Studies in experimental spinal cord trauma: comparison of treatment with steroids, low molecular weight dextran, and catecholamine blockade. J Neurosurg 40: 44–51
Hogan EL, Banik NL (1985) Biochemistry of the spinal cord. Handbook Neurochem 10: 285–337
Homma S, Hayashi K, Nakajima Y (1984) Unidimensional latency-topography of the evoked potentials in the cat. In: Homma S, Tamaki T (eds) Fundamentals and Clinical Applications of Spinal Cord Monitoring. Saikon, Tokyo, pp 23–32
Homma S, Tamaki T (1984) Fundamentals and Clinical Applications of Spinal Cord Monitoring. Saikon, Tokyo, pp 1–450
Hökfelt T, Elde R, Johansson O, Ljungdahl Å, Schultzberg M, Fuxe K, Goldstein M, Nilsson G, Pernow B, Terenius L, Ganten D, Jeffocote SL, Rehfeld J, Said S (1978) Distribution of peptide containing neurons. In: Lipton MA, DiMascio A, Killam KF (eds) Psychopharmacology: A Generation of Progress. Raven Press, New York, pp 39–66

Hökfelt T, Johansson O, Holets V, Meister B, Melander T (1987) Distribution of neuropeptides with special reference to their coexistence with classical neurotransmitters. In: Melter HY (ed) Psychopharmacology: The third Generation of Progress. Raven Press, New York, pp 401–416

Hsu CY, Hogan EL, Gadsen RH, Spicer KM, Shi MP, Cox RD (1985) Vascular permeability in experimental spinal cord injury. J Neurol Sci 70: 275–282

Hughes P, Dragunow M (1995) Induction of immediate-early genes and the control of neurotransmitter-regulated gene expression within the nervous system. Pharmacol Rev 47: 133–178

Hughes J, Gasser HS (1934) Some properties of the cord potentials evoked by a single afferent volley. Am J Physiol 108: 295–306

Hurlbert RJ, Koyanagi I, Tator CH (1993) Sensory evoked potentials for selective monitoring of the rat spinal cord: a cerebellar evoked potential to assess ventral cord integrity. J Nerotrauma 10: 181–194

Ikata T, Iwasa K, Morimoto K, Tonai T, Taoka Y ((1989) Clinical considerations and biochemical basis of prognosis of cervical spinal cord injury. Spine 14: 1096–1101

James HE, Marshall RF, Reulen H-J, Baethmann A, Marmarou A, Ito U, Hoff JT, Kuroiwa T, Czernicki Z (1997) Brain Edema X. Acta Neurochir [Suppl] (Wien) 70: 1–298

Kakulas BA (1984) Pathology of spinal injuries. Cent Nerv Syst Trauma 1: 117–129

Kakulas BA, Taylor JR (1991) Pathology of injuries to the vertebral column and the spinal cord. In: Vinken PJ *et al.* (eds) Handbook of Clinical Neurology no 61 (revised series 17) Spinal Cord Trauma (Frankel HL), pp 21–51

Kaufmann SHE (1992) Heat shock proteins in health and disease. Int J Clin Lab Res 21: 221–226

Kiewerski J (1991) The influence of mechanism of cervical spine injury on the degree of the spinal cord lesion. Paraplegia 29: 531–536

Kimura H, Steinbusch HWM (1996) The role of nitric oxide in the central nervous system. From basic to therapeutic aspect. J Chem Neuroanat 10: 179–322

Klee M, Rall W (1977) Computed potentials of cortically arranged populations of neurons. J Neurophysiol 40: 647–666

Kobrine AI, Doyle TF, Martins AN (1975) Local spinal cord blood flow in experimental traumatic myelopathy. J Neurosurg 42: 144–149

Kochanek P, Schoettle R, Uhl M, Magargee MJ, Nemoto E (1991) Platelet-activating factor antagonists do not attenuate delayed posttraumatic cerebral edema in rats. J Neurotrauma 8: 19–25

Kríz N, Syková E, Ujec E, Vyklicky (1974) Changes of extracellular porassium concentration induced by neuronal activity in the spinal cord of the cat. J Physiol 238: 1–15

Landau WM, Freygang WH, Rowland LP, Sokoloff L, Kety SS (1955) The local circulation of the living brain: Values in unanesthetized and anesthetized cat. Trans Am Neurol Ass 80, p 125

Levy W, McCaffrey M, York D (1986) Motor evoked potential in cats with spinal cord injury. Neurosurgery 19: 9–19

Lewin MG, Hansebout RR, Pappius MM (1974) Chemical characteristics of spinal cord edema in cats: Effects of steroids on potassium depletion. J Neurosurg 40: 65–75

Lindqvist S (1986) The heat shock response. Ann Rev Biochem 55: 1151–1191

Lindsberg PJ, Hallenbeck JM, Feuerstein G (1991) Platelet-activating factor in stroke and brain injury. Ann Neurol 30: 117–129

Lipton SA (1993) Molecular mechanisms of trauma-induced neuronal degeneration. Curr Opin Neurol Neurosurg 6: 588–596

Liu D (1994) An experimental model combining microdialysis with electrophysiology, histology and neurochemistry for studying excitotoxicity in spinal cord injury. Effect of NMDA and kainate. Mol Chem Neuropathol 23: 77–92

Liu D, Thangnipon W, McAdoo, DJ (1991) Excitatory amino acids rise to toxic levels upon impact injury to the rat spinal cord. Brain Res 547: 344–348

Lorente de Nó R (1947) A study of nerve physiology. In: Studies from the Rockefeller Institute for Medical Research. Rockefeller Institute for Medical Research, New York, pp 384–477

Lundh H, Thesleff S (1977) The mode of action of 4-aminopyridine and guanidine on transmitter release from motor nerve terminals. Eur J Pharmacol 42: 411–412

Malmstadt HV, Enke CG, Crouch SR (1981) Electronics and instrumentation for scientists. Benjamin/Cummings, Reading, pp 105–128

McCord JM (1987) Oxygen-derivated radicals: a link between reperfusion injury and inflammation. Federation Proc 46: 2402–2406

Misra HP, Fridovich I (1972) The role of superoxide anion in the auto-oxidation of epinephrine and a simple assay for superoxide dismutase. J Biol Chem 247: 3170–3175

Mohanty S, Dey PK, Sharma HS, Ray AK (1985) Role of serotonin in brain edema. In: Mohanty S, Dey PK (eds) Bhargava Bhushan Press. Banarus Hindu University, Varanasi, pp 25–36

Mohanty S, Dey PK, Sharma HS, Singh S, Chansouria JPN, Olsson Y (1989) Role of histamine in traumatic brain edema. An experimental study in the rat. J Neurol Sci 90: 87–97

Molgo J, Lundh H, Thesleff S (1980) Potency of 3,4-diaminopyridine and 4-aminopyridine on mammalian neuromuscular transmissions and the effect of pH changes. Eur J Pharmacol 61: 25–34

Mustafa A, Sharma HS, Olsson Y, Gordh T, Thorén P, Sjöquist P-O, Roos P, Adem A, Nyberg F (1995) Vascular permeability to growth hormone in the rat central nervous system after focal spinal cord injury. Influence of a new antioxidant compound H 290/51 and age. Neurosci Res 23: 185–194

Nemecek S (1978) Morphological evidence of microcirculatory disturbances in experimental spinal cord trauma. Adv Neurol 20: 395–405

Nemeth W, Reger SI, Henry DT, Wang GJ, Schnell MD, Stamp WG (1978) Spinal cord evoked potentials recorded from vertebral elements. Proc 24th Annual Orthop Res Soc 3, p 26

Nicholson C (1973) Theoretical analysis of field potentials in anisotropic ensembles of neuronal elements. IEEE Trans Biomed Eng 20: 278–288

Nicoletti F, Cavallaro S, Bruno V, Virgili M, Catania MV, Contestabile A, Canonico PL (1989) Gangliosides attenuate NMDA receptor-mediated excitatory amino acid release in cultured cerebellar neurons. Neuropharmacology 28: 1283–1286

Nyberg F, Sharma HS, Wisenfeld-Hallin Z (1995) Neuropeptides in the Spinal Cord. Prog Brain Res. Elsevier, Amsterdam 104: 1–450

Okumura R, Asato R, Shimada T, Kusaka H, Mizutani T, Miki Y, Konishi J (1992) Degeneration of the posterior columns of the spinal cord. Postmortem MRI and histopathology. J Comput Assist Tomogr 16: 865–867

Olsson Y, Sharma HS, Pettersson CÅV (1990) Effects of p-chlorophenylalanine on microvascular permeability changes in spinal cord trauma. An experimental study in the rat using ^{131}I-sodium and lanthanum tracers. Acta Neuropathol (Berl) 79: 595–603

Olsson Y, Sharma HS, Pettersson Å, Cervós-Navarro J (1992) Endogenous release of neurochemicals may increase vascular permeability, induce edema and influence on cell changes in trauma to the spinal cord. Prog Brain Res 91: 197–203

Olsson Y, Sharma HS, Nyberg F, Westman J (1995) The opioid receptor antagonist naloxone influences the pathophysiology of spinal cord injury. Prog Brain Res 104: 381–399

Osterholm JL, Mathews GJ (1972) Altered norepinephrine metabolism following experimental spinal cord injury. Part 2: protection against traumatic spinal cord hemorrhagic necrosis by norepinephrine synthesis blockade with alpha methyl tyrosine. J Neurosurg 36: 395–401

Osterholm JL, Alderman JL, Northrup BE (1987) Acute experimental spinal cord injury. In: Ghista DN, Frankel HL (eds) Spinal Cord Injury Medical Engineering. Thomas, Springfield, pp 5–46i

Perot PL (1973) The clinical use of somatosensory evoked potentials in spinal cord injury. Clin Neurosurg 20: 367–369
Panter SS, Yum SW, Faden AI (1990) Alteration in extracellular amino acids after traumatic spinal cord injury. Ann Neurol 27: 96–99
Purpura DP (1959) Nature of electrocortical potentials and synaptic organization in cerebral and cerebellar cortex. Int Rev Neurobiol 1: 47–163
Rall W (1977) Core conductor theory and cable properties of neurons. In Handbook of Physiology: Section 1. The Nervous System, Vol I, Cellular Biology of Neurons. American Physiology Society, Bethesda, pp 39–97
Rasmussen H (1986) The calcium messenger system (first of two parts). N Engl J Med 314: 1094–1101
Reger S, Henry DT (1979) Physiologic aspects of noise in spinal cord monitoring. In: Nash CL (ed) Proc Spinal Cord Monitoring Workshop, Data Acquisition and Analysis. St Louis Monograph 2, pp 189–203
Reger SI, Wang GJ, Anderson CP, Henry DT, Stamp WG (1980) The quantitative correlation of spinal cord damage with spinal cord evoked potentials. Trans 26th Ann Orthop Res Soc 5: 194–197
Reger SI, Andersson CP, Henry DT, Wang CJ, Whitehill R, Stamp WG (1987) Long-term quantitative correlation of spinal cord evoked potentials with cord injury. In: Ghista DN, Frankel HL (eds) Spinal Cord Injury Medical Engineering. Charles C Thomas, Springfield, pp 223–243
Renshaw B (1940) Activity in the simplest spinal reflex pathways. J Neurophysiol 9: 240–263
Salzman SK (1990) Neural Monitoring. The Prevention of Intraoperative Injury. Humana Press, New Jersey
Salzman SK (1991) Comparison of a serotonin antagonist, opioid antagonist, and TRH analog for the acute treatment of experimental spinal trauma. J Neurotrauma 8: 193–203
Salzman SK, Mendez AA, Sabato S, Lee WA, Ingersoll EB, Choi IH, Fonseca AS, Agresta CA, Freeman GM (1990) Anesthesia influences the outcome from experimental spinal cord injury. Brain Res 521: 33–39
Sandler AN, Tator CH (1976a) Effect of acute spinal cord compression injury on regional spinal cord blood flow in primates. J Neurosurg 45: 600–676
Sandler AN, Tator CH (1976b) Review of the effect of spinal cord trauma on the vessel and blood flow in the spinal cord. J Neurosurg 45: 638–646
Schmidt RF (1971) Presynaptic inhibition in the vertebrate nervous system. Ergeb Physiol 63: 20–101
Schramm J, Krause R, Shigeno T, Brock M (1984) Relevance of spinal cord evoked potentials for monitoring. In: Homma S, Tamaki T (eds) Fundamentals and Clinical Applications of Spinal Cord Monitoring. Saikon, Tokyo, pp 113–124
Schramm J, Jones S (1985) Spinal Cord Monitoring. Springer, Berlin Heidelberg, pp 1–350
Schwab ME, Bartholdi D (1996) Degeneration and regeneration of axons in the lesioned spinal cord. Phys Rev 76: 319–370
Sharma HS (1982) Blood-Brain Barrier in Stress. Banaras Hindu University Press, Varanasi, pp 1–85
Sharma HS (1987) Effect of captopril (a converting enzyme inhibitor) on blood-brain barrier permeability and cerebral blood flow in normotensive rats. Neuropharmacology 26: 85–92
Sharma HS (1988) Neurobiology of the nitric oxide in the nervous system. Basic and clinical perspectives (Editorial). Amino Acids 14: 83–85
Sharma HS, Dey PK (1982) Correlation of spinal cord tissue 5-HT with edema development following surgical spinal cord trauma in rats. Indian J Physiol Pharmacol 26 [Suppl] I: 8–9
Sharma HS, Dey PK (1986) Influence of long-term immobilization stress on regional blood-brain barrier permeability, cerebral blood flow and 5-HT level in conscious normotensive young rats. J Neurol Sci 72: 61–76

Sharma HS, Dey PK (1987a) Influence of long-term acute heat exposure on regional blood-brain barrier permeability, cerebral blood flow and 5-HT level in conscious normotensive young rats. Brain Res 424: 153–162
Sharma HS, Dey PK (1987b) Increased blood-brain and blood-csf barrier permeability following long-term immobilization stress in conscious rats. Wissenschaftliche Zeitschrift Karl-Marx Universitat Leipzig (Mathematisch-Naturwissenschaftliche Reihe) 36: 104–106
Sharma HS, Dey PK (1988) EEG changes following increased blood-brain barrier permeability under long-term immobilization stress in young rats. Neurosci Res 5: 224–239
Sharma HS, Dey PK, Olsson Y (1989) Brain edema, blood-brain barrier permeability and cerebral blood flow changes following intracarotid infusion of serotonin: modification with cyproheptadine and indomethacin. In: Krieglstein J (ed) Pharmacology of Cerebral Ischemia 1988. CRC Press. Boca Raton, Florida, pp 317–323
Sharma HS, Cervós-Navarro J (1990) Nimodipine improves cerebral blood flow and reduces brain edema, cellular damage and blood-brain barrier permeability following heat stress in young rats. In: Krieglstein J, Oberpichler H (eds) Pharmacology of Cerebral Ischemia 1990. CRC Press. Boca Raton, Florida, pp 303–310
Sharma HS, Olsson Y (1990) Edema formation and cellular alteration in spinal cord injury in the rat and their modification with p-chlorophenylalanine. Acta Neuropathol (Berl) 79: 604–610
Sharma HS, Olsson Y, Dey PK (1990a) Early accumulation of serotonin in rat spinal cord subjected to traumatic injury. Relation to edema and blood flow changes. Neuroscience 36: 725–730
Sharma HS, Westman J, Olsson Y, Johansson O, Dey PK (1990b) Increased 5-HT immunoreactivity in traumatized spinal cord. An experimental study in the rat. Acta Neuropathol (Berl) 80: 12–17
Sharma HS, Nyberg F, Olsson Y, Dey PK (1990c) Alteration in substance P in brain and spinal cord following spinal cord injury. An experimental study in the rat. Neuroscience 38: 205–212
Sharma HS, Olsson Y, Dey PK (1990d) Blood-brain barrier permeability and cerebral blood flow following elevation of circulating serotonin level in the anaesthetized rats. Brain Res 517: 215–223
Sharma HS, Winkler T, Stålberg E, Olsson Y, Dey PK (1991) Evaluation of traumatic spinal cord edema using evoked potentials recorded from the spinal epidural space. An experimental study in the rat. J Neurol Sci 102: 150–162
Sharma HS, Ceróvs-Navarro J, Gosztonyi G, Dey PK (1992a) Role of serotonin in traumatic brain injury. An experimental study in the rat. In: Globus MY-T and Dietrich WD (eds) The Role of Neurotransmitters in Brain Injury. Plenum Press, New York, pp 147–152
Sharma HS, Nyberg F, Olsson Y (1992b) Dynorphin A content in the rat brain and spinal cord after a localized trauma to the spinal cord and its modification with p-chlorophenylalanine. An experimental study using radioimmunoassay technique. Neurosci Res 14: 195–203
Sharma HS, Nyberg F, Cervós-Navarro J, Dey PK (1992c) Histamine modulates heat stress induced changes in blood-brain barrier permeability, cerebral blood flow, brain oedema and serotonin levels: An experimental study in conscious young rats. Neuroscience 50: 445–454
Sharma HS, Westman J, Olsson Y (1992d) Increased heat shock protein (HSP-70 kD) immunoreactivity following acute spinal cord trauma in the rat. Clin Neuropathol 11 (4), p 174
Sharma HS, Olsson Y, Cervós-Navarro J (1993a) Early perifocal cell changes and edema in traumatic injury of the spinal cord are reduced by indomethacin, an inhibitor of prostaglandin synthesis. Acta Neuropathol (Berl) 85: 145–153
Sharma HS, Olsson Y, Nyberg F, Dey PK (1993b) Prostaglandins modulate alterations of microvascular permeability, blood flow, edema and serotonin levels following spinal cord injury. An experimental study in the rat. Neuroscience 57: 443–449

Sharma HS, Olsson Y, Cervós-Navarro J (1993c) p-Chlorophenylalanine, a serotonin synthesis inhibitor, reduces the response of glial fibrillary acidic protein induced by trauma to the spinal cord. Acta Neuropathol (Berl) 86: 422–427

Sharma HS, Nyberg F, Thörnwall M, Olsson Y (1993d) Met-Enkephalin-Arg6-Phe7 in spinal cord and brain following traumatic injury of the spinal cord: Influence of p-chlorophenylalanine. An experimental study in the rat using radioimmunoassay technique. Neuropharmacology 32: 711–717

Sharma HS, Olsson Y, Nyberg F (1993e) Edema and cell changes in the traumatised spinal cord are reduced by topical application of 5-HT antibodies. An experimental study in the rat. IXth Noordwijkerhout-Camerino Symposium on Trends in Drug Research, p 90

Sharma HS, Olsson Y (1994a) Neurochemical mediators of increased microvascular permeability, vasogenic oedema and cell changes after trauma to the spinal cord. Experimental observations in the rat. J Physiol (Lond) 480, p 13

Sharma HS, Olsson Y (1994b) Nimodipine attenuates brain edema abd cell changes in rats following spinal cord trauma, Brain Pathol 4 [Suppl] 1, p 34

Sharma HS, Nyberg F, Olsson Y (1994c) Topical application of dynorphin-A antibodies reduced edema and cell changes in traumatised rat spinal cord. Reg Pept [Suppl] 1: S91–S92

Sharma HS, Alm P, Gordh T, Nyberg F, Olsson Y, Westman J (1994d) Involvement of nitric oxide in microvascular permeability disturbances and edema formation following injury to the rat spinal cord. Nitric oxide in the central nervous system, Satellite Symposium of IUPHAR, Montreal, Canada (abstract)

Sharma HS, Olsson Y, Nyberg F (1995a) Influence of dynorphin-A antibodies on the formation of edema and cell changes in spinal cord trauma. Prog Brain Res 104: 401–416

Sharma HS, Olsson Y, Pearsson S, Nyberg F (1995b) Trauma induced opening of the blood-spinal cord barrier is reduced by indomethacin, an inhibitor of prostaglandin synthesis. Experimental observations in the rat using ^{131}I-sodium, Evans blue and lanthanum as tracers. Restor Neurol Neurosci 7: 207–215

Sharma HS, Olsson Y, Westman J (1995c) A serotonin synthesis inhibitor, p-chlorophenylalanine reduces the heat shock protein response following trauma to the spinal cord. An immunohistochemical and ultrastructural study in the rat. Neurosci Res 21: 241–249

Sharma HS, Olsson Y, Dey PK (1995d) Serotonin as a mediator of increased microvascular permeability of the brain and spinal cord. Experimental observations in anaesthetised rats and mice. In: Greenwood J, Begley D, Segal M, Lightman S (eds) New Concepts of a Blood-Brain Barrier. Plenum Press, New York, pp 75–80

Sharma HS, Fonnum F, Olsson Y, Gordh T, Westman J (1995e) p-Chlorophenylalanine, a serotonin synthesis inhibitor reduces the alterations of GABA and glutamate levels following a focal spinal cord injury. An experimental study in the rat using immunohistochemical and HPLC technique. Amino Acids 9, p 33

Sharma H S, Westman J, Cervós-Navarro J, Dey PK, Nyberg F (1995f) Alterations of amino acid neurotransmitters following heat stress. Amino Acids 9: 33–34

Sharma HS, Olsson Y, Thóren P, Sjöquist P-O (1995g) H 290/51, a new inhibitor of lipid peroxidation reduces edema and microvascular permeability changes following trauma to the rat spinal cord. J Neurotrauma 12, p 453

Sharma HS, Alm P, Gordh T, Nyberg F, Olsson Y, Westman J (1995h) Involvement of nitric oxide in the pathophysiology of spinal cord injury in the rat. J Neurotrauma 12, p 370

Sharma HS, Westman J, Olsson Y, Alm P (1996a) Involvement of nitric oxide in acute spinal cord injury: an immunohistochemical study using light and electron microscopy in the rat. Neurosci Res 24: 373–384

Sharma HS, Westman J, Olsson Y (1996b) Spinal cord injury induced *c-fos* expression is reduced by p-CPA, a serotonin synthesis inhibitor. Neuropathol Appl Neurobiol 22: 15–16

Sharma HS, Westman J (1997) Prostaglandins modulate constitutive isoform of heat shock protein (72 kD) response following trauma to the rat spinal cord. Acta Neurochir [Suppl] (Wien) 70: 134–137

Sharma HS, Nyberg F, Gordh T, Alm P, Westman J (1997a) Topical application of insulin like growth factor-1 reduces edema and upregulation of neuronal nitric oxide synthase following trauma to the rat spinal cord. Acta Neurochir [Suppl] (Wien) 70: 130–133

Sharma HS, Westman J, Alm P, Sjöquist P-O, Cervós-Navarro J, Nyberg F (1997b) Involvement of nitric oxide in the pathophysiology of acute heat stress in the rat. Ann NY Acad Sci 813: 581–590

Sharma HS, Alm P, Westman J (1997c) Upregulation of hemeoxygenase-II in the rat spinal cord following heat stress. In: Johannsen Nielsen B, Nielsen R (eds) Thermal Physiology, The August Krogh Institute, pp 135–138

Sharma HS, Westman J, Nyberg F (1997d) Topical application of 5-HT antibodies reduces edema and cell changes following trauma to the rat spinal cord. Acta Neurochir [Suppl] (Wien) 70: 155–158

Sharma HS, Westman J, Cervós-Navarro J, Dey PK, Nyberg F (1997e) Opioid receptor antagonists attenuate heat stress-induced reduction in cerebral blood flow, increased blood-brain barrier permeability, vasogenic brain edema and cell changes in the rat. Ann NY Acad Sci 813: 559–571

Sharma HS, Alm P, Westman (1998a) Nitric oxide and carbon monoxide in the pathophysiology of brain functions in heat stress. In: Sharma HS, Westman J (eds) Brain Function in Hot Environment. Prog Brain Res 115: 297–336

Sharma HS, Nyberg F, Westman J, Alm P, Gordh T, Lindholm D (1998b) Brain derived neurotrophic factor and insulin like growth factor-1 attenuate upregulation of nitric oxide synthase and cell injury following trauma to the spinal cord. An immunohistochemical study in the rat. Amino Acids 14: 121–129

Sharma HS, Westman J, Nyberg F (1998c) Pathophysiology of brain edema and cell changes following hyperthermic brain injury. In: Brain Function in Hot Environment. Prog Brain Res 115: 351–412

Sharma HS, Nyberg F, Gordh T, Alm P, Westman J (1998d) Neurotrophic factors attenuate nitric oxide synthase upregulation, microvascular permeability disturbances, edema formation and cell injury in the spinal cord following trauma. An experimental study using topical application of BDNF and IGF-1 in the rat. In: Spinal Cord Monitoring. Basic principles, regeneration and pathophysiological aspects. Springer, Wien New York, pp 181–210

Shimoji K, Kano T, Higashi H, Morioka T, Henshel EO (1972) Evoked spinal electrograms recorded from epidural space in man. J Appl Physiol 33, p 468

Siesjö BK (1988) Historical overview: calcium, ischemia, and death of brain cells. Ann NY Acad Sci 522: 638–661

Singer JM, Russel GV, Coe JE (1970) Changes in evoked potentials after experimental cervical spinal cord injury in monkey. Exp Neurol 29: 449–461

Stokes BT, Fox P, Hollinden G (1985) Extracellular metabolites; their measurement and role in the acute phase of spinal cord injury. In: Dacey RG, Winn HR, Rimmel RW, Jane JA (eds) Trauma of the Central Nervous System. Raven, New York, pp 309–323

Stålberg E (1996) SFEMG, an update. In: Kimura J, Shibasaki H (eds) Recent Advances in Clinical Neurophysiology. Elsevier, Amsterdam, pp 48–60

Stålberg E, Grimby G (1995) Dynamic electromyography and biopsy changes in a 4 year follow up: study of patients with history of polio. Muscle Nerve 18: 699–707

Sypert GW (1990) Stabilization and management of cervical injuries. In: Pitt LH, Wagner FC (eds) Craniospinal Trauma. Thieme, New York, pp 363–370

Tamaki T, Noguchi T, Takano H, Tsuji H, Dincer MD (1984) The effects of hypovolemic hypotension and hypoxia on the jeopardised spinal cord. In: Homma S, Tamaki T (eds) Fundamentals and Clinical Applications of Spinal Cord Monitoring. Saikon Publishing Co, Tokyo, pp 145–154

Tator CH (1991) Review of experimental spinal cord injury with the emphasis on the local and systemic circulatory effects. Neurochirurgie 37: 291–302

Tator CH, Edmonds VE (1979) Acute spinal cord injury: analysis of epidemiological factors. Can J Surg 121: 1453–1464

Tator CH, Fehlings MG (1991) Review of the secondary injury theory of acute spinal cord trauma with emphasis on vascular mechanisms. J Neurosurg 75: 15–26

Thörnwall M, Sharma HS, Olsson Y, Thóren P, Sjöquist P-O, Nyberg F (1995) Dynorphin converting enzyme in spinal cord trauma. Influence of a new anti-oxidant H 290/51. Pharmacol Toxicol 76 [Suppl] IV, p 27

Thörnwall M, Sharma HS, Gordh T, Sjöquist P-O, Nyberg F (1997) Substance P endopeptidase activity in the rat spinal cord following injury. Influence of a new anti-oxidant compound H 290/51. Acta Neurochir [Suppl] (Wien) 70: 212–215

Wagner JFC, Stewart WB (1981) Effect of trauma dose on spinal cord edema. J Neurosurg 54: 802–806

Wall PD (1958) Excitability changes in afferent fibre terminations and their relation to slow potentials. J Physiol 142: 1–21

Walker JG, Yates RR, Yashon D, O'Neill JJ (1977) Effect of experimental trauma on dog spinal cord energy state. J Neurosurg 29, p 929

Westman J, Sharma HS (1998) Heat shock protein response in the CNS following heat stress. In: Brain Function in Hot Environment. Prog Brain Res 115: 207–240

Willis WD Jr (1980) Spinal cord potentials, In: W F Windle (ed) Spinal cord and its reaction to traumatic injury. Marcel Dekker, New York, pp 159–187

Willis WD Jr (1984) Evoked spinal cord potentials in the cat and monkey: use in the analysis of spinal cord function. In: S Homma, T Tamaki (eds) Fundamentals and Clinical Applications of Spinal Cord Monitoring. Saikon, Tokyo, pp 3–20

Willis WD, Weie MA, Skinner ED, Bryan RN (1980) Differential distribution of spinal cord field potentials. Exp Brain Res 17: 169–179

Winkler T (1994) Evaluation of spinal cord injuries using spinal cord evoked potentials. An experimental study in the rat. Acta Univ Ups 467: 1–45

Winkler T, Sharma HS, Stålberg E, Olsson Y (1992) ß-adrenergic receptor antagonist propranolol attenuates alteration in spinal cord evoked potentials and reduces edema formation following spinal cord injury in rats. 7th International Catecholamine Symposium, Amsterdam, abstracts, p 354

Winkler T, Sharma HS, Stålberg E, Olsson Y (1993) Indomethacin, an inhibitor of prostaglandin synthesis attenuates alteration in spinal cord evoked potentials and edema formation after trauma to the spinal cord. An experimental study in the rat. Neuroscience 52: 1057–1067

Winkler T, Sharma HS, Stålberg E, Olsson Y, Nyberg F (1994a) Opioid receptors influence spinal cord electrical activity and edema formation following spinal cord injury. Experimental observations using naloxone in the rat. Neurosci Res 21: 91–101

Winkler T, Sharma HS, Stålberg E, Olsson Y, Nyberg F (1994b) Naloxone reduces alterations in evoked potentials and edema in trauma to the rat spinal cord. Acta Neurochir [Suppl] (Wien) 60: 511–515

Winkler T, Sharma HS, Stålberg E, Olsson Y, Nyberg F (1995a) Role of histamine in spinal cord evoked potentials and edema following spinal cord injury. Experimental observations in the rat. Inflamm Res 44 [Suppl] 1: S44–S45

Winkler T, Sharma HS, Stålberg E, Olsson Y, Dey PK (1995b) Impairment of blood-brain barrier function by serotonin induces desynchronisation of spontaneous cerebral cortical activity. Experimental observations in the anaesthetised rat. Neuroscience 68: 1097–1104

Winkler T, Sharma HS, Stålberg E, Westman E (1997) Benzodiazepine receptors influence spinal cord evoked potentials and edema following trauma to the rat spinal cord. Acta Neurochir [Supp] (Wien) 70: 216–219

Winkler T, Sharma HS, Stålberg E, Badgaiyan RD, Alm P, Westman J (1998) Spinal cord evoked potentials and edema in the pathophysiology of rat spinal cord injury. Involvement of nitric oxide. Amino Acids 14: 131–139

Young W (1985) Blood flow, metabolic and neurophysiological mechanisms in spinal cord injury. In: Becker PB, Povlishock JT (eds) Central Nervous System Trauma Report. National Institute of Health, Bethesda, pp 463–473

Young W, Flamm ES (1982) Effect of high-dose corticosteroid therapy on blood flow, evoked potentials, and extracellular calcium in experimental spinal injury. J Neurosurg 57: 667–673

Correspondence: Dr. H. S. Sharma, Laboratory of Neuroanatomy, Department of Medical Cell Biology, Biomedical Centre, Post Box 571, Uppsala University, S-751 23 Uppsala, Sweden.

Spinal Cord Monitoring
Eds: E. Stålberg, H. S. Sharma, Y. Olsson

Chapter 14

Spinal cord blood flow and evoked potentials as outcome measures for experimental spinal cord injury

S. Li and **C. H. Tator**

Division of Neurosurgery and Canadian Paraplegic Association Spinal Cord Injury Research Laboratory, Toronto Hospital and University of Toronto, Toronto, Canada

There are two mechanisms of damage to the spinal cord after acute spinal cord injury(SCI): the primary mechanical injury and a secondary injury due to one or more additional damaging processes initiated by the primary injury (Collins, 1983; Sandler and Tator, 1976a,b). Many pathophysiological processes are involved in the secondary injury mechanisms including vascular, electrolyte and biochemical changes, edema and loss of energy metabolism (Tator and Fehlings, 1991). Of the several postulated mechanisms, the vascular theory has considerable supporting evidence based on biochemical, pathological, angiographic, blood blow and therapeutic studies. Acute spinal cord trauma causes numerous vascular changes which may be divided into systemic and local effects. Several studies have shown that acute SCI produces immediate damage to the microvasculature of the cord followed by a secondary injury to these vessels, and this combination induces spinal cord ischemia which may be progressive (Tator, 1991 and 1994; Dohrmann *et al.*, 1971; Hall *et al.*, 1989). One of the most direct methods of observing the spinal cord ischemia caused by SCI is to measure spinal cord blood flow (SCBF).

Efforts to find treatment strategies for SCI have necessitated the development of objective assessments of spinal cord function. In human, functional outcome has been assessed mainly by standard necrologic tests based on voluntary motor capability and sensory discrimination. In studies involving experimental SCI, behavioral tests involving general sensory and motor function have been devised including gross observation of general locomotor function, the inclined plane test, grid walking, footfall analysis, responses to pain and pressure, withdrawal reflexes, beam walking, and forelimb/hindlimb grasping (Gruner, 1993). Such tests may be useful indicators of motor function, but there are significant limitations (Goldberger *et al.*, 1990; Kunkel-Bagden *et al.*, 1993). For example, behavioral tests do not provide specific information regarding the physiologic mechanisms of treatment effects, such as changes in segmental excitability, axonal dysfunction, neuronal death, or even muscle atrophy or hypertrophy (Blight and Someya, 1985; Blight and Young, 1989). There is a definite need for objective, quantifiable techniques of assessing the physiological integrity of neurons and axons within the mammalian spinal cord in both clinical and experimental settings (Fehlings and Tator, 1988). This chapter outlines our experience with evoked potentials (EPs) and SCBF as outcome measures in experimental SCI. We have found that EP and SCBF monitoring are objective and quantitative methods for evaluating experimental SCI.

Spinal cord blood flow in trauma to the spinal cord

Techniques of SCBF measurements

The blood supply to the spinal cord has been defined in humans and animals, and the alterations in the vascular supply to the spinal cord after experimental cord injury have been assessed qualitatively and quantitatively. Gray matter blood flow estimated by H_2 clearance typically ranges from 40–100 ml/100/gm/min and white matter flow clusters around 15–20 ml/100 gm/min (Young, 1980). The following is a discussion of blood flow techniques used for the study of the spinal cord (see Chapter 1, in this volume).

Hydrogen clearance method. The utility of hydrogen gas in the measurement of blood flow was first described by Auckland *et al.* (1964) and represents an adaptation of the Fick principle by which blood flow is linearly related to the local hydrogen concentration and obtained by an initial slope technique

$$BF = 0.693/t_{1/2}$$

where BF is blood flow in ml / minute / g and $t_{1/2}$ is the time taken for a 50% reduction in observed current. This technique has become the most commonly used method for measuring SCBF because it possesses distinct advantages over other blood flow monitoring methods. First, H_2 clearance can be determined in any tissue in which a small platinum electrode can be inserted. Second, unlike the microsphere or autoradiographic techniques, it allows repeated measurement of flow. Third, blood flow can be estimated from the clearance rate of H_2 independently of the absolute amplitude of the H_2 signal (Young, 1980). However, there are also technical limitations for this method. For example, in the normal cord, the desaturation curves are bi-exponential. One could assume a bicompartmental model of the cord represented by gray and white matter and obtain separate SCBF values for gray and white matter by calculating the coefficients for the fast and slow components of the desaturation curve. However, after cord injury, the hydrogen clearance method cannot reliably distinguish gray from white matter SCBF in the rat because of disruption of the central gray matter, with loss of the gray/white interface, and hence a bicompartmental model would no longer be valid. Thus, most investigators use the initial slope index method to analyze the polarographic desaturation curves, because the initial slope of the logarithmically transformed hydrogen clearance curve represents a linear combination of flows in all tissue compartments. This method allows direct comparison of preinjury and postinjury SCBF and has been successfully used for cerebral blood flow in many species from primates to rats (Branston *et al.*, 1984; Guha *et al.*, 1987). The volume measured is related to the separation distance of the electrode elements and is generally less than 1 mm^3 (Corbally and Brennan, 1990).

Microsphere technique. The use of microspheres in the determination of SCBF in the rat after clip compression injury has been established in our laboratory (Dolan and Tator, 1982; Wallace and Tator, 1987). Radioactive and nonradioactive microspheres have been used for blood flow measurement (Ross and Tator, 1994). The standard deviations with the microsphere technique are considerable and small changes in regional SCBF could be missed, but the method is sufficiently sensitive to detect a marked reduction in SCBF after severe cord injury. However, in the rat we found that the technique required sampling of cord segments at least 1.5 cm long and allowed only two flow studies per animal. The injection of more microspheres to enable sampling of smaller segments or more than two flow studies per animal caused irreversible hemodynamic changes (Tator and Fehlings, 1991). In addition, the precise regional distribution of the decreased SCBF noted after clip compression injury, such as differences between white and gray matter flow, could not be determined with the microsphere technique (Wallace and Tator, 1986c).

Autoradiography. Basing on the uptake of a tracer, the autoradiographic technique has been used to measure blood flow in experimental studies with tracers such as ^{14}C-antipyrine, ^{14}C-iodoantipyrine, xenon-133(^{133}Xe), ^{14}C-ethanol, ^{3}H-water, tri-fluoro-iodomethane ^{131}I(CF_3^{131}I)(Ross and Tator, 1994; Sandler and Tator, 1976a,b; Dolan *et al.*, 1980; Cawthon *et al.*, 1980). Much work has been done using ^{14}C-antipyrine for measuring SCBF in SCI in rats and other species. This technique is one the best techniques for the spatial resolution of flow. However, it requires sacrifice of the experimental animal for each measurement of flow, and does not allow repeated measurements in the same animal (Ross and Tator, 1994).

Laser Doppler flowmetry. Laser Doppler flowmetry represents considerable technologic progress in the measurement of flow in small regional volumes. It offers the potential to optically measure flow in small regional volumes continuously. It is ideally suited to measure surface flow on skin, mucosa or following surgical exposure. This technique couples the Doppler principle in detecting the frequency shift of laser light imparted by moving red blood cells in the blood stream (Stern, 1977). Incident coherent light is carried to the tissue by fiberoptic cables, where it is scattered by the moving red blood cells. Blood flow is measured because it is linearly related to the mean frequency shift of the scattered light. The minimum tissue volume assessed is approximately 0.8 to 1 mm^3, although it is not known precisely (Rouen, 1976). This method is being applied widely in SCI because of the availability of a small probe, high spatial resolution, rapid response, and lack of tissue contact if required (Kristensen *et al.*, 1994; Yanase *et al.*, 1995). However, it does not provide absolute flow values (Corbally and Brennan, 1990) and measures only the peripheral rim of the spinal cord.

Nuclear magnetic resonance. The application of nuclear magnetic resonance (NMR) to the study of blood flow offers an exciting potential for noninvasive flow measurement. The technique is in the development stage and is not yet widely available. The NMR signal is derived from the nuclei of the tissue sample, no radionuclide tracer is necessary, and the procedure is totally noninvasive. Two different techniques are used to measure venous and arterial flow. Arterial flow is pulsatile and linked to cardiac systole. The signal intensity varies with the amplitude, and the flow rate can be obtained from the variation in amplitude detected by a receiving coil (Roughton, 1982; Mills *et al.*, 1984). Venous flow is measured by temporarily exciting the nuclei at some point in the venous blood and detecting these excited or tagged nuclei at a point downstream. Flow velocity is then measured by the transit time between the two stations (Roughton, 1982; Wedeen *et al.*, 1985). However, when the velocity of flow increases, there is a high signal loss and difficulty in quantifying regional blood flow. We have begun work in our laboratory to measure SCBF in the rat by NMR after a clip compression cord injury.

In addition, ^{133}Xe and stable Xe-enhanced computed tomography and positron emission tomography scanning are attractive techniques for measuring SCBF because they are noninvasive to neural tissue and allow serial measurements in the same animal, but they are not suitable for most experimental studies because of poor resolution. Thus, absolute blood flow in the spinal cord can be measured by a variety of techniques. Selection of the appropriate method for a given experiment is based on the volume of tissue measured, the need for precise region and the specific limitations of each method. The advent of laser Doppler and local hydrogen techniques make serial SCBF measurement comparatively simple and easy.

Several angiographic techniques have been used to investigate the vascular changes in the traumatized spinal cord, including perfusion with colloidal barium suspension (Fairholm and Turnbull, 1971; Dohrmann *et al.*, 1973), colloidal carbon (Wallace *et al.*, 1986), silicone rubber (Koyanagi *et al.*, 1993c) and injection of fluorescent dyes (Dohrmann and Allen, 1975; Dohrmann *et al.*, 1973). Microangiography provides excellent anatomical information about regional vascularity, but has the major shortcoming of permitting only one determination of vascularity per animal (Sandler and Tator, 1976a,b).

SCBF changes after trauma to the spinal cord

SCBF in compressive SCI. It has long been known that marked reductions in SCBF may be associated with traumatic SCI in experimental animals (Sandler and Tator, 1976). A decline in SCBF after SCI has been documented in most SCI models, although hyperaemia has also been found (Kobrine *et al.*, 1976), usually after milder injury. Investigators have used a variety of methods to measure SCBF after acute cord trauma and almost all have found a major reduction of blood flow in the spinal cord following severe injury (Hall, 1988; Hall and Wolf, 1986; Young *et al.*, 1981; Tator, 1991 and 1994). One of the most interesting findings has been the progressive worsening of posttraumatic ischemia during the first few hours after SCI, which suggests that at least a portion of the ischemia might be preventable if treated early (see Chapters 1 and 13, in this volume). Our laboratory demonstrated a severe decrease in posttraumatic SCBF with the C^{14}-antipyrine autoradiographic technique in three different models of spinal trauma: the extradural cuff compression injury in monkeys (Sandler and Tator, 1976a,b), the acute distraction injury in cats (Dolan *et al.*, 1980), and the acute clip compression model in rats (Rivlin and Tator, 1978). With a microcomputer system for on-line determination of flow by the H_2 clearance technique (Piper *et al.*, 1987), we consistently found that severe cord trauma caused profound and persistent ischemia of the cord. After acute clip compression injury in the rat, ischemia began almost immediately, similar to our previous findings in monkeys after acute extradural cuff compression. In the rat, there was a direct linear, dose-response relationship between the severity of injury as represented by clip force and the ischemic response (Fehlings *et al.*, 1989a; Guha and Tator, 1988), and even mild injuries (1.5 g clip compression for 1 min) resulted in a significant reduction in SCBF at the lesion site. In another study, we showed that mild SCI caused a 45% decline in SCBF while severe SCI caused a 70% decrease (Guha *et al.*, 1989). The ischemia in rats worsened during the first 3 hours after cord injury, if left untreated, and persisted for at least 24 hours (Fehlings *et al.*, 1989b; Rivlin and Tator, 1978). Similarly, in monkeys, the ischemia lasted for at least 24 hours after severe injury. Kato *et al.*, (1985) also noted a progressive decline in SCBF with increasing compression due to expanding epidural tumours until a critical level, above which there was an accelerated deterioration in both SCBF and neurologic function.

Several angiographic methods have also been used to study the microvasculature of the cord after experimental cord trauma and all have shown a major reduction in the microcirculation and lack of perfusion (Fairholm and Turnbull, 1971; Fried and Goodkin, 1971). In our laboratory, we used colloidal carbon angiography with a colloidal carbon suspension perfused into the blood stream of rats at 15 minutes, 2 hours, or 24 hours after severe trauma. Serial 200-μm sections of the cord showed major areas with lack of filling of the arterioles, capillaries, and venules both at the injury site and for a considerable distance cephalad and caudad in the cord.

The ischemic zones encompassed a large portion of the gray matter and the surrounding white matter, and were especially severe in white matter areas adjacent to hemorrhages in the gray matter. The anatomical distribution and temporal sequence of the ischemic zones could not be entirely explained on the basis of mechanical damage to vessels caused by the primary injury, and therefore we postulated a secondary injury to the microcirculation such as thrombosis or vasospasm of arterioles traversing the gray matter to supply the white matter (Tator and Fehlings, 1991; Tator, 1991 and 1994). Lack of perfusion of the microcirculation at the injury site has also been demonstrated by microangiographic techniques in other models of cord trauma (Fairholm and Tunbull, 1971; Fried and Goodkin, 1971).

SCBF in contusive SCI. In the past two decades, alterations of SCBF in response to trauma have also been investigated in contusive injury models, such as the weight-drop method. These models cause damage predominantly in the dorsal-medial areas of the spinal cord. Even mild to moderate trauma may cause marked declines in gray matter blood flow. With more severe injuries, there is a progressive and sustained reduction of white matter blood flow that begins centrally and progresses to involve more peripheral white matter (Ducker *et al.*, 1978; Sandler and Tator, 1967). A 300- or 500-g × cm weight-drop SCI model in the canine cord produced 30% and 60% declines in SCBF, respectively (Griffiths *et al.*, 1979). Ducker *et al.* documented no decrease and a 66% decrease in SCBF in paraparetic and paraplegic monkeys, respectively (1978).

Hyperaemia in SCI. There has been the occasional report that spinal cord trauma with the weight-drop model can cause hyperaemia in the white matter (Kobrine *et al.*, 1975). One possible reason is the variability in the severity of injury with this model. We have never observed hyperaemia at the injury site after severe injury in any species, although we found hyperaemia at the injury site in monkeys 24 hours after a mild injury (Sandler and Tator, 1976b), and at sites adjacent to a mild injury in rats in experiments on autoregulation when the mean systemic arterial pressure was elevated to hypertensive levels (Guha *et al.*, 1989). Chehrazi found no reduction in SCBF in cats after cord injury produced by the weight-drop technique (Chehrazi *et al.*, 1989). It appears that some of the controversy regarding posttraumatic ischemia may be due to several factors, including variations in injury severity, sampling site in the cord, and arterial pressure.

The cause of posttraumatic ischemia

Elucidation of the etiology of posttraumatic ischemia would be of major importance for improving the treatment of cord injury. Unfortunately, the exact cause of posttraumatic ischemia remains unknown although numerous theories have been postulated such as vasospasm due to mechanical damage or to the release of a vasoactive amine or other vasoconstrictor (see Chapter 13). In our laboratory, the results of microangiography after acute SCI in rats indicated that the pathogenesis of progressive ischemia may be mainly due to arterial obstruction, especially in the anterior sulcal arteries (Koyanagi *et al.*, 1993a,b). We have also shown that vasospasm occurs in the anterior sulcal arteries and have documented contraction of the smooth muscle cells in the walls of these blood vessels in rats (Anthes *et al.*, 1996). Furthermore, posttraumatic hemorrhage in the spinal cord was also clearly demonstrated by silicone rubber and colloidal carbon angiography (Koyanagi *et al.*, 1993c; Wallace *et al.*, 1986). We found that hemorrhage predominated in the gray matter and in the posterior and lateral white columns at the injury site and extended into the posterior columns far from the injury site. The hemorrhages were associated with zones of ischemia of the white matter in the distribution of the vessels coursing through the haemorrhagic gray matter and, thus hemorrhage may promote ischemia. Direct endothelial cell damage or edema, and extracellular edema causing vascular compression are additional possible mechanisms of ischemia. Other studies have suggested that the ischemia may result from thrombosis or platelet aggregation due to an agent such as thromboxane A_2(De al Torre, 1981). In addition, it has also been postulated that excitotoxic amino acids (mainly glutamate and aspartate) may be involved in the secondary vascular injury (Jørgensen and Diemer, 1982). For example, glutamate receptor activation may play a key role in the development of ischemic damage in cerebral tissue. In ischaemic-hypoxic models of brain damage, the excitotoxic effect of glutamate has been documented extensively both *in vivo* and *in vitro* (Meldrum, 1985; Rothman and Olney, 1986). Ischemia or trauma of the brain and spinal cord causes a massive increase in the extracelluar concentration of glutamate at the injury site (Demediuk *et al.*, 1988 and 1989). The excitotoxic mechanism is discussed below in greater detail.

Strategies to improve SCBF after SCI

To characterize the posttraumatic changes in axonal function in the motor and somatosensory tracts of the cord and to correlate these changes with SCBF, the relationships among the severity of cord injury, motor and somatosensory evoked potentials, and SCBF were examined in our laboratory (Fehlings and Tator, 1989). A linear relationship ($r = -0.89$) was found between the severity of cord injury and the reduction in blood flow at the injury site, and linear discriminant analysis revealed that the motor and somatosensory evoked potentials were significantly related to the severity of cord injury, and the reduction in amplitude of the evoked potentials was significantly correlated with the reduction in posttraumatic blood flow. Multiple regression analysis revealed that both the severity of cord injury and the degree of posttraumatic ischemia were significantly related to posttraumatic axonal dysfunction. Therefore, these studies provide quantitative evidence linking posttraumatic ischemia to axonal dysfunction following acute cord trauma.

Studies of the correlation between white matter ischemia, axonal conduction, and the severity of trauma in the acute phase (Tator and Fehlings, 1991; Sandler and Tator, 1976a; Lohse *et al.*, 1980) suggest that the degree of functional recovery is dependant on the return of blood flow. Indeed, counteracting or preventing the systemic or local vascular effects, and preservation of regional SCBF by various therapeutic modalities has been associated with improved recovery of spinal cord conduction (Guha *et al.*, 1985 and 1987; Fehlings and Tator, 1989b; Wallace and Tator, 1986a,b). Although a definite causal relationship has still not been established between posttraumatic spinal cord ischemia and outcome (Tator, 1991 and 1994), there is increasing evidence (see Chapters 1 and 13).

Calcium channel blockers. Calcium ions play a key role in the pathogenesis of neural injury after trauma or ischemia (Hall and Wolf, 1986; Hogan *et al.*, 1986; Young and Koreh, 1986), and intracellular calcium influx has been termed the final common pathway of toxic cell death in the nervous system (Cheung *et al.*, 1986). For example, disruption of cell membranes due to trauma, depolarization induced by trauma or ischemia, or excitotoxin-mediated activation of calcium channels are associated with an intracellular shift of calcium ions. The latter has been shown to promote contraction of vascular smooth muscle resulting in vasospasm and ischemia. In neurons or axons excess intracellular calcium can impair mitochondrial function, promote axonal degeneration, activate neutral proteases and lipases, and increase the synthesis of toxic eicosanoids. Several studies have examined the changes in calcium ions after acute SCI, and support the role of calcium ions in the pathophysiology of acute SCI (Tator and Fehlings, 1991).

Calcium channel blockers have been studied extensively in the brain where they produce dilation of cerebral vessels and an increase in cerebral blood flow in normal and pathological states including cerebrovascular diseases (Mohamed *et al.*, 1984; Nowicki *et al.*, 1982; Steen *et al.*, 1983; Takakura, 1992; Weinstein, 1995). Since the vascular physiology of the spinal cord and brain are similar with respect to autoregulation, the blood-central nervous system barrier, and CO_2 reactivity, it has been postulated that calcium channel blockers may be helpful in overcoming posttraumatic ischemia of the cord. Our laboratory conducted extensive studies of nimodipine, one of the dihydropyridine group of calcium channel blockers, on SCBF after experimental SCI. In normal rats, we measured SCBF, arterial pressure and cardiac output during intravenous infusion of nimodipine at doses ranging from 0.001 to 1.0 mg/kg. The optimum dose, 0.05 mg/kg, produced a 40% increase in blood flow with only a 25% reduction in mean arterial pressure, but larger doses caused more marked systemic hypotension (Guha *et al.*, 1985 and 1987). This optimum (0.05 mg/kg) dose was then administered to rats with clip compression injury. However, due to the hypotension caused by both cord trauma and nimodipine, it was necessary to administer either whole blood, angiotensin, or adrenaline to counteract the hypotension and maintain the posttraumatic mean arterial pressure at 100 to 120 mm Hg. The combination of nimodipine and adrenaline was the most effective and produced an increase in posttraumatic blood flow. In this experiment, the pressor agent was administered first, followed by nimodipine. In a subsequent study, nimodipine and adrenaline were administered simultaneously and produced a 60% increase in posttraumatic blood flow, whereas adrenaline alone failed to improve blood flow (Guha *et al.*, 1989). It is noteworthy that the combination of nimodipine and adrenaline did not significantly increase the amount of intramedullary hemorrhage at the injury site as determined by image analysis. Thus, nimodipine increased posttraumatic SCBF if the arterial pressure was maintained. Nevertheless, it is more important to determine whether an increase in posttraumatic blood flow would improve axonal function. In another study, we counteracted systemic hypotension with dextran (Fehlings and Tator, 1989b), which had previously been shown to increase posttraumatic SCBF, cardiac output, and arterial pressure and to produce hypervolemic hemodilution (Wallace and Tator, 1987). The combination of nimodipine and dextran, but not dextran alone, produced significant elevation of posttraumatic SCBF and improved axonal function as measured by motor and somatosensory evoked potentials. Moreover, the improvement in axonal function was highly correlated with the improvement in SCBF. These findings support the hypotheses that posttraumatic ischemia is of major importance in the pathogenesis of axonal dysfunction after acute SCI, and that reversal of ischemia can promote axonal function after SCI.

At least two other groups have studied nimodipine in experimental SCI. In an ischemic model in rabbits (Faden *et al.*, 1984) and in a modified weight-drop model in cats (Ford and Malm, 1985), nimodipine did not decrease the extent of pathological damage or improve clinical recovery, although SCBF was not measured. Holtz *et al.* (1989) gave nimodipine to rats for 4 hours after acute cord compression and found no change in SCBF or function on the inclined plane 4 days after SCI. Black *et al.* (1988) gave nicardipine, another dihydropyridine calcium channel antagonist, to rats injured by the weight-drop technique and found no improvement in behavioral function, but blood flow was not measured.

Steroids. There is evidence that large doses of glucocorticoid steroids improve posttraumatic SCBF and microvascular perfusion (Anderson *et al.*, 1982; Young *et al.*, 1982) as well as clinical neurological recovery after experimental SCI. In addition, clinical studies showed that high doses of methylprednisolone given within the first 8 hours of acute cord injury improved neurological function (Bracken *et al.*, 1990). These findings have raised optimism about treatment strategies with steroids, although the improvement was modest.

The exact mechanisms of glucocorticoid treatment in acute SCI are not known. It has been suggested that glucocorticoid can beneficially affect the injured cord through influence on at least three mechanisms including facilitation of neuronal excitability and impulse conduction, improved blood flow (Anderson *et al.*, 1982), and perhaps most importantly, the preservation of cord ultrastructure through a reduction of injury-induced, free radical-catalyzed lipid peroxidation (Hall *et al.*, 1982). Ducker and Hamit (1969) first suggested that the beneficial effect of glucocorticoid was based on a maintenance of microvascular integrity and blood flow within the injured spinal segment. Some experimental studies have provided direct or indirect evidences that glucocorticoid may enhance blood flow in SCI by interfering with the actions of vasoconstrictor amines such as serotonin, norepinephrine (Altura BM and Altura BT, 1974; Dow-Edwards *et al.*, 1980) or prostaglandins within the injured spinal segment (Jonsson and Daniell, 1976).

Opiate antagonists. There is evidence that administration of naloxone (a nonselective opioid receptor antagonist) improve posttraumatic hypotension, SCBF (Faden *et al.*, 1982; Young *et al.*, 1981) and clinical recovery (Flamm *et al.*, 1982) after SCI by the weight-drop technique in cats and rats. However, our laboratory evaluated naloxone treatment in rats injured by clip compression and found no improvement in SCBF, arterial pressure, cardiac output (Wallace and Tator, 1986b) or neurological function as assessed by the inclined plane test (Wallace and Tator, 1986a). There are also other similar negative reports with naloxone in experimental SCI (Black *et al.*, 1986a,b; Haghighi and Chehrazi, 1987, for details see Chapter 13 in this volume).

Glutamate antagonists. The excitatory amino acid (EAA) neurotransmitter glutamate is considered to play a pivotal role in the pathogenesis of neuronal death due to ischemia or trauma in the central nervous system (Muir and Lees, 1995). There is increasing evidence that EAA-mediated processes contribute significantly to the progressive neuronal loss in these conditions. The mechanisms of EAA-mediated injury to neurons including cellular ionic alterations have been determined in *in vitro* and *in vivo* preparations (Choi, 1987; Garthwaite, 1986; Rothman and Olney, 1986; Tymianski *et al.*, 1993; Tymianski and Tator, 1996). For example, glutamate causes early intracellular accumulation of sodium and chloride ions which result in acute cytotoxic edema, and early intracellular influx of extracelluar calcium ions resulting in delayed damage to neurons (Choi, 1987). There are several subtypes of glutamate receptors and among these, the N-methyl-D-aspartic acid (NMDA), 1-amino-3-hydroxy-5-methyl-4-isoxazole-propionic acid (AMPA) and kainic acid (KA) receptors are especially important in mediating neurotoxicity (Meldrum and Garthwaite, 1990). Since 1985, it has been known that there are NMDA and AMPA receptors in the spinal cord and that local damage is induced when sufficiently high concentrations of glutamate or its agonists are injected into the spinal subarachnoid space (Davis and Watkins, 1983; Ganong *et al.*,

1983; Kwak and Nakamura, 1985; Monaghan and Gotman, 1985). SCI alters the tissue levels of EAA (glutamate and aspartate) within the injury region (Demediuk *et al.*, 1988 and 1989).

The glutamate receptor antagonist, MK801, produces a non-competitive use-dependent blockade of the NMDA receptor, which is one of the main mediators of the lethal intracellular influx of calcium responsible for the neurotoxic effects of glutamate in cultures of spinal cord neurons (Tymianski *et al.*, 1993). Treatment with MK801 has produced variable results in SCI models. MK801, administrated systemically before or after SCI produced by the weight drop method, improved long-term neurologic recovery and reduced pathological changes and alterations in certain tissue cations (Faden *et al.*, 1988; Gomez-Pinilla *et al.*, 1989). In contrast, MK801 did not improve neurologic function or SCBF after compressive injury (Holtz and Gerdin, 1991). In our laboratory, the effects of pre- and posttreatment with MK801 were assessed in normal and SCI rats by three outcome measures: axonal function measured by somatosensory evoked potentials (SSEPs) and cerebellar evoked potentials (CEPs); SCBF determined by the hydrogen clearance technique; and cord water content determined by the wet-dry method. We found that MK801 inhibits evoked potentials, and does not improve SCBF or cord edema in normal and SCI rats. In the same SCI model, an AMPA antagonist, NBQX, gave similar results. Our findings indicate that NMDA or AMPA receptor antagonists produce a blockade of glutamate excitatory transmission in afferent pathways and retard the somatosensory conduction function in SCI rats during a four hour observation period. The reduced evoked responses may be evidence that NMDA or AMPA receptor antagonists attenuate glutamate neurotoxicity and provide neuroprotection after SCI in rats (Li and Tator, unpublished results).

Elevation of systemic pressure. Since the injured spinal cord loses its ability for autoregulation, the systemic hypotension resulting from acute SCI should be treated to prevent worsening of posttraumatic ischemia. The aim of treating posttraumatic hypotension should be restoration of mean arterial pressure to normotensive levels, to avoid the possibility of enhancing intramedullay hemorrhage by hypertension and hyperaemia (Tator and Fehlings, 1991). However, we showed that restoration of normotension is insufficient to restore posttraumatic SCBF and function (Fehlings *et al.*, 1989b; Guha *et al.*, 1987 and 1989), and that additional measures are required to improve the local microcirculation in the cord, such as preventing vasospasm. We improved posttraumatic SCBF by a variety of agents including blood transfusion and dopamine, and the combinations of nimodipine with either adrenaline or dextran, the latter combination also produced functional improvement. Other investigators showed that systemic hypertension produced by noradrenaline or restoration of normotension improved somatosensory evoked potentials after SCI in cats (Hardy *et al.*, 1972; Brodkey *et al.*, 1972).

Evoked potentials in trauma to the spinal cord

A critical aspect of SCI is functional neurological assessment after either experimental or clinical SCI. The functional assessment must provide consistent and reproducible results by different observers and different laboratories. Behavioral tests have been developed and widely used in both clinical and experimental studies of SCI, and provide direct and objective functional information. However, they cannot quantitatively assess cord function after SCI. Over the past two decades, evoked potentials have become a powerful tool in the assessment of SCI, principally in evaluating the conduction of white matter, although its success has been more apparent in experimental injury than in clinical trials.

Principles and basis of evoked potentials

Somatosensory evoked potentials (SSEPs). SSEPs have been widely studied in experimental SCI. In our laboratory, SSEPs have are a frequently employed electrophysiological technique for evaluating axonal function after clip compression SCI in rats (Fig. 1). Under anesthesia with chloralose-urethane or halothane, electrical stimulation is applied to the sciatic nerve from a bipolar needle electrode and SSEPs are recorded epidurally over the contralateral sensorimotor cortex. The SSEP waveforms recorded consist of negative and positive peaks at approximately 11 msec (N_{11}) and 13 msec (P_{13}) followed by a large, broad negative wave with an approximate peak latency of 25 msec (N_{25}). The N_{11} activity is widely distributed over the cerebral cortex, while the P_{13} and N_{25} peaks are sharply localized to a small area. The pattern of distribution of each peak is consistent in all rats, although the SSEP amplitude is variable (Koyanagi and Tator, 1997) and several factors can affect the SSEP waveforms. For example, the SSEP complex can be markedly affected by anesthesia and cortical stimulation. The recording site over the sensory cortex and the depth of anesthesia are critical for reliable SSEP monitoring of the rat spinal cord. Also, evoked potential monitoring should be carefully designed if both SSEPs and motor evoked potentials are recorded from the same animal, because the sensory cortex which responds to sciatic nerve stimulation is vulnerable to electrical stimulation employed to produce motor evoked potentials (MEPs, Koyanagi and Tator, 1996).

Motor evoked potentials. The SSEP principally reflects activity in large fibre ascending afferents in the dorsal columns and dorsal spinocerebellar tracts, and therefore does not accurately reflect motor function (Fehlings *et al.*, 1988 and 1989a). The inability of the SSEPs to reliably monitor motor tracts of the spinal cord has prompted interest in developing other neurophysiological means to do so. The MEP is primarily conducted along type A myelinated efferents in the pyramidal and non-pyramidal tracts of the cord (Fehlings *et al.*, 1988 and 1989a) and MEPs have a potential

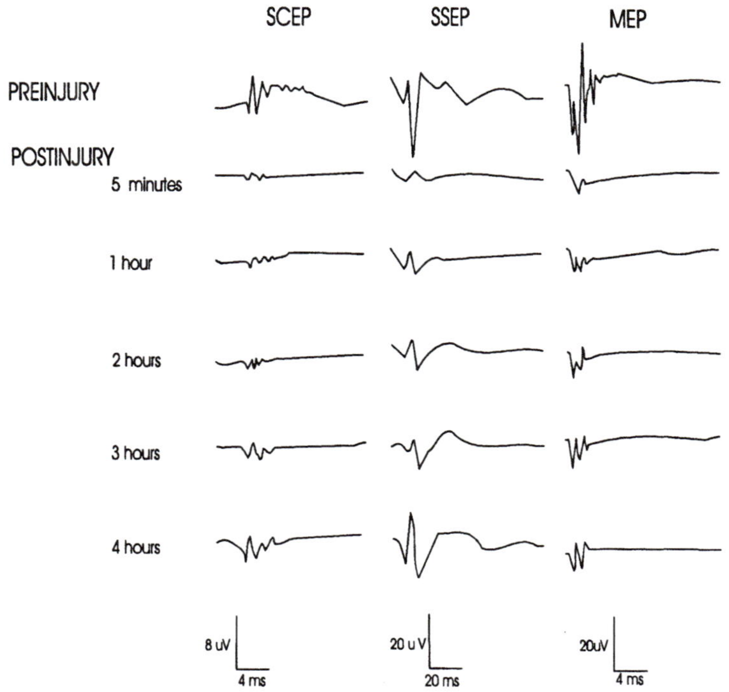

Fig. 1. Electrophysiological recordings from a rat receiving a 21g clip compression injury of the cord at T1. SCEP and SSEP were recorded extradurally from the spinal cord at C6 and from the sensorimotor cortex (*SMC*), respectively, after sciatic nerve stimulation. MEP was recorded extradurally from the spinal cord at T11 after SMC stimulation. The amplitude of the SCEP, SSEP and MEP was decreased significantly 5 minutes after SCI, but partly recovered from 1 hour to 4 hours after SCI

useful role in this area. In 1954, Patton and Amassian characterized the responses evoked by directly stimulating the motor cortex of primates and recorded from microelectrodes in the bulbar pyramids and the lateral funiculus of the cervical spinal cord. Unfortunately, the clinical application of MEPs was largely hindered by the inability to stimulate painlessly the motor cortex through the intact skull in awake patients (Gualtierotti and Paterson, 1954). In 1980, the report of transcutaneous, transcranial stimulation of the motor cortex with electromyographic recording from limb muscles was a critical contribution to this development (Merton and Morton, 1980). Levy *et al.* (1984a,b) took this approach to cats in an experimental setting and then to clinical application in humans. They found it to be a more reliable predictor of motor function than the SSEP and a more reliable indicator of anterior spinal cord integrity.

There are two kinds of MEP stimulations: electrical and magnetic. Electrical transcranial stimuli can produce direct activation of corticospinal axons (Kalkman *et al.*, 1993), while magnetic stimulation activates motor pathways transsynaptically. Electrical stimulation including monopolar and bipolar stimulus configurations has been widely used to elicit MEPs from the motor cortex in experimental SCI investigations. Monopolar stimulation consists of an anode electrode placed over the motor cortex and the cathode positioned at a distance, usually on the palate. The stimulus has been delivered either through the intact skull (transcranial) or with the anode electrode lying directly on the motor cortex (transcortical). In contrast, other studies have attempted to limit the stimulus spread by positioning both electrodes directly over the motor cortex (bipolar, Zappulla and Wang, 1993). Transcranial magnetic stimulation allows contactless stimulation of the motor cortex by inducing a current in brain tissue underlying the magnetic coil (Kalkman *et al.*, 1993) and is mainly used in humans.

The predominant efferent pathways traversing the spinal cord are composed of the pyramidal system arising from the motor cortex and the brainstem efferents derived from various brainstem nuclei. A third efferent pathway arises from the limbic system (locus coeruleus, raphe nuclei, and hypothalamic, nuclei). The first two efferent systems are involved in controlling and integrating muscle contraction and body movement, and mostly likely contribute to the generation of the MEPs (Zappulla and Wang, 1993). Bipolar stimulation of the cortex produces a specific activation of pyramidal tracts with a much slower conduction speed. In contrast, the more frequently used monopolar cortical stimulation for recording MEPs in experimental SCI, activates the faster and more powerful extrapyramidal pathways, and reflects activity in the extrapyramidal system. However, this extrapyramidal MEP is actually very useful, since corticospinal inputs to the spinal cord and the control of hindlimb movements in quadrupeds is extremely limited (Blight, 1994).

Transcranial and transcortical stimulations result in depolarization of cortical neurons or axons, which produce a descending volley comprised of an initial D (direct) wave and the later I (indirect) waves. The former is believed to represent stimulation of the axons of the large corticospinal tract neurons, while the latter are a series of descending impulses resulting from repetitive synaptic bombardment of pyramidal tract neurons through chains of interneurons (Cracco, 1987; York, 1987; Amassian *et al.*, 1987; Fehlings *et al.*, 1987). Because of the unidirectional nature of synapses, no signal is propagated via sensory axons to the spinal cord. However, Su *et al.* (1992) found that the neurogenic MEP signal is not entirely specific for the motor tracts, since it contains components produced by antidromic conduction in sensory tracts.

SSEPs in trauma to the spinal cord

Effects of SCI on the SSEPs. To characterize the changes in axonal function in somatosensory tracts of the cord after acute SCI, SSEPs were recorded after different degrees of cord injury caused by clip compression in rats (Figs. 1, 2). There were no significant abnormalities in the SSEPs when the rats received a 1.5 g compression SCI, but the 20 g compression injury produced a significant reduction in amplitude and a slight prolongation in latency. In contrast, the SSEPs were abolished in most rats when the injury was increased to 56 g (Fehlings *et al.*, 1989a).

It has been reported that it is possible to define the onset of the phase of second damage using SSEPs (Blight, 1994). Transient abnormalities in the SSEPs may occur at the third to sixth day after SCI and may suggest the presence of spinal cord edema. There is an almost complete loss of SSEPs immediately after weight-drop contusion injury in cats, a partial recovery in some animals within the first hour, but a secondary loss within the first day, followed by a slow and variable recovery over the succeeding weeks (Blight and Young, 1989 and 1990). Similarly, a study in rhesus monkeys showed that a transient recovery of sural nerve SSEPs 1–2 hours after SCI caused by weight-drop method, and then a secondary loss 1–7 days after injury (Martin and Bloedel, 1973). The abnormalities of SSEPs after SCI have included a reduction of amplitude, latency shifts of the early scalp components and the disappearances of SSEP peaks. Young (1982) suggested that amplitude changes predominated with little latency change in the acute phase (1 week), but in the more chronic phase (6 weeks), the SSEPs showed both amplitude and latency changes.

Relationship among SSEPs, pathology and functional recovery in SCI. The SSEP is well suited to the study of spinal conduction through the injury site (York *et al.*, 1983). Serial recordings after SCI provide information concerning the extent of the lesion and are an index of changes in spinal cord function of predictive value regarding outcome. Early experience with the SSEPs in experimental SCI demonstrated that the absence or presence of the SSEPs predicted eventual neurological recovery (Martin and Bloedel, 1973). SSEPs were useful in patients with acute SCI (Grundy and Griedman, 1987) and as an intraoperative spinal cord monitoring technique (Møller, 1988; Young, 1987). Even with the advent of other EP techniques including MEPs and cerebellar evoked potentials, SSEPs continue to be one of the most widely used methods to monitor cord function (Nash *et al.*, 1989). SSEP measurements several hours after acute SCI were reasonably good predictors of final behavioral recovery, which can take 4 weeks or more to become stable (Perot and Vera, 1982; Young, 1985). In contrast, chronic recovery of the SSEPs is not well correlated with behavioral outcome (Young, 1985) because relatively complete recovery of the SSEP can occur with very small proportions (2–5%)of the myelinated axons surviving in the dorsal half of the thoracic spinal cord (Blight and Young, 1990). It has been shown that the relation between SSEP amplitude and spinal input is extremely nonlinear, and cortical SSEPs mediated by the dorsal column and lateral tracts in cats can reach 100% amplitude with only 20–40% of the afferent volley in the peripheral nerve (Mark and Steiner, 1958). A practically normal cortical SSEP can be recorded by hindlimb nerve stimulation even when the thoracic spinal cord is largely transected, leaving only the dorsal column (Blight, 1994).

Thus, there are very definite limitation on the use of SSEPs in SCI. Although the utility of SSEPs to predict surviving axons or as a measure of returning neurologic competence is still open to question, the SSEPs can provide information of considerable prognostic significance after SCI. The presence of well-preserved SSEPs in the first 24 hours after injury indicates the best prognosis for recovery of motor and sensory function, whereas the absence of SSEPs is associated with a poor prognosis for useful recovery. However, if there is no recovery of the SSEPs in the first day, it is still possible that some recovery of motor function will occur over the succeeding weeks. Thus, SSEP is best used as a monitor of the acute stage of axonal injury in experimental SCI (Salzman, 1990).

Spinal cord evoked potentials. Spinal cord evoked potentials (SCEPs) are the evoked electrical responses elicited directly from the spinal cord by stimulation of a peripheral nerve (Fig. 1). They are usually composed of two types of electrical events, propagating action potentials present in the spinal roots and tracts, and stationary action potentials reflecting postsynaptic activity in the spinal gray matter. Several studies reported the usefulness of SCEPs in assessing spinal cord function after experimental SCI. It has been proven that SCEPs are more stable and reliable than SSEPs recorded from the cortex (Sharma *et al.*, 1991), and can be used to examine the level or extent of transverse lesions of the spinal cord (Sarnowski *et al.*, 1975). Tamaki *et al.* (1985) reported SCEPs evoked by direct spinal cord stimulation are more sensitive than potentials evoked by peripheral nerve stimulation. Several investigators have correlated the early alterations in SCEPs with behavioral outcome and pathological findings after experimental SCI (Homma and Tamaki, 1984). However, detailed relationships between altered SCEP and late pathological changes are still obscure since the late morphological changes in the cord are complicated by factors such as edema. Changes in SCEPs are present within a few minutes after incomplete SCI, but they can not reflect the effects of cell necrosis or profound edema (Sharma *et al.*, 1991) because the secondary cord damage caused by the disturbances in microcirculation and by several biochemical factors begins to occur within a few minutes after SCI and lasts 24 hours or more (Tator and Fehlings, 1991). However, when combined with other methods of assessment of residual spinal cord function, such as MEPs, SCEPs can be of benefit in establishing criteria for selection and application of methods in restoration of spinal cord function (Dimitrijevic *et al.*, 1992). Furthermore, SCEP changes after SCI could pharmacologically be modified with drugs known to alter serotonin metabolism (Sharma *et al.*, 1991). Pretreatment with indomethacin (a potent inhibitor of endogenous prostaglandins synthesis) markedly attenuated the decrease in the amplitude of rostral SCEPs and edema occurring in untreated rats, but did not affect the increases of SCEP latency (Winkler *et al.*, 1993).

Spinal cord evoked injury potentials. After experimental SCI, typical changes in the spinal evoked potentials have been described, consisting of a monophasic positive potential that is sometimes followed by a negative afterwave. Our laboratory was one of the first to describe these potentials (Deecke and Tator, 1973) which have labelled "killed end potential" or "final potential" (Schramm *et al.*, 1983). Similar potentials have been demonstrated in humans (Katayama *et al.*, 1988; Prestor *et al.*, 1993). These frequently reported positive-directed killed end potentials are one of the major electrical diagnostic signals indicating lesions within the spinal cord. However, there is no distinct correlation between the frequency of appearance of the positive-directed killed end potentials and histopathological changes in the acute phase of SCI (Baba *et al.*, 1989; Fehlings *et al.*, 1988 and 1989b). It has been shown that these altered evoked potentials indicate a major cord injury (Fehlings *et al.*, 1988 and 1989b), although they may occur in reversible as well as irreversible (transection) spinal cord lesions (Schramm *et al.*, 1983 and 1984).

The precise generator of this injury potential is unknown. It is quite likely that the use of a monopolar recording montage is important for displaying this potential (Winkler *et al.*, 1993). It has been postulated that this evoked potential may be generated (according to the volume conduction theory) by the interruption of a part of the afferent volley conduction measured by an electrode distal to it in the injured axons of the cord somatosensory system. Ionic changes in the extra- and intracellular spaces at the site of an acute cord lesion could also be responsible for this potential (Winkler *et al.*, 1993). This potential occurs when an impulse approaches, but does not arrive at a recording electrode because the fibres conducting the impulses terminate before reaching the electrode. In the case of incomplete blockade of spinal cord afferent tracts, the blocked afferent volleys generate a marked positivity, whereas the afferent volleys passing by along intact fibres cause the phase reversals (Schramm *et al.*, 1984).

MEPs in trauma to the spinal cord

Since the ascending sensory pathways are anatomically separate and have a different blood supply from the descending motor pathways, there is an obvious limitation to the use of SSEPs after SCI. Therefore, it is not surprising that patients have shown postoperative motor deficits when there has been no change in the SSEPs during surgery (Lesser *et al.*, 1986). Also, there are cases with preservation of motor function despite a loss of the SSEPs (Singer *et al.*, 1977; Hahn and Latchaw, 1983). Thus, SSEPs do not precisely predict the clinical state, especially the efferent conduction function, at the time of examination in acute SCI. In contrast, MEPs are extremely sensitive and specific for motor tract injury (Konrad *et al.*, 1987; Fehlings *et al.*, 1989a,b and 1991).

Effects of SCI on the MEPs. In SCI produced by the weight-drop method in rats, some investigators found that MEPs correlated well with the initial degree of neurological motor function, the extent of recovery and the ultimate histological damage (Simpson and Baskin, 1987; Levy *et al.*, 1987). An acute reduction in pyramidal MEP amplitude (up to 67% of preinjury values) was noted when rats were subjected to impacts up to 30 g-cm which produced no clinical deficits at 1 week postinjury. The pyramidal MEP had a residual peak 1 and its amplitude was reduced to 10 to 20% of the baseline when ataxia was evident in rats with 40 g-cm SCI. In contrast, the MEP components were absent when the SCI was 50 g-cm or more (Shiau *et al.*, 1992). Shiau *et al.* concluded that MEPs were a sensitive indicator of disruption of the spinal efferent pathway, an accurate prognosticator of postinjury motor activity, and more accurate than either the Tarlov scale or the inclined plane in predicting the severity of SCI as demonstrated by histological changes (Shiau *et al.*, 1992).

In our laboratory, we also found a relationship between MEP changes and the severity of SCI (Figs. 1, 2). After a 1.5 g injury, the latency of the

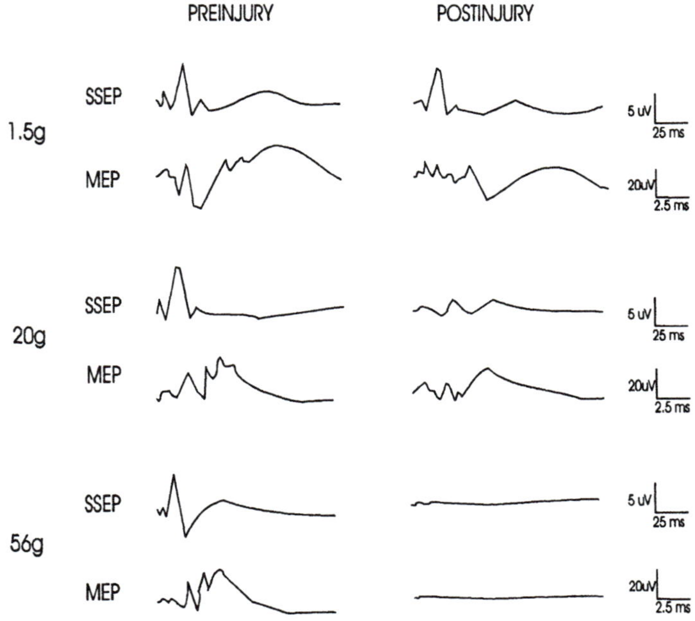

Fig. 2. SSEP and MEP recorded from three rats before and 30 minutes after 1.5 g, 20 g or 56 g clip compression injury of the cord at C8. The methods of SSEP recordings were the same as in Fig. 1. The MEP was recorded from the spinal cord at T10 by needle electrodes (Fehlings *et al.*, 1989a). There was a slight change in the MEP waveforms after the 1.5 g SCI, but a severe alteration in both after the 20 g SCI. In contrast, the 56 g SCI caused the complete disappearance of the SSEP and MEP waveforms

indirect peaks of the MEPs recorded from cord and sciatic nerve was delayed. A 20 g injury resulted in reduction of the amplitude in the direct peak and a prolongation of latency in both the direct and indirect peaks. A 56 g injury abolished the MEPs recorded from both the cord and sciatic nerve. These results indicate that alterations in MEPs are significantly related to the severity of cord injury (Fhelings *et al.*, 1989a).

Changes in MEP amplitude are a reliable and sensitive indicator for detecting SCI, and alterations in MEP from baseline levels may be more accurate than the absolute peak amplitude in assessing SCI. In contrast, changes in latency do not appear to be good indicators of the severity of SCI (Shiau *et al.*, 1992). There is evidence that MEPs recorded directly from the spinal cord are less sensitive to SCI or ischemia than MEPs recorded from the peripheral nerve or muscle (Fehlings, 1988 and 1989a,b; Kai *et al.*, 1995; Levy *et al.*, 1987). Machida *et al.* (1988) showed that cross-clamping of the thoracic aorta in dogs above the level of the artery of Adamkiewicz resulted in the rapid disappearance of the myogenic response to thoracic cord stimulation, but left the spinal cord response intact, and in another experiment, the myogenic response was more sensitive to over-distraction with or without derotation than the response recorded from the spinal cord in cats (Machida *et al.*, 1989). In addition, our laboratory recently found that the combined recordings of MEPs and evoked potentials with cerebellar stimulation (CEPs) provided complementary information regarding axonal integrity in the mammalian spinal cord (Fehlings *et al.*, 1991; Hurlbert *et al.*, 1992).

Similar to the killed end potentials noted with SSEPs, injury motor evoked potentials may also be recorded from animals with cord injury (Ross and Tator, 1991). These potentials are characterized by monophasic positive deflections, probably do not represent signal conduction down the spinal cord, and may result from an impulse travelling toward but not arriving at an electrode because of a block in conduction (Ross and Tator, 1991).

MEPs and functional recovery in SCI. MEPs are an important electro-physiological parameter for monitoring spinal cord functional integrity and functional recovery in experimental SCI (Fehlings *et al.*, 1989a,b; Shiau *et al.*, 1992), although there are still some problems to be solved about the application of MEPs. For example, in patients with acute SCI, there is often "spinal shock" lasting several weeks, and it is difficult to distinguish this from complete spinal cord transection. Macdonell and Donnan (1995) found that MEPs recorded from patients within 6 hours of acute SCI did not provide useful information regarding motor recovery. Caution must be exercised in generalizing conclusions concerning the MEP across species, because prediction of motor outcome based on changes in the MEP following experimental SCI must take into account the various efferent spinal pathways that may contribute to the generation of the MEPs (Zappulla and Wang, 1993).

Comparison of SSEPs and MEPs in SCI

It has been found that both the SSEP and MEP are affected by moderate contusive SCI produced by the weight-drop method (Baskin and Simpson, 1987; Simpson and Baskin, 1987). Levy *et al.* (1986) found the MEPs to be more sensitive than the SSEPs in detecting blunt cord injury using the same model, and concluded that MEPs may offer a valuable and more reliable indication of motor function than the SSEPs (Levy *et al.*, 1984a,b). Similarly, in our compressive SCI model, the MEP was extremely sensitive to mild cord injury, but it did not distinguish between mild and moderate injuries as readily as the SSEP (Fehlings *et al.*, 1989a). However, cortical SSEPs were superior to MEPs for detecting surviving sensory pathways in severe SCI (Blight, 1994). Thus, both methods appear valuable, without one being superior to the other. They assess the integrity of different spinal cord tracts, and both are significantly related to the severity of cord injury. The recording of these modalities in combination enhances the accuracy of the neurophysiological assessment and may be a more accurate neurophysiological technique to assess posttraumatic spinal cord function (Fehlings *et al.*, 1989a,b).

Conclusions

1. There is a marked reduction in the blood supply to the spinal cord after experimental SCI and the most direct method of observing post-traumatic ischemia is to measure SCBF. Techniques for measuring SCBF include hydrogen clearance, microsphere, autoradiography, laser Doppler flowmetry and nuclear magnetic resonance. There is a direct linear relationship between the severity of injury and changes of SCBF. Post-traumatic ischemia is presumably caused by vasospasm due to mechanical damage or the release of a vasoactive amine or other vasoconstrictors. Preservation of regional SCBF, by counteracting or preventing the systemic or local vascular effects of SCI by various therapeutic modalities, may improve the recovery of spinal cord function.

2. SCI causes significant abnormalities in the SSEPs which include a reduction in amplitude, a prolongation in latency, or complete loss of signal. SSEPs recorded directly from the spinal cord are more stable and reliable than those recorded from the cortex. The occurrence of positive-directed killed end potentials is a useful diagnostic signal indicating lesions within the spinal cord. SSEP measurements in the first day after cord injury can provide considerable prognostic information. The presence of well-preserved SSEPs in the acute stage of axonal injury forecasts a good recovery of motor and sensory function. In contrast, the MEP is an extremely sensitive and specific indicator of disruption of the spinal efferent pathways following experimental SCI, although the relationship between the MEP and the recovery of motor function is still controversial.

3. SSEPs and MEPs are significantly related to both the severity of cord injury and the reduction in SCBF. Thus, axonal dysfunction in SCI correlates with the degree of cord injury and posttraumatic ischemia.

References

Altura BM, Altura BT (1974) Peripheral vascular actions of glucocorticoids and their relationship to protection in circulatory shock. J Rest Pharmacol Exp Ther 190: 300–315

Amassian VE, Stewart M, Quirk GJ, Rosenthal JL (1987) Physiological basis of motor effects of a transient stimulus to cerebral cortex. Neurosurgery 20: 74–93

Anderson DK, Means ED, Waters TR, Green ES (1982) Microvascular perfusion and metabolism in injured spinal cord after methylprednisolone treatment. J Neurosurg 56: 106–113

Anthes DL, Theriault E, Tator CH (1996) Ultrastructural evidence for arteriolar vasospasm after spinal cord trauma. Neurosurgery 39: 804–814

Auckland K, Bower BF, Berliner RW (1964) Measurements of local blood flow with hydrogen gas. Circ Res 14: 164–187

Baba H, Tomita K, Umeda S, Kawahara N, Nagata S (1989) Experimental ascending evoked potentials in spinal cord injury. In: Ducker TB, Brown RH (eds) Neurophysiology and Standards of Spinal Cord Monitoring. Springer, Berlin Heidelberg New York Tokyo, pp 46–51

Baskin DS, Simpson RK (1987) Corticomotor and somatosensory evoked potential evaluation of acute spinal cord injury in the rat. Neurosurgery 20: 871–87

Black P, Markowitz RS, Finkelstein SD, McMonagle-Strucko K, Gillespie JA (1988) Experimental spinal cord injury: effect of a calcium channel antagonist (nicardipine). Neurosurgery 22: 61–66

Black P, Markowitz RS, Keller S, Wachs K, Gillespie JA, Finkelstein SD (1986a) Naloxone and experimental spinal cord injury. Part 2: Megadose treatment in a dynamic load injury model. Neurosurgery 19: 909–913

Black P, Markowitz RS, Keller S, Wachs K, Gillespie JA, Finkelstein SD (1986b) Naloxone and experimental spinal cord injury. Part 1: High dose administration in a static load compression model. Neurosurgery 19: 905–908

Blight AR (1994) Electrophysiological approaches to mechanisms of central nervous system trauma. In: Salzman SK, FadenAI (eds) The Neurobiology of Central Nervous System Trauma. Oxford University Press, Oxford, pp 41–56

Blight A, Young W (1990) Axonal morphometric correlates of evoked potentials in experimental spinal cord injury In: SK Salzman(ed) Neural Monitoring. Humana Press, Chilton, pp 87–114

Blight A, Young W (1989) Central axons in injured cat spina cord recover electrophysiological function following remyelination by Schwann cells. J Neurol Sci 91: 15–34

Blight A, Someya S (1985) Depolarizing afterpotentials in myelinated axons of mammalian spinal cord. Neuroscience 15: 1–12

Bracken MB, Shepard MJ, Collins WF, Holford TR, Young W, Baskin DS, Eisenberg HM (1990) A randomized, controlled trial of methyl-prednisolone or naloxone in the treatment of acute spinal-cord injury. Results of the secondary national acute spinal cord injury study. N Engl J Med 322: 1405–1411

Branston NM, Ladds A, Symon L, Wang AD (1984) Comparison of the effects of ischemia on early components of the somatosensory evoked potential in brainstem, thalamus and cerebral cortex. J Cereb Blood Flow Metab 4: 68–81

Brodkey JS, Richards DE, Blasingame JP (1972) Reversible spinal cord trauma in cats. Additive effects of direct pressure and ischemia. J Neurosurg 37: 591–593

Cawthon DF, Senter HJ, Stewart WB (1980) Comparison of hydrogen clearance and ^{14}C-antipyrine autoradiography in the measurement of spinal cord blood flow after severe impact injury. J Neurosurg 52: 801–807

Chehrazi BB, Scremin O, Decima EE (1989) Effect of regional spinal cord blood flow and central control in recovery from spina cord injury. J Neurosurg 71: 747–753
Cheung JY, Bonventre JV, Malis CD, Leaf A (1986) Calcium and ischemic injury. N Engl J Med 314: 1670–1676
Choi DW (1987) Ionic dependence of glutamate neurotoxicity. J Neurosci 7: 369–379
Collins WF (1983) A review and update of experimental and clinical studies of spinal cord injury. Paraplegia 21: 204–219
Corbally MT, Brennan MF (1990) Noninvasive measurement of regional blood flow in Man. Am J Surg 1990 160: 313–321
Cracco RQ (1987) Evaluation of conduction in central motor pathways: Techniques, pathophysiology, and clinical interpretation. Neurosurgery 20: 199–203
Davies J, Watkins JC (1983) Excitatory amino acid receptors in mono- and polysynaptic excitation in the cat spinal cord. Exp Brain Res 49: 280–290
De la Torre JC (1981) Spinal cord injury: Review of basic and applied research. Spine 6: 315–335
Deecke L, Tator CH (1973) Neurophysiological assessment of afferent and efferent conduction in the inured spinal cord of monkeys. J Neurosurg 39: 65–74
Demediuk P, Daly MP, Faden AI (1989) Effect of impact trauma on neurotransmitter and nonneuro-transmitter amino acids in rat spinal cord. J Neurochem 52: 1529–1536
Demediuk P, Daly MP, Faden AI (1988) Free amino acid levels in laminectomized and traumatized rat spinal cord. Trans Am Soc Neurochem 19, p 176
Dimitrijevic MR, Hsu CH Y, McKay WB (1992) Neurophysiological assessment of spinal cord and head injury. J Neurotrauma 9 [Suppl] 1: S293–S300
Dohrmann GJ, Wich KM, Bucy PC (1973) Spinal cord blood flow patterns in experimental traumatic paraplegia. J Neurosurg 38: 52–58
Dohrmann GJ, Wagner FC Jr, Bucy PC (1971) The microvasculature in transitory traumatic paraplegia. An electron microscopic study in the monkey. J Neurosurg 35: 263–271
Dohrmann GJ, Allen WE III (1975) Microcirculation of traumatized spinal cord. A correlation of microangiography and blood flow patterNS in transitory and permanent paraplegia. J Trauma 15: 1003–1013
Dolan EJ, Tator CH (1982) The effect of blood transfusion, dopamine and gamma hydyoxybutyrate on post-traumatic ischemia of the spina cord. Neurosurgery 56: 350–358
Dolan EJ, Transfeldt EE, Tator CH (1980) The effect of spinal distraction on regional spinal cord blood flow in cats. J Neurosurg 53: 756–764
Dow-Edwards D, DeCrescito V, Tomasula JJ, Flamm ES (1980) Effect of aminophylline and isoproterenol on spinal cord blood flow after impact injury. J Neurosurg 53: 385–390
Ducker TB, Salzman M, Lucas JJ, Garrison WB, Perot PL (1978) Experimental spinal cord trauma II: Blood flow, tissue oxygen, evoked potentials in both paraparetic and paraplegic monkeys. Surg Neurol 10: 64–70
Ducker TB, Hamit HF (1969) Experimental treatment of acute spinal cord injury. J Neurosurg 30: 693–697
Faden AI, Jacobs TP, Holaday JW (1982) Comparison of early and late naloxone treatment in experimental spinal injury. Neurology 34: 677–681
Faden AI, Jacobs TP, Smith MT (1984) Evaluation of the calcium channel antagonist nimodipine in experimental spinal cord ischemia. J Neurosurg 60: 796–799
Faden AI, Lemke M, Simon R, Noble L (1988) N-methyl-D-aspartate antagonist MK801 improves outcome following traumatic spinal cord injury in rats: behavioral, anatomic and neurochemical studies. J Neurotrauma 5: 33–45
Fairholm DJ, Turnbull IM (1971) Microangiographic study of experimental spinal cord injuries. J Neurosurg 35: 277–286
Fehlings MG, Tator CH, Linden RD (1989a) The relationships among the severity of spinal cord injury, motor and somatosensory evoked potentials and spinal cord blood flow. Electroencephalogr Clin Neurophysiol 74: 241–259

Fehlings MG, Tator CH, Linden RD (1989b) The effect of nimodipine and dextran on axonal function and blood flow following experimental spinal cord injury. J Neurosurg 71: 403–416

Fehlings MG, Tator CH, Linden RD, Piper IR (1988) Motor and somatosensory evoked potentials recorded from the rat. Electroencephalogr Clin Neurophysiol 69: 65–78

Fehlings MG, Hurlbert RJ, Tator CH (1991) The electrophysiological assessment of the pyramidal and non-pyramidal tracts of spinal cord of rats. Electroencephalogr Clin Neurophysiol [Suppl] 43: 287–296

Fehlings MG, Tator CH, Linden RD, Piper IR (1987) Motor evoked potentials recorded form normal and spinal cord-injured rat. Neurosurgery 20: 125–130

Fehlings MG, Tator CH (1988) A review of models of acute experimental spinal cord injury. In: Illis L(ed) Spinal Cord Dysfunction Assessment. Oxford University Press, Oxford, pp 3–43

Flamm ES, Young W, Demopoulos HB, DeCrescito V Tomasula J (1982) Experimental spinal cord injury: Treatment with naloxone. Neurosurgery 10: 227–231

Ford RW, Malm DN (1985) Failure of nimodipine to reverse acute experimental spinal cord injury. Cent Nerv Syst Trauma 2: 9–17

Fried LC, Goodkin R (1971) Microangiographic observations of the experimentally traumatized spinal cord. J Neurosurg 35: 709–714

Ganong AG, Lanthorn TH, Cotman CK (1983) Kynurenic acid inhibits synaptic and acidic amino acid-induced responses in the rat hippocampus and spinal cord. Brain Res 273: 170–174

Garthwaite G, Garthwaite J (1986) Neurotoxicity of excitatory amino acid receptor agonists in rat cerebellar slices: dependence on calcium concentration. Neurosci Lett 66: 193–198

Goldberger ME, Bregman BS, Bierck CJ Jr, Brown M (1990) Criteria for assessing recovery of function after spinal cord injury: behavioral methods. Exp Neurol 107: 113–117

Gomez-pinilla F, Tram H, Cotman CW, Nieto-Sampder M(1989) Neuroprotective effect of MK801 and U-50488H after contusive spinal cord injury. Exp Neurol 104: 11–24

Griffiths IR, Trench JG, Crawford RA (1979) Spinal cord blood flow and conduction during experimental cord compression in normotensive and hypotensive dogs. J Neurosurg 50: 353–360

Grundy BL, Friedman W (1987) Electrophysiological evaluation of the patients with acute spinal cord injury. Crit Care Clin 3: 519–549

Gruner JA, Wade CK, Menna G, Stokes BT (1993) Myoelectric evoked potentials versus locomotor recovery in chronic spinal cord injured rats. J Neurotrauma 10: 327–346

Gualtierotti J, Paterson AS (1954) Electrical stimulation of the unexposed cerebral cortex. J Physiol (Lond) 125: 278–291

Guha A, Tator CH, Smith CR Piper I (1989) Improvement in post–traumatic spinal cord blood flow with a combination of a calcium channel blocker and a vasopressor. J Trauma 29: 1440–1447

Guha A, Tator CH (1988) Acute cardiovascular effects of experimental spinal cord injury. J Trauma 28: 481–490

Guha A, Tator CH, Piper I (1987) Effect of a calcium channel blocker on posttraumatic spinal cord blood flow. J Neurosurg 66: 423–430

Guha A, Tator CH, Piper I (1985) Increase in rat spinal cord blood flow with the calcium channel blocker, nimodipine. J Neurosurg 63: 250–259

Guha A, Tator CH, Rochon J (1989) Spinal cord blood flow and systemic blood pressure after experimental spinal cord injury in rats. Stroke 20: 372–377

Haghighi SS, Chehrazi B (1987) Effect of naloxone in experimental acute spinal cord injury. Neurosurgery 20: 385–388

Hahn JF, Latchwa JP (1983) Evoked potentials in the operating room. Clin Neurosurg 31: 389–403

Hall ED (1988) Effects of the 21-aminosteroid U74006F on posttraumatic spina cord ischemia in cats. J Neurosurg 68: 462–465

Hall ED, Braughler JM (1982) Glucocorticoid mechanisms in acute spinal cord injury: a review and therapeutic rationale. Surg Neurol 18: 320–327

Hall ED, Wolf DL (1986) A pharmacological analysis of the pathophysiologic mechanisms of posttraumatic spinal cord ischemia. J Neurosurg 64: 951–961

Hall ED, Yonkers PA, Horan KL (1989) Correlation between attenuation of post-traumatic spinal cord ischemia and preservation of tissue vitamin E by the 21-aminosteroid U7400F: evidence for an in vivo antioxidant mechanism. J Neurotrauma 6: 169–176

Hardy RW, Brodkey JS, Richards DE, Nulsen FE (1972) Effect of systemic hypertension on compression block of spinal cord. Surg Forum 23: 434–435

Hogan EL, Hsu CY, Banik NL (1986) Calcium-activated mediators of secondary injury in the spinal cord. Cent Nerv Syst Trauma 3: 175–179

Holtz A, Gerdin B (1991) MK801, an OBS N-methyl-D-aspartate channel blocker, does not improve the functional recovery nor spinal cord blood flow after spinal cord compression in rat. Acta Neurol Scand 84: 334–338

Holtz A, Nystrom B, Gerdin B (1989) Spinal cord injury in rats: inability of nimodipine or anti-neutrophil serum to improve spinal cord blood flow or neurologic status. Acta Neurol Scand 79: 460–467

Homma S, Tamaki T (1984) Fundamentals and Clinical Application of Spinal Cord Monitoring. Saikon, Tokyo, pp 23–32

Hurlbert RJ, Tator CH, Fehlings MG, Niznik G, Liden RD (1992) Evoked potentials from direct cerebellar stimulation for monitoring of the rodent spinal cord. J Neurosurg 76: 280–291

Jonsson HT, Daniell HB (1976) Altered levels of PGF in cat spinal cord tissue following traumatic injury. Prostaglandins 11: 51–61

Jørgensen MB, Diemer NG (1982) Selective neuron loss after cerebral ischemia in the rat: possible role of transmitter glutamate: Acta Neurol Scand 66: 536–546

Kai Y, Owen JH, Allen BR, Dobras M, Davis C (1995) Relationship between evoked potentials and clinical status in spinal cord ischemia. Spine 20: 291–296

Kalkman CJ, Been HD, Ongerboer de Visser BW (1993) Intraoperative monitoring of spinal cord function. Acta Orthop Scand 64: 114–123

Katayama Y, Tsubokawa T, Yamamoto T, Hirayama T, Maejima S (1988) Preoperative determination of the level of spinal cord lesions from the killed end potential. Surg Neurol 29: 91–94

Kato A, Ushio Y, Hayakawat T, Yamada K, Ikeda H, Mogami H (1985) Circulatory disturbance of the spinal cord with epidural neoplasm in rats. J Neurosurg 63: 260–265

Kobrine AJ, Doyle TF, Rizzolin HV (1976) Spinal cord blood flow was affected by changes in systemic arterial blood pressure. J Neurosurg 44: 12–15

Kobrine AI, Doyle TF, Martins AN (1975) Local spinal cord blood flow in experimental traumatic myelopathy. J Neurosurg 42: 144–149

Konrad PE, Tacker WA, Levy WJ, Reedy DP, Cook JR, Geddes SL (1987) Motor evoked potentials in the dog: effects of global ischemia on spinal cord and peripheral nerve signals. Neurosurgery 20: 117–124

Koyanagi I, Tator CH, Lea PJ (1993a) Three-dimensional analysis of the vascular system in the rat spinal cord with scannign electron microscopy of vascular corrosion casts. Part 1: Normal spinal cord. Neurosurgery 33(2): 277–283

Koyanagi I, Tator CH, Lea PJ (1993b) Three-dimensional analysis of the vascular system in the rat spinal cord with scannign electron microscopy of vascular corrosion casts. Part 2: Acute spinal cord injury. Neurosurgery 33(2): 285–291

Koyanagi I, Tator CH (1993c) Theriault E: Silicone rubber microangiography of acute spinal cord injury in the rat. Neurosurg 32: 260–268

Koyanagi I, Tator CH (1996) The effects of cortical stimulation, anesthesia and recording site on somatosensory evoked potentials in the rat. Electroencephalogr Clin Neurophysiol 101: 534–542

Kristensen JD, Karlsten R ND, Gordh T (1994) Laser-Doppler evaluation of spinal cord blood flow after intrathecal administration of an N-methyl-D-aspartate antagonist in rats. Anesth Analg 78: 925–931
Kunkel-Bagden E, Dai HN and Bregman BS (1993) Methods to assess the development and recovery of locomotor function after spinal cord injury in rats. Exp Neurol 119: 154–164
Kwak S, Nakamura R (1985) Acute and late neurotoxicity in the rat spinal cord in vivo induced by glutamate receptor agonists. J Neurol Sci 129: 99–103
Lesser RP, Raudzens P, Luders H, Nuwer MR, Goldie WD (1986) Postoperative neurologic deficits may occur despite unchanged intraoperative somatosensory evoked potentials. Ann Neurol 19: 22–25
Levy WJ, McCaffrey M, Hagichi S (1987) Motor evoked potential as a predictor of recovery in chronic spinal cord injury. Neurosurgery 20: 138–142
Levy WJ, York DL, McCaffrey M, Tanzer F (1984a) Motor evoked potentials from transcranial stimulation of the motor cortex in humans. Neurosurgery 15: 287–302
Levy WJ, McCaffrey M, York DL, Tanzer F (1984b) Motor evoked potentials from transcranial stimulation of the motor cortex in cats. Neurosurgery 15: 214–227
Levy WJ, McCaffrey M, York DL (1986) Motor evoked potentials in cats with acute spinal cord injury. Neurosurgery 19: 9–19
Lohse DC, Senter HJ, Kauer JS, Wohns R (1980) Spinal cord blood flow in experimental transient traumatic paraplegia. J Neurosurg 52: 335–345
Macdonell RAL, Donnan GA (1995) Magnetic cortical stimulation in acute spinal cord injury. Neurology 45: 303–306
Machida M, Weinstein SL, Imamura Y, Usui T, Yamada T, Kimura J, Toriyama S (1989) Compound muscle action potentials and spinal evoked potentials in experimental spine manoeuvre. Spine 14: 687–691
Machida M, Weinstein SL, Yamada T, Kimura J, Toriyama S (1988) Dissociation of muscle action potentials and spinal somatosensory evoked potentials after ischemic damage of spinal cord. Spine 13: 1119–1124
Mark RF, Steiner J (1958) Cortical projection of impulses in myelinated cutaneous affernt nerve fibres of the cat. J Physiol (Lond) 142: 544–562
Martin SH, Bloedel JR (1973) Evaluation of experimental spinal cord injury using cortical evoked potentials. J Neurosurg 39: 75–81
Meldrum B, Garthwaite J (1990) Excitatory amino acid neurotoxicity and neurodegenerative disease. TiPs 11: 379–387
Meldurm B (1985) Excitatory amino acids and anoxic/ischemic brain damage. Trends Neurosci 8: 47–48
Merton PA, Morton HB (1980) Stimulation of the cerebral cortex in the intact human subject. Nature 285: 227
Mills CM, Brant-Zawadzki M, Crooks LE (1984) Nuclear magnetic resonance: principles of blood flow imaging. AJR 142: 165–170
Mohamed AA, McCulloch J, Mendelow AD, Teasdale GM, Harper AM (1984) Effect of the calcium antagonist nimodipine on local cerebral blood flow: relationship to arterial pressure. J Cereb Blood Flow Metab 4: 206–211
Møller AR (1988) Evoked Potentials in Intraoperative Monitoring. Williams & Wilkins, Baltimore, pp 67–88
Monaghan DT, Cotman CW (1985) Distribution of N-methyl-D-aspartate-sensitive L-[H^3]glutamate binding sites in rat brain. J Neurosci 5: 2909–2919
Muir KW, Lees KR (1995) Clinical experience with excitatory amino acid antagonist drugs. Stroke 26: 503–513
Nash CL, Brown RB (1989) Current Concepts Review: Spinal Cord Monitoring. J Bone Joint Surg 71-A: 627–630
Nowicki JP, Mackenzie ET, Young AR (1982) Brain ischaemia, calcium and calcium antagonists. Pathol Biol (Paris) 30: 282–288
Patton HD, Amassian VE (1954) Single multiple unit analysis of cortical stage of pyramidal tract activation. J Neurophysiol 17: 345–357

Perot PL, Vera CL (1982) Scalp-recorded somatosensory-evoked potentials to stimulation of nerves in the lower extremities and evaluation of patients with spinal cord trauma. Ann NY Acad Sci 388: 359–368
Piper I, Guha A, Tator CH (1987) A microcomputer system for on-line collection of blood flow and related physiological data. Comput Biol Med 4: 279–291
Prestor B, Zgur T, Dolenc VV (1993) Incomplete spinal cord evoked injury potential in man. Spine 18: 252–256
Rivlin AS, Tator CH (1978) Regional spinal cord blood flow in rats after severe cord trauma. J Neurosurg 49: 844–853
Ross IB, Tator CH (1991) Further studies of nimodipine in experimental spinal cord injury in the rat. J Neurotrauma 8: 229–238
Ross IB, Tator C (1994) Review of experimental methods for studying spinal cord or cerebral blood flow. In: Salzman SK, Faden AI (eds) The Neurobiology of Central Nervous System Trauma. Oxford University Press, Oxford, pp 57–67
Rothman SM, Olney JW (1986) Glutamate and the pathophysiology of hypoxic-ischemic brain damage. Ann Neurol 19: 105–111
Rouen LR (1976) Physiological testing in clinical angiology. Angiology 27: 57–64
Roughton JE (1982) Non-invasive measurement. Instrument Sci Tech 15: 1207–1269
Salzman SK (1990) Neural Monitoring. Humana Press, Clifton
Sandler AN, Tator CH (1967) Review of the effect of spinal cord trauma on the vessels and blood flow in the spinal cord. J Neurosurg 5, p 638
Sandler AN, Tator CH (1976a) Review of the effect of spinal cord trauma on the vessels and blood flow in the spinal cord. J Neurosurg 45: 638–646
Sandler AN, Tator CH (1976b) Effect of acute spinal cord compression injury on regional spina cord blood blow in primates. J Neurosurg 45: 660–676
Sarnowski RJ, Cracco RQ, Vogel HB, Mount F (1975) Spinal cord response in the cat. J Neurosurg 43: 329–336
Schramm J, Drause R, Shigeno T, Brock M (1984) Relevance of spinal cord evoked injury potential for spinal cord monitoring. In: Homma S, Tamaki T, Shimoji K, Kurokawa T (eds) Fundamentals and Clinical Application of Spinal Cord Monitoring. Saikon, Tokyo, pp 113–124
Schramm J, Drause R, Shigeno T, Brock M (1983) Experimental investigation on the spinal cord injury evoked potential. J Neurosurg 59: 485–492
Sharma HS, Winkler T, Stalberg E, Olsson Y, Dey PK (1991) Evaluation of traumatic spinal cord edema using evoked potentials recorded from the spinal epidural space. An experimental study in the rat. J Neurol Sci 102: 150–162
Shiau JS, Zappulla RA, Nieves J (1992) The effect of graded spinal cord injury on the extrapyramidal and pyramidal motor evoked potentials of the rats. Neurosurgery 30: 76–84
Simpson RK, Baskin DS (1987) Corticomotor evoked potentials in acute and chronic blunt spinal cord injury in the rat: correlation with neurological outcome and histological damage. Neurosurgery 20: 131–137
Singer PA, Prockop LD, Anderson DK (1977) Somatosensory cortical evoked response after feline experimental spina cord injury. Paraplegia 15: 160–165
Steen PA, Newberg LA, Milde JH, Michenfelder JD (1983) Nimodipine improves cerebral blood flow and neurologic recovery after complete cerebral ischemia in the dog. J Cereb Blood Flow Metab 3: 38–43
Stern MD (1977) Continuous measurement of tissue blood flow by laser-Doppler Spectroscopy. Am J Physiol 232: H441–H448
Su CF, Haghighi SS, Oro JJ, Gaines RW (1992) "Backfiring" in spinal cord monitoring. High thoracic spinal cord stimulation evoked sciatic response by antidromic sensory pathway conduction, not motor tract conduction. Spine 17: 504–508
Takakura S, Satoh Y, Satoh H, Mori J, Kohsaka M (1992) Effects of nilvadipine on regional cerebral blood flow and skin blood flow in anesthetized cats. Arch Internati Pharmacodym Therapie 319: 38–48

Tamaki T, Takano H, Takakuwa K (1985) Spinal cord monitoring: Basic principles and experimental aspects. CNS Trauma 2: 137–149

Tator C (1994) Ischemia as a secondary neural injury. In: Salzman SK, Faden AI (eds) The Neurobiology of Central Nervous System Trauma. Oxford University Press, Oxford, pp 209–215

Tator CH (1991) Review of experimental spinal cord injury with emphasis on the local and systemic circulatory effects. Neurochirurgie 37: 291–302

Tator CH and Fehlings MG (1991) Review of the secondary injury theory of acute spinal cord trauma with emphasis on vascular mechanisms. J Neurosurg 75: 15–26

Tymianski M, Charlton MP, Carlen PL, Tator CH (1993) Source specificity of early calcium neurotoxicity in cultured embryonic spinal neurons. J Neurosci 13: 2085–2104

Tymianski M, Tator CH (1996) Normal and abnormal calcium homeostasis in Neurons: a basis for the pathophysiology of traumatic and ischemic central nervous system injury. Neurosurgery 38: 1176–1195

Wallace MC, Tator CH (1986a) Failure of blood transfusion or naloxone to improve clinical recovery after experimental spinal cord injury. Neurosurgery 19: 489–494

Wallace MC, Tator CH (1986b) Failure of naloxone to improve spinal cord blood flow and cardiac output after spinal cord injury. Neurosurgery 19: 489–494

Wallace MC, Tator CH (1986c) Spinal cord blood flow measured with microspheres following experimental spinal cord injury in the rat. Can J Neurosci 13: 91–96

Wallace MC, Tator CH (1987) Successful improvement of blood pressure, cardiac output, and spinal cord blood flow after experimental spinal cord injury. Neurosurgery 20: 710–715

Wallace MC , Tator CH, Frazee P (1986) Relationship between posttraumatic ischemia and hemorrhage in the injured rat spinal cord as shown by colloidal carbon angiography. Neurosurgery 18: 433–439

Wedeen FJ, Reto AM, Edilman RR (1985) Projective imaging of pulsatile flow with magnetic resonance. Science 230: 946–948

Weinstein GB (1995) Effects of nimodipine and verapamil on cerebral blood flow and cerebrovascular reactivity in conscious rabbits. Ann NY Acad Sci 765: 334–335

Winkler T, Sharma HS, Stalberg, Olsson Y (1993) Indomethacin, an inhibitor of prostaglandin synthesis attenuates alteration in spinal cord evoked potential and edema formation after trauma to the spinal cord: an experimental study in the rat. Neuroscience 52: 1057–1067

Yanase M, Sakou T, Fukuda T (1995) Role of N-methyl-D-aspartate receptor in acute spinal cord injury. J Neurosurg 83: 884–888

York DH, Watts C, Raffensberger M, Spagnolia T, Joyce C (1983) Utilization of somatosensory evoked cortical potentials in spinal cord injury. Spine 8: 832–839

York DH (1987) Review of descending motor pathways involved with transcranial stimulation. Neurosurgery 1987 20: 70–73

Young W (1982) Correlation of somatosensory evoked potentials and neurological findings in spinal injury. In: Tator CH (ed) Early Management of Acute Spinal Cord Injury. Raven Press, New York, pp 153–165

Young W, Em V, Blight A (1982) Extracelluar calcium ionic activity in experimental spinal cord contusion. Brain Research 235: 105–130

Young W, Flamm ES, Demopoulos HB, Tomasula JJ, Decrescito V (1981) Effect of naloxone on posttraumatic ischemia in experimental spinal contusion. J Neurosurg 55: 209–219

Young W (1980) H_2 clearance measurement of blood flow: a review of technique and polarographic principles. Stroke 11: 552–564

Young W, Flamm ES (1982) Effect of high-dose corticosteroid therapy on blood flow, evoked potentials, and extracelluar calcium in experimental spinal injury. J Neurosurg 57: 667–673

Young W, Koreh I (1986) Potassium and calcium changes in injured spinal cords. Brain Res 365: 42–53

Young W (1985) Somatosensory-evoked potentials (SEP) in spinal cord injury. In: Schramm J , Jones SJ (eds) Spinal Cord Monitoring. Springer, Berlin Heidelberg New York Tokyo, pp 127–142
Zappulla RA, Wang W (1993) Neural origins of the motor evoked potential. Experimental approach. Adv Neurol 63: 51–60

Correspondence: Dr. C. H. Tator, Division of Neurosurgery, the Toronto Hospital, Western Division, 399 Bathurst street, Toronto, Ontario, M5T 2S8 Canada.

Section V

Spinal cord monitoring in clinic: pathology and disease processes

Spinal Cord Monitoring
Eds: E. Stålberg, H. S. Sharma, Y. Olsson

Chapter 15

White matter changes in human spinal cord injury

B. A. Kakulas, R. L. Lorimer, and **A. D. Gubbay**

Department of Neuropathology, Royal Perth Hospital and the Australian Neuromuscular Research Institute and the University of Western Australia, Perth, W. Australia

Introduction

The neurological disability in human spinal cord injury (SCI) is not determined by the local gray matter changes but by the interruption of the long white matter pathways. Typically in flexion, rotation or extension injuries of the cervical spine there is central haemorrhagic necrosis of the gray matter and a variable degree of disruption of the surrounding white matter depending on the severity of the trauma. Although the central gray matter is destroyed at the level of the lesion it does not usually cause a serious clinical problem due to the overlapping segmental innervation of myotomes and dermatomes. Within the white matter there may be involvement only of the propriospinal pathways or the lesion may extend more widely and affect the long tracts. However it is unusual for the spinal cord to be completely transected in road trauma, industrial or sporting accidents, the usual causes of SCI. More often there is some preservation of the white matter in human SCI.

The residual neurological state of the paraplegic or tetraplegic patient is determined by the number of white matter fibres which have escaped the injury. Descending corticospinal tracts for voluntary motor control are in the lateral and anterior columns and the afferent, sensory pathways lie in the posterior and lateral columns of the spinal cord. It is notable that a significant number of patients who are clinically complete, lacking all voluntary motor and sensory functions below the level of the lesion, retain some degree of continuity of nerve fibres at the level of the lesion. Such preserved ascending or descending pathways may also be recognized by neurophysiological methods, a syndrome for which the term "discomplete" has been applied (Dimitrijevic *et al.*, 1983; Sherwood *et al.*, 1992). Such discomplete patients form a special category of human SCI and pose a challenge to the restorative neurologist who seeks to enhance residual functions and thereby improvement of the patient's rehabilitative status.

In order to fully appreciate the white matter changes in human SCI a detailed study of axons and myelin was undertaken in cases selected from the Royal Perth Hospital (RPH) SCI data and tissue bank (Woods *et al.*, 1991). Our investigation employed a recently described sensitive and specific method for assessing axonal damage which is dependent upon β amyloid precursor protein (βAPP) immunocytochemistry (Sherrif *et al.*, 1994; Gentleman *et al.*, 1995). Myelin was investigated by morphological methods which rely on tinctorial differences between central, oligodendrocytic and peripheral, Schwann cell myelin. In addition quantitative information concerning the residual white matter in established SCI was undertaken for the purpose of clinicopathological correlation.

Overview of neuropathological reactions in human SCI

Our experience is derived from a database and tissue bank of 542 cases of spinal cord trauma, consisting of 338 "hyperacute" cases with the victim dead on arrival (DOA) at hospital, 120 "acute" survivors of less than 6 months and 84 "chronic" SCI survivors of 6 months or more. A feature of the "hyperacute" forensic cases is the large number of patients with severe vertebral injuries and who have an intact spinal cord. For instance of the 338 who were DOA, the spinal cord was intact to the naked eye at the level of the vertebral lesion in 131. In general these patients survived for less than one hour so that it is not possible to exclude the presence of submicroscopic damage which would have become evident if the patient had survived. Spinal cord transection is rare, of the 120 "acute" injuries with fractures, the spinal cord was transected in 14 and of 84 "chronic" survivors only 21 showed spinal cord transection. The spinal cord lesions were carefully studied at all stages and the following account of the neuropathology of SCI derived from this source.

The neuropathological events which follow SCI are stereotyped (Kakulas and Bedbrook, 1976; Kakulas and Taylor, 1992). At first the cord is swollen, bruised and soft. In cross sections central haemorrhagic necrosis extending over several segments is the rule. Microscopically there is hyperaemia with free red cells within the disrupted tissues. A variable number of polymorphs appear first, followed by macrophages which gradually remove the dead tissue. Axonal damage can be detected within hours by βAPP immunocytochemistry. Activated astrocytes appear in 5 or 6 days when glial scar formation begins. Typically a proportion of white matter at the periphery of the cord is preserved.

White matter lesions in human SCI may be divided into early, intermediate and late. Immediately following injury there is complete disruption of a proportion of axis cylinders and myelin sheaths and partial injury of others. Depending on the degree of injury some of the axonal changes are reversible while others will go on to complete necrosis.

Traumatic demyelination occurs at this early stage. As the lesion progresses debris is removed by macrophages and healing takes place. Remyelination of bare but viable axis cylinders occurs. Scar tissue in the form of astrogliosis, heteromorphic at the level of the lesion and isomorphic above and below in association with Wallerian degeneration. There may also be collagenous fibrosis to a variable degree particularly near the leptomeninges which are always thickened. The central area of haemorrhagic necrosis is usually replaced by a multilocular cyst resulting from macrophage removal of dead tissue and red cells. Occasionally a solid collagenous and astrocytic scar replaces the central gray matter. Thus the intermediate stage consists of reactive changes. The cellular and tissue changes continue until the "end stage" which consists of glial lined, multilocular cysts with a variable degree of preserved white matter at the periphery. Late changes consist of continuing Wallerian and possible transsynaptic degenerations involving the afferent and efferent pathways. Axis cylinders which have escaped the initial injury are remyelinated by Schwann cells and nerve root proliferations are present in the established lesions. Late complications such as post-traumatic syringomyelia, cord compression due to osteoarthrosis or ischaemic changes secondary to atherosclerosis of segmental arteries may aggravate the neurological state of the patient many years after the initial injury. For detailed information concerning the molecular events the reader is referred to the comprehensive review of the pathophysiology and pathology of acute spinal cord injury by Tator (1995).

The concept of discomplete

The concept of "discomplete" lesions was first introduced by Dimitrijevic *et al.* 1983 to explain electrophysiological transmission of impulses across the lesion in patients who were clinically complete. We have examined neuropathologically many patients who were clinically complete but in whom intact nerve fibres transversing the lesion have been found. These are "anatomically discomplete". Of the 120 acute SCI patients 21 were classified as "discomplete" having no neurological functions below the level of the lesions yet with anatomical continuity of the spinal cord at the level of injury. Eighteen were clinically and anatomically incomplete and 14 were both clinically and anatomically complete. The status of the remaining 67 of the 120 cases was unknown. Of the 84 chronic SCI patients 31 were "discomplete" 23 incomplete, 21 complete and 9 were unknown. The finding that a large proportion of patients have anatomical continuity of the spinal cord at the level of the lesions (the incomplete and "discomplete" groups) emphasises the relative sparing of white matter even in the most severe spinal injury caused by road trauma, sporting or industrial accidents but not in gunshot wounds where there may be total disruption of the spinal cord tissue.

Early axonal changes

Despite their importance, little information is available concerning the early axonal changes in human SCI. For this reason we undertook a detailed study of axonal reactions in selected cases from the database. We also sought to confirm the presence of traumatic demyelination similar to that observed in experimental conditions (Blight, 1985). Following the report that the β-amyloid precursor protein (βAPP) immunochemistry was of value for the detection of early axonal changes in head injury (Sherrif *et al.*, 1994) we applied this method to the spinal cord to determine whether it would also be a useful marker of axonal damage in SCI. βA4 is the main constituent of amyloid in the senile plaque of Alzheimer's disease (Maggio and Mantyh, 1996) and is derived from a larger transmembrane (amyloid) precursor protein (APP).

The materials studied were from patients who had survived spinal injuries from a few minutes to 72 hours. They were divided into two groups, those in whom the cord was obviously contused or disrupted and those in whom the cord appeared to be intact on macroscopic examination, the solid cord injury of Bunge *et al.* (1993)

Of the 14 cases in whom the cord was obviously disrupted 11 showed axonal damage with the βAPP stain. In this group of 11 patients survival was from 30 mins. to 72 hours. In the sections taken at the site of injury, βAPP stained many irregular and swollen axons in the white matter. Similar changes were also present in sections taken a segment above and below the lesion in 7 members of this group. In 5 patients who survived for more than 24 hours, 2 showed axonal swellings and 3 axonal spheroids or "retraction bulbs". Spheroids were not found in silver stained or βAPP sections of patients who survived less than 24 hours. However irregular "varicose" enlargement of axis cylinders was frequently observed. Three out of 14 of the patients in whom the spinal cord was obviously disrupted by the trauma were negative with the βAPP immunocytochemistry. However, in 2 of the cases survival was very short being less than 1 hour. The other case showed advanced autolysis.

Of the 17 SCI patients with an intact spinal cord on macroscopic and microscopic sections, 7 stained negatively for the βAPP. Three of these cases had short survival ie. 1 hour or less and hence possibly inadequate time had elapsed for the axonal response to occur. It is possible that in the other 4, although the vertebral column was traumatised, the spinal cord escaped injury. The 10 remaining cases with an intact spinal cord were all βAPP positive showing axonal swellings and irregular enlargements. The range of survival was from 30 minutes to 24 hours post injury. Swollen and irregular positively stained axons were grouped focally in the white matter in 5 cases and more widespread in the other 5. This group of patients were all involved in severe multitrauma and were in a deep coma in the Emergency or Intensive Care Units of the Royal Perth Hospital.

Staining characteristics with βAPP in our SCI study differed in its appearance from the pericentral zone of white matter damage evident with βAPP in prolonged terminal hypotension of critically ill patients (Ohgami *et al.*, 1993). In the traumatic group βAPP positivity was not confined to the pericentral zone and was clearly associated with regions of spinal cord damage. Eight control spinal cords without trauma were negative with the βAPP method. These patients died from a number of "medical" causes such as bleeding gastric ulcer, cerebral haemorrhage, septicaemia, drowning, carbon monoxide poisoning and asthma. Others had neurological conditions such as multiple sclerosis, old ischaemic damage to the spinal cord and old cervical disc herniation.

In common with other reports (Ahlgren *et al.*, 1996) our study demonstrated βAPP immunocytochemistry to be a very specific and a sensitive detector of early axonal injury in the spinal cord. However the earliest survivor in their group was 10 days and βAPP positivity was reported to persist for at least a month after injury. The accumulation of βAPP at the site of axonal injury in a quantity which can be detected immunocytochemically is due to disruption of fast axoplasmic transport mechanisms. The small quantity of βAPP present in a normal axon is insufficient for detection by immunocytochemistry so that any staining at

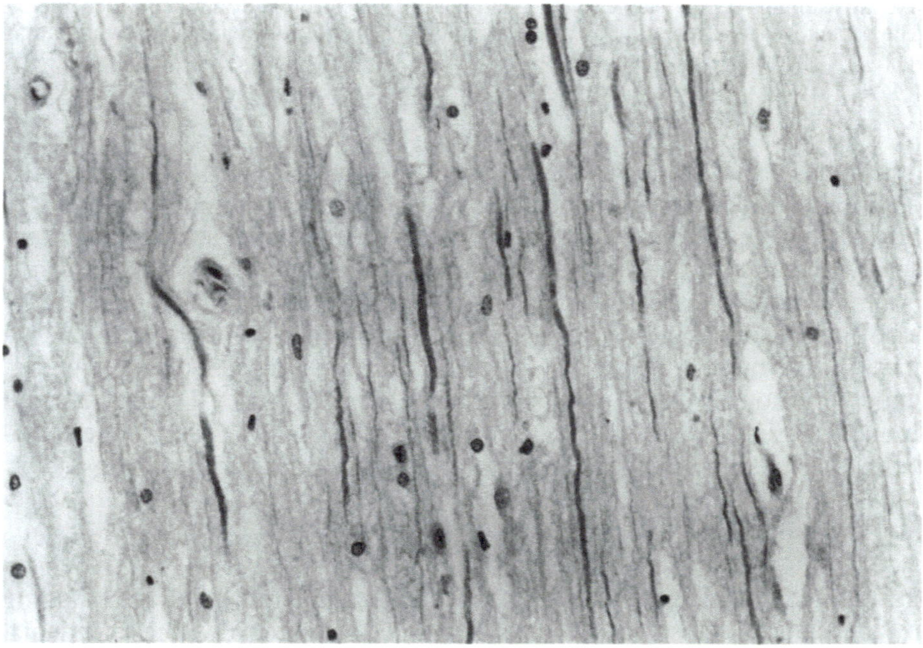

Fig. 1. X93/99 × 2000 Spinal cord white matter in longitudinal section below the level of injury at C5 showing many disrupted irregular axons stained with β-amyloid precursor protein immunocytochemistry. The patient was a 68 year old man who suffered a spinal injury at the C4 level and survived for 30 hours

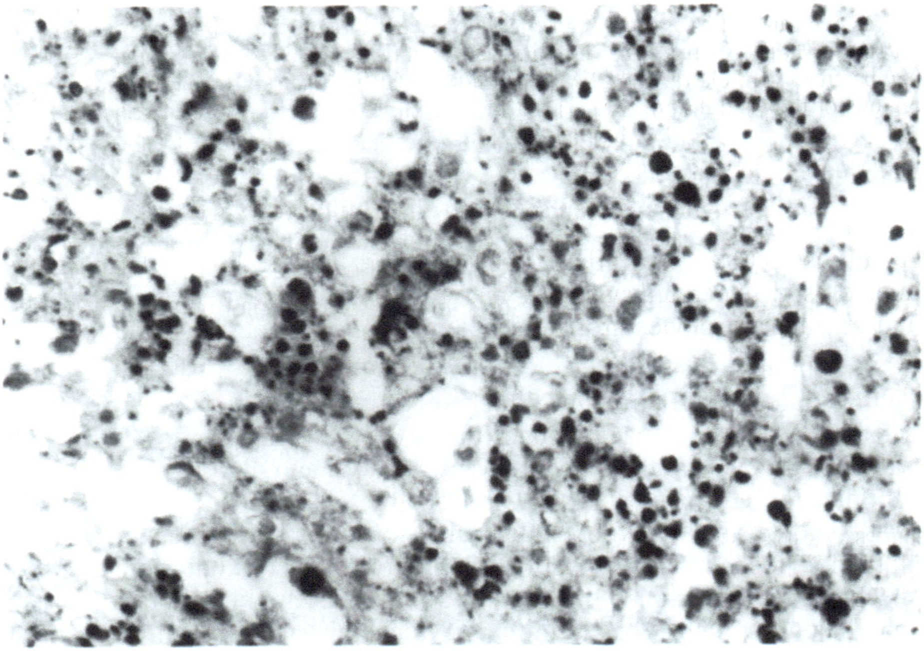

Fig. 2. 84/730 × 2000 Cross section at the level of spinal cord injury showing many axonal swellings deeply stained with β-amyloid precursor protein immunocytochemistry. The patient was a 12 year old boy who was injured when struck by a motor car while riding a bicycle. The spinal injury was at C2-3 and he lived for only 5 hours

all implies accumulation of βAPP due to slowing of axonal transport. Thus, a period of short survival is necessary following axonal injury for βAPP to be detected. In the brain βAPP has been shown to be an accurate and very sensitive means of detecting diffuse axonal injury in patients who survive 3 hours or more (Gentleman *et al.*, 1995). It was reported in this head injury study that axonal bulb function was not seen under 15 hours survival, but that a large number of patients showed axonal varicosities with βAPP in under 15 hours. Our findings suggest that the term "Diffuse Axonal Injury" (Strich, 1961; Adams *et al.*, 1991) may also be applied to the spinal cord. Axonal spheroids and irregular swellings are depicted in Figs. 1, 2.

Myelin changes

Early myelin changes consist of swollen vacuolated and faint staining myelin sheaths. This appearance is usually considered to be due to oedema. If the patient survives, macrophages appear in about 48 hours to approximately 72 hours post injury and ingest myelin debris converting it to neutral fat. Neutral fat is abundant and readily visible in oil-red-o stained frozen sections at the level of the lesion. Bare axis cylinders, representing traumatic demyelination is an uncommon finding in human SCI. Demyelination is strictly defined as loss of the myelin sheath but with preservation of the axis cylinder. Demyelination of central axons is typically found in multiple sclerosis (MS). In trauma, demyelination is best recognized under experimental conditions (Blight, 1985; Waxman, 1989). On the other hand disruption of the axis cylinder is inevitably followed by breakdown of the myelin sheath. This process is known as Wallerian degeneration and proceeds distally from the point of injury away from the neuronal cell body in a centrifugal direction. As the axis cylinder progressively dies presumably due to failure of axonal transport, the myelin sheath disintegrates. Wallerian degeneration is the common result of all types of injury of the white matter in the CNS and peripheral nerves.

In demyelination of whatever cause the axon may die as a result of a continuation of the noxious influence and/or if neuronal transport mechanisms fail. True demyelination is not common in human SCI because of concomitant damage to the axis cylinder as well as the myelin sheath as a result of the trauma. However the axon is capable of surviving without a myelin heath and remyelination may take place later. In MS remyelination is due to migration of the oligodendrocytes into the affected area. Oligodendrocytes proceed to remyelinate the axon by enwrapping their cell processes around the axon as in normal myelination during development. Remyelination by oligodendrocytes can be detected by shortened and irregularly spaced internodal lengths.

Remyelination of central spinal tract axons has been observed in our studies of human SCI, not of the central oligodendrocytic type, but peripheral myelin produced by Schwann cells. Schwann cells form a basement membrane (lamina) and oligodendrocytes do not. These and other differences between the 2 cell types are best detected by electron microscopy (EM). Unfortunately EM cannot be undertaken on post mortem tissues because of artefacts created by autolysis. Nevertheless a method developed many years ago for paraffin embedded tissues in MS research serves the purpose quite well (Feigen and Cravioto, 1961; Feign-Popoff, 1966). By combining the luxol fast blue (LFB) stain for myelin with the periodic acid Schiff (PAS) method for basement membranes central oligodendrocytic myelin stains green whereas peripheral Schwann cell myelin stains blue. Moreover in well prepared sections the basement membrane of peripheral myelin can be seen by light microscopy. Remyelination by Schwann cells is unequivocal evidence of previous demyelination in human SCI.

It has been shown that demyelination may be followed by remyelination in experimental SCI (Gledhill *et al.*, 1973). Electrophysiological studies have shown recovery of function in animal models following remyelination of central axons by Schwann cells and that newly formed nodes of Ranvier develop membrane properties similar to those in normal fibres. (Blight and Young, 1989).

Our observations of human central axonal remyelination in human SCI were taken from a study of 27 suitable cases selected from the RPH collection. These patients survived at least six months after sustaining their spinal injury. Tissues were taken from the area of disrupted cord and sections stained with LFB/PAS. Of the 27 cases, 16 showed central remyelination by peripheral Schwann myelin in the LFB/PAS stained sections. The age range of these 16 patients showing remyelination was from 12 to 52 years. Post injury survival ranged form 1 to 38 years. Thirteen were cervical injuries (C5–C7) and 3 thoracic (T4, T6, T12/L1).

All tissue sections were at the level of the lesion. A cystic cavity due to extensive loss of normal cord parenchyma with gliotic walls containing residual axons and their myelin sheaths among vascular bundles form the common lesion. Thus a cystic cavity was present in 14 out of the 16 cases showing remyelination and 9 of 11 of the non-remyelinated cases. The remyelinated axons are morphologically different to peripheral nerve root regenerations and readily distinguishable. Central remyelinated axons occur in clusters or as single remyelinated axons generally in the centre of the lesion, in close proximity to the cavity, but occasionally are found near dorsal roots and in the subpial region. In longitudinal sections it is possible to find many axis cylinders surrounded by peripheral type myelin several millimetres above or below the lesion as well as at the site of injury. Some cases show peripherally myelinated axons encircling a new blood vessel within the reactive changes. Remyelinated axons are often tortuous following the deformity of the cord parenchyma at the level of the lesion. All cases with remyelination show the presence of distinct nerve root regenerations and we presume that the Schwann cells are derived from the nerve roots.

Blight and Young (1989) have shown in animal experiments that central axons remyelinated by Schwann cells may conduct electrical impulses. It is implied therefore that in human SCI this type of remyelination may be associated with function. However what effect this may have on the clinical picture is uncertain. Our attempt to determine whether or not central remyelination was associated with any particular clinical feature such as reflex contractions (muscular spasms), pain, phantom sensations, autonomic disturbances or late recovery of function did not reveal any positive correlations.

It has long been known that Schwann cells will allow regeneration of central axons (Aguayo *et al.*, 1979), an observation which has been repeatedly confirmed and extensively investigated. The question therefore is posed as to whether central regeneration of Schwann cell myelinated axons has occurred in human SCI. The finding of many axons in continuity coated by oligodendrocytic as well as peripheral myelin suggests such a possibility. Oligodendrocytic myelin is at a distance from the lesion and the Schwann cell myelin is at the site of the injury in these axons. Unfortunately this intriguing possibility cannot be proven or disproven by current methods in human SCI and must remain conjectural.

Quantification of residual white matter

Analysis of the residual state of the spinal cord in human SCI after the reactive changes have subsided provides further information for useful clinicopathological correlation. At the level of injury there are glial lined multilocular cysts, traversed by gliovascular bundles, regenerated nerve roots, collagenous fibrosis associated with thickened leptomeninges and a variable amount of preserved white matter representing the residual pathways. This is the stereotyped appearance of the "end stage" lesion of which the constituent elements were quantified by Kaelan (1993) within his doctoral project. His results are summarized below.

In 11 patients who were clinically complete no residual white matter was present in 8 while 3 did show some preservation of white matter tracts. These 3 may be considered "anatomically discomplete". The amount of white matter which was present in the 3 discomplete patients was 1.09 mm^2, 1.12 mm^2 and 3.89 mm^2 (or 8.43%, 2.36% and 27.4% of the total area of the cord at the level of the lesion). In 10 clinically incomplete patients the amount of white matter at the level of the lesion ranged from 1.17 mm^2 to 13.89 mm^2 (2.1% to 40.4%). Most were associated with gliosis 8.83 mm^2 to 24.08 mm^2 (34.5% to 68.3%), 7 had cavities 1.51 mm^2 to 9.4 mm^2 (5.9% to 29.7%) and 4 showed nerve root regeneration 0.87 mm^2 to 10.51 mm^2 (2.1% to 37.2%). In the incomplete patients the amount of preserved white matter did not directly correlate with the degree of retention of volitional motor control or with the distribution and modality of sensation below the level of the lesion. For instance one of the complete patients had a greater amount of white matter (3.89 mm^2) than 4 of the clinically incomplete cases (1.17 mm^2, 3.17 mm^2, 1.94 mm^2, 1.73 mm^2). Therefore the location and quality of the preserved white matter may be more important than the quantity in determining the neurological status of the SCI patient.

Axonal counts were undertaken in order to determine the minimum number of residual axons which may be necessary to conduct voluntary motor and/or sensory signals across the lesion. The minimum number of descending and/or ascending fibres associated with retained conscious functions in SCI was estimated using a Cambridge Quantimet 520 Image Analysis System. Six clinically incomplete patients and 4 complete were examined. Three of the 6 incomplete cases were used for the investigation of motor and 4 for sensory functions. They were also compared with normal controls. All were cervical injuries. The descending fibres in the lateral pyramidal tracts below the level of the lesion were counted at T4 and ascending axonal fibres in the posterior columns were counted above the level of the lesion at C4. Normal controls had an average of 41472 nerve fibres at T4. The lowest number of fibres estimated in the lateral pyramidal tracts in a clinically incomplete patient with retained motor function below the level of the lesion was 3173 axonal fibres on the right side. This patient had dorsal and plantar flexion of the right foot and flickering movement of the toes. The average number of fibres in 4 complete SCI cases counted on both right and left sides was 2113. All of these were anatomically complete. The findings indicate that very few intact axons are required to subserve voluntary motor functions.

The lowest number of fibres counted in the posterior columns of a sensory clinically incomplete patient with a level of injury at C5 was 117 359 fibres. This patient had preservation of light touch and vibration sense in the right arm and pinprick and touch perianally. The normal controls had an average of 452 480 axonal fibres in the posterior columns. It should be emphasized that the quantification of preserved axons using these techniques is very approximate and is limited by the resolution of the image analysis system of 0.5μ. The axons were counted after immunochemical staining with the neurofilament 200 kD antibody method.

Conclusion

In this Chapter the principal changes which occur in the white matter at various stages in the natural history of human SCI have been described. The lack of a supporting stroma renders the white matter extremely vulnerable to injury as it consists primarily of millions of nerve fibres, the majority unmyelinated. The white matter of the spinal cord has no ability to repair itself but is subject to a stereotyped progression of well recognized pathological reactions common to the CNS. These reactions consist of necrosis due to direct mechanical injury or to a lesser extent poor perfusion resulting from compression by oedema and haemorrhage and disruption of blood vessels. As it consists of complex lipids CNS tissue is firm at body temperature. However in necrosis the compound lipids breakdown to smaller molecules which are liquid at room or body temperatures. This process is termed myelomalacia. If the patient survives the SCI these broken down lipids are further converted to neutral fat by macrophages after phagocytosis. Necrotic axis cylinders and myelin sheaths are removed by the phagocytes. Extravasated red cells are also removed by macrophages. Astrocytes migrate into the lesion and multiply in 5 to 10 days forming a gliotic scar surrounding the necrotic area. Haemorrhage may also stimulate collagenous fibrosis. After some weeks a multilocular cavity replaces the softened area. Most commonly in human SCI caused by motor vehicle or sporting accidents there is some continuity of the cord parenchyma across the lesion. Correspondingly a certain amount of white matter survives in the long term even in clinically complete patients. Wallerian degeneration above the lesion in the afferent pathways and below in the efferent systems becomes apparent 6 to 12 months after the injury and continues throughout life. Tract degeneration is associated with fine longitudinally orientated astrogliosis (isomorphic gliosis). Although transsynaptic degeneration is suggested neurophysiologically it has not been possible to demonstrate a loss of motor neurons in the gray matter below the level of the lesion by morphological methods (Kaelan *et al.*, 1988).

Apart from the sequence of reactions outlined above there are a few special features applicable to the white matter in human SCI. These are axonal swellings and retraction bulbs (spheroids), traumatic demyelination, remyelination by Schwann cells and nerve root regenerations. A recent advance in the investigation of early axonal changes has been the use of βAPP immunocytochemistry. By this technique axonal changes are revealed within hours and consist of irregular swellings made visible because of accumulation of βAPP resulting from failure of neuronal transport. Axonal spheroids were only found in survivors of 24 hours or more and maximal at 48 hours. It seems that there are reversible and irreversible axonal changes (Gentleman *et al.*, 1995). Spheroids or retraction bundles are clearly irreversible. However varicose swellings are compatible with survival although a number of these clearly continue to degenerate. It is not possible to distinguish which will go on to die. Surviving axons show traumatic demyelination. These surviving axons are probably those which show remyelination by Schwann cells in long term survivors. Whether Schwann cells also encourage central regeneration in human SCI is yet to be determined.

In considering the white matter changes in human SCI an attempt was made to quantify the lesions with counts of surviving nerve fibres for the purpose of clinicopathological correlation. Although this quantification was undertaken by the most accurate methods available the data recorded here should be regarded as approximate. There are many variables all of which cannot be controlled. The lack of neurological detail in the records is also a factor. Nevertheless there is some correspondence between our results and those of others in spinal cord studies of patients who have had cordotomies for pain (Windle, 1980; Nathan, 1994). The information provided of the white matter lesions described here including the quantification, the only data of this type currently available, is likely to be of value for the clinician, neurophysiologist, restorative neurologist, and especially the neurobiologist who is concerned with arresting the inexorable progression of neuropathological events in SCI and with the possibility of bringing about central regeneration.

Acknowledgments

The assistance of Dr. Vicki Fabian, Neuropathologist, Royal Perth Hospital and Mr. Brett Tizard, Research Technologist is gratefully acknowledged. Mr. Russell Johnsen, Laboratory Scientist, performed the immunocytochemistry. The work was supported by the Sir George Bedbrook Award provided by the International Spinal Research Trust of the UK.

References

Adams JH, Graham DI, Gernarealli TA, Maxwell WL (1991) Diffuse axonal injury in non-missle head injury. J Neurol Neurosurg Psychiatry 54: 481–583

Aguayo AJ, Bray GM, Perkins GS, Duncan ID (1979) Axon-sheath cell interactions in peripheral and central nervous system transplants. In: Ferrendelli JA (ed) Aspects of developmental neurobiology 4: 361–383

Ahlgren S, Li GL, Olsson Y (1996) Accumulation of β-amyloid precursor protein and ubiquitin in axons after spinal cord trauma in humans: immunohistochemical observations on autopsy material. Acta Neuropathol 92: 49–55

Blight AR (1985) Delayed demyelination and macrophage invasion: a candidate for secondary cell damage in spinal cord injury. Cent Nerv Syst Trauma 2: 299–315

Blight AR, Young W (1989) Central axons in injured cat spinal cord recover electrophysiological function following remyelination by Schwann cells. J Neurol Sci 91: 15–34

Bunge BP, Puckett WR, Becerra JL, Marcillo A, Quencer RM (1993) Observations on the pathology of human spinal cord injury: a review and classification of 22 new cases with details from a case of chronic cord compression with extensive focal demyelination. Adv Neurol 59: 75–89

Dimitrijevic MR, Faganel J, Lehmkuhl D, Sherwood A (1983) Motor control mechanisms in health and diseases. Raven Press, New York, pp 915–926

Feigin I, Cravioto H (1961) A histochemical study of myelin: A difference in the solubility of the glycolipid components in the central and peripheral nervous systems. J Neuropath Exp Neurol 20: 245–253

Feigin I, Popoff N (1966) Regeneration of myelin in multiple sclerosis: the role of mesenchymal cells in such regeneration and in myelin formation in the peripheral nervous system. Neurology 16(4): 364–372

Gentleman SM, Roberts GW, Gennarelli TA, Maxwell WL, Adams JH, Kerr S, Graham DI (1995) Axonal injury: a universal consequence of fatal closed head injury. Acta Neuropathol 89: 537–543

Gledhill RF, Harrison BM, McDonald WI (1973) Demyelination after acute spinal cord compression. Exp Neurol 38: 472–487

Kaelan C (1993) Neuropathology and clinical neurobiology of human spinal cord injury (SCI): Quantitation of tissue elements in human SCI and the comparison of experimental animal SCI and human SCI. Thesis presented for the degree of Doctor of Philosophy, Department of Pathology, University of Western Australia

Kaelan C, Jacobsen PF, Kakulas BA (1988) An investigation of possible transynaptic neuronal degeneration in human spinal cord injury. J Neurol Sci 86: 231–237

Kakulas BA, Bedbrook GM (1976) Chapter 3: Pathology of injuries of the vertebral column with emphasis on the macroscopical aspects. In: Vinken PJ, Bruyn GW (eds) Handbook of Clinical Neurology, Chapter 25. Injuries of the Spine and Spinal Cord, Part 1. North-Holland, Amsterdam/Elsevier, New York, pp 27–42

Kakulas BA, Taylor JR (1992) Chapter 2: Pathology of injuries of the vertebral column and spinal cord. In: Frankel HL (ed) Handbook of Clinical Neurology, Vol 17. Spinal Cord Trauma. Elsevier, Amsterdam, pp 21–51

Maggio JE, Mantyh PW (1996) Brain amyloid – a physiochemical perspective. Brain Pathol 6: 147–162

Nathan PW (1994) Effects on movement of surgical inclusions into the human spinal cord. Brain 117: 337–346

Ohgami T, Kitamoto T, Tateishi J (1993) Alzheimer's amyloid precursor protein accumulates within axonal swelling in human brain lesions. Neurosci Lett 160: 139–144

Sherrif FE, Bridges LR, Sivaloganathian S (1994) Early detection of axon injury after human head trauma using immunohistochemistry for beta-amyloid precursor protein. Acta Neuropathol 87: 55–62

Sherwood AM, Dimitrijevic MR, McKay WB (1992) Evidence of subclinical brain influence in clinically complete spinal cord injury: discomplete SCI. J Neurol Sci 110: 90–98

Strich SJ (1961) Shearing of nerve fibers as a cause of brain damage due to head injury. Lancet 2: 443–448

Tator CH (1995) Update on the pathophysiology and pathology of acute spinal cord injury. Brain Pathol 5: 407–413
Waxman SG (1989) Demyelination of spinal cord injury. J Neurol Sci 91: 1–14
Windle WF (1980) Neurotropism. Editorial commentary. Exp Neurol 67: 251–256
Woods A, Gaekwad UH, Kakulas BA, Smith ER (1991) Establishment of a clinicopathological database for traumatic human spinal injury. Paraplegia 29: 149–155

Correspondence: Dr. B. A. Kakulas, Royal Perth Hospital, Wellington St., Perth, W. Australia.

Spinal Cord Monitoring
Eds: E. Stålberg, H. S. Sharma, Y. Olsson

Chapter 16

Motor control in human spinal cord injury

M. R. Dimitrijevic

Baylor College of Medicine, Division of Restorative Neurology and Human Neurobiology, Houston, Texas, USA and the Ludwig Boltzmann Institute for Restorative Neurology, Austria, Vienna

Introduction

There are two major accomplishments in the practice of contemporary medicine of spinal cord injury (SCI): the first consists of significant new developments in clinical criteria for the assessment of spinal cord functions (Ditunno *et al.*, 1994), and the second of laboratory methods involving significant technological developments in neuroradiology (Weirich *et al.*, 1990; Bonduranti *et al.*, 1990) and human neurophysiology (Dimitrijevic and Halter, 1995).

Spinal cord monitoring of the long ascending and descending conducting pathways, as well as the occasional evaluation of electrical events while processing spinal cord mechanisms of the gray matter, have been widely used for the prevention of SCI during surgical procedures for the correction of spine malformation, such as scoliosis (Brown and Nash, 1991). Later, similar monitoring procedures were expanded during surgical interventions to the spinal cord for a variety of conditions of extra and intra-medullar compression, vascular malformation, and other surgically corrected spinal neurological disorders. However, it is a surprising paradox that the technique widely used to monitor spinal cord functions in tertiary care medical centers has not yet become a part of the protocol for the treatment and management of acute post-traumatic SCI for it can further contribute to the prevention of additional injuries to the already injured spinal cord during surgical stabilization procedures. It is of even greater value to assess the effect of systemic or locally applied neuroprotective drugs, and can provide potential guidance in developing strategies for the local pharmacological treatment of the injured portion of the spinal cord (Halter *et al.*, 1989).

In this chapter, we will briefly describe the clinical criteria for SCI syndromes and propose a neurophysiological classification of spinal cord injury. Supportive neuropathological evidence for the neurophysiological classification of human SCI will be mentioned together with an outline of motor control of the human spinal cord. We will finally discuss neurophysiological evidence for a human model of chronic SCI based on Brain Motor Control Assessment (BMCA) studies of residual volitional control of motor units in paretic and paralyzed muscles in addition to the already accepted model of the "discomplete" spinal cord injury, in which descending residual pathway activity only *influences* segmental motor control in chronic SCI patients.

In this chapter, we will define *motor control* as the voluntary, reflex, or automatic activity of the central nervous system integrated with internal and external feedback signals: i.e., the motor task which leads to the performance of a motor act. These motor acts are the ultimate expression of motor control processes whose features are characterized by their patterns of EMG activity.

SCI syndromes and neurophysiological classification

Medical terminology describes SCI according to the level of the lesion and whether or not the lesion is complete. When the lesion is incomplete, the neurological criteria cover a wide range of possible impairments of spinal cord functions. These are clinical syndromes of complete and incomplete, diffuse or patchy focal spinal cord lesion, and of anterior, posterior, hemisection, central cord, conus and cauda equina syndromes (Eidelberg, 1987). These categories are dependent upon the presence of clinical evidence for the degree of preservation of motor and sensory functions. This means the ability to elicit volitional muscle contraction or to perceive sensory stimuli below the level of the injury (Fig. 1).

Physicians specialized in the care of SCI patients have found it useful to standardize the criteria of location and extent of SCI by carefully examining sensory dermatome functions and volitional motor activity in their corresponding myotomes. According to the American Spinal Injury Association (ASIA) standards for the neurological assessment of spinal

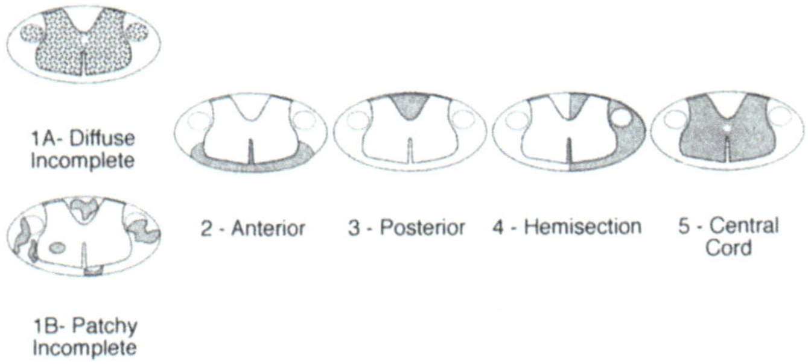

Fig. 1. A sketch of 5 clinical neurological syndromes based on clinically defined partially preserved brain motor control and sensory perception

injury patients, by examining dermatomes and myotomes as part of a properly conducted neurological examination, one can determine "the segments involved and, therefore, the location and extent of the spinal cord injury" (Ditunno *et al.*, 1994). The criteria for complete SCI are the absence of sensory and motor functions below the neurological level of the injury. However, if impaired sensory or motor functions are present more than the 3 segments below the level of the spinal cord lesion, then such an injury is incomplete. Such a standard neurological assessment in SCI can be useful to monitor recovery and acquire knowledge to predict the functional outcome (Ditunno *et al.*, 1997).

Another practical classification of the degree of incompleteness is known as the Frankel classification consisting of the A, B, C, D, and E categories (Frankel *et al.*, 1969). A is used for complete spinal cord lesions when all motor and sensory functions are absent below the "Zone of Partial Preservation". Category B describes spinal cord lesions which are incomplete with preserved sensation, excluding phantom sensation, and lack volitional motor functions. Category C covers those spinal cord lesions in which patients show some preserved, minimal and non-functional, voluntary motor activity (less than grade 3, fair movement according to the motor grading score). Category D is used to describe the spinal cord lesion with preserved, useful voluntary motor function, meaning that the majority of the key muscles tested score 3 or above. E is the category of patients showing complete return of all motor and sensory functions, but who may show some abnormal reflex response.

Standardized criteria for location and extent of spinal cord lesions are also useful in prospective studies defining early prognostic factors for the evolution of SCI. It has been shown on several occasions that if SCI patients are admitted to the hospital on the first day of injury, approximately one third will die within the first three months, another third will become independent of assistance for daily living within 18 months, leaving one third who will remain dependent on assistance for the rest of their lives (Daverat *et al.*, 1995; Dimitrijevic, 1996). This last group of SCI patients is looking forward to new developments that will upgrade the functional status of motor performance. It is crucial, therefore, that we should start to evaluate even "useless" motor and sensory activities below the level of the lesion. More in depth knowledge of the features of segmental motor activity when it is still under residual suprasegmental brain influence and peripheral input will allow us to learn which residual functions are available within the evolution of the spinal cord lesion for the application of restorative neurological procedures (Dimitrijevic, 1990) (Fig. 2).

In the neurologically intact subject, spinal motor neurons and interneuron pools are functionally organized by suprasegmental control according to the motor task. These interneuronal circuits, once thought to serve merely as relay stations, are now known to serve as integrators of activity converging from a variety of peripheral and central origins

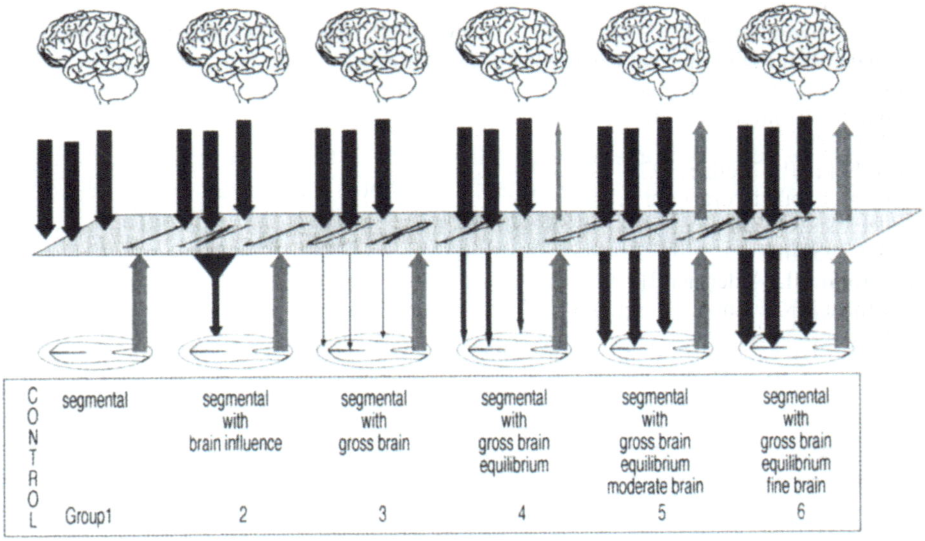

Fig. 2. Clinical neurophysiological classification of spinal cord injured individuals

(Hultborn *et al.*, 1976; Jankowska and Edgley, 1993). The ultimate role of these neurons is to translate descending commands and afferent volleys into a pattern of activation that will control the contraction of muscle fibers to accomplish a particular motor act (Davidoff, 1992). In the paretic individual such organization is altered but still present to an extent which can be characterized in terms of patterns of EMG activity.

On the basis of clinical and neurophysiological studies of 642 SCI patients with established spinal cord lesion (Dimitrijevic *et al.*, 1990), we were able to describe the following 6 distinct categories of spinal cord functions (Fig. 2).

In the *first category* there is complete absence of brain influence below the spinal cord injury zone. These individuals have phasic (tendon) stretch reflexes that are exaggerated but habituate rather rapidly if repeated with constant strength and rate. These exaggerated tendon jerks do not irradiate to other muscle groups. If clonus is present, it is unsustained. Cutaneomuscular, plantar withdrawal reflexes are present, brisk, short lasting, and do not irradiate to the other limbs; when repetitively elicited they habituate rapidly. Muscle tone is slightly increased, the manual passive stretch has a phasic feature and tonic reflexes are absent (Dimitrijevic and Lenman, 1980; Dimitrijevic and Faganel, 1985; Dimitrijevic *et al.*, 1983).

In the *second category*, there is complete paralysis below the level of the lesion, with subclinically present brain influence through a small population of brain originated axons that mediate excitatory and inhibitory influence to segmental reflex activity. Thus, proprio- and exteroceptive reflexes also become tonic, sustained and irradiate to other limbs. Spasticity increases markedly and acquires tonic features. This kind of clinically complete lesion with neurophysiological evidence of residual brain influence on segmental reflexes is known as "discomplete" spinal cord lesion (Dimitrijevic, 1988; Dimitrijevic, 1988; Dimitrijevic *et al.*, 1984; Cioni *et al.*, 1986).

The *third category* consists of individuals who can volitionally control flexion, extension, flexion-extension patterned movements dependently or independently from the contralateral limb. Segmental reflex activity is similar to that in category 2. Although they have volitional control of gross movement, they are unable to stand or walk. Muscle tone is severely increased and flexor and extensor spasms can be triggered by peripheral stimulation or by suprasegmental maneuvers and during volitionally induced movement (Dimitrijevic, 1987; Dimitrijevic, 1992).

In the *fourth category* are individuals who, even though they suffer from increased segmental activity and spasticity, can volitionally control patterned multi- and single-joint movements. They are able to stand and initiate a few stepping movements when they have the appropriate support (walker) (McKay *et al.*, 1993).

The *fifth category* consists of subjects who, in addition to all the previously described features of segmental and suprasegmental motor activity,

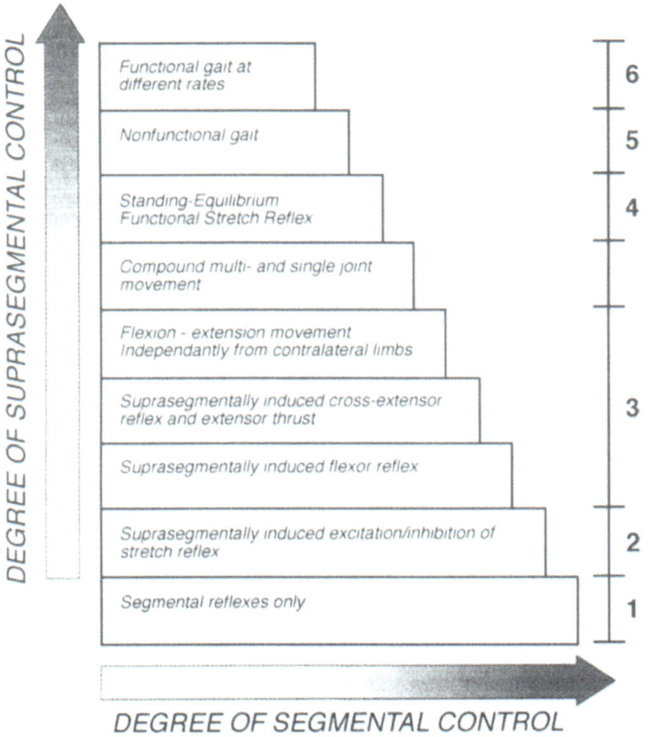

Fig. 3. Degrees of segmental and suprasegmental control and their relationship in 6 clinical neurophysiological categories

can stand and partially control their equilibrium. In addition, they are able to initiate non functional gait with assistance or assistive devices over limited distances (Dimitrijevic, 1988).

In the *sixth category* are individuals who have moderately increased muscle tone and segmental reflex activity. They are able to initiate gait which is functional, even if restricted in speed and endurance, with or without orthotic and assistive devices (McKay *et al.*, 1993).

The above described clinical neurophysiological classifications of spinal cord injury and their degrees of segmental and suprasegmental control are illustrated in Fig. 3.

Supportive neuropathological evidence for the neurophysiological classification of human spinal cord injury

Under the mentorship of Professor Kakulas, Dr. Kaelan carried out a quantitative study of tissue elements in human spinal cord injury in his thesis on Neuropathology and Clinical Neurobiology of Human Spinal Cord Injury (Kaelan, 1993; Kakulas and Taylor, 1992). The material for the study was obtained from the tissue bank of spinal cords within the Department of Neuropathology, Royal Perth Hospital. From this extensive tissue bank of 1106 human spinal cord specimens, 78 cases had been chronic complete and incomplete SCI patients who had survived from 12 months onward. Microscopic examination of specimens of the 78 chronic cases revealed that 18 of them had totally transected spinal cords. Thus, the totally transected spinal cord is not a common finding. Of these 78 cases, the author selected 21 suitable for the quantification of tissue elements according to the criteria of completeness of clinical data and availability of tissue blocks or wet specimens of the spinal cord at the level of the lesion, above and below. Of these 21 cases, 11 were clinically complete and 10 incomplete. The complete cases were further subdivided into complete and "discomplete" on the basis of anatomical findings or presence of nervous tissue at the level of the lesion.

The term "discomplete" was proposed on the basis of neurophysiological studies of the apparently isolated spinal cord in which clinical signs indicated complete motor paralysis, whereas the neurophysiological evaluation documented the presence of *supraspinal influence* below the level of the injury on spinal reflex activity, or even volitionally induced facilitation and suppression of motor unit activity (Dimitrijevic *et al.*, 1984; Cioni *et al.*, 1986; Sherwood *et al.*, 1992).

Quantitative measurements of the group of complete cases showed that the preserved tissue elements at the level of the lesion consisted mostly of connective tissue, fibrosis and gliosis with some cases showing nerve root regeneration. No white or gray matter was found in the complete cases entirely agreeing with the clinical condition. However, in the group of "discomplete" cases, in addition to fibrosis, connective tissue, gliosis and nerve root regeneration, there was also preservation of gray and white matter. Moreover, in the "discomplete" group the connective tissue was not as dense as it was in the complete group. In one case of "discomplete", a preserved white matter area was even greater than in the group of incomplete cases. Finally, in "discomplete" cases, the preserved white matter was mainly distributed in the anterior columns. Thus, this condition of "discomplete", which was proposed on neurophysiological grounds (Dimitrijevic, 1988; Dimitrijevic, 1995), showed microscopic preservation of white and grey matter (Kakulas and Taylor, 1992). This finding supported the hypothesis that the absence of volitional motor activity does not preclude the existence of CNS axons traversing the injury site and influencing the activities below the level of the lesion (Dimitrijevic, 1994).

In the group of incomplete cases, which were anatomically and clinically incomplete, the results of quantification of tissue elements showed similarities with the complete and "discomplete" group, such as gliosis and fibrosis. However, the most distinct finding in this category was the preservation of nerve tissue, and the clinical picture in incomplete SCI patients was determined by the neuropathology in each case (Kakulas and Taylor, 1992). Tissue elements in the group of incomplete cases are responsible for the newly established profile of the residual brain descending system and its integration with the spinal network. Moreover, in our studies on motor control in chronic SCI patients with multi-channel surface polyelectromyography, we showed that brain control reduced by spinal cord injury results in the establishment of new, distinct residual brain control (Dimitrijevic, 1994). This is also the case for the quality of gait performance which depends upon the extent of residual suprasegmental brain influence and brain control (McKay *et al.*, 1993; Tang *et al.*, 1994).

The systematic use of a comprehensive protocol for multi-channel surface EMG recording, which we developed as a method for brain motor control assessment (BMCA) (Sherwood *et al.*, 1996), was found to be very useful to describe the characteristics of motor control features in persons with SCI. Key information is contained in the overall temporal pattern of motor unit activity, observed in the EMG envelope. In paralysis, a rudimentary form of suprasegmental control of tonic and phasic reflexes can be demonstrated. EMG patterns evoked by voluntary and passive maneuvers and by volitional modulation of reflex responses reveal features of motor control not apparent in the clinical examination. Such subclinical findings may explain paradoxically different responses in apparently similar SCI subjects (Sherwood *et al.*, 1996), and may be used to monitor spontaneous or induced changes (Clifton *et al.*, 1996).

Model of human spinal cord injury with reduced and altered motor control

Systematic studies of motor units activity from the lower thoracic, lumbar and upper sacral segments of the spinal cord during reflex, automatic and volitional motor tasks in the chronic SCI subjects, led us to describe in this

Chapter a clinical neurophysiological classification for spinal cord injury. Furthermore, we also reviewed supportive anatomical evidence for the proposed clinical neurophysiological classification and we especially addressed newly reported clinical-pathomorphological evidence for the "discomplete" model of the human spinal cord.

In this study of the model of the human spinal cord with "reduced and altered motor control" we applied the identical approach used in the "discomplete" model. This means that we carried out a study of the qualitative features of motor control recorded by BMCA in subjects with residual volitional control of motor units in paretic and paralyzed muscles below the spinal cord injury (Dimitrijevic *et al.*, 1997). This "reduced functional anatomy of spinal cord injury" is just the opposite of the "discomplete" SCI model in which residual motor control is absent and is replaced by the brain's facilitatory and suppressive influence on segmental motor output.

Results were obtained from 91 chronic SCI patients and 15 neurologically healthy individuals. Of these 91 SCI, 23 were women and 68 were men, with ages ranging from 11 to 77 years, with a mean of 33.6 $\pm$ 1.7 (mean $\pm$ SE). All suffered from post-traumatic closed SCI between C2 and T10. The time between injury and the recording of motor control ranged from 3 months to 27 years with a mean duration of 78.5 $\pm$ 10.9 months. Twenty-one of these 91 patients showed no signs of volitional movement or volitionally induced motor unit activity below the level of injury, 21 could move voluntarily but could not ambulate, and 48 were ambulatory with a variety of assistive devices.

We analyzed the EMG patterns during gait and 2 discrete volitional motor tasks in the supine position: the first consisted of volitional efforts to perform full flexion and extension of the lower limbs and the second consisted of dorsal and plantar flexion of single ankle movements.

The main characteristic of integrated EMG activity during gait is the repetitive pattern of burst discharges followed by silent periods. Another general feature of EMG activity is that EMG bursts are of lower amplitude and shorter duration in proximal and trunk muscles than in the leg muscles. The third general feature is the reciprocal relationship of bursts of EMG activity between antagonistic leg and thigh muscle groups during the stance and swing phases of gait. Bearing in mind these general EMG features for the adult patient with an intact nervous system, it is easy to recognize alterations in motor output (Dimitrijevic *et al.*, 1997).

Unlike the semi-automatic and stereotyped motor control during steady gait, if we record EMG activity in a SCI patient during a volitional motor task of single- or multi-joint movements in the supine position, it is possible to expand observations on motor control in paralyzed patients (Dimitrijevic *et al.*, 1997).

We have found in EMG studies of motor control during volitional motor tasks that there is a consistency in the different features of motor unit activity. This was the case in 11 wheelchair-bound and 23 ambulatory patients. In this group of 11 and 23 SCI patients, we analyzed the percentage of the presence of volitional multi-joint flexion, isolation of single- from multi-joint movement, and isolation of agonist from antagonist EMG activity. All 3 features indicate discrete motor control and are observed more frequently in ambulatory patients than in the wheelchair-bound population. On the other hand, diffuse patterns of motor output and delayed onset are much more frequent in the wheelchair-bound than in the ambulatory group.

The fact that we can recognize different features of motor control during single-joint and multi-joint volitional motor tasks and that these features are consistent, repetitive, and related to the severity of the lesion led us to hypothesize that a new organization of spinal circuits is established and that this new structural and functional relationship can generate definite patterns of residual motor functions.

In incomplete SCI, the organization of brain motor control depends on the profile of the residual brain descending system and its integration with the spinal network. In the past, our understanding was that motor activity after incomplete SCI was inconsistent and considered abnormal because it differed from normal motor activity. However, we have shown that locomotor patterns in SCI people are consistent. Thus, the functional properties of these newly established sensory motor integration mechanisms are expressed in the repeatable neurophysiological characteristics of motor unit activity during single- and multi-joint volitional tasks. A thorough understanding of this new motor control will play an important role in establishing the scientific basis for neurobiological interventions to restore impaired SCI functions.

Human SCI models based on studies with surface EMG polyelectromyography (BMCA) play an important role in assessing and understanding the underlying mechanisms of newly developing protocols for the modifications of the already established clinical output in chronic SCI subjects. An example of this is found in locomotor training with or without FES for the improvement of walking speed and capacity (Ladouceur *et al.*, 1997).

Furthermore, the new spinal cord organization established as a result of spinal cord injury can develop different potentials for neuroplasticity and, as a result of this activity-dependent capacity, the CNS can respond by increasing motor performance (Muir and Steeves, 1997). Moreover, another factor which will influence the capacity of neuroplasticity will be the profile of residual supraspinal control. Such a residual profile can have components of postural, locomotor and volitional activity, or only one of these components can prevail. It has been shown, for instance, that interactive locomotor training using body weight support can improve walking capabilities in patients and not change the capacity for volitional muscle contraction (Wernig *et al.*, 1995).

Thus, the assessment of motor control in SCI subjects can reveal subclinical brain influence below the level of the lesion in otherwise clinically complete SCI, or it can describe and measure the relation between different components of residual suprasegmental control in incomplete SCI patients. This new possibility will help in defining clinical and laboratory inclusion criteria of SCI patients for a variety of available restorative procedures and neurorehabilitation protocols (Dimitrijevic, 1990).

Summary and conclusions

In this Chapter, we have discussed that the surface EMG polyelectromyographic recording, carried out according to the protocol for the assessment of segmental and suprasegmental residual functions below the spinal cord lesion (BMCA), allows us to distinguish six distinct categories of spinal cord functions (Fig. 2, Fig. 3). The novelty of this proposed clinical neurophysiological classification is that motor control is defined by the subclinical features of motor unit activity and their characteristic temporal and spatial plurisegmental output as revealed by the presence or absence of clinical neurological findings for volitional movement. Furthermore, we have discussed clinical-neuropathological evidence for the proposed clinical neurophysiological classification, and reviewed evidence for the structural basis of the newly described syndrome of "discomplete" spinal cord lesion. In the condition of "discomplete" syndrome, residual conduction of excitation and inhibition can be mediated through the uninjured axons, and the brain's influence below the level of the lesion can be demonstrated by modifying segmental reflex behavior, or activating motor unit activity during reinforcement maneuvers, but not during volitional motor tasks.

In conclusion, there are three basic human SCI models: the first is the model with no continuity of spinal cord tissue; the second is the "discomplete" SCI model, and in the third model partially present residual descending functions are expressed during postural, locomotor and volitional motor tasks.

These SCI models are defined by means of the BMCA protocol and can be carried out as a part of routine practice. This information is essential for the selection of patients for existing restorative procedures if they are to become a part of the protocol for the enhancement of residual spinal cord functions.

References

Ditunno JF, Young W, Donovan WH, Creasey G (1994) The international standards booklet for neurological and functional classification of spinal cord injury. Paraplegia 32: 70–80

Weirich SD, Cotler HB, Narayana PA, Hazle JD, Jackson FE, Coupe KJ, McDonald CL, Langford LA, Harris JH (1990) Histopathological correlation of magnetic resonance imaging signal patterns in a spinal cord injury model. Spine 15: 630–638

Bonduranti FJ, Cotler HB, McArdle CB, Harris JH (1990) Acute spinal cord injury: A study using physical examination and magnetic resonance imaging. Spine 15: 161–168

Dimitrijevic MR, Halter JA (1995) Atlas of Human Spinal Cord Evoked Potentials. Butterworth-Heinemann, Boston

Brown RH, Nash CL Jr (1991) Intraoperative spinal cord monitoring. In: Frymoyer JW (ed) The Adult Spine: Principles and Practice. Raven Press, New York, pp 557–558
Halter JA, Haftek I, Sarzynska M, Dimitrijevic MR (1989) Spinal evoked injury potentials in patients with acute spinal cord injury. J Neurotrauma 6: 231–245
Eidelberg E (1987) Spinal Cord Syndromes. In: Davidoff RA (ed) Handbook of the Spinal Cord. Marcel Dekker, New York, pp 1–17
Ditunno JF Jr, Graziani V, Tessler A (1997) Neurological assessment in spinal cord injury. In: Seil FJ (ed) Advances in neurology, Vol. 72: Neuronal Regeneration, Reorganization, and Repair. Lippincott-Raven, Philadelphia, pp 325–333
Frankel HL, Hancock DO, Hyslop G *et al.* (1969) The value of postural reduction in the initial management of closed injuries of the spine with paraplegia and tetraplegia, Part I. Paraplegia 7: 179–192
Daverat P, Petit H, Kemoun G, Dartigues JF, Barat (1995) The long term outcome in 149 patients with spinal cord injury. Paraplegia 33: 665–668
Dimitrijevic MR (1996) Motor recovery in humans after spinal cord injury. Neurol Neurochir Pol 30 [Suppl] 1: 37–47
Dimitrijevic MR (1990) Neurobiology of recovery and practice of restorative neurology. In: Chopra JS, Jagannathan K, Sawhney IMS (eds) Advances in Neurology. Excerpta Medica International Congress Series 883. Elsevier, Amsterdam, pp 375–384
Hultborn H, Illert M, Santini M (1976) Convergence on interneurones mediating the reciprocal inhibition of motoneurons. 3. Effects from supraspinal pathways. Acta Physiol Scand 96: 368–391
Jankowska E, Edgley S (1993) Interaction between pathways controlling posture and gait at the level of spinal interneurons in the cat. Prog Brain Res 97: 161–171
Davidoff RA (1992) Skeletal muscle tone and the misunderstood stretch reflex. Neurol 42: 951–963
Dimitrijevic MR, Lissens MA, McKay WB (1990) Characteristics and extent of motor activity recovery after spinal cord injury. In: Seil FJ (ed) Advances in Neurology, Vol 72. Neuronal Regeneration, Reorganization, and Repair. Lippincott-Raven, Philadelphia
Dimitrijevic MR, Lenman JAR (1980) Neural control of gait in patients with upper motor neurone lesions. In: Feldman R, Young R, Koella WP (eds) Spasticity: Disordered Motor Control. Symposia Specialists, Miami
Dimitrijevic MR, Faganel J (1985) Motor control of the spinal cord. In: Eccles J, Dimitrijevic MR (eds) Recent achievements in restorative neurology, vol I: Upper motor neuron functions and dysfunctions. Karger, Basel, pp 150–162
Dimitrijevic MR, Faganel J, Lehmkuhl LD, Sherwood AM (1983) Motor control in man after partial or complete spinal cord injury. In: Desmedt JE (ed) Advances in Neurology, Vol 39. Motor Control Mechanisms in Health and Disease. Raven Press, New York, pp 915–926
Dimitrijevic MR (1988) Residual motor functions in spinal cord injury. In: Waxman SG (ed) Advances in Neurology, vol. 47: Functional Recovery in Neurological Disease. Raven Press, New York, pp 139–155
Dimitrijevic MR (1988) Model for the study of plasticity of the human nervous system: features of residual spinal cord motor activity resulting from established post-traumatic injury. In: Evered D, Whelan J (eds) Plasticity of the Neuromuscular System (CIBA Foundation Symposium 138). Wiley, New York, pp 227–239
Dimitrijevic MR, Dimitrijevic M, Faganel J, Sherwood AM (1984) Suprasegmentally induced motor unit activity in paralyzed muscles of patients with established spinal cord injury. Ann Neurol 16: 216–221
Cioni B, Dimitrijevic MR, McKay WB, Sherwood AM (1986) Voluntary supraspinal suppression of spinal reflex activity in paralyzed muscles of spinal cord injury patients. Exp Neurol 93: 574–583
Dimitrijevic MR (1987) Neurophysiology in spinal cord injury. Paraplegia 25: 205–208
Dimitrijevic MR (1992) Development of neurophysiological aspects of the spinal cord during the past ten years. Paraplegia 30: 92–95

McKay WB, Verhagen Metman L, Dimitrijevic MM, Sherwood AM, Dimitrijevic MR (1993) Locomotor patterns in humans with impaired spinal cord functions. In: Kraft GH, Shahani B (eds) Motor Control Disorders, Physical Medicine and Rehabilitation Clinics of North America, Vol. 4, No. 4. Saunders, Philadelphia, pp 707–730

Kaelane (1993) Neuropathology and clinical neurobiology of human spinal cord injury. Thesis presented for the Degree of Doctor of Philosophy of the University of Western Australia, Department of Pathology

Kakulas BA, Taylor JR (1992) Pathology of injuries of the vertebral column and spinal cord. In: Frankel HL (ed) Handbook of Clinical Neurology, Vol. 17. Spinal Cord Trauma. Elsevier, Amsterdam, pp 27–42

Sherwood AM, Dimitrijevic MR, McKay WB (1992) Evidence of subclinical brain influence in clinically complete spinal cord injury: discomplete SCI. J Neurol Sci 110: 90–98

Dimitrijevic MR (1995) Clinical aspects of traumatic injury to central nervous system axons. In: Waxman S, Kocsis J, Stys P (eds) The Axon. Oxford University Press, New York, pp 669–679

Dimitrijevic MR (1994) Motor control in chronic spinal cord injury patients. Scan J Rehab Med [Suppl] 30: 53–62

Tang SFT, Tuel SM,·McKay WB, Dimitrijevic MR (1994) Correlation of motor control in the supine position and assistive device used for ambulation in chronic incomplete SCI persons. Am J Phys Med Rehab 73: 268–274

Sherwood AM, McKay WB, Dimitrijevic MR (1996) Motor control after spinal cord injury: assessment using surface EMG. Muscle Nerve 19: 966–979

Clifton GL, Donovan WH, Dimitrijevic MR *et al.* (1996) Omental transposition in chronic spinal cord injury. Spinal Cord 34: 193–203

Dimitrijevic MR, McKay WB, Sherwood AM *et al.* (1997) Motor control physiology below spinal cord injury: residual volitional control of motor units in paretic and paralyzed muscles. In: Seil FJ (ed) Advances in Neurology, Vol. 72. Neuronal Regeneration, Reorganization, and Repair. Lippincott-Raven, Philadelphia

Ladouceur M, Pépin A, Norman KE, Barbeau H (1997) Recovery of walking after spinal cord injury. In: Seil FJ (ed) Advances in Neurology, Vol. 72. Neuronal Regeneration, Reorganization, and Repair. Lippincott-Raven, Philadelphia, pp 249–255

Muir DG, Steeves JD (1997) Sensorimotor stimulation to improve locomotor recovery after spinal cord injury. TINS 20: 72–77

Wernig A, Müller S, Nanassy A, Cagol E (1995) Lufband therapy based on "rules of spinal locomotion" is effective in spinal cord injury persons. Eur J Neurosc 7: 823–829

Dimitrijevic MR (1990) Neurological rehabilitation and restorative neurology in spastic syndromes. In: Benecke R, Emre M, Davidoff RA (eds) The Origin and Treatment of Spasticity. Parthenon Press, Casterton, pp 113–124

Correspondance: Dr. M. R. Dimitrijevic, Baylor College of Medicine, Division of Restorative Neurology and Human Neurobiology, One Baylor Plaza (Suite 800), Houston TX 77030, USA.

Spinal Cord Monitoring
Eds: E. Stålberg, H. S. Sharma, Y. Olsson

Chapter 17

Intraoperative neurophysiology of the corticospinal tract

V. Deletis and **K. Kothbauer**

Institute for Neurology and Neurosurgery, Beth Israel Medical Center North Division, New York, USA

Introduction

Interventions that directly or indirectly put the spinal cord at risk are considered as particularly challenging for the neurosurgeon, the interventional neuroradiologist, the orthopedic surgeon, the vascular surgeon and the clinical neurophysiologist. In this context the objective of intraoperative neurophysiology is threefold: first, to prevent injury to the spinal cord, second, to document time and nature of injury any, and third, to "guide" the surgeon during the procedure according to continuously updated neurophysiologic data.

So far, a large body of knowledge has been gathered on the anatomy and neurophysiology of the corticospinal tract (Amassian *et al.*, 1987; Coulter *et al.*, 1977; Kuypers, 1964; Patton *et al.*, 1954; Penfield *et al.*, 1937; Philips *et al.*, 1964). However, most of the work has been done in vertebrates and nonhuman primates. Aside for Penfield's groundbreaking work of mapping the cerebral cortex which was performed in the surgical setting simple and pragmatic methods to study motor tract physiology in man were not available for a long time. Since 1980 Merton and Morton described transcranial electrical stimulation of the human motor cortex (Merton *et al.*, 1980) and 1985 when transcranial magnetic stimulation was introduced (Barker *et al.*, 1985), the study of the motor system in health and disease has undergone a quantum leap.

This chapter will discuss some of the theoretical aspects of transcranial stimulation of the motor cortex and our nine years of experience with intraoperative monitoring the functional integrity of the spinal cord.

Theoretical and methodological aspects of motor evoked potentials

Motor evoked potentials after transcranial electrical stimulation

Single stimulus technique

A strong electrical stimulus to the scalp over the motor cortex can break the impedance barrier of the skull (Merton and Morton, 1980) and allow a portion of the current (approximately 10 to 20%) to reach the motor cortex and activate motor neurons of the corticospinal tract. Synchronized volleys of so activated fast corticospinal axons can be recorded along the spinal cord as traveling waves. Early work of Patton and Amassian (Patton and Amassian, 1954) showed evidence that these traveling waves belong to the same generator which is activated in two different ways yielding two distinct types of traveling waves. The first wave is elicited by threshold current intensity applied through a focal anode, a setting which appears to stimulate axons directly. The first of these volleys was thus called "D-wave". The subsequent waves after increasing the current are generated by indirect excitation of axons via synaptic connections ending on the corticospinal tract motor neurons within the motor cortex. These subsequent volleys were then called "I-waves". Depressive influence on synaptic activity such as anesthesia as well as cooling or ischemia of the motor cortex significantly affects I-waves (Patton and Amassian, 1954) .

We regard only the D-waves as a reliable parameter to monitor the functional integrity of the corticospinal tract because: (1) they are relatively robust under anesthesia, (2) they represent the population of fast conducting fibers of the corticospinal tract and (3) they have turned out to correlate with postoperative motor function.

The methods of intraoperative transcranial electrical stimulation in man were described by a number of groups (Boyd *et al.*, 1986; Burke *et al.*, 1992; Katayama *et al.*, 1988). This is basically the original method of Merton and Morton (Merton and Morton, 1980), but modified for intraoperative use. Initially Cz was used for the stimulating anode and a point 6 cm in front as the cathode (Deletis, 1993). After we found that stimulating at C3/C4 or C1/C2 yields higher amplitudes of D-waves with a lower stimulation intensity and smaller stimulus artifacts, what is obviously an advantage for intraoperative monitoring, we routinely use this latter electrode montage (Fig. 1).

Stimuli are square wave impulses of 500 µs duration with a stimulus intensity of 15–160 mA. For intraoperative monitoring purposes the stimulus intensity is increased until it does not further increase the D-wave amplitude. D-wave and *epidural MEP* will be used as synonyms in this chapter.

Monitoring MEP from the spinal epidural space. This method requires the single stimulus method for stimulation. For recording of D-waves, we use a catheter-electrode (Type JX-330, Arrow Int., Reading, PA) with three

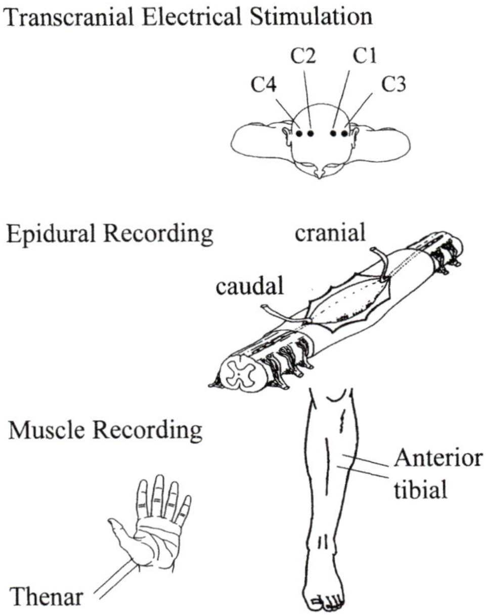

Fig. 1. Anodal transcranial electrical stimulation at C3/C4 or C1/C2 with recording either over the spinal cord via an electrode in the epi- or subdural space or from representative limb muscles

platinum-iridium recording cylinders, each 3 mm long and 1 mm in diameter, separated by 18 mm.

After laminectomy, the electrodes are inserted into the epidural or subdural space at the upper and lower edges of the opening (Fig. 1). Recordings from the catheter electrode rostral to the tumor serve as a monitoring control of the incoming signal (Fig. 2). Caudal recording then allows monitoring of the functional integrity of the corticospinal tract through the level of the lesion.

For spine surgery without laminectomy the catheter-electrode can be inserted through a flavectomy. If the spine is not exposed at all, as for supratentorial or brainstem surgery, an epidural catheter-electrode is inserted percutaneously through a 14 gauge Tuohy needle. (T 466LNRH, Becton Dickinson, Franklin Lakes, NJ). The angular tip of the Tuohy needle allows safe advancement of the catheter. We have reported 47 operations with use of this percutaneous placement of the recording electrode (Canter *et al*., 1995). In one case the electrode formed a loop when being advanced in the epidural space, in two cases the subarachnoid space was entered without further negative effects. There were no complications such as bleeding or infection.

In our institution *epidural MEP* monitoring has attained a standard-of-treatment status. In every procedure involving the spinal cord, monitoring

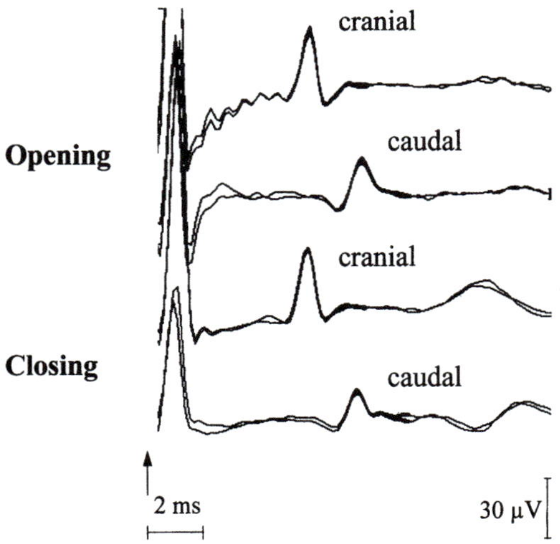

Fig. 2. Epidural MEP cranially of and caudally to a midthoracic intramedullary spinal cord tumor at the beginning (upper two traces) and the end (lower two traces) of complete removal of the lesion

of *epidural MEPs* is performed. It has been employed in over 300 procedures without any reported complications.

The same electrode placed cranially to the lesion can be used to record spinal somatosensory evoked potentials (SEPs) after tibial nerve stimulation.

There are, however, also limitations to the application of *epidural MEP* monitoring. The reasons for unmonitorability of epidural MEPs are due to anatomy and pathology: If an operated lesion involves the lumbar spinal cord, there is no corticospinal tract to record from caudally. Therefore in surgery of the conus medullaris epidural MEP monitoring is not feasible. As a rule of thumb we expect lesions below the bony level of Th 10 not to be amenable to epidural recording. Aside for this anatomical reason there is a pathological reason for unmonitorability concerning mainly intramedullary spinal cord tumors. About one third of these procedures are unmonitorable with this technique. The tumor itself may alter the conducting properties in a way that no D-wave can be recorded. Evidence points to a kind of desynchronization of the D-wave (Fig. 3), i.e., the fast conducting fibers of the pyramidal tract are dispersed in their conduction velocities by the biological effect of the lesion. The final result of extreme desynchronization is unmonitorability of the *epidural MEP* (Fig. 4a). Not surprisingly this is more likely to occur in large lesions that involve more segments and have significant surrounding edema as well as in patients after radiation therapy involving the spinal cord. It should be emphasized that in the case of absent epidural MEPs caudally to the tumor, muscle MEP can be recorded with the use of the multiple stimuli technique (see below Fig. 4), if the patients do not have a significant motor deficit.

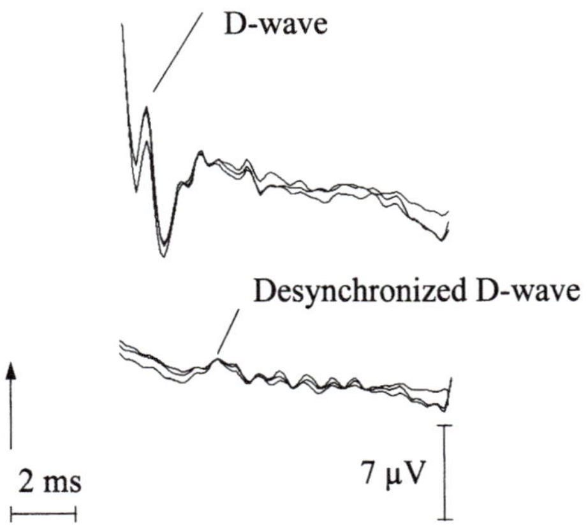

Fig. 3. MEP desynchronization shown as the D-wave caudally (lower trace) to an intramedullary spinal cord tumor as opposed to the normal D-wave cranially (upper trace)

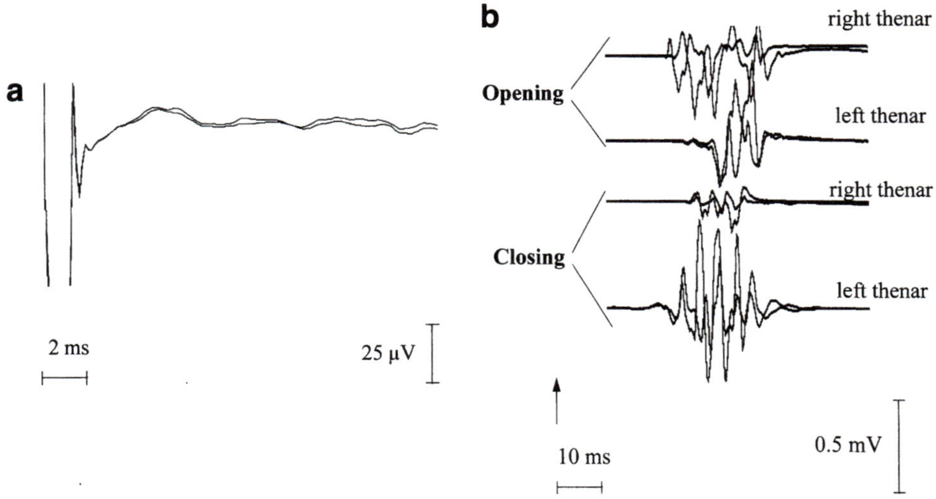

Fig. 4. Absent epidural MEPs caudal to the lesion as extreme desynchronization (**a**) in a patient with a high cervical intramedullary spinal cord tumor. Note, that the muscle MEPs elicited with a short train of six stimuli in the upper extremities were present (**b**)

In a few patients who had a previous operation extensive dural scarring prevented placement of the epidural electrode.

The D-wave amplitude serves as a reliable indicator for the functional integrity of the corticospinal tract. So far a 50% – decrease of the *epidural MEP* amplitude is considered significant and is used as a point to stop surgical dissection. This is however an arbitrary measure, taken from the experience in SEP monitoring. The clinical evidence suggests that a patient with a 50% deterioration of epidural MEP amplitude is at risk for immediate postoperative deficit but is likely to recover. This is consistent with Nathan's report that a patient with a hemichordotomy including the lateral corticospinal tract fully recovered motor function (Nathan *et al.*, 1973). It is also consistent with experimental studies in cats showing that as little as 10% of the corticospinal tract fibers appear to be sufficient to support locomotion (Blight, 1983a,b).

The single stimulation technique does not produce muscle twitches in limb muscles. Some twitching of cranial muscles due to direct stimulation of cranial nerves may occur.

Multiple stimuli technique

Despite early enthusiasm for the new technique of transcranial stimulation of the motor cortex in the late eighties, the single stimulus method described in the previous section turned out to be inadequate to elicit MEPs from limb muscles under general anesthesia. Early results have been found difficult to reproduce. However, the preliminary reports, applying single stimuli (Levy *et al.*, 1984; Taniguchi *et al.*, 1991; Zentner, 1989) have shown the potential of the technique for intraoperative monitoring of the motor system and triggered efforts for technical refinement. More recent reports confirm that motor cortex stimulation with a short train of stimuli is necessary to reliably record muscle MEPs in a routine surgical environment (Jones *et al.*, 1996; Pechstein *et al.*, 1996; Rodi *et al.*, 1996; Taniguchi *et al.*, 1993a).

The development of transcranial electrical train stimulation has proved to be a step forward in establishing reliable and reproducible motor tract monitoring. In order to optimize the stimulation technique and to minimize the current delivered to the brain it is essential to determine the optimal stimulation parameters. Understanding the neurophysiologic mechanism of the multiple stimuli technique is a crucial factor in obtaining optimal *muscle MEP*. One critical parameter is the interstimulus interval (ISI) in the stimulus train.

Recovery time of the D-wave. It has been shown in the previous section that the D-wave is a kind of neurogram of the corticospinal tract. No synapses are involved in its generation and propagation. Therefore after a double stimulus applied transcranially the recovery time of the D-wave generated by the second stimulus (D2-wave) can be calculated.

At the left side of Fig. 5 traces with D1- and D2-waves elicited by ISIs from 1.6 to 8.2 ms are presented. The recordings were performed during anesthesia without the use of muscle relaxants so that the epidural electrode also recorded the muscle response in the paraspinal muscles (M-wave). The maximal amplitude of the M-wave corresponded to the complete recovery of the D2-wave as shown in the diagram on the right side of Fig. 5. It is evident that the ISI between 2.9 and 4 ms is the most successful in eliciting the M-wave. ISIs shorter than 2.9 ms and longer than 4 ms were less successful. The unsuccessful result of the shorter ISI was most likely due to

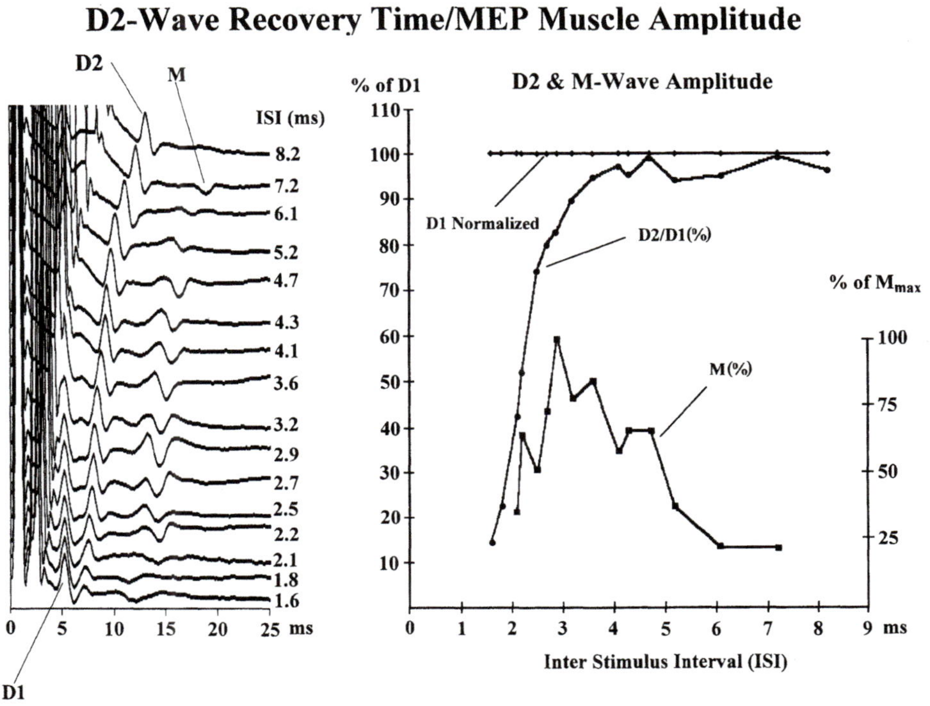

Fig. 5. D-wave recovery time. Two transcranial stimuli given at interstimulus intervals from 1.6 ms to 8.2 ms generated two D-waves. The traces with D1 and D2 waves show the generation of a muscle response (M-wave, "muscle artifact") in the paraspinal muscles. The maximal amplitude of the M-wave corresponds to near complete recovery of the D2-wave amplitude. This is at interstimulus intervals between 2.9 and 4 ms

the incomplete recovery of the D2-wave amplitude while in the longer intervals it was probably due to a lack of temporal summation at the spinal α-motorneuron.

Therefore the ISI in the train of electrical stimuli should not be set up arbitrarily. If a single electrical stimulus elicits a single D-wave without additional I-waves, the ISI should be long enough to allow a complete recovery of each consecutive D-wave generated by the train of stimuli. The ISI should not be too long in order to ensure temporal summation at the α-motorneuron. This can be achieved by adjusting the ISI in the stimulus train between 3 and 4 ms (333–250 Hz).

Facilitation of I-waves. Our previous work (Rodi *et al.*, 1996) described another neurophysiological phenomenon related to the train stimulation technique. When three or more electrical stimuli were applied to the motor cortex during relatively light anesthesia, they elicited not only a matching number of D-waves but also additional I-waves (Fig. 6a). We believe that

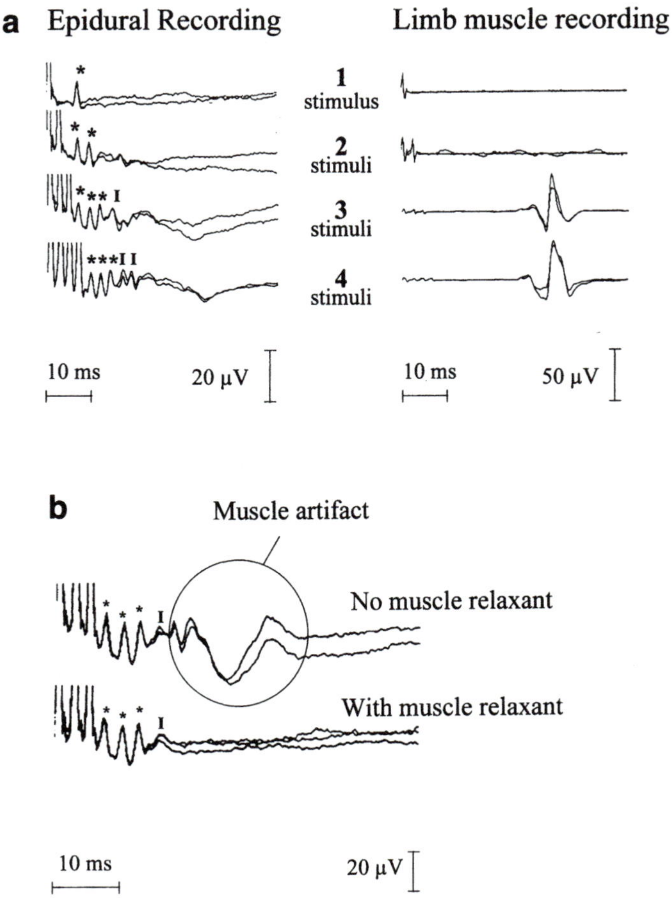

Fig. 6. Facilitation of I waves by repetitive stimulation. Single and double transcranial stimuli elicit single and double D-waves (*) in the epidural recording without a limb muscle response. After three stimuli, an additional I-wave (*I*) is generated and a corresponding peripheral muscle MEP is produced. After four stimuli the effect is even more pronounced (**a**). The administration of a muscle relaxant did not change the epidurally recorded D- and I-waves but wiped out the muscle artifact from the paraspinal muscles thus depicting the I-waves more clearly (**b**). Modified from (Rodi *et al.*, 1996) with permission

these additional I-waves are generated due to a facilitatory effect of multiple stimuli on the corticomotoneuronal synapses within the motor cortex. We termed this phenomenon the "facilitation" of I-waves. These epidurally recorded I-waves do not disappear after administration of a muscle relaxant, a muscle artifact from the paraspinal muscle thus being excluded (Fig. 6b).

Facilitation of I-waves is an important contributing factor in generating additional descending volleys that end on the spinal α-motorneuron because it is the number of volleys arriving at the α-motorneuron, regardless D or I, that is essential for bringing it beyond firing threshold.

The left side of Fig. 6a shows D-waves recorded after single, double, triple and quadruple transcranial stimuli were given with an ISI of 2 ms. The right side shows the corresponding *muscle MEPs* from the anterior tibial muscle. Muscle MEPs appeared when an additional I-wave was generated.

Cortical and spinal depression by anesthetics. As outlined in the previous sections there are two main reasons why a train of stimuli rather than a single stimulus appears to be necessary for reliably eliciting *muscle MEPs* during general anesthesia. These are the suppression of I-waves (Burke *et al.*, 1993; Hicks *et al.*, 1992) and the blocking of the spinal α-motorneuron by anesthetic agents.

More than one wave, D or I, in the corticospinal tract is necessary to generate an excitatory postsynaptic potential (EPSP) at the α-motorneuron that is high enough to reach its firing level (Philips and Porter, 1964). The single D-wave after a single transcranial stimulus mostly does not suffice. In addition many anesthetic agents have a depressive effect on the excitability of the α-motorneuron, thus making more corticospinal volleys necessary for its excitation. The latter has recently been documented by measuring the amplitude decrease of F-waves and H-reflexes in relation to the dose of anesthetic given (Zhou *et al.*, 1997). Certainly this does not exclude that the depression of the α-motorneuron is an indirect result of the loss of excitatory descending drive due to the anesthetic effect on the brain.

Consequently it appears imperative to overcome the blocking of the cortical motorneuron and the α-motorneuron by applying more transcranial stimuli than just one. Since each of the single stimuli in a short train elicits one D-wave in the corticospinal tract, the α-motorneurons of the spinal cord are hit by a train of D-waves. Thus the EPSPs finally reach firing level and a muscle potential is generated. At least three D-waves appear to be necessary to elicit limb muscle responses in a patient under our standard anesthesia management (Fig. 6a) (Rodi *et al.*, 1996).

The anesthetic effects on the excitability of both the motor cortex and the α-motorneuron however are profound and have long hampered efforts to monitor MEPs from muscle. These difficulties have led to systematic investigations of agents used in anesthesia and their effect on both epidural and muscle motor evoked potentials (Burke *et al.*, 1993; Kalkman *et al.*, 1995; Yamada *et al.*, 1995). The aim was to manage surgical anesthesia safely while still keeping the central nervous system excitable enough for muscle MEP recording.

In our institution a routine anesthesia management for all procedures requiring intraoperative neurophysiological monitoring has been established: propofol is given in a constant infusion of about 100 μg/kg/min, fentanyl (1 μg/kg/h) is added for analgesia and the patient is ventilated with oxygen and nitrous oxide up to 50 Vol. %. Bolus injections of both iv.

agents are avoided whenever possible. Halogenated anesthetics are avoided and muscle relaxants are not given after intubation. Most groups use partial neuromuscular blockade (Lang *et al.*, 1996; Yang, *et al.*, 1994) in order to minimize the twitches of masticatory, shoulder and paraspinal muscles that may interfere with surgery. We have managed anesthesia without relaxation and have found muscle twitches after train stimulation rarely to be an insurmountable problem.

These theoretical aspects are the basis for successful monitoring of *muscle MEPs* during general anesthesia.

Monitoring MEPs from limb muscles. Muscle MEP monitoring utilizes the multiple stimuli technique.

Muscle MEPs have a certain advantage over *epidural MEPs* for intraoperative monitoring in that they cover the entire motor tract from the motor cortex to beyond the neuromuscular junction. Thus to monitor muscle MEPs, the integrity of the corticospinal tract, the α-motorneurons, the peripheral nerves, the neuromuscular junctions, and the muscles is required. This is of particular interest for intramedullary procedures in the cervical and the lumbar spinal cord as well as for procedures involving the cauda equina and endovascular procedures jeopardizing the segmental spinal cord vasculature. Particularly in the cervical and lumbar segments tumor dissection, mechanical trauma or ischemia may damage the gray matter of the anterior horn or the nerve roots directly, a damage that would not be reflected by the *epidural MEP* recording. Furthermore, as has been noted above, the lumbar intumescence is not amenable to *epidural MEP* recording, – *muscle MEPs*, however, do cover this part of the motor system.

There is no consensus on which criteria should be applied for intraoperative monitoring of muscle MEPs. This is all the more so since each group uses a particular anesthetic as well as stimulation technique. Most data have been published using the single stimulus technique. But for both the single and the multiple stimuli technique a tremendous variability of the latency and particularly the amplitude of muscle responses has been reported (Jones *et al.*, 1996; Lang *et al.*, 1996; Pechstein *et al.*, 1996; Woodforth *et al.*, 1996; Yang *et al.*, 1994; Zentner, 1989). The only consistent finding in these reports is the correlation of a total loss of muscle responses to a new or increased postoperative motor deficit. This supports our own experience. Utilizing the multiple stimuli technique for recording *muscle MEPs* we have *never* observed an increased or new postoperative motor deficit when muscle MEPs were preserved at the end of surgery *regardless* their latency or amplitude. Our experience covers well over 800 procedures, including spinal cord tumor surgery, brainstem surgery, orthopedic spine surgery as well as endovascular procedures involving the spinal cord vasculature. The only significant finding appears to be a *complete loss* of *muscle MEPs* which was associated with a motor deficit postoperatively. The relationship between these early deficits and the definite neurological outcome is described later.

For recording muscle MEPs from limb muscles we use stainless steel EEG needle electrodes inserted into the thenar and anterior tibial muscles bilaterally (Fig. 1). As an alternative the forearm flexors (Taniguchi *et al.*, 1993a) may be used. The α-motorneurons of these muscles have abundant corticospinal tract innervation. Utilizing the above outlined technique of stimulating the motor cortex with a short train of stimuli a compound muscle action potential is recorded (Fig. 4b). Depending on the individual situation, the C1/C2 montage is tried first and in a number of cases appears to be sufficient to elicit muscle MEPs in all four extremities. If the threshold intensities for the lower extremities, which are usually higher than those for the arms, exceed 160 mA, the C3/4 montage is used. This appears to have a lower threshold for stimulation of the entire corticospinal tract than C1/C2 and thus more readily elicits muscle MEPs in the legs as well. But, as our experience shows, the C3/4 stimulation produces stronger muscle twitches of axial and masticatory muscles, often allowing muscle recording only when the surgeon briefly stops to dissect. However, since no averaging is needed for recording this appears to be a minor problem for fast and continuous data update.

In addition to the occurrence of muscle twitches there are drawbacks to the technique of *muscle MEP* monitoring. Since two synapses are involved, one between the corticospinal neuron and the α-motorneuron and one at the neuromuscular junction, the technique is vulnerable to anesthetics and muscle relaxants.

The combination of *epidural MEP* monitoring and *muscle MEP* monitoring is an important improvement of intraoperative spinal cord monitoring. Furthermore, *muscle MEPs* yield not only information about the α-motorneuron but about the elaborated and integrated command the stimulated motor cortex sends to the muscle as well.

The presence of *muscle MEPs* at the end of surgery indicates preserved integrity of the corticospinal tract. Its absence however is of long term predictive value only in conjunction with *epidural MEPs* (see transient paraplegia section).

Motor evoked potentials after transcranial magnetic stimulation

Single stimulus technique

MEPs after transcranial magnetic stimulation of the motor cortex are used clinically to assess the function of the motor system in a number of disorders (Cros *et al.*, 1993; Ferbert *et al.*, 1992; Hallett, 1996; Macdonell *et al.*, 1995; Maertens de Noordhout *et al.*, 1991; Rösler *et al.*, 1989; Rossini *et al.*, 1991; Xing *et al.*, 1990; Ying *et al.*, 1992).

The basic principle of magnetic stimulation is that a time-varying magnetic field induces an electrical field in any medium through which it passes (Barker *et al.*, 1987). The distinct advantage of magnetic stimulation over electrical stimulation in the awake subject is that it is not painful.

It was generally accepted that the current induced in the cerebral cortex by magnetic stimulation excites the cortical motor neurons transynaptically (Rothwell, 1991). There are a few lines of evidence that direct or transynaptic activation of cortical motor neurones depends on the orientation of the induced current in the brain. In monkeys a vertex-tangential orientation of a round-shaped coil excites cortical motor neurones transynaptically while lateral sagittal orientation excites them directly (Amassian *et al.*, 1990). In man Kaneko *et al.* (Kaneko *et al.*, 1996) showed by using a figure-of-eight coil that current induced in the brain in a latero-medial direction excites cortical motor neurones directly while a current induced in postero-anterior direction excites them transynaptically.

Not only the direction of the induced current in the brain, which depends upon the technical specifications of the coils and stimulators used, but also the anesthetic regimes used add to the variability of responses. A number of both experimental and clinical studies have been performed to define an anesthetic regime that would allow intraoperative muscle response recording (Edmonds *et al.*, 1989; Ghaly *et al.*, 1990; Kalkman *et al.*, 1992; Schmid *et al.*, 1992; Schönle *et al.*, 1989). The results of these studies are not comparable due to the previously mentioned variables. For example propofol appears to be a feasible anesthetic agent when using the Cadwell stimulator (Cadwell Laboratories Inc., Kennewick, WA) (Edmonds *et al.*, 1989) but not when using the MagStim (MagStim Co. Ltd., Spring Gardens, Dyfed, UK) (Schmid *et al.*, 1992; Taniguchi *et al.*, 1993b). Only two agents have consistently been reported to allow muscle recording after single transcranial magnetic stimulation. Etomidate was reported to be feasible in a surgical environment utilizing the MagStim stimulating device (Herdmann *et al.*, 1993; Yang *et al.*, 1994). Ketamine appears to be the best suited anesthetic agent to allow reproducible muscle response recording (Kalkman *et al.*, 1994; Kothbauer *et al.*, 1993; Yang *et al.*, 1994). Unfortunately due to its cardiovascular effects and particularly the postanesthesia emergence reactions (White *et al.*, 1982) many anesthesiologists are reluctant to use it routinely.

In addition, the use of magnetic stimulators in the operating room is inconvenient because these devices are usually bulky, the coils are difficult to keep in place and have a tendency to overheat. Not to mention that the machinery is also expensive.

For pre- and postoperative objective assessment of corticospinal tract integrity however transcranial magnetic stimulation is of great value.

Multiple stimuli technique

A stimulator can generate a train of up to four independent magnetic stimuli with variable interstimulus intervals (Quadropulse, MagStim Co. Ltd., Spring Gardens, Dyfed, UK) . This stimulator has only recently become available, therefore not much experience has been gathered so far. According to the previous theoretical data and our limited

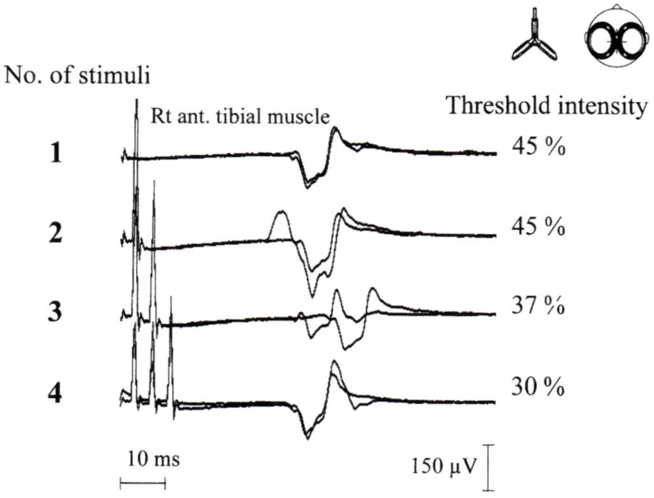

Fig. 7. Thresholds for eliciting motor responses to transcranial magnetic stimulation with single stimuli and a short train of two, three and four stimuli respectively. The interstimulus interval was 4 ms. Note the significantly lower threshold of output intensity when a train of four stimuli is applied

experience this stimulator enhances our ability to elicit muscle MEPs in the lower extremities which is always difficult, particularly under general anesthesia.

With the new stimulator the threshold of muscle MEPs in the anterior tibial muscles was determined. With a double-cone coil the threshold for eliciting muscle MEPs was significantly lower with the application of four stimuli than a single stimulus (45% of maximal compared to 30%) (Fig. 7).

The preliminary report published recently by Amassian *et al.* (Amassian *et al.*, 1996) demonstrates that when the stimulating coil was positioned in a specific orientation for mainly trans-synaptic activation of cortical motorneurons, the optimal interstimulus interval in the short train of magnetic stimuli was between 1.6 and 1.7 ms. According to the authors their data support the hypothesis of a "clock-like" time quantizer in the motor cortex. In other words magnetic trans-synaptic train stimulation coincides with a spontaneous interneuron chain activity at ~600 Hz. This spontaneous activity has a much higher frequency than the activity of corticomotorneurons during voluntary movement (Amassian *et al.*, 1987). In the awake subject the optimal interstimulus interval that elicits maximal MEPs is 1.6 to 1.7 ms when the stimulating coil was positioned to elicit only I-waves.

Practical aspects of direct monitoring of the corticospinal tract

The functional integrity of the spinal cord in surgery of intramedullary tumors

In surgery of intramedullary spinal cord tumors the conventional cortical and spinal SEP monitoring techniques were used because no direct motor tract monitoring was available for a long time. SEPs however, have several drawbacks for their intraoperative use: no instant feedback is possible because of the need for averaging, baseline recordings are frequently poor due to the effect of the lesion, and in many cases the response disappears early, usually at myelotomy. The latter appears often inevitable because the cord needs to be opened in order to approach a lesion that does not reach its surface but it leaves the surgeon without neurophysiological feedback during the essential part of the procedure. In a recently analyzed series of 100 patients operated upon for removal of an intramedullary spinal cord tumor in two thirds it was possible to record *epidural MEPs*, and SEPs could be recorded in approximately as many cases. *Muscle MEPs* could be elicited in all patients who did not have a significant preoperative motor deficit (Kothbauer *et al.*, 1998a).

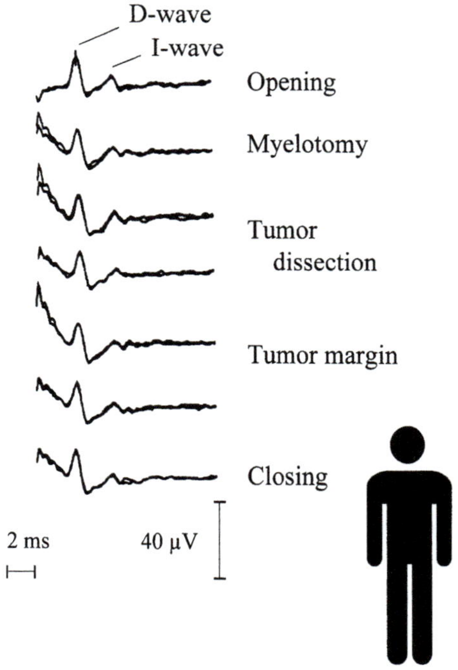

Fig. 8. Epidural MEP stable throughout an intramedullary spinal cord tumor removal. The presence of epidural MEP at the end of the procedure predicts preserved motor function postoperatively

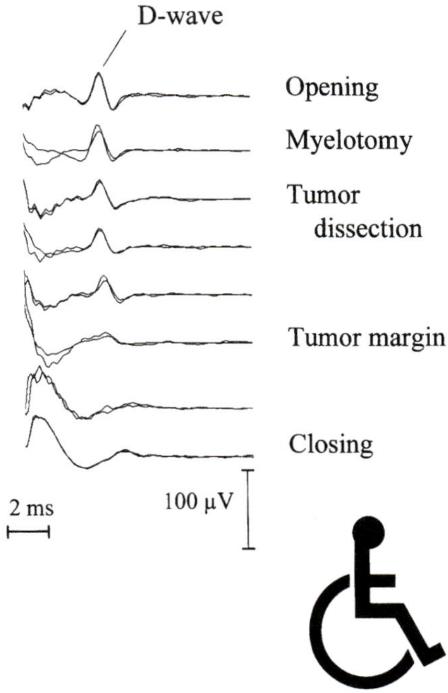

Fig. 9. Rapid loss of epidural MEP during removal of an intramedullary spinal cord tumor. In this instance the patient became quadriparetic after surgery

SEPs were lost during myelotomy in over half of the patients while MEPs remained intact. The postoperative motor function showed no association whatsoever with the intraoperative SEP data.

Patients with preserved epidural MEPs at the end of the operation had normal motor function postoperatively (Fig. 8). Two patients with significant decrease or loss of *epidural MEPs* (Fig. 9) had postoperative deterioration of motor function (Morota *et al.*, 1997). Patients with lost muscle MEPs at the end of surgery were paraplegic or severely paraparetic and recovered within weeks. We have come to call this latter phenomenon the phenomenon of transient paraplegia (see next section). In two of these patients the documented persistence of *epidural MEPs* is evidence for preservation of the corticospinal tract (Fig. 10). Only the preservation of epidural MEPs predicts recovery from postoperative paraplegia. In other words, *epidural and muscle MEPs* are the core of monitoring spinal cord surgery, since they selectively represent the motor system and are shown to correlate with postoperative motor function.

These data showed that intramedullary spinal cord surgery warrants selective monitoring of both the sensory and motor pathways. Not only one of the two should be monitored as representative for the functional integrity of the entire spinal cord. Particularly monitoring SEPs as a general measure of overall spinal cord integrity is not advisable in intramedullary spinal

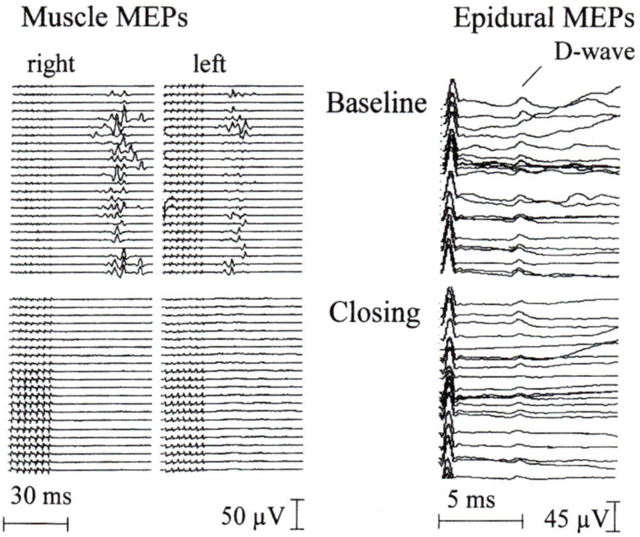

Fig. 10. Preserved epidural MEP with loss of muscle MEP at the end of surgery of an intramedullary spinal cord tumor. This patient was transiently paraplegic, but recovered within a week

cord surgery. It would primarily lead to a premature warning to the surgeon about neurophysiological decline in spinal cord integrity. Thus in a substantial proportion of cases tumors would only be partially removed when resection could actually have been safely completed had motor tract monitoring been performed. SEPs often have a mere documentary function, since they disappear early during myelotomy, and are thus only of value when they persist. Furthermore their disappearance does not correlate with a change in motor function. Sensory function, particularly joint position sense, however, does to some extent correlate with the disappearance of SEPs, but any detailed and extensive studies on the effect of intramedullary surgery on sensory function are not available so far.

MEP and the phenomenon of transient paraplegia

Transient or reversible paraplegia is a poorly recognized neurological entity. It is observed after spinal cord surgery where a certain number of patients are paraplegic immediately after surgery, only to recover in a relatively short period of time, i.e. days or a few weeks. The neurosurgeon observes a similar sort of transient paraplegia frequently after decompression of the spinal cord from metastatic tumors when some patients may be paraplegic for a brief period of usually only hours after decompression and then improve rapidly.

In the recent series of 100 patients undergoing surgery for intramedullary spinal cord lesions (Kothbauer *et al.*, 1998) and about 80 patients undergoing interventional radiology for spinal cord vascular malformations

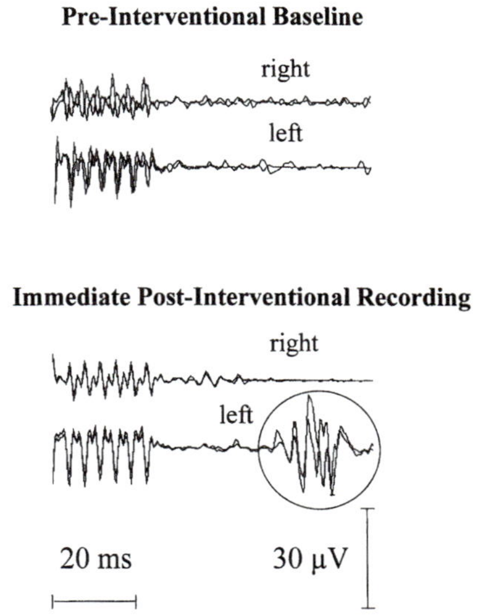

Fig. 11. Appearance of a previously absent left anterior tibial muscle MEP after embolization of a spinal dural arteriovenous malformation. The patient reported immediate improvement in motor function after the procedure and significantly improved in the subsequent weeks

38 patients with transient paraplegia have been documented. The surgical patients were ambulatory before surgery and paraplegic or deeply paraparetic immediately after surgery. They regained preoperative motor control within hours or upto 1–2 weeks thereafter. Muscle MEPs in these patients disappeared during surgery. In those cases where epidural MEPs were recordable, they remained present throughout the procedure (Fig. 10).

Thus intraoperatively preserved *epidural MEPs* indicate basically intact motor function and predict rapid recovery even in the absence of *muscle MEPs* at the end of the procedure.

The patient with the dural AVM was paraplegic before and improved immediately after embolization. Before embolization, *muscle MEPs* of the lower extremities, and also SEPs, were absent. Within minutes after embolization of the feeding artery, while the patient was still on the table, the previously absent *muscle MEPs* appeared (Fig. 11), but not SEPs. This corresponded to rapid improvement of motor function thereafter. The patient was able to move her legs on the next day. Motor function did significantly improve in the subsequent days and months and the patient regained the ability to walk and control of bladder function (Kothbauer *et al.*, 1998b).

It has been described before, that patients with spinal dural AVMs may substantially recover after endovascular or surgical treatment even though after long-standing paraplegia. This is believed to be due to venous congestion which leads to a congestive ischemia of the spinal cord that may impair function but not survival of the neural elements involved.

Practical aspects of indirect monitoring of the corticospinal tract

The potentially selective damage to certain tracts in the spinal cord is obvious in microsurgical procedures that go *into* the cord tissue. This selectivity is less pronounced when procedures involve the spinal cord vasculature and may cause ischemia or when they involve the spine by exerting pressure or traction upon the spinal cord.

Embolization of vascular lesions of the spinal cord and its coverings, or cardiovascular surgery that significantly lowers the blood pressure in the descending aorta may cause ischemia of the cord. Already over ten years ago the effect of injecting sodium amytal or even only contrast solution into the anterior spinal artery has been determined. Angiographic contrast solution can lead to a transient decrease in the amplitude of cortically recorded SEPs (Berenstein *et al.*, 1984). The same applies for the injection of sodium amytal (Fig. 12) (Young *et al.*, 1989). Since the anterior spinal artery anatomically supplies the anterior two thirds of the cord this effect of contrast and amytal on SEPs documents that the vascular connections to the posterior spinal arteries are sufficient to affect the posterior third as well. Thus ischemia is much less likely than intramedullary surgery to affect *only* the motor tract. Therefore the "indirect" monitoring of the functional integrity of the corticospinal tract by monitoring only the sensory *or* the motor part may be sufficient.

For example, during repair of a thoracoabdominal aneurysm, after cross-clamping the aorta the distal pressure has to be monitored very closely. The case presented in Fig. 13 illustrates that even a slight decrease of distal pressure may bring the spinal cord perfusion below the critical point, and be it only on a certain segmental level. Under these circumstances we are convinced that in this particular case failure to elevate the distal blood pressure would have resulted in paraplegia because a selective ischemia of the dorsal columns would be hard to conceive.

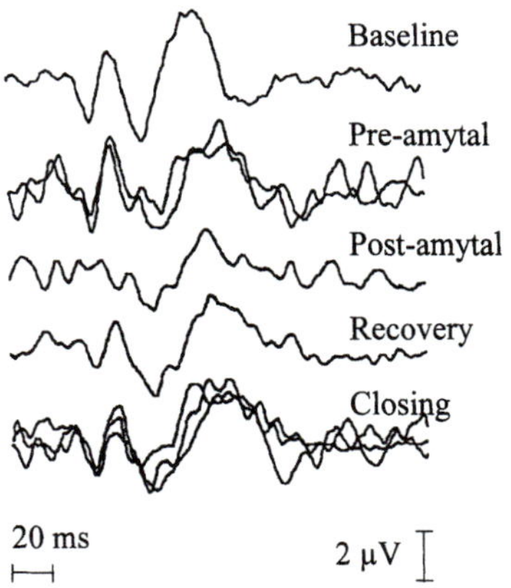

Fig. 12. SEP changes in angioembolization during and after sodium amytal injection into the anterior spinal artery (Young, Mollin, 1989, with permission)

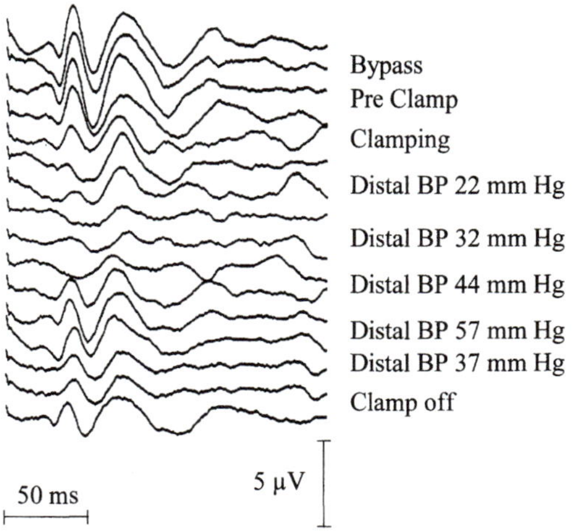

Fig. 13. Transient loss of tibial nerve SEP during cross clamping of the aorta and an episode of distal hypotension (distal blood pressure 22 mm Hg). No postoperative neurologic deficits were noted

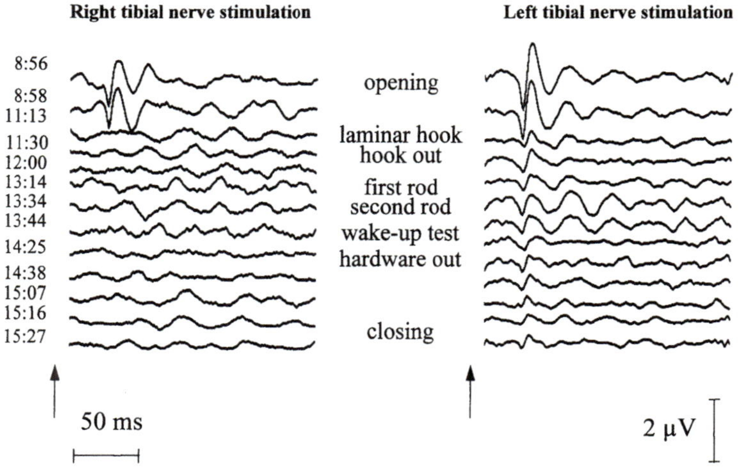

Fig. 14. Cortical somatosensory evoked potentials. SEP changes during correction of scoliosis. The right sided loss of tibial SEPs correlated with a right monoplegia at the wake-up test. All laminar hooks were removed. The patient recovered to normal motor function within a week postoperatively

The selectivity of damage in spinal surgery such as decompressive procedures, stabilizations, instrumentation to repair scoliosis and the like may be somewhere between the vascular and the intramedullary procedure. From this field after all came the first reports of "false negatives" for intraoperative SEP recording (Ginsburg *et al.*, 1985; Lesser *et al.*, 1986). However the examples provided in Figs. 14 and 15 illustrate the occurrence

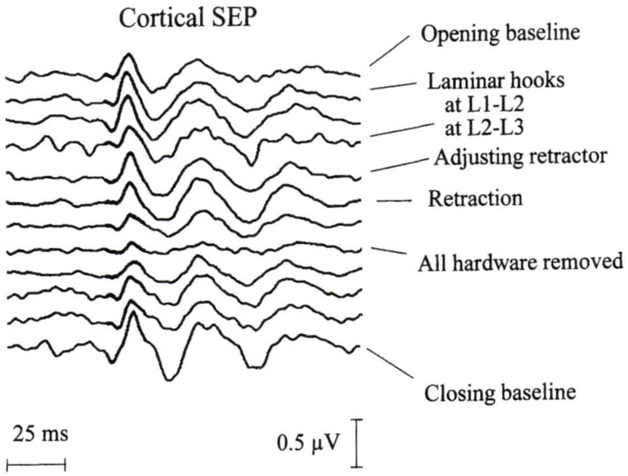

Fig. 15. SEP changes during spinal stabilization that required removal of all hardware. This patient had normal postoperative motor function

of SEP changes in spinal surgery and their correlation to motor function. In one case (Fig. 14) the unilateral loss of cortical SEPs was noted. A wake-up test confirmed an ipsilateral monoplegia, the hardware was removed and the patient was unable to move this leg after surgery. Fortunately, the patient recovered full motor function within a matter of days and underwent subsequent surgery without any further complications. The other case (Fig. 15) required removal of all stabilization hardware because bilateral loss of SEPs had occurred. The potentials recovered and in the light of the previous experience it was not doubted that this patient would have been paraplegic without monitoring. In both cases however, only SEP monitoring was performed. As confirmed with wake-up test in one case this did show the motor deficit, thus supporting the concept of "indirect" monitoring, i.e. taking SEPs as indicative not only for sensory but also for motor function.

Since spinal orthopedic procedures are performed in such great numbers, any sort of monitoring, SEP only or MEP plus SEP is certainly desirable. This was recently underlined by a multicenter-study that showed a clear benefit from the use of monitoring (Nuwer *et al.*, 1995).

Conclusion

The development of new methods for intraoperative monitoring of the corticospinal tract significantly enhances our ability to protect the motor tract within the spinal cord from damage. Even more with these neurophysiological data we can predict the patient's long term motor outcome better than with the clinical picture a few hours after surgery. We hope that these methods will be widely used. Their application can encourage surgeons to be more aggressive during surgery and at the same time to operate more safely.

References

Amassian V, Rothwell JC, Redding M, Priori A, Jalinous R (1996) A high frequency resonant circuit in human motor cortex. Soc Neurosci (abstract), p 658

Amassian VE, Quirk GJ, Stewart M (1990) A comparison of corticospinal activation by magnetic and electrical stimulation of monkey motor cortex. Electroencephalogr Clin Neurophysiol 77: 390–401

Amassian VE, Stewart M, Quirk GJ, Rosenthal JL (1987) Physiological Basis of motor effects of a transient stimulus to cerebral cortex. Neurosurgery 20: 74–93

Barker AT, Freeston IL, Jalinous R, Jarratt JA (1987) Magnetic Stimulation of the human brain and peripheral nervous system: an introduction and the results of an initial clinical evaluation. Neurosurgery 20: 100–109

Barker AT, Jalinous R, Freeston IL (1985) Non-invasive magnetic stimulation of the human motor cortex. Lancet 1: 1106–1107

Berenstein A, Young W, Ransohoff J, Benjamin V, Merkin H (1984) Somatosensory evoked potentials during spinal angiography and therapeutic transvascular embolization. J Neurosurg 60: 777–785

Blight AR (1983a) Axonal physiology of chronic spinal cord injury in the cat: intracellular recording in vitro. Neuroscience 10: 1471–1486

Blight AR (1983b) Cellular morphology of chronic spinal cord injury in the cat: analysis of myelinated axons by line sampling. Neuroscience 10: 521–543

Boyd SG, Rothwell JC, Cowan JMA, Webb PJ, Morley T, Asselman P, Marsden CD (1986) A method of monitoring function in corticospinal pathways during scoliosis surgery with a note on motor conduction velocities. J Neurol Neurosurg Psychiatry 49: 251–257

Burke D, Hicks R, Gandevia SC, Stephen J, Woodforth I, Crawford M (1993) Direct comparison of corticospinal volleys in human subjects to transcranial magnetic and electrical stimulation. J Physiol (Lond) 470: 383–393

Burke D, Hicks R, Stephen J (1992) Anodal and cathodal stimulation of the upper-limb area of the human motor cortex. Brain 115: 1497–1508

Canter M, Deletis V (1995) Spinal epidural electrode catheter for intraoperative recording of evoked potentials and injection of drugs. Sixth International Symposium on Spinal Cord Monitoring, New York, p 50 (abstract)

Coulter JD, Jones EG (1977) Differential distribution of corticospinal projections from individual cytoarchitectonic fields in the monkey. Brain Res 129: 335–340

Cros D, Chiappa KM (1993) Clinical applications of motor evoked potentials. In: Devinsky O, Beric A, Dogali M (eds) Electrical and Magnetic Stimulation of the Brain and Spinal Cord. Raven Press, New York, pp 179–185

Deletis V (1993) Intraoperative monitoring of the functional integrity of the motor pathways. In: Devinsky O, Beric A, Dogali M (eds) Electrical and Magnetic Stimulation of the Brain and Spinal cord. Raven Press, New York, pp 201–214

Edmonds HL, Paloheimo MPJ, Backman MH, Johnson JR, Holt RT, Shields CB (1989) Transcranial magnetic motor evoked potentials (tcMMEP) for functional monitoring of motor pathways during scoliosis surgery. Spine 14: 683–686

Ferbert A, Vielhaber S, Meincke U, Buchner H (1992) Transcranial magnetic stimulation in pontine infarction: correlation to degree of paresis. J Neurol Neurosurg Psychiatry 55: 294–299

Ghaly RF, Stone JL, Levy WJ, Roccaforte P, Brunner EB (1990) The effect of etomidate on motor evoked potentials induced by transcranial magnetic stimulation in the monkey. Neurosurgery 27: 936–942

Ginsburg HH, Shetter AG, Raudzens PA (1985) Postoperative paraplegia with preserved intraoperative somatosensory evoked potentials. J Neurosurg 63: 296–300

Hallett M (1996) Transcranial magnetic stimulation: a useful tool for clinical neurophysiology. Ann Neurol 40: 344–345 (editorial)

Herdmann J, Lumenta CB, Huse KOW (1993) Magnetic stimulation for monitoring of motor pathways in spinal procedures. Spine 18: 551–559

Hicks RG, Woodforth IJ, Crawford MR, Stephen JPH, Burke DJ (1992) Some effects of isoflurane on I waves of the motor evoked potential. Br J Anesth 69: 130–136

Jones SJ, Harrison R, Koh KF, Mendoza N, Crockard HA (1996) Motor evoked potential monitoring during spinal surgery: responses of distal limb muscles to transcranial cortical stimulation with pulse trains. Electroenceph Clin Neurophysiol 100: 375–383

Kalkman CJ, Drummond JC, Patel PM, Sano T, Chesnut RM (1994) Effects of droperidol, pentobarbital, and ketamine on myogenic transcranial magnetic motor-evoked responses in humans. Neurosurgery 35: 1066–1071

Kalkman CJ, Drummond JC, Ribberink AA, Patel PM, Sano T, Bickford RG (1992) Effects of propofol, etomidate, midazolam and fentanyl on motor evoked responses to transcranial electrical or magnetic stimulation in humans. Anesthesiology 76: 502–509

Kalkman CJ, Ubags LH, Been HD, Swaan A, Drummond JC (1995) Improved amplitude of myogenic motor evoked responses after paired transcranial electrical stimulation during sufentanil/nitrous oxide anesthesia. Anesthesiology 83: 270–276

Kaneko K, Kawai S, Fuchigami Y, Morita H, Ofuji A (1996) The effect of current direction induced by transcranial magnetic stimulation on the corticospinal excitability in human brain. Electroencephalogr Clin Neurophysiol 101: 478–482

Katayama Y, Tsubokawa T, Maemjima S, Hirayama T, Yamamoto T (1988) Corticospinal direct response in humans: identification of the motor cortex during intracranial surgery under general anesthesia. J Neurol Neurosurg Psychiatry 51: 50–59

Kothbauer K, Schmid UD, Liechti S, Rösler KM (1993) The effect of ketamine anesthetic induction on muscle responses to transcranial magnetic cortex stimulation studied in man. Neurosci Lett 154: 105–108

Kothbauer KF, Deletis V, Epstein FJ (1998a) Motor evoked potential monitoring for intramedullary spinal cord tumor surgery: correlation of clinical and neurophysiological data in a series of 100 consecutive procedures. Neurosurg Focus 4, Article 1

Kothbauer K, Pryor J, Berenstein A, Setton A, Deletis V (1998b) Motor evoked potentials predicting recovery from paraparesis after embolization of a spinal dural arteriovenous fistula. Intervent. Neuroradiol 4: 81–84

Kuypers HGJM (1964) The descending pathways to the spinal cord, their anatomy and function. In: Eccles JC, Schade JP (eds) Progress in Brain Research. Organization of the Spinal Cord. Elsevier, Amsterdam, pp 178–202

Lang EW, Beutler AS, Chesnut FM, Patel PM, Kennelly NA, Kalkman CJ, Drummond JC, Garfin SR (1996) Myogenic motor-evoked potential monitoring using partial neuromuscular blockade in surgery of the spine. Spine 21: 1676–1686

Lesser RP, Raudzens P, Lüders H, Nuwer MR, Goldie WD, Morris HH, Dinner DS, Klem G, Hahn JF, Shetter AG, Gisnburg HH, Gurd AR (1986) Postoperative neurological deficits may occur despite unchanged intraoperative somatosensory evoked potentials. Ann Neurol 19: 22–25

Levy WJ, York DH, McCaffrey M, Tanzer F (1984) Motor evoked potentials from transcranial stimulation of the motor cortex in humans. Neurosurgery 15: 287–302

Macdonell RAL, Donnan GA (1995) Magnetic cortical stimulation in acute spinal cord injury. Neurology 45: 303–306

MaertensdeNoordhout A, Remacle JM, Pepin JL, Born JD, Delwaide PJ (1991) Magnetic stimulation of the motor cortex in cervical spondylosis. Neurology 41: 75–80

Merton PA, Morton HB (1980) Stimulation of the cerebral cortex in the intact human subject. Nature 285, p 227

Nathan PW, Smith MC (1973) Effects of two unilateral cordotomies on the motility of the lower limbs. Brain 96: 471–494

Nuwer MR, Dawson EG, Carlson LG, Kanim LE, Sherman JE (1995) Somatosensory evoked potential spinal cord monitoring reduces neurologic deficits after scoliosis surgery: results of a large multicenter study. Electroencephalogr Clin Neurophysiol 96: 6–11

Patton HD, Amassian VE (1954) Single-and multiple unit analysis of cortical state of pyramidal tract activation. J Neurophysiol 17: 345–363

Pechstein U, Cedzich C, Nadstawek J, Schramm J (1996) Transcranial high-frequency repetitive electrical stimulation of recording myogenic motor evoked potential with the patient under general anesthesia. Neurosurgery 39: 335–344

Penfield W, Boldrey E (1937) Somatic motor and sensory representation in the cerebral cortex of man as studied by electrical stimulation. Brain 60: 389–443

Philips CG, Porter R (1964) The pyramidal projection to motoneurones of some muscle groups of the baboon's forelimb. In: Eccles JC, Schadé JP (eds) Progress in Brain Research. Elsevier, Amsterdam, pp 222–243

Rodi Z, Deletis V, Morota N, Vodusek DB (1996) Motor evoked potentials during brain surgery. Pflüger's Archiv. Eur J Physiol 431: R291–292

Rösler K, Hess C, Schmid U (1989) Investigation of facial motor pathways by electrical and magnetic stimulation: sites and mechanisms of excitation. J Neurol Neurosurg Psychiatry 52: 1149–1156

Rossini PM, Desiato MT, Lavaroni F, Caramia MD (1991) Brain excitability and electroencephalographic activation: non-invasive evaluation in healthy humans via transcranial magnetic stimulation. Brain Res 567: 111–119

Rothwell JC (1991) Physiological Studies of electric and magnetic stimulation of the human brain. In: Levy WJ, Cracco RQ, Barker AT, Rothwell J (eds) Magnetic Motor Stimulation: Basic Principles and Clinical Experience. Electroenceph Clin Neurophysiol [Suppl] 43: 29–35

Schmid UD, Boll J, Liechti S, Schmid J, Hess CW (1992) Influence of some anesthetic agents on muscle responses to transcranial magnetic cortex stimulation: A pilot study in man. Neurosurgery 30: 85–92

Schönle PW, Isenberg C, Crozier TA, Dressler D, Machetanz J, Conrad B (1989) Changes of transcranially evoked motor responses in man by midazolam, a short acting benzodiazepine. Neurosci Lett 101: 321–324

Taniguchi M, Schramm J, Cedzich C (1991) Recording of myogenic motor evoked potentials under general anesthesia. In: Schramm J, Møller ÅR (eds) Intraoperative Neurophysiologic Monitoring in Neurosurgery. Springer, Berlin Heidelberg New York Tokyo, pp 72–87

Taniguchi M, Cedzich C, Schramm J (1993a) Modification of cortical stimulation for motor evoked potentials under general anesthesia: technical description. Neurosurgery 32: 219–226

Taniguchi M, Nadstawek J, Langenbach U, Bremer F, Schramm J (1993b) Effects of four intravenous anesthetic agents on motor evoked potentials elicited by magnetic transcranial stimulation. Neurosurgery 33: 407–415

White PF, Way WL, Trevor AJ (1982) Ketamine – its pharmacology and therapeutic uses. Anesthesiology 56: 119–136

Woodforth IJ, Hicks RG, Crawford MR, Stephen JP, Burke DJ (1996) Variability of motor-evoked potentials recorded during nitrous oxide anesthesia from the tibialis anterior muscle after transcranial electrical stimulation. Anesth Analg 82: 744–749

Xing J, Katayama Y, Yamamoto T, Hirayama T, Tsubokawa T (1990) Quantitative evaluation of hemiparesis with corticomyographic motor evoked potential recorded by transcranial magnetic stimulation. J Neurotrauma 7: 57–64

Yamada H, Transfeldt EE, Tamaki T, Nishiura H, Taylor BA, Torres R, Iaizzo PA (1995) General anesthetic effects on compound muscle action potentials elicited by single or dual spinal cord stimulation. J Spin Disord 8: 157–162

Yang L-H, Lin S-M, Lee W-Y, Liu C-C (1994) Intraoperative transcranial electrical motor evoked potential monitoring during spinal surgery under intravenous ketamine or etomidate anesthesia. Acta Neurochir (Wien) 127: 191–198

Ying Z, Schmid UD, Schmid J, Hess CW (1992) Motor and somatosensory evoked potentials in coma: analysis and relation to clinical status and outcome. J Neurol Neurosurg Psychiatry 55: 470–474

Young W, Mollin D (1989) Intraoperative somatosensory evoked potentials monitoring of spinal surgery. In: Desmedt JE (eds) Neuromonitoring in Surgery. Elsevier, Amsterdam, pp 165–173

Zentner J (1989) Noninvasive motor evoked potential monitoring during neurosurgical operations in the spinal cord. Neurosurgery 24: 709–712

Zhou HH, Mehta M, Leis AA (1997) Spinal cord mononeuron excitability during isoflurane and nitrous oxide anesthesia. Anesthesiology 86: 302–307

Correspondence: Dr. V. Deletis, Institute for Neurology and Neurosurgery, Beth Israel Medical Center North Division, 170 East End Avenue, NY 10128.

Spinal Cord Monitoring
Eds: E. Stålberg, H. S. Sharma, Y. Olsson

Chapter 18

Corticospinal volleys evoked by transcranial electrical and magnetic stimulation

D. Burke and **R. Hicks**

Prince of Wales Medical Research Institute and Department of Neurology, Prince of Wales Hospital, Randwick, Australia

Monitoring corticospinal function during surgery is now feasible and a number of different techniques have been implemented by different authorities, as discussed below. Based on recordings of somatosensory volleys using recording electrodes inserted into the epidural space, as pioneered by Jones and colleagues (1982, 1983), and the first report of corticospinal volleys recorded using similar electrodes (Boyd *et al.*, 1986), our unit has developed a recording system that allows the simultaneous recording of descending corticospinal volleys and ascending somatosensory volleys from the spinal cord (Fig. 1), and has used it extensively, primarily during scoliosis surgery (Hicks *et al.*, 1991; Burke *et al.*, 1992b; Stephen *et al.*, 1996). This chapter addresses the nature of the corticospinal volleys set up by electrical and magnetic stimulation of the human motor cortex through the scalp, as recorded using this technique. The data are relevant not only for those who monitor spinal cord function during surgery but also for those who use transcranial stimulation of the motor cortex as a diagnostic or research tool.

There a number of theoretical advantages in being able to monitor accurately the function of corticospinal pathways in the brain and spinal cord rather than the function of somatosensory pathways only. First, inadvertent damage during scoliosis surgery or unavoidable damage during neurosurgery is usually more devastating to the patient when the result is weakness rather than anaesthesia. Secondly, somatosensory evoked potentials monitor the function of primarily the posterior quadrants of the spinal cord and may not detect damage confined to the anterior quadrants. Thirdly, vascular lesions involving the anterior spinal artery impair the lateral columns and spinal grey matter, largely sparing the posterior columns.

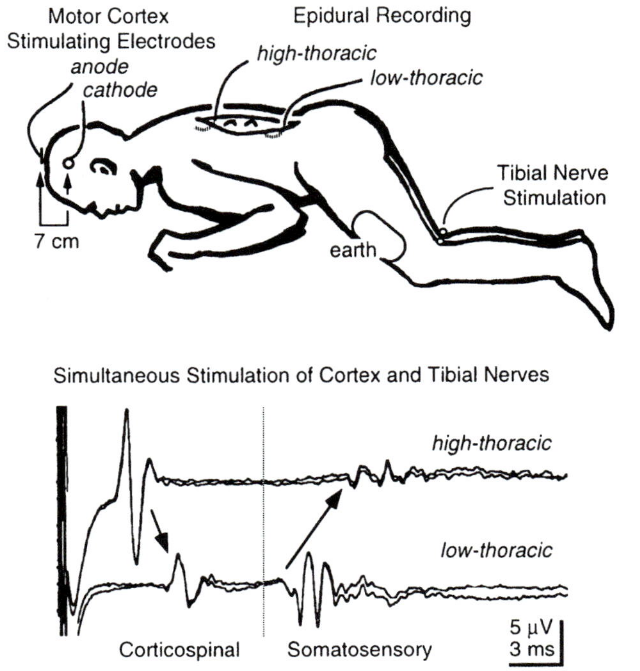

Fig. 1. The technique used for recording simultaneously descending corticospinal volleys in response to anodal electrical stimulation of the motor cortex and ascending somatosensory volleys in response to stimulation of the tibial nerves in the popliteal fossa. The stimuli are delivered simultaneously, and recorded at two levels from the spinal cord as illustrated in the lower traces. The descending corticospinal volley has a shorter latency and propagates down the spinal cord while the ascending somatosensory volley has a longer latency and propagates up the spinal cord. In the lower panel the traces are duplicate averages of 10 sweeps. In these traces, negativity for the corticospinal volley is shown as an upward deflection and negativity for the somatosensory volley as a downward deflection, reflecting the fact that the volleys approach the bipolar recording electrodes from opposite directions

Factors influencing choice of technique for intraoperative monitoring of corticospinal function

Corticospinal function can be studied in a number of different ways: by electrical or magnetic stimulation of motor cortex, or by electrical stimulation of the pathways in the brainstem or the spinal cord. With motor cortex and brainstem stimulation, the evoked corticospinal volleys can be recorded either directly from the spinal cord uncontaminated by antidromic somatosensory activity (due to intervening synapses in somatosensory pathways), or after overcoming the corticospinal-motoneurone synapse as an evoked EMG potential from an innervated muscle. With direct stimulation of the spinal cord, the discharge of motoneurones can be recorded as a "motor evoked potential" (MEP) from the peripheral nerve ("neurogenic MEP", see Owen *et al.*, 1988) or as an EMG potential from muscle.

There are advantages and disadvantages to each technique, no one technique being optimal for all monitoring purposes. For example, where surgery is on the most caudal segments of the cord, the conus medullaris or nerve roots, to record a corticospinal volley from the spinal cord proximal to the lesion would be of little value. When complete muscle relaxation is desirable, recording a "neurogenic MEP" may be more useful than attempting to record an EMG potential.

Complete avoidance of muscle relaxants during surgery is not necessary when recording the MEP as an EMG potential, and the precise dosage of relaxant can be titrated using trains of repetitive stimuli to motor nerves. However, stimulation of the motor cortex or spinal cord will still produce a twitch contraction of paraspinal and limb muscles and this, even though relatively weak, can be surgically disturbing. Only by recording the evoked activity proximal to the neuromuscular junction can this problem be avoided, and that implies either directly from the spinal cord (before the corticospinal-motoneurone synapse) or from the motor nerve as a neurogenic MEP (after the synapse).

The corticospinal-motoneurone synapse represents a major barrier to translating the evoked corticospinal volley into a discernible MEP. Even in awake but relaxed subjects, not all motoneurones in a target pool discharge in response to the volleys set up by intense transcranial stimuli, and this is more so with the relatively weak magnetic stimulus than the potentially stronger electrical (see below). In the anaesthetised patient, choice of anaesthetic is critical, as is the use of additional agents such as midazolam. Anaesthetic considerations are of lesser importance when the corticospinal volley itself is recorded rather than the motoneurone discharge, though they still represent a limitation when the transcranial stimulus is magnetic rather than electrical (see below). One method to overcome the barrier presented by the corticospinal-motoneurone synapse is to use pairs of stimuli or trains of stimuli, to either the motor cortex or the spinal cord (Taylor *et al.*, 1993; Kalkman *et al.*, 1995).

For many purposes, for example scoliosis surgery, to monitor corticospinal and somatosensory pathways at the same time would be preferable to monitoring only corticospinal function, simply because of the larger volume of cord studied. However, if the somatosensory evoked potential is recorded from the scalp, transcranial stimulation cannot be used at the same time. Similarly, the stimuli to limb nerves required for somatosensory studies can interfere with the ability to record the motor evoked potential from the peripheral nerve (as a neurogenic MEP) or from the muscle as an EMG potential. Epidural recording obviates these problems, such that ascending and descending volleys converge on the spinal cord recording site(s), and this implies that the peripheral nerve and transcranial stimuli need not be delivered in separate sequences. Indeed, as in the technique used in this Unit for scoliosis surgery, the stimuli can be delivered simultaneously (Fig. 1, see Hicks *et al.*, 1991; Burke *et al.*, 1992b).

The amplitude and variability of the monitored waveform determine the number of trials that must be averaged for reliable results. For example, the response to direct electrical stimulation of corticospinal axons may be quite large, commonly >10 μV, sufficient to be clearly visible in single trials, and can be highly reproducible from trial to trial (Burke *et al.*, 1995), but the volley produced by magnetic stimulation is smaller, often more complex and usually more variable (Burke *et al.*, 1993). As an EMG potential, the MEP recorded from muscle may be larger but it is highly variable in single sweeps because only a small but variable number of motoneurones (if any) discharge in each trial (Woodforth *et al.*, 1996).

Corticospinal volleys evoked by transcranial electrical stimulation of the motor cortex

As established in the cat and monkey (Lance and Manning, 1954; Patton and Amassian, 1954; Kernell and Wu, 1967), electrical stimulation of motor cortex can evoke a complex descending volley with early and late components due respectively to "direct" stimulation of the corticospinal neurone or its axon (hence "D" wave) and "indirect" stimulation (hence "I" waves), the latter involving stimulation of corticocortical interneurones (or their axons) which then activate the corticospinal neurone transynaptically. These animal studies used a ball stimulating electrode placed directly on the pia, but the findings are remarkably comparable when transcranial stimuli are applied in human subjects (see below, and the single-motor-unit studies of Day *et al.*, 1989; for review see Rothwell *et al.*, 1991), even though the stimulus is more remote from the cortex and the orientation of the current flow to pyramidal neurones is more problematic.

The response at threshold. The threshold for the corticospinal volley is lower with anodal stimulation than cathodal, presumably because, with anodal stimulation, current that entered the vertically orientated apical dendrite of the pyramidal cell would exit at the axon such that depolarization would occur in the axon, presumably at or near the initial segment of the neurone (Fig. 2). With cathodal stimulation, a reversed direction of current flow might depolarize the apical dendrite but it would also tend to hyperpolarize the axon impeding impulse initiation (Fig. 2, see Phillips and Porter, 1977).

At threshold, the response consists of a single wave that has the same latency with both stimulus polarities, and is presumably due to direct stimulation of the corticospinal neurone or its axon, close to the cell body (Burke *et al.*, 1990, 1992a).

The effects of increasing stimulus intensity. With increasingly strong electrical stimuli, both anodal and cathodal, the evoked volley becomes more complex as waves appear before and after the component of lowest

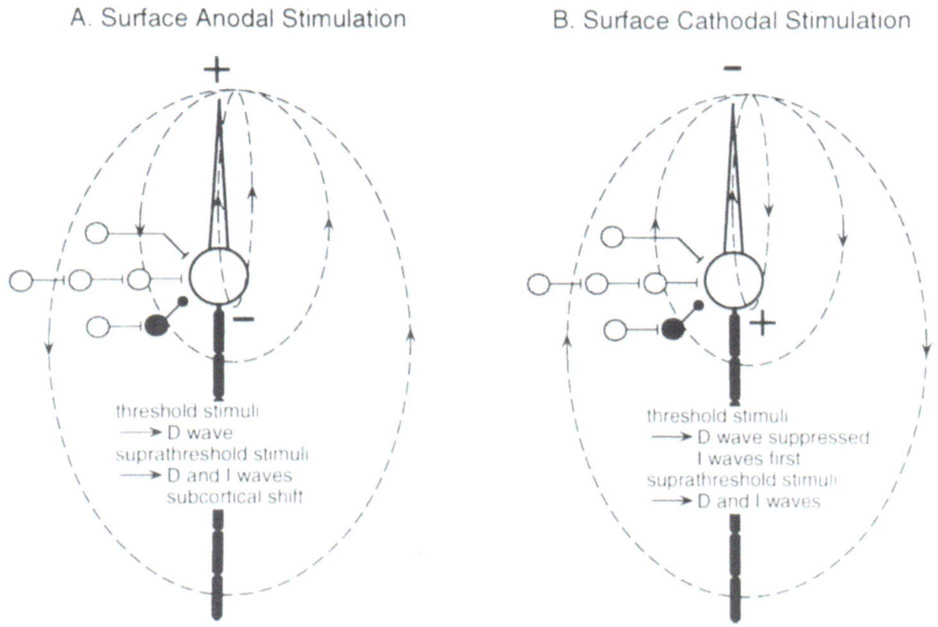

Fig. 2. The elements in the motor cortex believed to be activated by transcranial electrical stimulation. With relatively weak anodal stimuli the direction of current flow favours depolarisation at the initial segment or proximal nodes of Ranvier of the corticospinal neurone, thus producing the "D wave". With stronger stimuli, the corticospinal neurone will be activated directly at deeper nodes of Ranvier and, in addition, the axons of cortico-cortical interneurones projecting to the corticospinal neurone will be stimulated. Those cortical interneurones with excitatory projections can reactivate the corticospinal neurone, thus setting up a number of volleys in the corticospinal axon. These later volleys are known as "I waves". With surface cathodal stimulation, the threshold for the D wave is higher because the arc of current flow favours hyperpolarization at the initial segment and proximal nodes of Ranvier. Theoretically, I waves could appear at lower threshold than the D wave with cathodal stimuli, but in practice this does not occur (see text)

threshold (Fig. 3). The waves of longer latency are probably "I waves", as discussed below. The components with shorter latencies result from direct stimulation of the corticospinal axon at subcortical sites, as also discussed below. In other words, they constitute components of the D-wave volley.

Origin of early waves. The precise site of origin of the component of lowest threshold (termed "D_1") is not certain, even though all other components in the complex descending volley are referred to it. The lowest-threshold component has a latency at the low-cervical region that would be compatible with initiation at or near the initial segment of the corticospinal neurone, and such an origin is consistent with the fact that this low-threshold component is quite sensitive to volatile anaesthetics (Hicks *et al.*, 1992b). In addition, a wave of identical latency can be produced by transcranial magnetic stimulation (Burke *et al.*, 1993; see below).

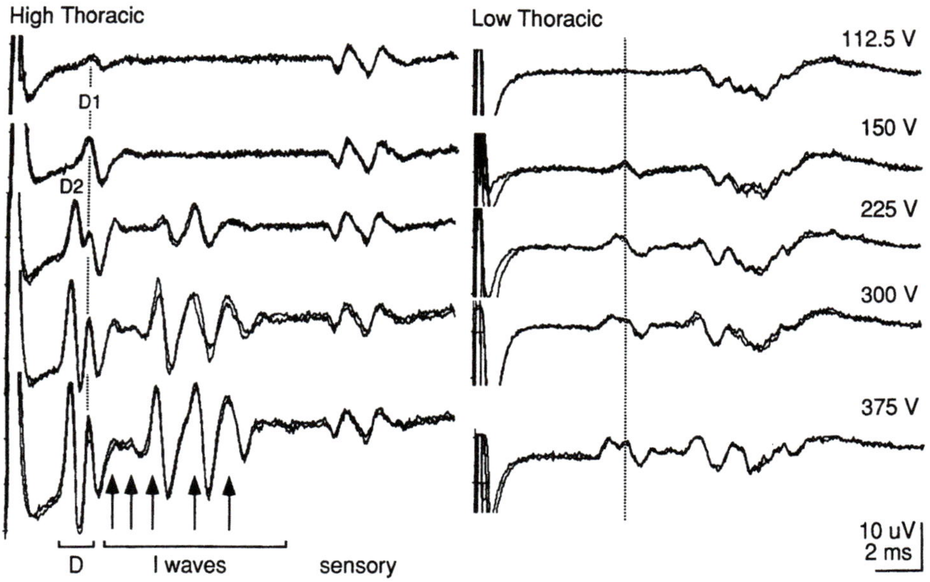

Fig. 3. The corticospinal volleys produced by increasingly strong anodal transcranial electrical stimulation after withdrawal of isoflurane. With stimuli of 112.5 V and 150 V, the corticospinal volley consists of a simple D wave (D_1) with a latency to peak (indicated by the dotted vertical line) of 3.5 ms at the high-thoracic site and 6.5 ms at the low-thoracic site. At 225 V, the D wave is followed by I waves at both sites, and the D wave is becoming bifid. At 375 V, the high-threshold short-latency D_2 component occurs 0.7–0.8 ms in advance of the initial low-threshold D_1 component. I waves (I_1, I_2, I_3, I_5 and I_6) are indicated by vertical arrows, and are particularly prominent because of the earlier withdrawal of isoflurane. At the low-thoracic site the I waves are superimposed on the somatosensory potential. The somatosensory potential begins at 9.9 ms in the low-thoracic recording and at 13.7 ms in the high-thoracic recording. Duplicate averages of 25 responses from a female patient aged 11 years. From Hicks *et al.* (1992a), with permission

The two progressively higher-threshold and earlier components (first "D_2" and then "D_3") arise from stimulation of corticospinal axons deep to cortex. Accordingly, they have the same conduction velocity as the low-threshold component and, with increasingly intense stimuli, they gradually replace the lower threshold components, until only the component of shortest-latency and highest threshold (D_3) remains (Fig. 4). As D_2 and D_3 emerge, their latencies initially shorten slightly but, when they are well-established, they appear consistently in successive steps, with D_2 0.8–1.0 ms in advance of D_1 and D_3 0.8–1.0 ms in advance of D_2 (Fig. 4). These fixed latency intervals suggest that there are two sites at which corticospinal axons are preferentially stimulated within the cranium (in addition to the site of lowest threshold) when transcranial electrical stimuli are applied using our standard stimulating electrode placement.

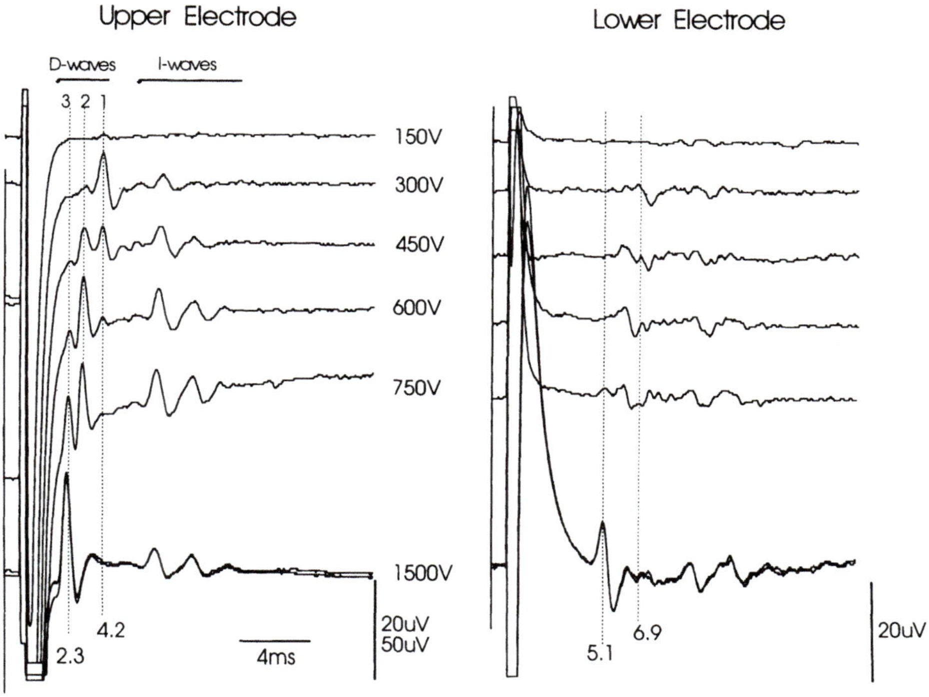

Fig. 4. The corticospinal volleys evoked by increasingly strong anodal stimulation of the motor cortex. Epidural recordings from the low-cervical and low-thoracic levels (from a female patient, aged 14). The stimulus artefact begins 1 ms after the start of the sweep. Two averages are superimposed for the 1500 V stimulus. Note the gradual recruitment of earlier D wave components as the stimulus intensity is increased, and the progressive increase in the size and number of I waves. In the recordings from the upper electrode, the 50 μV calibration applies only to the responses at 1500 V. The dotted lines indicate successive D-wave components and their respective peak latencies. The latency shift from D_1 to D_3 is the same at the two levels. Note that D_1 is attenuated from 450 V upwards, disappearing at 750 V. D_2 is abolished at 1500 V. These findings confirm that the different components of the D wave reflect activity arising at different levels of the neuraxis in the same population of axons, rather than activity in non-corticospinal axons. Note also that the interval between the D wave and the I waves is the same at the two recording sites, best seen with stimuli of 1500 V, reflecting the fact that I waves are not due to activity in slowly conducting axons. From Rothwell *et al.* (1994), with permission

Initially it was suggested that these sites could be the internal capsule and the cerebral peduncle (Burke *et al.*, 1990), but transmission from a near-cortex site to the pyramidal decussation in the brainstem probably takes only about 2 ms (Thompson *et al.*, 1991; Ugawa *et al.*, 1991). The latency of D_3 to the spinal epidural recording sites is the same as that of the volley produced by direct brainstem stimulation (Fig. 5, Rothwell *et al.*, 1994). Such stimulation probably activates the corticospinal system at the pyramidal decussation (Ugawa *et al.*, 1991). D_2 must then arise between

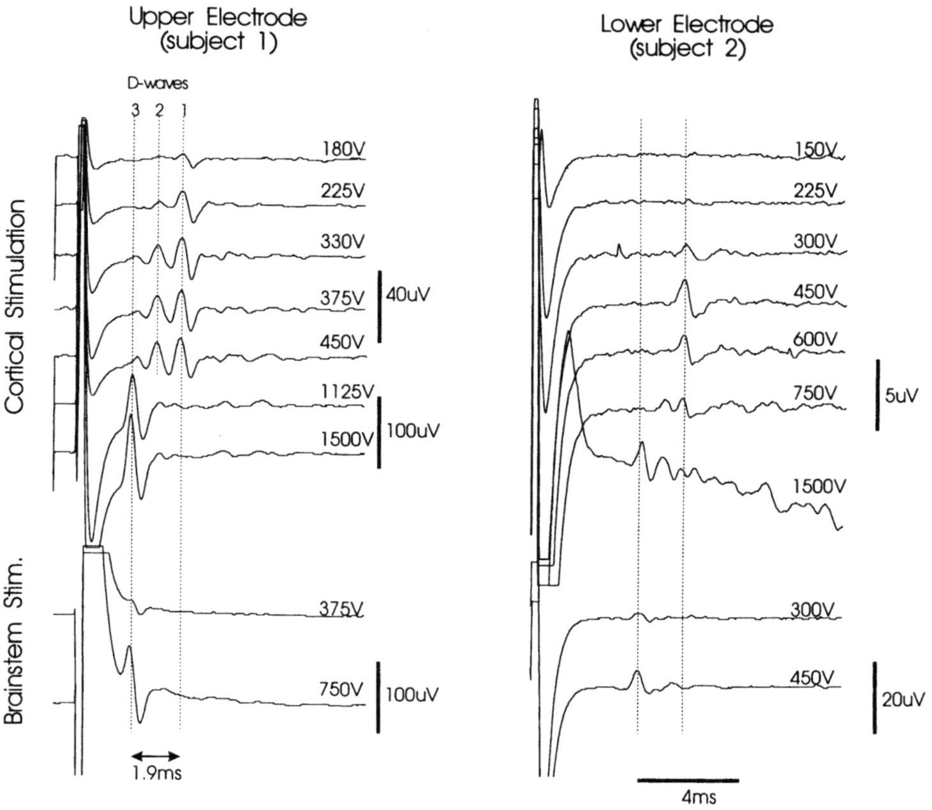

Fig. 5. Comparison of the responses to transcranial electrical stimulation and to electrical stimulation of the brainstem. Data from two subjects are illustrated, from the low-cervical level in a female aged 14 and from the low-thoracic level in a male aged 12. For the traces on the left, the stimulus was delivered 1 ms after the onset of the sweep. For the traces on the right, the stimulus was delivered at the onset of the sweep. As in Fig. 4, gradually increasing the stimulus intensity resulted in a progressive shift of the site of activation of corticospinal neurones to two subcortical sites. The latency to peak of D_3 is the same as the latency of the volley produced by brainstem stimulation, using electrodes on the mastoid processes. Each trace is the average of 10 responses. From Rothwell *et al.* (1994), with permission

cortex and decussation, presumably at the cerebral peduncle where (like the pyramidal decussation) a bend in the axon would probably make them more susceptible (Amassian *et al.*, 1992; Maccabee *et al.*, 1993).

These shifts in latency have importance for both monitoring and reflex studies. For monitoring purposes, the amplitude variability of the individual components of a complex 3-component D wave volley is higher than that of a large simple D wave, i.e., D_1 alone (Burke *et al.*, 1995). For motor control studies, the fact that D_2 begins to appear with stimuli of 225–375 V and D_3 with 300–600 V implies that some axons are stimulated deep to cortex even when quite modest electrical stimuli are used. It would be unwise to assume that the latency of the EMG potential produced by such stimuli reflected transmission from the cortical level.

Origin of late waves. The evidence that the waves of longer latency than that of lowest threshold (i.e., the D wave) are indeed I waves is as follows. First, many of the late waves can be identified at two recording sites (low-cervical and low-thoracic), and their latency after the D wave is not greater in the caudal recording (Fig. 4). In other words, these I waves result from activity in axons that have the same conduction velocity as those contributing to the D wave. Presumably, then, their longer latency must result from their initiation after the D wave. Late waves in the rostral recording without counterparts in the caudal recording may be attributed to activity directed to intervening segments (e.g., low cervical and thoracic segments) and to the lower signal-to-noise ratio at the caudal site. Secondly, the I waves are very sensitive to anaesthetic agents, such as isoflurane (Fig. 6, Hicks *et al.*, 1992a,b), as would be expected for activity dependent on secure synaptic transmission. Thirdly, the I waves increase in size even with very intense stimuli (750–1500 V, see Fig. 4), stimuli that are sufficient to activate all corticospinal axons in the brainstem thus occluding the lower-threshold more rostrally originating components (see above). Fourthly, as discussed later, the corticospinal volley evoked by transcranial magnetic stimulation is biased towards these I waves (as predicted by Day *et al.*, 1989), though it still contains D-wave activity even when the stimulus intensity is low (Burke *et al.*, 1993).

Variability of individual components. The variability of the components of the corticospinal volley at the low-cervical level has been assessed by measuring their amplitude and latency in 100 consecutive single sweeps without averaging, when there was no operative interference and when anaesthetic conditions, blood pressure and temperature were stable (Burke *et al.*, 1995). The variability in latency of the D- and I-wave components of the corticospinal volley was quite low in single sweeps, with coefficients of variation of $<3\%$. The amplitude variability was high for I waves and the components of complex D waves (coefficients of variation of about 30%), but were low for large simple D waves of amplitude 10–20 μV (coefficients of variation of 3–8%).

The intrinsically low variability of the D wave at the low-cervical level could be sufficient for single sweeps to be satisfactory for some monitoring purposes but averaging is recommended as a routine because (i) the signal-to-noise ratio is usually poorer at the low-thoracic site, (ii) monitoring is often performed in less-than-ideal conditions and must continue in the face of operative and anaesthetic interference, and (iii) the D wave may not be as well defined in patients with pre-existing pathology. In the authors' experience, averages of 10–25 sweeps usually provide satisfactory definition of the D wave at both levels, and this takes about 30–75 s when stimuli are delivered once every 3 s. Although it is conventional to set an amplitude decrease of 50% as a criterion of significant dysfunction, this

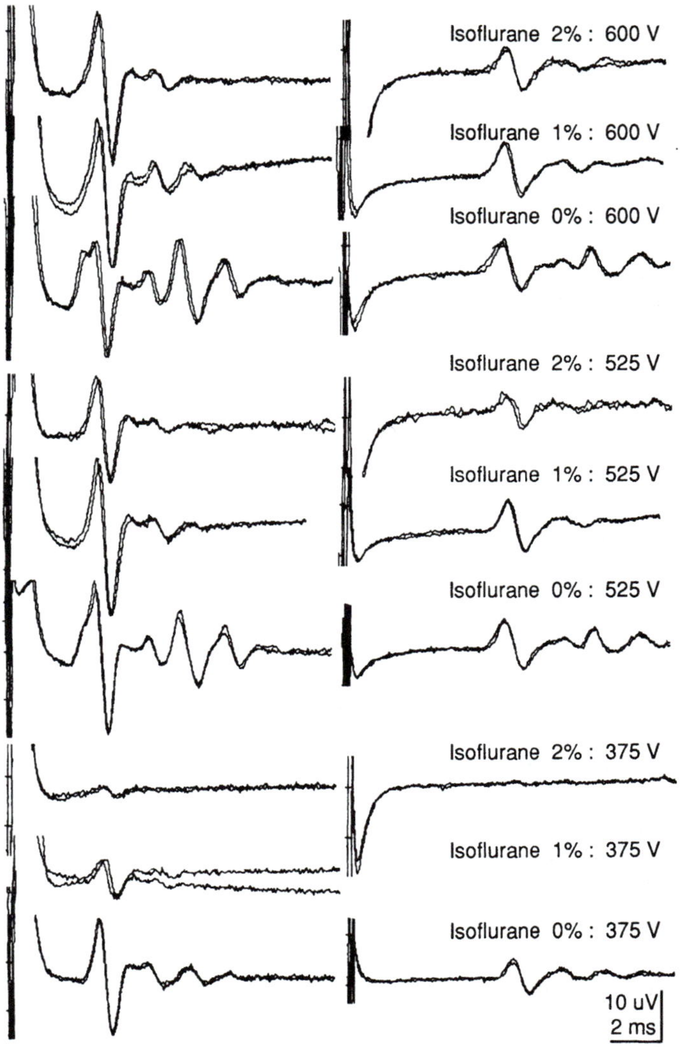

Fig. 6. The effect of isoflurane on the corticospinal volley evoked by transcranial anodal electrical stimulation in a female aged 15. The panels on the left show recordings from low-cervical/high-thoracic level and those on the right the corresponding low-thoracic recordings. I waves are prominent at each stimulus intensity the lower the isoflurane concentration. With the weakest stimulus intensity, 375 V, isoflurane has a dramatic effect on the D wave as well as the I waves. Duplicate averages of 25 responses. From Hicks *et al.* (1992a), with permission

criterion is not appropriate for corticospinal volleys as recorded using the present techniques for the following reasons. First, the reproducibility of the monitored waveform is much higher, particularly when a number of trials are averaged. Secondly, simultaneous recording of two volleys, ascending and descending, at two levels of the neuraxis reduces, if not eliminates, the usual reasons for "false-positive" findings due, for example, to displacement of a recording electrode or failure of a stimulator.

The higher variability of complex bifid and trifid D waves indicates that the optimal stimulus intensity is that which produces the largest simple D wave, not the most intense that can be used before stimulus artifact obscures the recording. The presence of I waves does not enhance the value of the recording for monitoring purposes: indeed, a sequence of I waves at the low-thoracic site can obscure the somatosensory volley when both modalities are tested simultaneously, as in our monitoring technique (see lower traces in Fig. 3, right panel). Moreover, I waves seem to have high intrinsic variability, so that they present no advantages over the D wave.

Effects of anaesthetic agents. There is usually little difficulty in recording corticospinal volleys to transcranial electrical stimulation even in the presence of high concentrations of volatile anaesthetics (e.g., isoflurane, end-tidal concentration 2%), though a higher stimulus intensity may be required to produce a satisfactory volley for monitoring purposes. All components are anaesthetic-sensitive, at least to some extent (Fig. 6), even those due to direct stimulation of the corticospinal axon (i.e., the D wave components), but the most sensitive are the I waves and threshold D waves, as mentioned above (Hicks *et al.*, 1992a,b). The anaesthetic sensitivity of threshold D waves could be because the initial segment and the immediately adjacent node(s) will reflect the excitability of corticospinal neurones. The sensitivity of I waves presumably reflects their dependence on synaptic transmission as well as the excitability of the corticospinal neurone and, in this regard, it is notable that, with increasing anaesthetic levels, late I waves are more affected than early I waves, as if there were more synapses in the chain producing late I waves.

Corticospinal volleys evoked by transcranial magnetic stimulation of the motor cortex

A number of studies has now demonstrated that transcranial magnetic stimulation can produce D waves as well as I waves (Berardelli *et al.*, 1990; Thompson *et al.*, 1991; Burke *et al.*, 1993; Nakamura *et al.*, 1995; Fujiki *et al.*, 1995). The responses to graded stimulus intensities have been compared for transcranial magnetic and electrical stimulation with and without volatile anaesthetics by Burke *et al.* (1993). The coil was the conventional round 9-cm coil of the Magstim 200, centred over the vertex. To the authors' knowledge, there are, as yet, no studies documenting in epidural recordings, the volleys evoked using different coil orientations, though Fujiki *et al.* (1995) used the more focal "figure-of-eight" coil centred on the upper-limb area and Nakamura *et al.* (1995) a double-cone coil. The coil and its orientation were chosen by Burke *et al.* (1993) to be the same as used by Day and colleagues (1989) in their influential studies

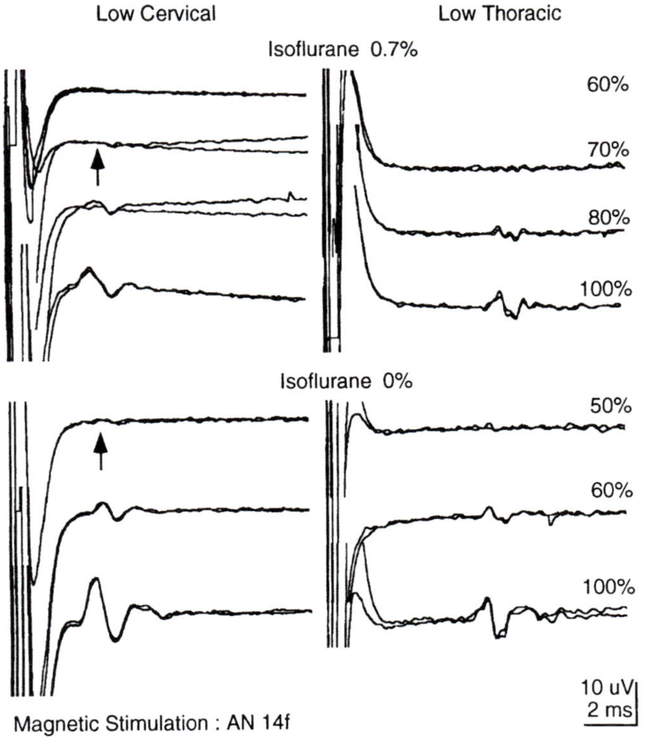

Fig. 7. Corticospinal volleys evoked by graded magnetic stimulation, before and after the withdrawal of isoflurane, in a female patient aged 14. Magnetic stimulation evoked D waves, which had the same latency to peak as anodal electrical stimulation (not illustrated). Following withdrawal of isoflurane in the lower panels, the threshold for the magnetic D wave was reduced from 70% to 50%. No I waves occurred at D-wave threshold. With magnetic stimuli of 100% of the output of the Magstim 200, the D wave was 7 μV when under isoflurane but greater than 14 μV after withdrawal of isoflurane. There were similar changes in the two I waves. The latency of the D wave decreased as it became larger (0.34 ms between threshold and 100% in the lower panel), but there was no evidence of a discrete high-threshold short-latency peak as occurs with electrical stimulation. The vertical arrows indicate the threshold D wave. Traces are duplicate averages of 8 responses. From Burke *et al.* (1993), with permission

comparing the responses of single motor units to transcranial magnetic and electrical stimulation.

Nature of the corticospinal volley. Magnetic stimulation evokes D waves at threshold with this coil and this coil orientation, even when volatile anaesthetics have been completely withdrawn (Figs. 7, 8). However, I waves are also seen at or just above threshold, a finding that is rare with electrical stimulation. Whether the D waves were directed to upper-limb and lower-limb segments cannot be decided from these recordings (though the more focal stimulation used by Fujiki *et al.* (1996) indicates that

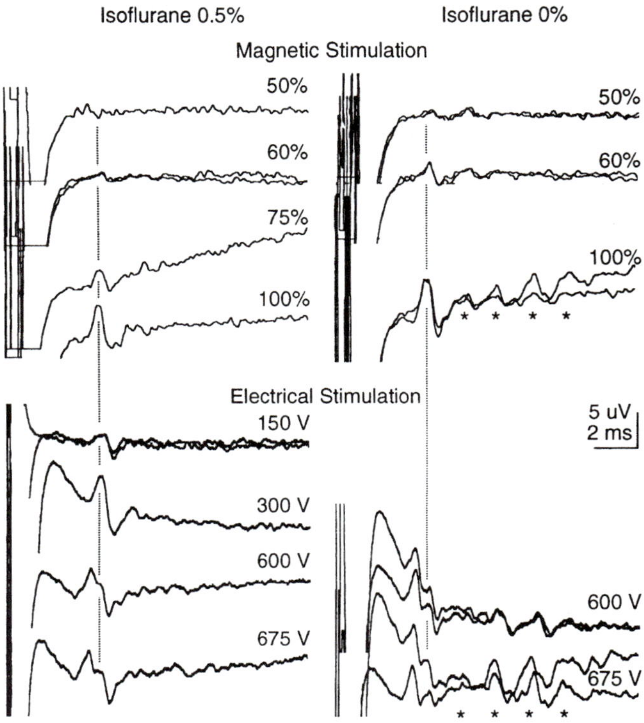

Fig. 8. Corticospinal volleys from the low-cervical region in a male patient aged 13, in response to transcranial magnetic stimulation (upper panels) and transcranial electrical stimulation (lower panels), when receiving isoflurane (left panels) and after with its withdrawal (right panels). Each trace is the average of 4 responses, duplicated for the traces on the right and for two traces on the left. Both types of stimulation produced D waves at threshold. With magnetic stimulation, threshold was 60% of the maximal stimulus intensity, and I wave activity was not seen until 100%. Following withdrawal of isoflurane, I waves were more prominent with the most intense stimuli (indicated by the stars) but an I wave was also present at D-wave threshold (50%). The D wave at 300 V when receiving isoflurane was of a size (7–8 μV) similar to the magnetic D wave at 100% when not receiving isoflurane. With electrical stimulation, but not with magnetic stimulation, a discrete short-latency peak (D_2) developed in advance of the threshold component of the D wave (D_1) as stimulus intensity was increased. From Burke *et al.* (1993), with permission

selective magnetic stimulation of the upper-limb area does activate corticospinal neurones directly). In the anaesthetised patient after withdrawal of isoflurane, the magnetic stimulus intensity necessary to produce a threshold volley (~50%) was higher than that necessary to produce a detectable effect on the active motoneurone pool of an awake subject (~30–40%). It is conceivable that this "depression" affected I waves more than the D wave so eliminating a normally lower threshold for I-wave generation. However, the threshold magnetic D wave was very sensitive to volatile anaesthetics much as had been the threshold electrical D wave (Hicks *et al.*, 1992b; see above), probably no less so than the I waves.

The amplitude and complexity of the evoked volley grew with increasing stimulus intensity, but even with 100% of the output (~1.8–2.0 Tesla) the latency of the D wave did not change greatly. This contrasts with electrical stimulation as described above, but the findings were entirely consistent with the data of Edgley *et al.* (1990) who compared the corticospinal volleys evoked in the monkey by transcranial electrical and magnetic stimulation. However, whatever the stimulus intensity, the volley contained more I-wave activity with magnetic than electrical stimulation for the amount of D-wave activity.

Variability of individual components. Formal studies of the trial-to-trial variability of magnetic volleys have not been undertaken, but the similarities between the magnetic D wave and just-suprathreshold electrical D wave and of the magnetic and electrical I waves suggest that they will share a similar high variability. There are two inter-dependent problems in using magnetic stimulation for monitoring purposes: first, all components in the volley are sensitive to volatile anaesthetics and, secondly, magnetic stimulation does not usually generate large simple D waves, the component of the electrical volley with lowest variability (and lowest anaesthetic sensitivity).

Conclusions

Choice between magnetic and electrical stimulation for monitoring purposes. Ultimately, magnetic and electrical stimulation both excite neural tissue electrically, the difference lying in the mechanism through which the current reaches the site of activation. The major clinical problem with electrical stimulation is pain, a problem that is not relevant in the anaesthetised or unconscious patient.

The coil for magnetic stimulation is cumbersome and requires special supports to secure it to the skull, and this can impede access to the face and larynx by anaesthetic staff. Alternatively, an anaesthetic mask and equipment can interfere with fixation of the coil. On the other hand the scalp electrodes for electrical stimulation are not physically intrusive and are stable once applied.

The D wave in volleys evoked by electrical stimulation are usually larger than with magnetic stimulation, have lower variability and are relatively anaesthetic-resistant.

Repetitive transcranial magnetic stimulation has been advocated recently to potentiate transmission across the corticospinal-motoneurone synapse so that larger and more reliable EMG potentials (or neurogenic MEPs) can be recorded (unpublished promotional literature). However, there is little point in choosing an intrinsically weaker stimulus methodology associated with a more cumbersome method of stimulus delivery, particularly when the necessary equipment is expensive. Repetitive stimulation can be performed more effectively (and more cheaply) using electrical stimulation, without compromising patient safety (Taylor *et al.*, 1993; Kalkman *et al.*, 1995).

The disadvantages of magnetic stimulation detailed above also apply to varying degrees in the awake patient, but are outweighed by the pain produced by the alternative. However, this does not mean that magnetic stimulation should be the preferred stimulus for all situations.

Limitations in extrapolating from the present data. It would be injudicious to extrapolate from the present data to the situation in awake subjects without retaining some reservations. The data recorded using the present technique is affected by filter settings, the choice of bipolar rather than "monopolar" (referential) recording and, most importantly, as discussed above, anaesthetics.

Bipolar recordings provide satisfactory definition of descending and ascending spinal cord volleys, at least as good as a monopolar epidural lead, and were chosen because there is then one recording lead less in the surgical field. A high-pass filter setting of 500 Hz is preferred in this Unit when monitoring corticospinal volleys from the spinal cord because it provides remarkable baseline stability and attenuates stimulus artifact which would otherwise present problems with the low-cervical recording. With it, monitoring is not interrupted by excessive operative artifacts, except when the diathermy is used. However, such a high filter setting effectively differentiates the descending volley, altering latency and attenuating the recorded activity. Its use, together with bipolar recording using electrodes 1–2 cm apart, results in clear definition of only synchronized travelling waves. Standing waves (such as the somatosensory potential generated in the lumbosacral cord) and activity that is desynchronized will be attenuated. For monitoring purposes, this is not a disadvantage, and the trade-off is the relatively artifact-free recording. It is likely that some descending corticospinal activity, particularly I wave activity, is not defined in our recordings. However, dispersed, poorly synchronized I wave activity could not produce discrete peaks in post-stimulus time histograms of the discharge of single motor units in awake subjects (see Day *et al.*, 1989), and this factor probably does not reduce the relevance of the findings described above for the interpretation of such studies.

Acknowledgments

This work was supported by the National Health and Medical Research Council of Australia.

References

Amassian VE, Eberle L, Maccabee PJ, Cracco RQ (1992) Modelling of magnetic coil excitation of human cerebral cortex with a peripheral nerve immersed in a brain-shaped volume conductor: the significance of fibre bending in excitation. Electroencephalogr Clin Neurophysiol 85: 291–301

Berardelli A, Inghilleri M, Cruccu G, Manfredi M (1990) Descending volley after electrical and magnetic transcranial stimulation in man. Neurosci Lett 112: 54–58

Boyd SG, Rothwell JC, Cowan JMA, Webb PJ, Morley T, Asselman P, Marsden CD (1986) A method of monitoring function in corticospinal pathways during scoliosis surgery, with a note on motor conduction velocities. J Neurol Neurosurg Psychiatry 49: 251–257

Burke D, Hicks RG, Stephen JPH (1990) Corticospinal volleys evoked by anodal and cathodal stimulation of the human motor cortex. J Physiol (Lond) 425: 283–299

Burke D, Hicks RG, Stephen JPH (1992a) Anodal and cathodal stimulation of the upper-limb area of the human motor cortex. Brain 425: 283–299

Burke D, Hicks R, Stephen J, Woodforth I, Crawford M (1992b) Assessment of corticospinal and somatosensory conduction simultaneously during scoliosis surgery. Electroencephalogr Clin Neurophysiol 85: 388–396

Burke D, Hicks R, Gandevia SC, Stephen J, Woodforth I, Crawford M (1993) Direct comparison of corticospinal volleys in human subjects to transcranial magnetic and electrical stimulation. J Physiol (Lond) 470: 383–393

Burke D, Hicks R, Stephen J, Woodforth I, Crawford M (1995) Trial-to-trial variability of corticospinal volleys in human subjects. Electroencephalogr Clin Neurophysiol 97: 231–237

Day BL, Dressler D, Maertens de Noordhout A, Marsden CD, Nakashima K, Rothwell JC, Thompson PD (1989) Electric and magnetic stimulation of human motor cortex: surface EMG and single motor unit responses. J Physiol (Lond) 412: 449–473

Edgley SA, Eyre JA, Lemon RN, Miller S (1990) Excitation of the corticospinal tract by electromagnetic and electrical stimulation of the scalp in the macaque monkey. J Physiol (Lond) 425: 301–320

Fujiki M, Isono M, Hori S, Ueno S (1996) Corticospinal direct response to transcranial magnetic stimulation in humans. Electroencephalogr Clin Neurophysiol 101: 48–57

Hicks RG, Burke D, Stephen JPH (1991) Monitoring spinal cord function during scoliosis surgery with Cotrel-Dubousset instrumentation. Med J Aust 154: 82–86

Hicks RG, Woodforth IJ, Crawford MR, Stephen JPH, Burke DJ (1992a) Some effects of isoflurane on I waves of the motor evoked potential. Br J Anaesth 69: 130–136

Hicks RG, Burke D, Stephen JPH, Woodforth I, Crawford M (1992b) Corticospinal volleys evoked by electrical stimulation of human motor cortex after withdrawal of volatile anaesthetics. J Physiol (Lond) 456: 393–404

Jones SJ, Edgar MA, Ransford AO (1982) Sensory nerve conduction in the human spinal cord: epidural recordings made during scoliosis surgery. J Neurol Neurosurg Psychiatry 45: 446–451

Jones SJ, Edgar MA, Ransford AO, Thomas NP (1983) A system for the electrophysiological monitoring of the spinal cord during operations for scoliosis. J Bone Joint Surg 65B: 134–139

Kalkman CJ, Ubags LH, Been HD, Swaan A, Drummond JC (1995) Improved amplitude of myogenic motor evoked responses after paired transcranial electrical stimulation during sufentanil/nitrous oxide anesthesia. Anesthesiology 83: 270–276

Kernell D, Wu C-P (1967) Responses of the pyramidal tract to stimulation of the baboon's motor cortex. J Physiol (Lond) 191: 653–672

Lance JW, Manning RL (1954) Origin of the pyramidal tract in the cat. J Physiol (Lond) 124: 385–399

Maccabee PJ, Amassian VE, Eberle LP, Cracco RQ (1993) Magnetic coil stimulation of straight and bent amphibian and mammalian peripheral nerve in vitro: locus of excitation. J Physiol (Lond) 460: 201–219

Nakamura H, Kitagawa H, Kawaguchi Y, Tsuji H, Takano H, Nakatoh S (1995) Intracortical facilitation and inhibition after paired magnetic stimulation in humans under anesthesia. Neurosci Lett 199: 155–157

Owen JH, Laschinger J, Bridwell K, Shimon S, Nielsen C, Dunlap J, Kain C (1988) Sensitivity and specificity of somatosensory and neurogenic-motor evoked potentials in animals and humans. Spine 13: 1111–1118

Patton HD, Amassian VE (1954) Single- and multiple-unit analysis of cortical stage of pyramidal tract activation. J Neurophysiol 17: 345–363
Phillips CG, Porter R (1977) Corticospinal neurones. Their role in movement. Academic Press, London
Rothwell JC, Thompson PD, Day BL, Boyd S, Marsden CD (1991) Stimulation of the human motor cortex through the scalp. Exp Physiol 76: 159–200
Rothwell J, Burke D, Hicks R, Stephen J, Woodforth I, Crawford M (1994) Transcranial electrical stimulation of the motor cortex in man: further evidence for the site of activation. J Physiol (Lond) 481: 243–250
Stephen JP, Sullivan MR, Hicks RG, Burke DJ, Woodforth IJ, Crawford MR (1996) Cotrel-Dubousset instrumentation in children using simultaneous motor and somatosensory evoked potential monitoring. Spine 21: 2450–2457
Taylor BA, Fennelly ME, Taylor A, Farrell J (1993) Temporal summation – the key to motor evoked potential spinal cord monitoring in humans. J Neurol Neurosurg Psychiatry 56: 104–106
Thompson PD, Day BL, Crockard HA, Calder I, Murray NMF, Rothwell JC, Marsden CD (1991) Intra-operative recording of motor tract potentials at the cervico-medullary junction following scalp electrical and magnetic stimulation of the motor cortex. J Neurol Neurosurg Psychiatry 54: 618–623
Ugawa Y, Rothwell JC, Day BL, Thompson PD, Marsden CD (1991) Percutaneous electric stimulation of corticospinal pathways at the level of the pyramidal decussation in man. Ann Neurol 29: 418–427
Woodforth IJ, Hicks RG, Crawford MR, Stephen JP, Burke DJ (1996) Variability of motor evoked potentials recorded during nitrous oxide anesthesia from the tibialis anterior muscle after transcranial electrical stimulation. Anesth Analg 82: 744–749

Correspondence: Dr. D. Burke, Prince of Wales Medical Research Institute, High Street, Randwick, N.S.W. 2031, Australia.

Section VI

Spinal cord monitoring in clinic: pharmacology, therapy and outcome measures

Spinal Cord Monitoring
Eds: E. Stålberg, H. S. Sharma, Y. Olsson

Chapter 19

Novel approaches to the treatment of post compressive spinal cord injury

W. F. Young, D. Dominique, U. S. Vasthare, and **R. F. Tuma**

Department of Neurosurgery, Temple University Hospital, Philadelphia, USA

Introduction

Spinal cord injuries are an important cause of longterm disability with current estimates of their frequency at a range of 30–50 cases per million population. Over the last decade, our knowledge of the pathophysiological mechanisms associated with spinal cord injury has increased dramatically. Clinical and basic science research have improved our ability to preserve and restore function following these devastating injuries. In this chapter, we will briefly review some of the important advancements that have been gained through basic science research, and in particular review our experimental findings utilizing a model of compressive spinal cord injury.

Mechanisms of spinal cord injury

Human spinal cord injuries stem from both primary mechanical injury and a cascade of secondary pathophysiologic mechanisms. Primary injury mechanisms which characterize clinical spinal cord injuries include acute compression, impact injury, lacerations, distraction forces and shear forces (Balentine, 1978; Cusick *et al.*, 1982; Taylor, 1951). As a result of the primary mechanical damage, a number of "secondary" mechanisms have been postulated to occur resulting in an exacerbation of the initial damage.

Allen in 1911, initially proposed the theory of secondary mechanisms of spinal cord injury when it was demonstrated that removal of post-traumatic hematomyelia by myelotomy improved neurologic function after a weight-drop injury (Allen, 1911, 1914). He also provided histologic evidence of a progressive increase in damage with time. Allen reported petechial hemorrhage in the spinal cord gray matter and edema of the white matter. Hemorrhagic regions in the gray matter increased after two hours and swollen neuron axis cylinders were seen after four hours. These findings provided the first evidence that damage after injury is progressive and is a result of an ongoing "secondary" process.

Table 1. Prosposed secondary injury mechanisms associated with spinal cord injury

1. *Biochemical mechanisms*
Arachidonic acid cascade
O_2 free radical formation/lipid peroxidation
Excess glutamate (neurotransmitter, excitotoxocity)
Eicosanoid production (prostaglandins)
Nitric oxide
2. *Vascular mechanisms*
Loss of autogregitation
Neurogenic shock
Alterations in microcirculation
Spinal cord hypoperfusion
vasospasm
thrombosis
3. *Cell-mediated mechanisms*
Leukocyte/endothelial interactions
Platelet/endothelial interactions
Mononuclear infiltration
4. *Electrolyte mechanisms*
Increased intracellular calcium
Increased extracellular potassium
Increased sodium permeability

A multitude of factors are responsible for secondary spinal cord injury which include formation of free radicals, excessive cellular calcium entry, excitatory neurotransmitter release and microcirculatory changes (Table 1) (Anderson *et al.*, 1985; Anderson and Hall, 1993; Anderson *et al.*, 1976; Balentine, 1978; Demopoulos *et al.*, 1978; Fairholm and Tunbull, 1970; Green and Wagner, 1973; Griffith *et al.*, 1978; Hall and Wolf, 1986; Kapadia, 1984; Osterholm, 1974; Sandler and Tator, 1976; Stewart and Wagner, 1979; Tanno *et al.*, 1992; Tator and Fehlings, 1991). Greater understanding of these mechanisms affords the opportunity to impact pharmacologically and thus ameliorate secondary damage.

Models of spinal cord injury

The first experimental model of spinal cord injury was developed by Allen in 1911. His technique of causing injury utilized a weight-drop method in which the magnitude of injury was dependent upon the amount of weight and distance of impact. For many years, this method dominated the literature. Recently, weight-drop methods have been improved to include technology which can monitor the acceleration associated with the weight drop (Behrmann *et al.*, 1994). These newer technologies offer the possibility of better quantifying the biomechanical aspects of the injury, and therefore improved standardization (see Chapter 13, in this volume).

Table 2. Models of experimental spinal cord injury

Rapid contusion models	Compression model	Compression/contusion
Weight drop	static weights	aneursym clip
Controlled acceleration models	balloon catheter	forceps
	inflatable cuffs	
	screws	
Closed impact	miscellaneous	
	distraction	
	ischemia	
	cutting	
	aspiration	

Most experimental methods involve performing a laminectomy prior to producing the injury. This allows for direct visualization of the cord, and the potential for more consistent injury production. A compression model of spinal cord injury was first developed by Tarlov (Tarlov *et al.*, 1953). A balloon catheter was used to produce graded compression. Other investigators have utilized a variety of techniques to produce a controlled compression including inflatable cuffs, screws and static weight (Al-Mefty *et al.*, 1993; Black *et al.*, 1986; Nurick, 1972). (Table 2). Still others have developed methods to produce brief periods of compression using aneurysm clips or forceps (Tarlov *et al.*, 1953; Tator and Decker, 1973; Wakefield and Eidelberg, 1975). A closed impact model has been developed in dogs which offers the opportunity to closely mimic clinical situations (Griffith and Miller, 1974).

For several years our laboratory has utilized a static weight compression device, which was modified after the technique first developed by Nystrom (Nystrom *et al.*, 1988; Nystrom and Berglund, 1988). The device used to compress the cord consists of a platform on top of a metal rod connected to a 5 mm by 2 mm foot plate which rests on the spinal cord. A micromanipulator is used to place the foot plate over the cord. The edges of the foot plate are beveled to conform to the shape of the spinal cord. Weights are then placed on top of the platform. The amount of injury is thus dependent on two variables:

1. the total weight resting on the cord and
2. the duration of time that the weights rest on the cord.

We have found this model to be relatively easy to perform and produces a consistent injury as measured by somatosensory evoked potentials, neurologic function and histologic analysis.

Pharmacological treatment of spinal cord injury

Experimental work over the last decade has done much to advance our understanding of the pathophysiological basis of secondary injury (see Chapters 8, 13 and 14, in this volume). In addition, it has afforded the opportunity to identify agents which can potentially ameliorate secondary injury. Currently, the best example of taking basic science development to clinical application in spinal cord injury is the use of high dose methlylprednisolone.

Extensive research in animal preparations provided evidence that high dose methylprednisolone is neuroprotective after spinal cord injury (Bracken *et al.*, 1990; Braughler *et al.*, 1987; Hall, 1992; Young and Flamm, 1982). This prompted the development of the National Acute Spinal Cord Injury Study or NASCIS (Bracken *et al.*, 1990). In a prospective randomized trial, methylprednisolone(MP 30 mg/kg) followed by infusion at 5.4 mg/kg/h for 23 hours was administered. The findings of this study demonstrated that patients with acute spinal cord injury benefited from treatment with methylprednisolone as characterized by improved neurologic function. The neuroprotective effects of MP are believed to result from scavenging oxygen free radicals. Experimental work suggested that high doses of this glucocorticoid inhibits lipid peroxidation (Demediuk *et al.*, 1985). In addition, MP promotes the release of anti-inflammatory peptides (lipocortins). Lipocortins inhibit calcium-activated phospholipase activity. They therefore inhibit the production of eicosanoids and thus prevent the formation of superoxide radicals by leukocytes (Beckman *et al.*, 1990; Williams and Higgs, 1988). In addition, glucocorticoids may regulate cytokine induced cyclooxygenase 2 expression. Through blockage of inflammatory substances, spinal cord blood flow is improved and cord edema is reduced. The understanding of neuroprotective pharmacology of MP led to the design of an analog which lacks the glucocorticoid receptor based actions, yet retain its protective inhibition of lipid peroxidative action (21 aminosteroids).

Since, oxygen radical mediated lipid peroxidation appears to be a critical factor in post traumatic neuronal degeneration numerous studies have evaluated neuroprotective agents with lipid antioxidant activity. Intensive pretreatment of animals with the endogenous lipid peroxyl radical scavenger alpha tocopherol (Vitamin E) results in attenuation of post-traumatic spinal cord ischemia and enhances chronic neurologic recovery (Hall *et al.*, 1989). The compounds that combine the amino functionality of the 21-amino steroids with the peroxyl radical scavenging chromanal portion of Vitamin E (2-methylaminochromans) have recently shown promise as neuroprotectants.

Free radical formation within neurons is greatly enhanced by excessive intracellular calcium. There is a large body of evidence suggesting that calcium may be the final common pathway of cell death. After injury, calcium may enter the neuron via a number of mechanisms including through disrupted cell membranes, through voltage dependent calcium channels or through receptor mediated transmembrane transport activated by excitatory transmitters such as glutamate. Currently, voltage dependent calcium channel antagonist such as Nimodipine has not yet proved effective in patients. However, Flunarizine has shown promise when used in combination with methylprednisolone experimentally. Experimental evidence firmly suggests that excess glutamate contributes greatly to secondary tissue damage after CNS injury. Its importance has been demonstrated in post-cerebral ischemia and after brain and spinal trauma. Glutamate excitotoxicity probably plays a major role in neuronal death. Experimental evidence also suggests that the N-methyl-D-aspartate (NMDA) receptor (glutamate binds to NMDA receptor) contributes to glutamate excitotoxicity by allowing calcium to enter neurons. Ketamine, dextromethorphan and MK801 inhibit glutamate receptors and are effective neuroprotective agents experimentally.

The understanding of the involvement of nitric oxide (NO) in both spinal cord and brain injury is complicated becuase of its multiple effects (see Chapter 8). NO can function as a neurotransmitter and as a vasodilator. NO is released from the substrate L-Arginine in the presence of the nitric oxide synthase (NOS) (Bastian and Hibb, 1994; Bereta *et al.*, 1994; Cho *et al.*, 1992; Choi, 1990; Hecker *et al.*, 1990). The effects of NO are variable, depending upon the circumstances stimulating its synthesis. NOS occurs in two forms calcium dependent NOS (constitutive NOS cNOS) which is present in brain & endothelial cells and calcium independent NOS (inducible NOS-iNOS) in macrophages and other cells. Under pathophysiological conditions such as ischemia and reperfusion injury or trauma to brain and spinal cord, inflammatory mediators such as cytokines trigger the induction of calcium independent NOS from various cells including endothelial cells, vascular smooth cells, and also from neutrophils. Once induced, iNOS continuously produces NO resulting in sustained relaxation of vascular smooth muscle, reduced responsiveness to vasoconstrictors and possibly tissue damage via its cytotoxic effects. NO produced through stimulation of cNOS under other circumstances may have beneficial effects such as enhancing blood flow, inhibition of platelet aggregation, inhibition of leukocyte adhesion. Therefore appropriate use of selective inhibitors of iNOS may attenuate the cytotoxic damage, while stimulating cNOS could in theory enhance the beneficial effects. Much of our understanding of the effects of NO is derived from experiments utilizing models of cerebral ischemia. It has been suggested that its effects may also be important in the pathophysiology of spinal cord injury. However, the role of NO in spinal cord injury remains to be elucidated (see Chapter 8 in this volume).

In this section, we have briefly reviewed important developments in the treatment of acute spinal cord injury. Many other agents have been studied, and to detail them is beyond the scope of this chapter. To date, MP is the only one shown to be effective clinically and has become the standard by which all future agents will be judged.

The use of hypertonic saline in experimental spinal cord injury

During the 1980s, it was suggested that infusion of hypertonic saline solutions is a better method of restoring cardiovascular homeostasis after hypovolemic shock. Investigators have demonstrated through experimental models of hemorrhagic shock that hypertonic saline infusion increases cardiac output, and improves capillary perfusion in a variety of organs (kidney, pancreas, gastrointestinal tract, skeletal muscle, and brain) when compared to isotonic saline solutions (Battistella and Wisner, 1991; Gazitua *et al.*, 1971; Nakayama *et al.*, 1984; Nemeck, 1978; Nolte *et al.*, 1992; Prough *et al.*, 1991; Tocha-e-Silva *et al.*, 1986; Schmoker *et al.*, 1991; Shackford *et al.*, 1994). These findings are thought to be accomplished at least in part through rapid mobilization of extravascular fluid volume into the vascular space and possibly through a direct positive ionotropic effect.

It has been hypothesized that the beneficial effects of hypertonic saline may be mediated through:
1. osmotically-induced relaxation of vascular smooth muscle
2. post-traumatic reduction in vascular endothelial swelling
3. anti-inflammatory effects

The post-traumatic reduction in endothelial swelling observed after hypertonic saline infusion was identified by Mazzoni *et al.* (1989, 1990). The anti-inflammatory effects of hypertonic saline may also be mediated through alterations in leukocyte/endothelial interactions (Bevilacqua, 1993; Granger and Kubes, 1994; Harlan, 1985; Johnson *et al.*, 1988; Weiss, 1989). It has been demonstrated that animals recuscitated following hemorrhage with hypertonic saline exhibited decreased leukocyte adhesion and rolling in skin blood vessels (Saetzler *et al.*, 1996).

Endothelial cell injury leading to leukocyte infiltration may be an important mechanism of secondary injury after spinal cord injury. Expression of cell adhesion molecules (CAMs) may be a primary mechanism for attracting leukocytes into areas of injury. (Albelda and Buck, 1990; Kubes *et al.*, 1994; Slogan *et al.*, 1992; Xu *et al.*, 1990; Zhuang *et al.*, 1993). Two important classes of adhesion molecules are integrins and selectins. Expression of these adhesion molecules and subsequent activation of leukocytes may be an important mechanism for hypoperfusion after spinal cord injury. (Bernofsky, 1991; Biagas *et al.*, 1992; Blight, 1985, 1992; Giulian and Robertson, 1990; Lewis and Granger, 1986). Down regulation of leukocyte and endothelial activation following injury could attenuate secondary inflammatory changes that would lead to a reduction in tissue perfusion.

Our laboratory was the first to evaluate the effect of 7.5% hypertonic saline administration following spinal cord injury. We have tested the efficacy of hypertonic saline (7.5%) on three studies utilizing animal models of spinal cord injury (acute and chronic). The first study (acute model 85) compared the effects of hypertonic saline with isotonic saline solutions on spinal cord blood flow and somatosensory evoked responses for upto 1 h following cervical spinal cord injury. The second study (acute model 70) evaluated the effect of hypertonic saline administration at various time periods following cervical spinal cord injury. The third study (chronic injury 72) evaluated hypertonic saline administration on neurologic function, spontaneous voiding and histopathologic changes following thoracic spinal cord injury.

Our method of inducing injury utilizing a compression plate is described earlier in the Chapter. We monitor spinal cord blood flow utilizing a TSI Laser Blood Perfusion Monitor (Model BPM 403a, TSI, St. Paul, MN, USA). A 0.84 mm diameter probe is fixed 1 mm above the spinal cord directly over the site of injury to monitor flow. Flow is monitored prior to injury and continuously throughout the experiment following the removal of compression. To briefly describe the measurement technique, when the fiberoptic probe is placed over a perfused tissue, the emitted photons collide with moving red blood cells and stationary tissue. Only those photons that are scattered by the moving red blood cells undergo a Doppler frequency shift. A photodetector that uses a fiber optic probe to collect the scattered photons, converts the reflected Doppler frequency spectrum into an analog voltage signal. This technique provides continuous on line measurement of flow but does not allow measurement of absolute flow values. However, the laser Doppler technique has been shown to correlate very closely with the radioisotope labeled microsphere technique in the assessment of changes in flow (Eyre *et al.*, 1988).

In the acute studies, somatosensory evoked potentials (SSEPs) are recorded utilizing a technique we have used for a number of years (Tator and Fehlings, 1991). Platinum needle electrodes are placed subcutaneously for field stimulation of the median nerve in the foreleg. Electrical stimuli are delivered at a frequency of 4 hz. with a duration of 0.1 ms and a voltage of 8–10 v. Electroencephalographic activity is recorded from C3 and Fz electrode position corresponding to the somatosensory area of the cortex on the contralateral side. The information was fed through an analog-digital (A–D) converter to a computer for analysis.

Hypertonic saline administration was found to have a significant impact, not only on spinal cord blood flow, but also on preserving somatosensory evoked potentials (acute studies). The magnitude of reactive hyperemia was increased in the animals receiving hypertonic saline. Since blood pressure did not differ significantly between the groups the increased blood flow seen following spinal cord injury was primarily the result of a decrease in spinal cord vascular resistance. Equivalent expansion of blood volume with isotonic saline did not produce a protective effect.

In the second study, we evaluated the effects of hypertonic saline infusion at various time periods following spinal cord injury. The magnitude of blood flow changes following hypertonic saline administration was related to the time of injury. The greatest increase in flow occured when hypertonic saline was administered immediately following injury. These results indicate that the effect of hypertonic saline administration depends on the pathologic state of the tissue (Tuma *et al.*, 1997). Interestingly, hypertonic saline administration did not increase flow in uninjured tissue. This might have greater significance in preventing steal syndrome. Moreover, we have demonstrated a sustained hyperosmolar effect after a single dose of hypertonic saline (as measured by serum osmolality).

Recently, we investigated the effects of hypertonic saline infusion post spinal cord injury in a chronic model (Tuma *et al.*, 1997). After thoracic spinal cord injury animals were allowed to recover from anesthesia, they were monitored for two weeks after injury with daily assessments of neurologic function. Rats treated with hypertonic saline after compression had much faster rate of recovery with respect to all parameters evaluated than rats treated with normal saline. None of the animals treated with normal saline had spontaneous voiding prior to the third day after injury. Forty percent of the hypertonic saline treated animals demonstrated spontaneous voiding the first day after injury. This number increased to 70% by the second day of injury. It was not until the seventh day of injury the number of animals with spontaneous voiding was approximately the same in both experimental groups. The mean number of days before the rats demonstrated spontaneous voiding was significantly less with hypertonic saline treated animals (2.5 days) than with animals treated with normal saline (5.6 days).

Motor function also returned more quickly in the animals treated with hypertonic saline. None of the animals treated with normal saline displayed any spontaneous hindlimb movement until five days after injury. This is in contrast to the group treated with hypertonic saline, where 40% of the animals had spontaneous movement during the first four days after injury. All of the hypertonic saline animals had spontaneous hindlimb movement by the eighth day after injury, whereas only 75% of the animals treated with normal saline had spontaneous hindlimb movement by the tenth day after injury. The neurologic score, as assessed by the Basso, Beattie and Bressnahan locomotor rating scale, was also better in the hypertonic treated group. The hypertonic saline treated group had significantly higher scores during each of the ten days of recovery.

In addition, through histologic analysis of spinal cords (control and hypertonic saline resuscitated animals) we demonstrated a better preservation of spinal cord architecture in animals treated with 7.5% hypertonic saline.

Table 3. Potential effects of hypertonic saline

Direct influence on vascular smooth muscle
Inhibiting free radical production
Direct free radical scavenging effects
Inhibition of cytokine production
Attenuation of endothelial swelling
Preventing endothelial dysfunction
Preventing leukocyte platelet activation
Preventing leukocyte endothelial activation
Influence on NO production

Hypertonic saline administration in the above mentioned models of spinal cord injury has resulted in improved spinal cord blood flow and better preservation of neurological function following injury. The mechanisms that are hypothesized to contribute to this effect are included in Table 3.

Future considerations

Hypertonic saline may have effect on one or more aspects of secondary injury mechanisms. These aspects of secondary injury will result in microcirculatory disturbances leading to hypoperfusion. Our laboratory is actively pursuing the mechanisms responsible for microcirculatory disturbances leading to hypoperfusion.

In order to better elucidate the mechanisms responsible for microcirculatory disturbances leading to hypoperfusion, we utilize in-vivo microscopy, to directly evaluate and perform experiments on the surface vasculature of the spinal cord.

Alter performing a laminectomy in our standard fashion, the dura is carefully opened as to not damage the cord. The spinal cord is observed using a Leitz Ortholux microscope using a 10x ocular and a 32x objective. The light source is provided by a ring light from Cole Palmer. A BG38 red suppression filter which will absorb the red wave length producing a dark image is used. The microscope is connected to a Zeiss color camera which is attached to a VCR where the images are taped for off-line analysis. The VCR is connected to a color television to observe the image during the experiment. The spinal cord is initially viewed and video-taped to obtain baseline parameters. An injury is induced in our standard manner. Microscopy images are then video-taped post injury.

In vivo microscopy has allowed us to measure spinal cord arteriole diameters and using fluorescent dyes, (which preferentially stain leukocytes), we can study the interaction of leukocytes and endothelial cells after injury. Preliminary data has demonstrated a significant increase in arteriole diameter of injured animals treated with hypertonic saline when compared to normal saline treated animals. Hypertonic saline seems to selectively effect vascular smooth muscle in regions of injury. We are also testing the hypothesis that hypertonic saline may inhibit the expression of adhesion molecules on leukocytes, endothelial cells and possibly platelets by using fluorescence microscopy. We therefore think that this technology will help us to further elucidate cellular and intravascular mechanisms after spinal cord injury.

References

Albelda S, Buck C (1990) Integrins and other cell adhesions molecules. FASEB J, pp 2869–2880

Allen AR (1911) Surgery of experimental lesion of spinal cord equivalent to crush injury of fracture dislocation of spinal column. JAMA 57: 878–890

Allen AR (1914) Remarks on the histopathological changes in the spinal cord due to impact: An experimental study. J Nerv Ment Dis 41: 141–147
Anderson DK, Demediuk P, Saunders RD, Dugan LL, Means ED, Horrocks LA (1985) Spinal cord injury and protection. Ann Emerg Med 14: 816–821
Anderson DK, Hall ED (1993) Pathophysiology of spinal cord trauma. Ann Emerg Med 22: 987–992
Anderson DK, Prockop LD, Means ED, Hartley LE (1976) Cerebrospinal fluid lactate following experimental spinal cord injury. J Neurosurg 49: 715–722
Al-Mefty O, Harkey HL, Marawi I, Haines DE, Peeler DF, Wilner HI, Smith RR, Holaday HR, Haining JL, Russell WF *et al.* (1993) Experimental cervical myelopathy effects of ischemia and compression of the canine cervical spine cord. J Neurosurg 79: 550–561
Balentine JD (1978) Pathology of experimental spinal cord trauma. 1. The necrotic lesion as a function of vascular injury. Lab Invest 39: 236–253
Balentine JD (1978) Pathology of experimental spinal cord trauma. ll. Ultrastructure of axons and myelin. Lab Invest 39: 254–266
Battistella FD, Wisner DH (1991) Combined hemorrhagic shock and head injury. Effects of hypertonic saline (7.5%) resuscitation. J Trauma 31: 182–188
Bastian NR, Hibb JB Jr (1994) Assembly and regulation of NADPH oxidase and nitric oxide synthase. Curr Opin Immunol 6: 131–139
Beckman JS, Beckman TW, Chen J *et al.* (1990) Apparent hydroxyl radical production by peroxynitrite: Implications for endothelial injury from nitric oxide and superoxide. Proc Natl Acad Sci USA 87: 1620–1624
Behrmann O, Bresnahan J, Beattie M (1994) Modeling of acute spinal cord injury in the rat: neurprotection and enhanced recovery with methyprednisone, 0–74006F and Ym-14673. Exp Neurol 126: 63–75
Bereta M, Bereta J, Georgoff I , Coffman FD, Cohen S, Cohen MC (1994) Methylxanthines and calcium-mobilizing agents inhibit the expression of cytokine-inducible nitric oxide synthase and vascular cell adhesion molecure-1 in murine microvascular endothelial cells. Exp Cell Res 212: 230–242
Bernofsky C (1991) Nucleotide chloramines and neutrophil-medicated cytotoxicity. FASEB 5: 295–300
Bevilacqua MP (1993) Endothelial-leukocyte adhesion molecules. Annu Res Immunol 11: 767–804
Biagas KV, Uhl MW, Schiding JK *et al.* (1992) Assessment of post-traumatic polymorphonuclear leukocyte accumulation in rat brain using tissue myeloperoxidase assay and vinblastine treatment. J Neurotrauma 9(4): 363–371
Black P, Markowitz R, Cooper V, Mechanic A, Kushner H, Damjanov I, Finkelstein SD, Wachs KC (1986) Models of spinal cord injury, Part 1. Neurosurgery 19: 752–762
Blazka ME, Harry GJ, Luster MI (1994) Effect of lead acetate on nitrite production by murine brain endothelial cell cultures. Toxicol Appl Pharmacol 126: 191–194
Blight AR (1985) Delayed demyelination and macrophage invasion: A candidate for secondary cell damage in spinal cord injury. Cent Nerv Syst Trauma 2: 299–315
Blight AR (1992) Macrophages and inflammatory damage in spinal cord injury. J Neurotrauma 9 [Suppl] 1: S83–S91
Bracken M, Shepard M, Collins W *et al.* (1990) A randomized controlled trial of methylprednisolone or naloxone in the treatment of acute spinal cord injury 322: 1405–1411
Braughler JM, Means E, Waters TR, Anderson DK (1987) Evaluation of an intensive methylprednisolone sodium succinate dosing regimen in experimental spinal cord injury. J Neurosurg 67: 102–105
Cho HJ, Xie O, Calacay J, Mumford RA, Swiderek KM, Lee TD, Nathan C (1992) Calmodulin is a subunit of nitric oxide synthase from macrophages. J Esp Med 176: 599–604
Choi DW (1993) Nitric oxide: Friend or foe to the injured brain? Proc Natl Acad Sci USA 90: 9741–9743
Cusick JF, Myklebust J, Zyvoloski M, Sances A Jr, Houterman C, Larson SJ (1982) Effects of vertebral column distraction in the monkey. J Neurosurg 57: 652–659

Demediuk P, Saunders RD, Anderson DK, Means ED, Horrocks LA (1985) Membrane lipid changes in laminectomized and traumatized cat spinal cord. Proc Natl Acad Sci 82: 7071–7075
Demopoulos HB, Yoder M, Gutman EG, Seligman ML, Flamm ES, Ransohoff J (1978) The fine structure of endothelial surfaces in the microcirculation of experimentally injured feline spinal cords. Scanning Electron Microsc 11: 677–682
Eyre JA, Essex TJ, Flecknell PA, Bartholomew PH, Sinclair JI (1988) A comparison of measurements of cerebral blood flow in the rabbit using laser Doppler spectroscopy and radionuclide labeled microspheres. Clin Phys Physiol Meas 9: 65–74
Fairholm D, Tunbull I (1970) Microangiographic study of experimental spinal injuries in dogs and rabbits. Surgical Forum 21: 435–455
Gazitua S, Scott JB, Swindall B, Haddy FJ (1971) Resistance responses to local changes in plasma osmolarity in three vascular beds. Am J Physiol 220: 284–391
Giulian D, Robertson C (1990) Inhibition of mononuclear phagocytes reduces ischemic injury in the spinal cord. Am Neurol 27: 33–42
Granger DN, Kubes P (1994) The microcirculation and inflammation: Modulation of leukocyte endothelial cell adhesion. J Leukoc Biol 5: 662–675
Green BA, Wagner FC (1973) Evolution of edema in the acutely injured spinal cord: A fluorescence microscopic study. Surg Neurol 1973: 98–101
Griffith IR, Miller R (1974) Vascular permeability to protein and vasogenic oedema in experimental concussive injuries to the canine spinal cord. J Neurol Sci 22: 291–304
Griffith IR, Burns N, Crawford AR (1978) Early vascular changes in the spinal gray matter following impact injury. Acta Neuropathol 41: 33–39
Hall ED (1992) The neuroprotective pharmacology of methylprednisolone. J Neurosurg 76: 13–22
Hall ED, Wolf DL (1986) A pharmacological analysis of the pathophysiological mechanisms of post-traumatic spinal cord ischemia. J Neurosurg 64: 951–961
Hall ED, Younkers PA, Horan KL, Baughler JM (1989) Correlation between attenuation of post-traumatic spinal cord ischemia and preservation of tissue Vitamin E by the 21-Aminosteroid U74006F. Evidence of an in vivo antioxidant mechanism. J Neurotrauma 6: 169–176
Harlan JM (1985) Leukocyte-endothelial interactions. Blood 65: 513–525
Hecker M, Sessa WC, Harris HJ, Anggard EE, Vane JR (1990) The metabolism of L-arginine and its significance for the biosynthesis of endothelium-derived relaxing factor: Cultured endothelial cells recycle L-citrulline to L-arginine. Proc Natl Acad Sci USA 87: 8612–8616
Johnson RJ, Alpers CE, Pritze P *et al.* (1988) Platelets mediate neurotrophil-dependent immune complex nephritis in the rat. J Clin Invest 82: 1225–1235
Kapadia SE (1984) Ultrastructural alteration in blood vessels of the white matter after experimental spinal cord trauma. J Neurosurg 61: 539–544
Kubes P, Kurose I, Granger DN (1994) NO donors integrin-induced leukocyte adhesion but P-selectin dependent rolling in postischemic venules. Am J Physiol 267: H931–937
Lewis RE, Granger HJ (1986) Neutrophil-dependent mediation of micro-vascular permeability. Fed Proc 45: 109–113
Mazzoni MC, Borgstrom P, Intaglietta M, Arfors KE (1990) Capillary narrowing in hemorrhage shock is rectified by hyperosmotic saline-dextran reinfusion. Circ Shock 31: 407–418
Mazzoni MC, Borgstrom P, Intaglietta M, Arfors KE (1989) Lumenal narrowing and endothelial cell swelling in skeletal muscle capillaries during hemorrhagic shock. Circ Shock 29: 27–39
Means ED, Anderson DK (1983) Neuronophagia by leukocytes in experimental spinal cord injury. Neuropathol Exp Neurol 42: 707–719
Moreno-Flores MT, Bovolenta P, Nieto-Sampedro M (1993) Polymorpho-nuclear leukocytes in brain parenchyma after injury and their interaction with purified astrocytes in culture. Glia 7(2): 146–157

Nakayama S, Sibley L, Gunther RA, Holcroft JW, Kramer GC (1984) Small-volume resuscitation with hypertonic saline (2,400 mOsm/Liter) during hemorrhagic shock. Circ Shock 13: 149–159
Nemecek S (1978) Morphological evidence of microcirculatory disturbances in experimental spinal cord trauma. Adv Neurol 20: 395–405
Hypertonic saline or lactated Ringer's (1983) (Effect on the pulmonary and systemic microcirculation). Circ Shock 10: 179–188
Nolte D, Bayer M, Lehr HA, Becker M, Kromback F, Kreimeier U, Messmer K (1992) Attenuation of postischemic microvascular disturbances in striated muscle by hyperosmolar saline dextran. Am J Physiol 263: H1411–H1416
Nurick S (1972) The pathogenesis of the spinal cord disorder associated with cervical spondylosis. Brain 95: 87–100
Nystrom B, Berglund JE, Bergquist E (1988) Methodological analysis of an experimental spinal cord compression model in the rat. Acta Neurol Scand 78: 460–466
Nystrom B, Berglund JE (1988) Spinal cord restitution following compression injuries in rats: Acta Neurol Scand 78: 467–472
Osterholm JL (1974) The pathophysiological response to spinal cord injury. The current status of related research. J Neurosurg 40: 5–33
Perry VH, Andersson P-B (1992) Glial symposium: The inflammatory response in the CNS. Neuropathol Appl Neurobiol 18: 454–459
Petersma A, De Jong N, Koster JF, Sluiter W (1994): Effect of hypoxia on adherence of granulocytes to endothelial cells in vitro. Am J Physiol 267: H874–879
Prough DS, Whitley JM, Toylor CL, Deal DD, DeWitt DS (1991) Regional blood flow following resuscitation fro hemorrhagic shock with hypertonic saline. Anesthesiol 75: 319–327
Tocha-e-Silva M, Negraes GA, Soares AM, Pontieri V, Loppnow L (1986) Hypertonic resuscitation from several hemorrhagic shock: Patterns of regional circulation. Circ Shock 19: 165–175
Ross I, Tator CH (1993) Spinal cord blood flow and evoked potential responses after treatment with Nimodipine or Myethylprednisolone in spinal cord-injured rats. Neurosurgery 33(3): 470–476
Saetzler RK, Badellino MM, Buckman RF, Eynon CA, Tuma RF, Arfors K-E (1996) Hypertonic saline attenuates leukocyte/endothelium and leukocyte/platelet interaction following hemorrhagic shock. J Trauma (in press)
Sandler AN, Tator CH (1976) Review of the effect of spinal cord trauma on the vessels and blood flow in the spinal cord. J Neurosurg 45: 638–646
Schoettle RJ, Kochanek PM, Magargee MJ, Uhl MW, Nemoto EM (1990) Early polymorpho-nuclear leukocyte addumulation correlates with the development of post-traumatic cerebral edema in rats. J Neurotrauma 7: 207–217
Schmoker JD, Zhuang J, Shackford SR (1991) Hypertonic fluid resuscitation improves cerebral oxygen delivery and reduces intracranial pressure after hemorrhagic shock. J Trauma 31: 1607–1613
Shackford SR, Schmoker JD, Zhuang J (1994) The effect of hypertonic resuscitation on pial arteriolar tone after brain injury and shock. J Trauma 37: 899–908
Slogan DJ, Wood MJ, Charlton HM (1992) Leucocyte recruitment and inflammation in the CNS. Trends Neurosci 15: 276–278
Smyth SS, Joneckis CC, Parise LV (1993) Regulation of vascular integrins. Blood 81: 2827–2843
Stewart WB, Wagner FC (1979) Vascular permeability changes in the contused feline spinal cord. Brain Res 169: 163–167
Tanno H, Nockels RP, Pitts LH, Noble LJ (1992) Breakdown of the blood-brain barrier after fluid percussive in the rat. Part 2. Effect of hypoxia on permeability to plasma proteins. J Neurotrauma 9: 335–347
Tarlov M, Klinger M, Vitale S (1953) Spinal cord compression studies. 1. Experimental techniques to produce acute and gradual compression. Arch Neurol Psychiatry 70: 813–819

Tator CH, Decker L (1973) Value of normothermic perfusion, hypothermic perfusion and durotomy in the treatment of experimental spinal cord trauma. J Neurosurg 39: 52–64
Taylor AR (1951) The mechanism of injury to the spinal cord in the neck without damage to the vertebral column. J Bone Joint Surg 33: 543–547
Tator CH, Fehlings MG (1991) Review of the secondary injury theory of acute spinal cord trauma with emphasis on vascular mechanisms. J Neurosurg 75: 15–26
Tuma RF, Vasthare U, Artors E, Young WF (1997) Hypertonic saline administration attenuates spinal cord injury. Vol 42, No 5. Trauma [Suppl] S54–60
Vasthare US, Rubin S, Riina H, Rosenwasser RH, Tuma RF (1991) Effect of fentanyl on recovery following incomplete global cerebral ischemia. Drug Dev Res 23: 227–232
Wakefield CL, Eidelberg E (1975) Electron microscope observations of the delayed effects of spinal cord compression. Exp Neurol 48: 637–646
Walter R, Schaffner A, Schoedon G (1994) Differential regulation of constitutive and inducible nitric oxide production by inflammatory stimuli in murine endothelial cells. Biochem Biophys Res Commun 202: 450–455
Weiss SJ (1989) Tissue destruction by neutrophils. NEJ Med 320: 365–376
Williams JI, Higgs GA (1988) Eicosanoids and inflammation. J Pathol 156: 101–110
Xu J, Hsu CY, Liu TH *et al.* (1990) Leukotriene and polymorphonuclear cell infiltration in spinal cord injury. J Neurochem 55: 907–912
Young WF, Flamm ES (1982) Effect of high-dose corticosteroid therapy on blood flow, evoked potentials, and extracellular calcium in experimental spinal cord. J Neurosurg 52: 667–673
Young WF, Rosenwasser RH, Vasthare US, Tuma RF (1994) Preservation of post-compression spinal cord function by infusion of hypertonic saline. J Neurosurg Anesth 6: 122–127
Zhuang J, Shackford SR, Schmoker JD, Anderson MI (1993) The association of leukocytes with secondary brain injury. J Trauma 35(3): 422–425

Correspondence: Dr. W. F. Young, Department of Neurosurgery, Temple University Hospital, 3401, N. Broad St., Philadelphia, PA 19140, USA.

Spinal Cord Monitoring
Eds: E. Stålberg, H. S. Sharma, Y. Olsson

Chapter 20

The multicenter outcome survey for spinal cord monitoring: monitoring reduces neurologic deficits

M. R. Nuwer

Department of Neurology, UCLA School of Medicine,
Department of Clinical Neurophysiology, UCLA Medical Center,
Los Angeles, California, USA

Introduction

Somatosensory evoked potential (SEP) monitoring can be used in surgery to identify early impending spinal cord damage from mechanical or ischemic causes. Scoliosis surgery is a major clinical setting for such monitoring.

The earliest monitoring techniques used electrodes placed in the epidural space (Shimoji *et al.*, 1971; Shimoji and Kano, 1975; Tamaki *et al.*, 1972, 1982; Jones *et al.*, 1982, 1983). This was soon supplemented by a scalp monitoring technique first popularized by Nash and Brown (Nash *et al.*, 1972, 1977). Scalp and neck techniques monitoring were studied subsequently to optimize the methods (Nuwer and Dawson, 1984a, 1984b). With these various techniques available, many individual teams have gone on to conduct monitoring and to publish their results. Previous publications include several excellent reviews of these various techniques (Nuwer, 1986; Ducker and Brown, 1986; Desmedt, 1989; Schramm and Møller, 1991; Jones *et al.*, 1994; Møller, 1995; Russell and Rodichok, 1995). Neurology and clinical neurophysiology organizations have endorsed monitoring and promulgated guidelines for its implementation (American Academy of Neurology 1990, International Federation of Clinical Neurophysiology 1993, American Electoencephalographic Society 1994).

It is important to know whether such surgical monitoring alters neurologic outcome. Many studies published by individual teams or surgeons describe their clinical experience (Engler *et al.*, 1978; Tamaki *et al.*, 1981, 1984; Leuders *et al.*, 1982; Jones *et al.*, 1982, 1983; LaMont *et al.*, 1983; Bradshaw *et al.*, 1984; Gonzalez *et al.*, 1984; Wilber *et al.*, 1984; Bieber *et al.*, 1986; Dinner *et al.*, 1986; Johnston *et al.*, 1986; Ryan and Britt, 1986; Whittle *et al.*, 1986; Mostegl *et al.*, 1987; Roy *et al.*, 1987; Machida *et al.*, 1988; Cheliout-Heraut *et al.*, 1989; Lubicky *et al.*, 1989; Tamaki,

1989; Abel *et al.*, 1990; Halonen *et al.*, 1990; Keith *et al.*, 1990; Albanese *et al.*, 1991; Apel *et al.*, 1991; Forbes *et al.*, 1991; Hicks *et al.*, 1991; Loder *et al.*, 1991; Fujioka *et al.*, 1994; Phillips *et al.*, 1995). The number of patients in these individual reports generally is too small to assess whether neurologic complications are reduced by using spinal cord monitoring. Therefore, assessing outcome is best done through a multicenter study among a large, well defined group of surgeons. Such an evaluation would study neurologic outcome for surgeons with much experience and frequent use of monitoring, compared to others who have made limited or no use of SEP spinal cord monitoring. Alternately, one could compare the neurologic outcomes rate before and after a group of surgeons begin using SEP spinal cord monitoring.

A recent multicenter survey of spinal cord monitoring outcome addressed these important clinical issues (Nuwer *et al.*, 1995; Dawson *et al.*, 1991). This survey was conducted through the Scoliosis Research Society (SRS), a group constituted primarily by surgeons and others with a special interest in scoliosis treatment. Most SRS members practice in the U.S. Each SRS member surgeon annually must submit surgical outcome data as a condition of continuing membership, making the SRS is a valuable resource for studying outcomes. Over several decades, the SRS has surveyed its members repeatedly about their rates of neurological deficits after scoliosis surgery. For the recent multicenter survey, SRS members were questioned about their neurologic postoperative deficits as well as the use of SEP monitoring. For each surgical practice, the affiliated monitoring clinical neurophysiologist provided many specific details of their SEP monitoring technique, equipment used, and criteria for determining when a change had occurred.

An international cohort of surgeons was also surveyed. This was done in conjunction with an international meeting of the SRS and the European Spinal Deformities Society. In this way, the surgeons from Europe and Japan were included in the international portion of this outcome survey. However, techniques of monitoring and general medical and surgical care may differ from nation to nation. To help control for this confounding factor, many elements studied in the multicenter survey focused on the US surgeons, i.e. those SRS members based in the US.

General results

Among the 173 US surgeons surveyed, 153 (88%) had used SEP spinal cord monitoring for some or all of their surgical cases in recent years (Nuwer *et al.*, 1995). The surgeons had used SEP typically for seven years (median) although some had used monitoring for up to 18 years. During those years in which the individual surgeons had used SEP monitoring, they had conducted 97,586 spinal surgery cases. Among those, 51,263 (53%) were done using SEP monitoring. The typical surgeon had used monitoring in 250 cases (median). The wake-up test, an older method of assessing intraoperative spinal cord function, was conducted in 18,830 of these patients. The vast majority of surgeons (74%) used the wake-up test in 10% or fewer of their cases after they began to use SEP monitoring. The majority of patients monitored (60%) had scoliosis, whereas smaller portions had kyphosis (6.5%) or other spinal disorders.

SEP monitoring techniques

Among the US teams, 85% used recordings from the scalp and the neck, with stimulation at the posterior tibial nerve at the ankle. The remaining teams used epidural monitoring or monitoring with interspinous ligament electrodes. Some teams used stimulation at the peroneal nerve in place of posterior tibial nerve. With most stimulating and recording techniques, new EP results could be obtained typically with 300 EP stimuli. When stimulating at 5 per second, this could produce a new EP tracing about once per minute. The median number of recording electrodes used was four, rather than recording just from one predetermined pair of electrodes.

Among teams using scalp and neck recording, 95% scored only the short latency peaks at 30–50 msec. The remaining 5% of teams looked at the middle or long latency SEP peaks. Most monitoring teams (90%) assessed both the latency and the amplitude to determine when a change from baseline had occurred. For EP amplitude monitoring, a 50% change from baseline amplitude was used as the criterion for "change" by 72% of surveyed teams. A smaller number of teams (21%) used a lower percentage of amplitude change, such as a 25%–40% decrease. Nearly one-half of the monitored teams (44%) used a 10% latency change as the criterion for significant "change." The other half of the teams (48%) identified a 5% to 8% latency change as being their criterion for "change."

Surgical outcome results

Surgical outcome is judged using validity measures such as the false negative and true positive rates. The clinical outcome is judged by whether the patient suffered new neurological deficits postoperatively. Some deficits were small in degree and transient, whereas others were definite major deficits, e.g. paresis or paraplegia. Neurologic outcome results were obtained both from the surgeon and from the neurophysiology monitoring team member. Some neurological deficits were very transient or questionable as to whether they should constitute a new "deficit" for our purposes. Such outcomes were considered "equivocal." Some neurological deficits first appeared several hours to several days postoperatively, without being present immediately postoperatively. Such outcomes were considered "delayed onset." All other outcomes were considered "definite."

False negative (FN) outcomes are those in which the SEP remained stable throughout the case but the patient awoke with a new postoperative neurologic deficit. A "stable" SEP is one that did not exceed that particular

Table 1. Neurologic outcome for sep monitoring in scoliosis surgery

Total procedures	51,263	(100%)
False negative (FN)	65	(0.127%)
False positive	774	(1.510%)
True positive (TP)	217	(0.423%)
Neurological deficits (FN plus TP)	282	(0.550%)
True negative	50,207	(97.94%)
Sensitivity	92%	
Specificity	98.9%	
Negative predictive value	99.93%	
Positive predictive value	42%	

From the multicenter outcome survey for spinal cord monitoring (Nuwer *et al.*, 1995).

team's criteria for change. In the U.S. segment of the multicenter survey, 51,263 patients were monitored. Among these, 65 (0.27%) had a false negative outcome. Among those with false negative outcome, 34 were considered definite, 13 were equivocal and 18 were delayed onset. Some of these cases are already individually reported in the existing literature that discusses false results using SEP monitoring (Harper *et al.*, 1988; Ben-David *et al.*, 1987; More *et al.*, 1988; Ginzburg *et al.*, 1985; Lesser *et al.*, 1986). Outcome rates are presented in Table 1.

True positive (TP) cases are those in which the SEP changed, the patient awoke with a new postoperative deficit despite the warning. Presumably, the surgical team attempted to investigate or intervene based on the EP changes but the intervention was not entirely successful. In some or many of these cases, the patient may have suffered a substantially less degree of deficit than would have occurred without monitoring, but the portion of such cases cannot be determined. In the U.S. portion of the multicenter survey, 217 patients (0.42%) had true positive monitoring outcome.

The neurological deficits rate was also assessed. This is the sum of the FN and TP rates. Among the U.S. multicenter survey patients, 282 (0.55%) had new postoperative neurologic deficits. These could be subdivided further. The rate of transient deficits was 0.24%, whereas persistent deficits occurred at a rate of 0.31%. The majority of patients with new deficits were characterized as minor, such as radiculopathy, sensory impairment without motor loss, postoperative bladder control problems or other lesser degrees of neurologic deficits. The rate of major deficits, i.e. paresis or paraplegia, was 0.24%.

These rates of persistent and major neurologic deficits were substantially lower than those historically encountered in this same group of surgeons. In Fig. 1, a comparison is made to the historical rates of deficits. This shows that persistent deficits dropped by one-third and major deficits dropped by about 60% among the patients operated on with SEP monitoring, in comparison to historical controls.

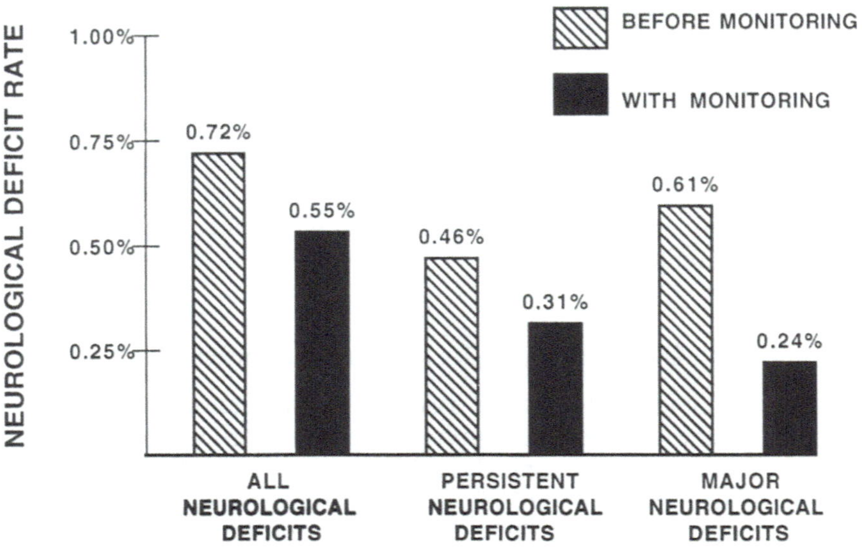

Fig. 1. Rates of all neurologic deficits, persistent neurologic deficits and major neurologic deficits are shown comparing the MacEwen *et al.* (1975) survey of S.R.S. members to the recent S.R.S. survey results, Nuwer *et al.* 1995, Dawson *et al.* 1991. Each rate is smaller in recent study (confidence interval test, $p<0.001$). Adapted from: Nuwer *et al.* (1995)

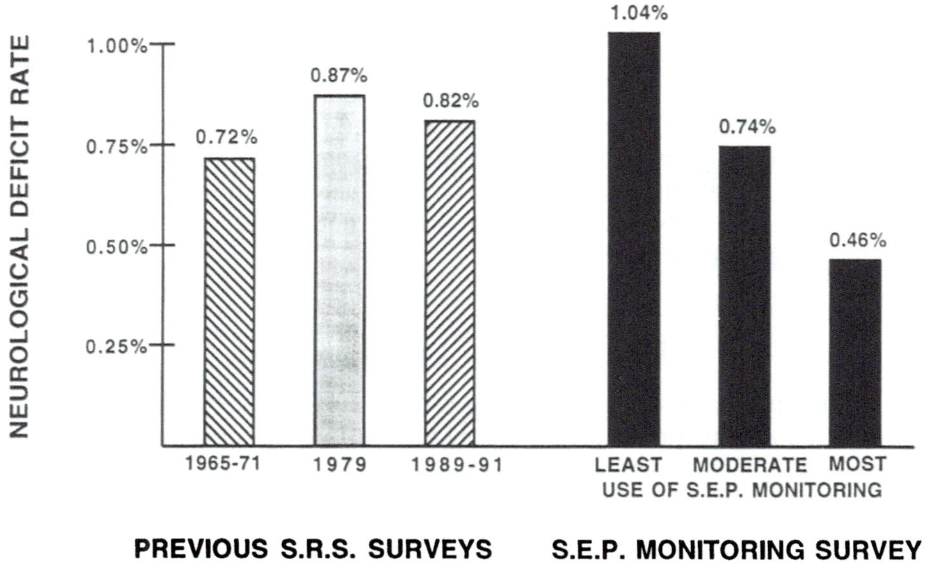

Fig. 2. Neurologic deficit rate over several previous S.R.S. surveys, compared to the recent survey. The recent survey data show less neurologic deficits among teams with the greatest SEP monitoring experience (chi-square, $p<0.01$). The neurologic deficits rate among the teams with the greatest SEP use was lower than that for teams with the least SEP use, by more than a factor of two. Adapted from: Nuwer *et al.* (1995)

Figure 2 shows a further comparison to historical data, including several surveys of SRS members over 15 years. That comparison subdivides the multicenter survey surgeons into those who have made extensive use of SEP monitoring, as opposed to those who have made only moderate or little use of SEP monitoring in their surgical practice. The data show clearly that the surgeons with the most extensive use of SEP monitoring have the lowest rate of neurological deficits, either in comparison to historical controls or in comparison to contemporaneous surgeons with less use of SEP monitoring.

Other factors affecting surgical outcome

A variety of other factors were assessed in the multicenter survey. The surgeons' years of experience with orthopedic practice, the frequency of wake-up test use, and technical SEP factors were studied. The years of experience with SEP monitoring was a predictor of a lower neurologic deficits rate ($r = 0.24$, $P < 0.005$). The years of orthopedic surgery practice also corresponded to a lower neurologic deficits rate ($r = 0.13$). Outcome rates were not significantly affected by how often the surgeon used the wake-up test. Technical factors in the way the SEP was monitored accounted for minor degrees of difference in outcome.

Geographical comparisons were made, assessing the outcomes in the European and Japanese monitoring groups in comparison to the U.S. subset in this survey (Nuwer and Carlson, 1994). Technically, surgeons in Japan and Europe, especially the U.K., used epidural monitoring extensively, whereas the surgeons in the U.S. used surface monitoring from the scalp and neck in general. However, the rates of neurological deficits and other validity measures of surgical outcome were comparable across geographic regions. This suggests that SEP monitoring can be done in a variety of ways, and these various technical approaches are all reasonable, satisfactory and similarly accurate in predicting neurologic outcome and performing the monitoring task.

Implications for clinical practice

After making some assumptions about the applications of monitoring in comparison to historical levels, the U.S. multicenter survey results can be used to calculate further validity measures. EP monitoring had a sensitivity of 92% and a specificity of 98.9% . In the same group, SEP monitoring had a high negative predictive value of 99.93%, i.e. SEP monitoring is highly likely to be correct when the SEPs remain stable. However, the test suffers from a mediocre positive predictive value of 42%. The latter value reflects a tendency toward false alarms, either technical in nature or due to true physiological changes which occur without high risk of new postoperative deficits.

The cost-effectiveness of a technique is sometimes used to judge a technique's value. The data in this multicenter survey suggest that the patient is spared a major new neurologic deficit at a rate about one in every 200 monitored cases. If the cost of monitoring is US$600, then it costs about US$120,000 to spare a patient from paraplegia or paraparesis. This amount, however great it may be, is substantially less than the cost of providing medical care, physical therapy and disability assistance for the remainder of the lifetime of a young adult with permanent paraparesis or paraplegia due to surgical complications. When viewed in this way, SEP spinal cord monitoring for scoliosis surgery appears to be a cost-effective monitoring intervention in this patient population.

Monitoring personnel

An important remaining question is who should staff the monitoring procedure. A variety of models have been tried. For the most part, surgeons themselves have found it difficult to oversee the SEP spinal cord monitoring directly, as their attention is focused on the principal surgical activity itself. In a similar way, anesthesiologists have tried to run the monitoring procedure themselves, but often have found that they are too distracted by other anesthesia responsibilities. As a result, it has become routine to have an additional person in the OR to conduct monitoring. This includes the activities of setting up the stimulating and recording electrodes, any necessary calibrations and preparation of equipment, establishing a suitable baseline recording early in the case, altering procedures as necessary to optimize interoperative recordings, and carrying out the principal monitoring itself. When interoperative monitoring problems arise, a variety of tasks must be conducted quickly to determine whether detected changes are due to technical, anesthetic or surgical factors.

A neurophysiologist conducting the monitoring needs to be knowledgeable about the SEP technique itself, with additional skills, knowlege and ability pertaining to the various factors encountered in an OR monitoring setting. In some hospitals, well-trained Ph.D. neurophysiologists directly conduct the monitoring. In other hospitals, highly skilled EEG technologists have been able to conduct monitoring when supervised by a knowledgeable physician clinical neurophysiologist. One approach to such monitoring involves electronic transmission of the monitoring data to the laboratory of a physician neurophysiologist who is not in the OR. Another version of this uses a physician clinical neurophysiologist to oversee several operating rooms, with EEG technologists manning individual rooms and calling for assistance as needed. In some medical centers, an EEG technologist reports directly to the anesthesiologist who is well-trained in clinical neurophysiology. Finally, in another model, the physician clinical neurophysiologist is in the operating room full-time throughout the entire case. These various models are still controversial. There are strongly held opinions that one or another of these is the proper model to follow.

Certainly, no EEG technologist should be left in the OR without a knowledgeable clinical neurophysiologist available on short notice to supervise the procedure, to intervene to assess problems encountered, and to provide quality control. The only EEG technologists who would be suitable for SEP OR monitoring are those who are highly skilled, who are already capable and well-trained in SEP techniques, and who have suitable patience and good judgement under pressure.

Experience has shown that some staffing models work poorly. One model to avoid is to allow an EEG technologist to perform monitoring without available supervision and quality control by a clinical neurophysiologist as needed. When an EEG technologist, or other less well-trained technical staff, answers directly to a surgeon or anesthesiologist who is not highly skilled in clinical neurophysiology, painful experience has shown that quality suffers, mistakes are made, and the monitoring technique is useless or counterproductive.

At the present time, it appears that the SEP spinal cord monitoring technique is likely to be effective and useful so long as at least one available member of the monitoring team is highly skilled in clinical neurophysiology and either in the OR throughout the case or available to interact with the OR monitoring team as needed during cases. This interesting and controversial staffing topic might best be settled in the future if data were available to compare outcomes and problems between the several common models.

Conclusions

SEP spinal cord monitoring and scoliosis surgery substantially reduces the postoperative neurologic deficits. The 60% reduction in major deficits, i.e. paraparesis and paraplegia, are a substantial factor in favor of using this monitoring on a wide scale. The monitoring appears to be cost-effective in the larger picture of medical care and the cost of treating major complications like paraplegia.

Until further advanced monitoring techniques become practical and readily available, SEP monitoring during scoliosis surgery should continue to be a clinically useful, valid procedure.

After all, the best way to treat paraplegia is to prevent it from occurring. SEP spinal cord monitoring in scoliosis surgery provides surgeons and neurophysiologists with a technique to achieve this goal.

References

Abel MF, Mubarak SJ, Wenger DR, Costello J, Hicks GE (1990) Brainstem evoked potentials for scoliosis surgery: a reliable method allowing use of halogenated anesthetic agents. J Pediatr Orthop 10: 208–213

Albanese SA, Spadaro JA, Lubicky JP, Henderson NA (1991) Somatosensory cortical evoked potential changes after deformity correction. Spine 16: S371–S374

American Academy of Neurology (1990) Intraoperative neurophysiology. Neurology 40: 1644–1646

American Electroencephalographic Society (1994) Guidelines for intraoperative monitoring of sensory evoked potentials. J Clin Neurophysiol 11: 77–87
Apel DA, Marrero G, King J, Tolo VT, Bassett GS (1991) Avoiding paraplegia during anterior spinal surgery. The role of somatosensory evoked potential monitoring with temporary occlusion of segmental spinal arteries. Spine 16: S365–S370
Ben-David B, Haller G, Taylor P (1987) Anterior spinal fusion complicated by paraplegia. A case report of a false-negative somatosensory evoked potential. Spine 12: 536–539
Bieber E, Tolo V, Uematsu S (1986) Spinal cord monitoring during posterior spinal instrumentation and fusion. Clin Orthop Rel Res 229: 121–124
Bradshaw K, Webb JK, Fraser AM (1984) Clinical evaluation of spinal cord monitoring in scoliosis surgery. Spine 9: 636–643
Cheliout-Heraut F, Vital JM, Pouliquen JC, Bitar Z, Heraut LA (1989) Surveillance peroperatoire des scolioses chez l'adulte jeune par les potentiels evoques somesthesiques: a propos de 33 cas. Neurophysiol Clin 19: 297–310
Dawson EG, Sherman JE, Kanim LEA, Nuwer MR (1991) Spinal cord monitoring results of the scoliosis research society and the European spinal deformity society survey. Spine 16: S361–S364
Desmedt JE (1989) Neuromonitoring in Surgery. Elsevier, Amsterdam
Dinner DS, Lüders H, Lesser RP, Morris HH, Barnett G, Klem G (1986) Intraoperative spinal somatosensory evoked potential monitoring. J Neurosurg 65: 807–814
Ducker TB, Brown RH (1986) Neurophysiology and Standards of Spinal Cord Monitoring. Springer, Berlin Heidelberg New York Tokyo
Engler GL, Spielholz NI, Bernhard WN, Danziger F, Merkin H, Wolff T (1978) Somatosensory evoked potentials during Harrington instrumentation for scoliosis. J Bone Joint Surg (Am) 60A: 528–532
Forbes HJ, Allen PW, Waller SC, Jones SJ, Edgar MA, Webb PJ, Ransford AO (1991) Spinal cord monitoring in scoliosis surgery. J Bone Joint Surg (Br) 73B: 487–491
Fujioka H, Shimoji K, Tomita S, Denda T, Takada T, Homma S, Uchiyama H, Takahashi H, Tobita T, Baba H (1994) Spinal cord potential recordings from the extradural space during scoliosis surgery. Br J Anesth 73: 350–356
Ginsburg HH, Shetter AG, Raudzens PA (1985) Postoperative paraplegia with preserved intraoperative somatosensory evoked potentials. J Neurosurg 6: 296–300
Gonzalez EG, Hajdu M, Keim H, Brand L (1984) Quantification of intraoperative somatosensory evoked potential. Arch Phys Rehabil 65: 721–725
Halonen JP, Jones SJ, Edgar MA, Ransford AO (1990) Multi-level epidural recordings of spinal SEPs during scoliosis surgery. New Trends Adv Tech Clin Neurophysiol 41: 342–347
Harper CM, Daube JR, Lichy WJ, Klassen RA (1988) Lumbar radiculopathy after spinal fusion for scoliosis. Muscle Nerve 11: 386–391
Hicks RG, Burke DJ, Stephen PH (1991) Monitoring spinal cord function during scoliosis surgery with Cotrel-Dubousset instrumentation. Med J Aust 154: 82–86
International Federation of Clinical Neurophysiology (1993) Neuromonitoring during surgery. Electroencephalogr Clin Neurophysiol 87: 263–276
Jones SJ, Boyd S, Hetreed M, Smith NJ (1994) Handbook of Spinal Cord Monitoring. Kluwer, London
Jones SJ, Edgar MA, Ransford AO (1982) Sensory nerve conduction in the human spinal cord: epidural recordings made during scoliosis surgery. J Neurol Neurosurg Psychiatry 45: 446–451
Jones SJ, Edgar MA, Ransford AO, Thomas NP (1983) A system for the electrophysiological monitoring of the spinal cord during operations for scoliosis. J Bone Joint Surg 65: 134–139
Johnston CE, Happel LT, Norris R, Burke S, King AG, Roberts JM (1986) Delayed paraplegia complicating sublaminar segmental spinal instrumentation. J Bone Joint Surg (Am) 68: 556–563

Keith RW, Stambough JL, Awender SH (1990) Somatosensory cortical evoked potentials: a review of 100 cases of intraoperative spinal surgery monitoring. J Spinal Dis 3: 220–226
LaMont RL, Wasson SL, Green MA (1983) Spinal cord monitoring during spinal surgery using somatosensory spinal evoked potentials. J Pediatr Orthop 3: 31–36
Lesser RP, Raudzens P, Lueders H, Nuwer MR, Goldie WD, Morris HH, Dinner DS, Klem G, Hahn JF, Shetter AG, Ginsburg HH, Gurd AR (1986) Postoperative neurological deficits may occur despite unchanged intraoperative somatosensory evoked potentials. Ann Neurol 19: 22–25
Leuders H, Gurd A, Hahn J, Andrish J, Weiker G, Klem G (1982) A new technique for intraoperative monitoring of spinal cord function. Spine 7: 110–115
Loder RT, Thomson GJ, LaMont RL (1991) Spinal cord monitoring in patients with nonidiopathic spinal deformities using somatosensory evoked potentials. Spine 16: 1359–1364
Lubicky JP, Spadaro JA, Yuan HA, Fredrickson BE, Henderson N (1989) Variability of somatosensory cortical evoked potential monitoring during spinal surgery. Spine 14: 790–798
MacEwen GD, Bunnell WP, Sriram K (1975) Acute neurological complications in the treatment of scoliosis. J Bone Joint Surg (Am) 57A: 404–408
Machida M, Weinstein SL, Yamada T, Kimura J, Itagaki T, Usui T (1988) Monitoring of motor action potentials after stimulation of the spinal cord. J Bone Joint Surg 70-A: 911–918
Møller AR (1995) Intraoperative neurophysiologic monitoring. Harwood, Luxembourg
More RC, Nuwer MR, Dawson EG (1988) Cortical evoked potential monitoring during spinal surgery: sensitivity, specificity, reliability, criteria for alarm. J Spinal Dis 1: 75–80
Mostegl A, Bauer R, Eichenauer M (1987) Intraoperative somatosensory potential monitoring. A clinical analysis of 127 surgical procedures. Spine 13: 396–400
Nash C, Brodkey J, Croft T (1972) A model for electrical monitoring of spinal cord function in scoliosis patients undergoing correction. J Bone Joint Surg (Am) 54A: 197–198
Nash C, Lorig R, Schatzinger L, Brown R (1977) Spinal cord monitoring during operative treatment of the spine. Clin Orthop 126: 100–105
Nuwer MR (1986) Evoked Potential Monitoring in the Operating Room. Raven Press, New York
Nuwer MR, Carlson LG (1994) A multicentre survey of spinal cord monitoring outcome. In: Jones SJ, Boyd S, Hetreed M, Smith NJ (eds) Handbook of Spinal Cord Monitoring. Kluwer, London, pp 78–87
Nuwer MR, Dawson E (1984a) Intraoperative evoked potential monitoring of the spinal cord: enhanced stability of cortical recordings. Electroencephalogr Clin Neurophysiol 59: 318–327
Nuwer MR, Dawson E (1984b) Intraoperative evoked potential monitoring of the spinal cord: a restricted filer, scalp method during Harrington instrumentation for scoliosis. Clin Orthop 3: 42–50
Nuwer MR, Dawson EG, Carlson LG, Kanim LEA, Sherman JE (1995) Somatosensory evoked potential spinal cord monitoring reduces neurologic deficits after scoliosis surgery: results of a large multicenter survey. Electroencephalogr Clin Neurophysiol 96: 6–11
Phillips LH, Blanco JS, Sussman MD (1995) Direct spinal stimulation for intraoperative monitoring during scoliosis surgery. Muscle Nerve 18: 319–325
Roy EP, Gutmann L, Riggs JE, Jones ET, Byrd A, Ringel RA (1987) Intraoperative somatosensory evoked potential monitoring in scoliosis. Clin Orthop Rel Res 229: 94–98
Russell GB, Rodichok LD (1995) Intraoperative Neurophysiologic Monitoring. Butterworth-Heinemann, Boston
Ryan TP, Britt RH (1986) Spinal and cortical somatosensory evoked potential monitoring during corrective spinal surgery with 108 patients. Spine 11: 352–361
Schramm J, Møller AR (1991) Intraoperative neurophysiologic Monitoring in Neurosurgery. Springer, Berlin Heidelberg New York Tokyo

Shimoji K, Kano T (1975) Evoked electrospinogram interpretation of origin and effects of anesthetics. Int Anesthesiol Clin 13: 171–189
Shimoji K, Kano T, Higashi H (1971) Epidural recording of spinal electrogram in man. Electroencephalogr Clin Neurophysiol 30: 236–239
Tamaki T, Yamashita T, Kobayashi H, Hirayama H (1972) Spinal cord monitoring. Jpn J Electroencephalogr Electromyogr 1, p 196
Tamaki T (1989) Spinal cord monitoring with spinal potentials evoked by direct stimulation of the spinal cord. In: Desmedt J (ed) Neuromonitoring in Surgery. Elsevier, Amsterdam, pp 139–149
Tamaki T (1994) Present status of spinal cord monitoring in Japan. In: Jones SJ, Boyd S, Hetreed M, Smith NJ (eds) Handbook of Spinal Cord Monitoring. Kluwer, London pp 89–91
Tamaki T, Noguchi T, Takano H, Tsuji H, Nakagawa T, Imai K, Inoue S (1984) Spinal cord monitoring as a clinical utilization of the spinal evoked potentials. Clin Orthop 184: 58–64
Tamaki T, Tsuji H, Inoue S, Kobayashi H (1981) The prevention of iatrogenic spinal cord injury utilizing the evoked spinal cord potential. Int Orthop 4: 313–317
Tamaki T, Yamashita T, Kobayashi H, Hirayama H (1972) Spinal cord monitoring. Jpn J Electroencephalogr Electromyogr 1, p 196
Whittle IR, Johnston IH, Besser M (1986) Recording of spinal somatosensory evoked potentials for intraoperative spinal cord monitoring. J Neurosurg 64: 601–612
Wilber RG, Thompson GH, Shaffer JW, Brown RH, Nash CL (1984) Postoperative neurological deficits in segmental spinal instrumentation. J Bone Joint Surg (Am) 66A: 1178–1187

Correspondence: Dr. M. R. Nuwer, Reed Neurological Research Center, 710 Westwood Plaza, Los Angeles, CA 90024, USA.

Spinal Cord Monitoring
Eds: E. Stålberg, H. S. Sharma, Y. Olsson

Chapter 21

Effects of anesthetics on somatosensory and motor evoked potentials in humans

T. Andoh and **F. Okumura**

Department of Anesthesiology, Yokohama City University School of Medicine, Yokohama, Japan

Importance of anesthetic effects on somatosensory evoked potentials

Somatosensory evoked potential (SEP) monitoring has been used to intraoperatively evaluate central nervous system (CNS) integrity during selected surgical procedures, such as carotid endoarterectomy, intracranial vascular surgery, thoracic aortic surgery, and spinal cord procedure (Fisher *et al.*, 1995; Nuwer *et al.*, 1995; Cunningham *et al.*, 1987; Gugino and Chabot, 1990). Several intraoperative factors including hypothermia, cooling of the stimulated extremities, hypoxia, hypercapnia, hypotension, and anesthetic agents, can influence the outcome of intraoperative SEP monitoring (Gugino and Chabot, 1990). Although the influence of factors other than anesthetic agents can be largely eliminated by careful control of body temperature and cardio-respiratory status, alteration in SEPs by anesthetic agents is unavoidable in patients undergoing procedures that require general anesthesia. Therefore, effects of anesthetics on SEPs should be considered when interpreting changes in SEPs. Different anesthetics have disparate effects, and different components of evoked potentials have distinct sensitivities to a given anesthetic. In general, early components (subcortical SEPs) are less sensitive to anesthetics than are late components (cortical SEPs). This is probably because anesthetics affect synaptic transmission more than axonal conduction – early components are due more to axonal conduction, whereas late components depend primarily on multi-synaptic transmission. For intraoperative cortical SEP monitoring, special attention is required to ensure that the agents and dose levels selected favorably preserve late waves.

Why different anesthetic have distinct effects on SEPs is unclear, but drug specific properties may reflect differences in electrophysiological effects. For example, synaptic transmission in a given region may have different sensitivities to different anesthetics. Therefore, comparison of effects of various anesthetics on SEPs may improve our understanding of the specific neuronal effects of individual agents.

Besides intraoperative use, SEP monitoring has also been employed to evaluate the severity of neural damage in comatose patients (Lindsay *et al.*, 1981; Newlon *et al.*, 1983). In such patients anesthetics or related agents are often used to control convulsions and to reduce intracranial pressure. The lack of a certain evoked potential may be due to anesthetic-induced depression or to neural injury itself. Therefore, it is also important to assess the reliability of SEP monitoring in comatose patients given anesthetic agents.

In this chapter, components of evoked responses are defined according to the American Electroencephalographic Society Guidelines (American Electroencephalographic Society, 1984). The following terminology was used for short-latency SEPs to median nerve stimulation at the wrist. The prominent negative peak in ipsilateral-contralateral Erb's point derivation is designated as Erb's point potential (Ep). N13 represents the most prominent peak recorded at the second or fifth cervical spinous process in neck-scalp derivation. In scalp(C3′ or C4′)-scalp(Fz) or contralateral earlobe derivation, N20 and P23 are normally measured. For posterior tibial nerve evoked short latency responses, the first positive and negative peak of scalp recordings are designated as P35 and N45. Central conduction time (CCT) is usually measured as the difference between the latencies of N13 and N20.

Intravenous anesthetics

Barbiturates

Subcortical SEPs. Although induction doses of thiopental do not affect the latency of N13 (Koht *et al.*, 1988; McPherson *et al.*, 1986), a large dose producing an isoelectric EEG reduces the amplitude of N13 by 33% without altering the latency (Drummond *et al.*, 1985). At both low and high doses pentobarbital has minimal effects on N13 latency and amplitude (Drummond *et al.*, 1987). In contrast to these reports, Shimoji *et al.* reported that clinical doses of thiamylal increased N1 amplitude and prolonged P2 latency in spinal somatosensory evoked potentials (spinal SEPs) recorded at the epidural space, suggesting that clinical doses of barbiturates may affect somatosensory responses at spinal level (Shimoji *et al.*, 1974). The discrepancy among the effects of barbiturates on N13 and spinal SEPs probably arises from differences in the origins of these responses.

Cortical SEPs. Induction doses of thiopental (4 mg/kg or 5 mg/kg plus 2 mg/kg/hr) increase the latency of N20 but do not reduce the amplitude of N20 (Koht *et al.*, 1988; McPherson *et al.*, 1986). At this dose level, thiopental increases the latency of the cortical response without changing the latency of the response recorded at cervical level, so that the central conduction time (CCT) is prolonged (Koht *et al.*, 1988).

At much higher doses (maximum 77.5 mg/kg) thiopental progressively depresses the amplitude of cortical responses; however, cortical SEP recording is still possible in patients with preexisting normal responses even after administration of doses far in excess of levels producing isoelectric EEG (Drummond *et al.*, 1985). Similar results have been reported for high doses of pentobarbital in patients undergoing elective neurosugery (Drummond *et al.*, 1987). In comatose patients the latency or amplitude of N13 and N20 are relatively unaffected by high doses of thiopental that produce isoelectricity, as compared with SEPs recorded in patients receiving lower doses of thiopental (Ganes and Lundar, 1983). Taken together, barbiturates suppress cortical responses in a dose-dependent manner but do not abolish such responses even at the high doses. Consequently, SEPs can be effectively monitored in patients treated with barbiturates. Absence, but not reduction, of the early cortical responses such as N20 implies neural damages of somatosensory pathways in patients receiving high doses of barbiturates, unless hypothermia is concurrently performed.

Benzodiazepines

Subcortical SEPs. Clinically used doses of midazolam do not affect N13 latency (Koht *et al.*, 1988).

Cortical SEPs. Midazolam reportedly prolongs the latency of N20 and CCT without changing the amplitude of N20 (Koht *et al.*, 1988). Diazepam (0.1 mg/kg) increases the latency and decreases the amplitude of early cortical SEPs (latency 20–120 msec) and abolishes late cortical responses (latency 200–400 msec) (Grundy *et al.*, 1979). Benzodiazepines given as premedication can influence cortical SEP recording before induction of general anesthesia. In patients receiving benzodiazepines as sedatives or anticonvulsants, drug-related effects on cortical SEPs should be considered.

Propofol

Subcortical SEPs. Propofol with or without low doses of fentanyl slightly prolongs N13 latency but does not change N13 amplitude significantly (Scheepstra *et al.*, 1989).

Cortical SEPs. Induction doses of propofol reportedly increase both the latency and amplitude of N20 (Scheepstra *et al.*, 1989). Propofol has also been used in combination with various agents to maintain anesthesia and has proved to be useful. Propofol combined with a low dose of fentanyl increases the latencies of N20 and CCT, but preserves the amplitude of N20 (Scheepstra *et al.*, 1989). Propofol combined with alfentanyl does not affect either the latency or amplitude of early cortical responses to posterior tibial nerve stimulation (Kalkman *et al.*, 1991). This combination preserves

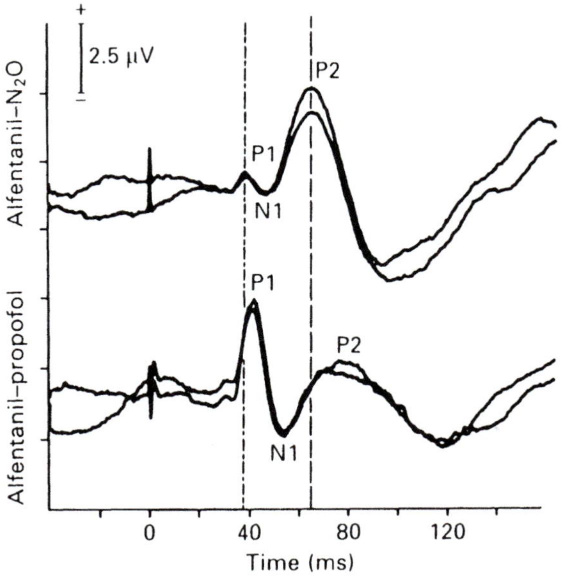

Fig. 1. Posterior tibial nerve somatosensory evoked potentials from one patient during alfentanil-nitrous oxide anesthesia and 15 min after substitution of nitrous oxide with propofol. *P1* first positive cortical peak; *N1* first negative cortical peak; *P2* second positive cortical peak. Note the increase in the amplitude of P1 by substitution of nitrous oxide with propofol. (From Kalkman *et al.*, 1991 with permission)

cortical potentials better than nitrous oxide with alfentanyl or nitrous oxide with 0.5 MAC of isoflurane and fentanyl (Kalkman *et al.*, 1991; Taniguchi *et al.*, 1992). Propofol combined with nitrous oxide and dextromoramide reportedly depresses the amplitude of early cortical responses in tibial nerve SEPs (Maurette *et al.*, 1988). It seems that propofol alone does not depress cortical SEPs and that total intravenous anesthesia with a combination of propofol and opioid has little effect on SEP recording (Fig. 1).

Etomidate

Subcortical SEPs. Etomidate reportedly causes no changes in Erb's point potential or N13 latency, but decreases N13 amplitude by 64% (Koht *et al.*, 1988; McPherson *et al.*, 1986). The magnitude of the depression in N13 amplitude seems larger than that induced by other anesthetics.

Cortical SEPs. Etomidate is known to markedly increase the amplitude of median nerve cortical SEPs (Koht *et al.*, 1988; McPherson *et al.*, 1986). Koht *et al.* reported a 4-fold increase in the amplitude of N20 by etomidate (Koht *et al.*, 1988), McPherson *et al.* noted a 2.7-fold increase (McPherson *et al.*, 1986). These increases in the amplitude are thought to be mediated by effects on somatosensory pathway above spinal level, because etomidate does not augment, but instead depresses, N13 amplitude. Prolongation of the latencies of N20 and CCT was observed in both studies. There are two conflicting opinions with regard to etomidate as an anesthetic for procedures requiring cortical SEP monitoring. Since an increase in cortical SEP amplitude is often an early sign of ischemia in the corresponding pathway and since etomidate changes SEP waveform, etomidate-induced alterations might interfere with real changes in neurological function, as assessed by SEP monitoring. (McPherson *et al.*, 1986). In contrast, etomidate can improve cortical SEP recording. One case report describes a patient anesthetized with continuous infusion of midazolam and fentanyl in whom recording of tibial nerve cortical SEP became possible after etomidate was additionally given (Sloan *et al.*, 1988). In this patient, stable SEP recording was enabled by the continuous infusion of etomidate. Therefore, etomidate is the agent of choice for patients in whom cortical responses are too small to be reliably monitored during anesthesia with other anesthetics.

Ketamine

Subcortical SEPs. Ketamine with or without nitrous oxide causes no change in the latency and amplitude of Erb's point potential or N13 (Schubert *et al.*, 1990). However, spinal SEPs recorded at the epidural space are reportedly altered by ketamine. Ketamine increases N1 amplitude and decreases P2 amplitude. (Kano and Shimoji, 1974).

Cortical SEPs. When ketamine is used alone, it enhances the amplitude of early cortical responses without affecting latency. Maximal increases in the amplitudes of N20 and P23 with ketamine alone have been reported as 15% and 88%, respectively, so that augmentation with ketamine is much smaller and less consistent than that with etomidate (Schubert *et al.*, 1990). Nitrous oxide added to ketamine reduces the cortical response amplitude to below control level. The amplitude of N20 decreased to about 60% of the control response when 70% nitrous oxide was added to ketamine, however, the amplitude appeared higher than expected with nitrous oxide alone, indicating that nitrous oxide with ketamine preserves early cortical SEPs.

Opioids

Subcortical SEPs. While medium and high doses of fentanyl slightly alter N13 (McPherson *et al.*, 1986; Schubert *et al.*, 1986), a low dose of fentanyl with droperidol significantly depresses the amplitudes of all components in spinal SEPs recorded at the epidural space (Kano and Shimoji, 1974). Morphine also depresses all components of spinal SEPs (Maruyama *et al.*, 1980).

Cortical SEPs. A medium dose of fentanyl alone (25 μg/kg) increased the latency of N20 and P23, without affecting the latency of N13 (McPherson *et al.*, 1986). The decrease in the amplitude of N20 was not significant in that study. A higher dose of fentanyl (mean dose 53 μg/kg) increased the latencies of N13, N20 and P23 and decreased the amplitude of N20 by about 36% in another study (Schbert *et al.*, 1986).

A low dose of fentanyl (2.5 μg/kg plus 1.5 to 2.5 μg/kg/hr) combined with 60% nitrous oxide consistently increased the latency of the first negative cortical response and decreased the peak amplitude measured by posterior tibial nerve stimulation (Pathak *et al.*, 1984). In the same report, an equianalgesic dose of morphine combined with 60% nitrous oxide also induced similar changes in tibial nerve cortical SEP.

Early cortical SEPs were readily recorded under all conditions, suggesting that high-dose opioid anesthesia as well as nitrous oxide anesthesia supplemented with a low dose of opioid are suitable for intraoperative SEP recording.

Inhalational anesthetics

Nitrous oxide

Subcortical SEPs. Nitrous oxide with halothane, enflurane, and isoflurane does not influence the latency of Erb's point potential or N13 (Peterson *et al.*, 1986).

Cortical SEPs. Nitrous oxide alone induces a dose-dependent decrease in the amplitude of N20 without affecting its latency (Sebel *et al.*, 1986). Treatment with 50% nitrous oxide reportedly decreases the amplitude of N20 by 44%. Nitrous oxide is usually combined with other intravenous or volatile anesthetics. When 50% nitrous oxide is added to high-dose fentanyl anesthesia, nitrous oxide decreases the amplitudes of early cortical responses of median and tibial nerve SEPs without altering the latencies of these responses. The effect of nitrous oxide on the amplitude of early cortical responses is greater than those of low concentrations of isoflurane and enflurane (McPherson *et al.*, 1985). Nitrous oxide (60%) induces further depression in N20 when combined with 1.5 MAC of halothane, enflurane, or isoflurane (Peterson *et al.*, 1986). Moreover, cessation of 60% nitrous oxide improved the recording of early cortical SEPs in some patients with poor recording under anesthesia with nitrous oxide and 1.5 MAC of isoflurane or enflurane. Therefore, the depressant effects of nitrous oxide on cortical SEPs seem to be additive to those of other anesthetics. These effects are not small despite the low potency of nitrous oxide as an anesthetic.

Halothane

Subcortical SEPs. Halothane with 60% nitrous oxide does not affect the latency of Erb's point potential or N13 (Peterson *et al.*, 1986). This combination has minimal effects on subcortical responses measured at the cervical or lumbar spinous process after posterior tibial nerve stimulation (Pathak *et al.*, 1989).

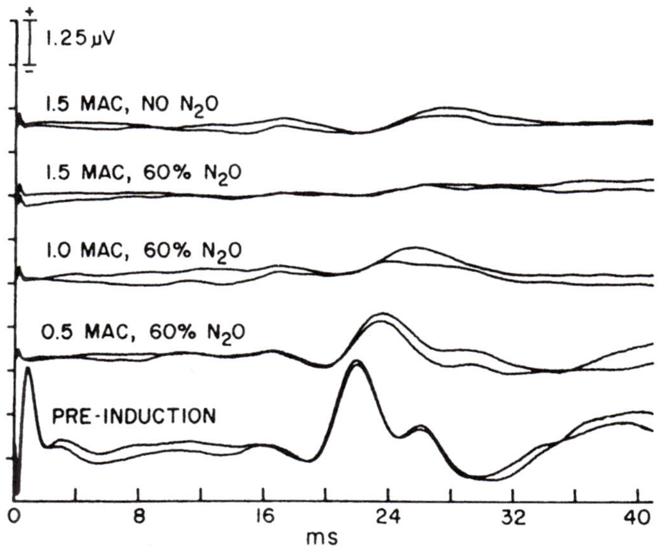

Fig. 2. Median nerve cortical somatosensory evoked potentials (C3′ or C4′ - Fz) at various MAC levels of isoflurane with 60% nitrous oxide (*N_2O*), and at 1.5 MAC without N_2O. (From Peterson *et al.*, 1986 with permission)

Cortical SEPs. Halothane combined with 60% nitrous oxide increases the latency and decreases the amplitude of N20 in a dose-dependent manner, resulting in a dose-dependent increase in CCT (Peterson *et al.*, 1986). In general, these changes are greater than those associated with clinical doses of intravenous anesthetics without inhalational agents. N20 was recordable with 1.5 MAC halothane and 60% nitrous oxide in all patients without previous neural deficits, and discontinuation of nitrous oxide decreased the latency and increased the amplitude of N20. In normal subjects, 1.0 MAC halothane with 60% nitrous oxide is considered compatible with effective median nerve SEP monitoring (Fig. 2).

Halothane has the qualitatively similar effects on cortical and subcortical tibial nerve SEPs (Pathak *et al.*, 1989). However, in contrast to median nerve SEPs, the first negative wave of cortical response (N45) was not detected at 1.0 MAC halothane in 60% nitrous oxide, and halothane decreased N45 to a greater extent than did isoflurane and enflurane.

Isoflurane

Subcortical SEPs. Isoflurane with 60% nitrous oxide does not influence subcortical responses elicited by median nerve or posterior tibial nerve stimulation (Peterson *et al.*, 1986; Pathak *et al.*, 1989). Isoflurane alone slightly prolongs the latencies of Erb's point potential and N13, but did not affect the amplitudes of these subcortical responses in another study (Samra *et al.*, 1987).

Cortical SEPs. Isoflurane in 60% nitrous oxide also increased the latency and decreased the amplitude of N20 in a dose-dependent fashion (Peterson *et al.*, 1986). While halothane preserved N20 in patients receiving 1.5 MAC with nitrous oxide, N20 disappeared in five of seven patients given 1.5 MAC isoflurane with 60% nitrous oxide. A combination of 0.5 MAC isoflurane with 60% nitrous oxide is considered suitable for SEP monitoring.

When given alone, an end-tidal isoflurane concentration of 0.5 to 2.0% has similar effects on N20, but the reduction in amplitude seems to be smaller for isoflurane alone than for isoflurane combined with 60% nitrous oxide (Samra *et al.*, 1987). Although N20 was recordable in 13 of 14 patients at 1.5% isoflurane, this peak was not detected in 30% of patients receiving 2.0% isoflurane, indicating that isoflurane, even when used as a sole agent, should be given only in low concentrations (less than 1.0% end-tidal concentration) for cortical SEP monitoring.

Isoflurane with 60% nitrous oxide affects cortical tibial nerve SEPs in a similar manner to median nerve SEPs (Pathak *et al.*, 1989). Isoflurane's effect on N45 amplitude was smaller than halothane's effect at 1.0 MAC. N45 was measurable at 1.0 MAC of isoflurane with 60% nitrous oxide, whereas it was not detectable at an equipotent dose of halothane. In contrast to this finding, the depressive effects of isoflurane on N20 are greater than those of halothane during nitrous oxide anesthesia, as described above (Peterson *et al.*, 1986). However, results among studies are inconsistent with respect as to which volatile anesthetic causes the greater depression of early cortical SEPs at the same MAC level (Peterson *et al.*, 1986; McPherson *et al.*, 1985; Pathak *et al.*, 1989).

Enflurane

Subcortical SEPs. Enflurane with 60% nitrous oxide does not affect the subcortical responses to median nerve or posterior tibial nerve stimulation (Peterson *et al.*, 1986; Pathak *et al.*, 1989).

Cortical SEPs. Enflurane with 60% nitrous oxide also causes dose-dependent increases in the latency and decreases in the amplitude of N20 and prolongs CCT (Peterson *et al.*, 1986). It has greater effects on cortical median nerve SEPs than do halothane and isoflurane. N20 was not recordable in four of seven patients given 1.0 MAC enflurane and all patient given 1.5 MAC enflurane. Therefore, for cortical SEP monitoring the concentration of enflurane should be limited to 0.5 MAC when combined with nitrous oxide.

The magnitude of the effects on tibial nerve SEPs of enflurane and nitrous oxide is similar to that of isoflurane and nitrous oxide (Pathak *et al.*, 1989). N45 was recordable at 1.0 MAC enflurane with 60% nitrous oxide.

Table 1. Effects of intravenous anesthetics on somatosensory evoked potentials

Anesthetics	Dose	Component measured	Latency change	Amplitude change	CCT	Loss of component	Reference
Thiopental	5 mg/kg+2mg/kg/hr	N13	no change	–	increase	no	Ganes and Lunder, 1983
		N20	+5%	no change		no	
Thiopental	4 mg/kg	N13	no change	–	increase	no	Grundy *et al.*, 1979
		N20	increase	no change		no	
Thiopental	1.25 mg/kg, 5min (total 6.25 mg/kg)	EP	no change	no change	+17%	no	Gugino and Chabot, 1990
		N13	no change	no change		no	
		N20	+5%	no change		no	
	1.25 mg/kg, 60min (total 77.5 mg/kg)	EP	no change	no change	+41%	no	
		N13	no change	−33%		no	
		N20	+15%	−57%		no	
Pentobarbital	0.6 mg/kg, 5min (total 3 mg/kg)	EP	no change	–	no change	no	Hicks *et al.*, 1992
		N13	+1.4%	no change		no	
		N20	no change	−20%		no	
	0.6 mg/kg, 33min (total 19.8 mg/kg)	EP	no change	–	+26%	no	
		N13	+2%	no change		no	
		N20	+9%	−45%		no	
Midazolam	0.3 mg/kg+2mg/kg/hr	N13	no change	–	increase	no	Ganes and Lundar, 1983
		N20	increase	no change		no	
Propofol with fentanyl	2.5 mg/kg+9 mg/kg/hr	N13	+3.8%	no change	+20.3%	no	Kalkman *et al.*, 1992
		N20	+8.3%	no change		no	
Propofol with alfentanyl	1 mg/kg	P37	no change	no change	–	no	Kalkman *et al.*, 1991
		N45	no change	no change		no	
Propofol with nitrous oxide and dextromoramide	2 mg/kg+3-6 mg/kg/hr	N45	no change	−66%	–	no	Kalkman *et al.*, 1995
Etomidate	0.3 mg/kg+2 mg/kg/hr	N13	no change	–	+16%	no	Ganes and Lunder, 1983
		N20	+7%	+300%		no	
Etomidate	0.4 mg/kg	EP	no change	no change	–	no	Grundy *et al.*, 1979
		N13	no change	−64%		no	
		N20	+5%	+170%		no	

Table 1. Effects of intravenous anesthetics on somatosensory evoked potentials (*continued*)

Anesthetics	Dose	Component measured	Latency change	Amplitude change	CCT	Loss of component	Reference
Ketamine	2 mg/kg+30 g/kg/hr	EP	no change	no change	no change	no	Kano and Shimoji, 1974
		N13	no change	no change		no	
		N20	no change	+16%		no	
Ketamine with nitrous oxide	2 mg/kg+30 g/kg/hr	EP	no change	no change	no change	no	Kano and Shimoji, 1974
		N13	no change	no change		no	
		N20	no change	-40%		no	
Fentanyl	25 g/kg	EP	no change	no change	–	no	Grundy *et al.*, 1979
		N13	no change	-28%		no	
		N20	+4%	no change		no	
Fentanyl	36–71 g/kg	EP	no change	–	+16%	no	Koht *et al.*, 1988
		N13	+3%	–		no	
		N20	+6%	−35%		no	
Fentanyl with nitrous oxide	2.5 g/kg+2 g/kg/hr	P37	+6%	−46%	–	no	Maruyama *et al.*, 1980
		N45	+11%		–	no	
Morphine with nitrous oxide	0.25 mg/kg +0.2mg/kg/hr	P37	+7%	−57%	–	no	Maruyama *et al.*, 1980
		N45	+9%			no	

Table 2. Effects of inhalational anesthetics on somatosensory evoked potentials

Anesthetics	Dose	Component measured	Latency change	Amplitude change	CCT	Loss of component	Reference
Nitrous oxide	10%	N20	no change	−25%	–	no	McPherson *et al.*, 1985
	30%	N20	no change	−30%	–	no	
	50%	N20	no change	−44%	–	no	
Halothane	0.5 MAC	N13	no change	–	+13%	no	Maurette *et al.*, 1988
with nitrous oxide		N20	+5%	−61%		no	
	1.0 MAC	N13	no change	–	+25%	no	
		N20	+8.4%	−70%		no	
	1.5 MAC	N13	no change	–	+42%	no	
		N20	+13%	−80%		no	
	1.5 MAC alone	N13	no change	–	+32%	no	
		N20	+12%	−74%		no	
Isoflurane	0.5 MAC	N13	no change	–	+22%	no	Maurette *et al.*, 1988
with nitrous oxide		N20	+7%	−54%		no	
	1.0 MAC	N13	no change	–	+36%	no	
		N20	+12%	−76%		no	
	1.5 MAC	N13	no change	–	+53%	no	
		N20	+20%	−95%		yes, 5/7	
	1.5 MAC alone	N13	no change	–	+49%	no	
		N20	+20%	−90%		yes, 3/7	
Enflurane	0.5 MAC	N13	no change	–	+20%	no	Maurette *et al.*, 1988
with nitrous oxide		N20	+6%	−52%		no	
	1.0 MAC	N13	no change	–	+64%	no	
		N20	+17%	−75%		yes, 4/7	
	1.5 MAC	N13	no change	–	–	no	
		N20	lost	lost		yes, 7/7	
	1.5 MAC alone	N13	no change	–	+94%	no	
		N20	+30%	−75%		yes, 4/7	

Table 2. Effects of inhalational anesthetics on somatosensory evoked potentials (*continued*)

Anesthetics	Dose	Component measured	Latency change	Amplitude change	CCT	Loss of component	Reference
Isoflurane	0.5%	N20	+11%	−49%	+25%	no	Am EES SOC, 1984
	1.0%	N20	+14%	−50%	+37%	no	
	1.5%	N20	+18%	−65%	+52%	no	
	2.0%	N20	+36%	−69%	+63%	no	
Isoflurane with nitrous oxide	0.5 MAC	N20	+3%	−42%	+6%	no	Nuwer *et al.*, 1995
	1.0 MAC	N20	+7%	−52%	+15%	no	
	1.5 MAC	N20	+10%	−64%	+25%	no	
Sevoflurane with nitrous oxide	0.5 MAC	N20	+2%	−38%	+2%	no	Nuwer *et al.*, 1996
	1.0 MAC	N20	+7%	−44%	+18%	no	
	1.5 MAC	N20	+10%	−51%	+22%	no	

Changes in latency and amplitude are % changes from preanesthetic control values, with the exception of data from reference 30, where control values were obtained after recovery from anesthesia. (Adapted from Gugino and Chabot,[4] with permission.) *CCT* central conduction time; *EP* Erb's point potential; *MAC* minimun alveolar concentration; – not measured.

Sevoflurane

Sevoflurane is a new volatile anesthetic which has been used clinically since 1992. Induction of and emergence from sevoflurane anesthesia is fast because of its low blood gas ratio.

Subcortical SEPs. Sevoflurane with 66% nitrous oxide does not alter N13 latency (Kameyama, 1994).

Cortical SEPs. Administration of 0.5 to 1.5 MAC of sevoflurane with 66% nitrous oxide causes dose-dependent increases in the latency and decreases in the amplitude of N20 (Kameyama, 1994). CCT is also increased in a dose-dependent manner. These effects on cortical SEP and CCT of sevoflurane are comparable to those of isoflurane (Kameyama, 1994). N20 was recordable in patients given 1.5 MAC of sevoflurane as well as in those given isoflurane with 66% nitrous oxide in that study.

Other drugs

Lidocaine

Lidocaine is administered intravenously, either as a supplement for anesthesia or as an antiarrythmic agent. Lidocaine 3 mg/kg followed by 4 mg/kg/hr during anesthesia with sufentanyl, nitrous oxide, and a low concentration of isoflurane affects N20, resulting in a slight prolongation of latency and a 26% reduction in the amplitude (Schubert *et al.*, 1992). Therefore, bolus administration of lidocaine should be curtailed during periods of cortical SEP monitoring.

Adenosine triphosphate

Adenosine triphosphate (ATP) and adenosine can be administered as adjunctive anesthetics, because these drugs augment anesthesia through purinergic receptor mechanisms (Seitz *et al.*, 1990). ATP, 100 and 200 μg/kg/min, added to nitrous oxide and a low concentration of isoflurane does not alter median nerve cortical and subcortical SEPs. ATP can be used during intraoperative SEP monitoring because it does not affect recording (Andoh *et al.*, 1993).

Anesthetic considerations for intraoperative SEP monitoring

For subcortical SEP monitoring, the anesthetic method used is not very important, because most anesthetics have minimal effects on subcortical responses recorded at the spinous process. However, depression of N13 amplitude by etomidate is rather large, as compared with other anesthetics (McPherson *et al.*, 1986). Barbiturates, ketamine and opioids have different effects on spinal SEPs recorded at the epidural space, suggesting that these anesthetics substantially alter spinal activity. When spinal function is monitored by spinal SEPs with the use of epidural electrodes, changes induced by these anesthetics should be considered when interpreting the data.

For cortical SEP monitoring, several methods have been reported to be suitable. These include propofol combined with alfentanyl (Kalkman *et al.*, 1991; Taniguchi *et al.*, 1992), nitrous oxide with a low dose of opioid (Kano and Shimoji, 1974), nitrous oxide with ketamine (Schubert *et al.*, 1990), high-dose opioid anesthesia with or without low concentrations of volatile anesthetics (Schubert *et al.*, 1986; McPherson *et al.*, 1985), nitrous oxide with low concentrations of volatile anesthetics (Peterson *et al.*, 1986), and balanced anesthesia with low doses of volatile anesthetics and opioids (Gugino and Chebot, 1990). Although all these methods probably preserve early cortical SEPs in patients without neural deficit, special caution is required in patients with preexisting neural deficit or abnormal SEP recording. Anesthetic regimens including nitrous oxide should be avoided, because this agent strongly depresses amplitude. As for effects on the amplitude of early cortical responses, total intravenous anesthesia with propofol and opioid seems to preserve the response best. High-dose opioid anesthesia also seems to be suitable for patients with neural deficit or abnormal SEP recording. Use of etomidate may help to restore cortical responses in patients with no detectable response, as described above.

Selection of the anesthetic regimen of choice should be based on careful evaluation of factors such as the condition of the patient, type and duration of surgery, and desirable postoperative condition. Nitrous oxide with ketamine, with a low dose of opioid, or with a low concentration of volatile anesthetics is unlikely to provide an adequate depth of anesthesia for major surgery, and is not suitable for patients with ischemic heart disease since cardiovascular responses to surgical stimuli might be difficult to control. Ketamine is not an option for patients with reduced intracranial compliance because it increases intracranial pressure. High-dose opioid anesthesia is likely to provide a stable circulatory condition in patients with heart disease or those undergoing major vascular surgery, but requires postoperative respiratory support because of prolonged respiratory depression. When early post-operative recovery is warranted, balanced anesthesia consisting of volatile anesthetics and opioids or total intravenous anesthesia with propofol plus opioid are suitable.

During critical periods for SEP monitoring, such as clamping of the blood supply to the CNS or correction of scoliosis, intravenous anesthetics should not be given by bolus injection, and the concentration of inhalational anesthetics should not be altered. Intravenous agents should be continuously infused throughout the surgical procedure to maintain a stable depth of anesthesia. It is also important to keep an adequate depth of anesthesia, because intense surgical stimulation can increase the amplitude of cortical responses when the depth of anesthesia is inadequate (Freye *et al.*, 1989). Increased amplitude of responses might be an early sign of ischemia in the corresponding pathway, so an inadequate depth of anesthesia can lead to false fluctuations in amplitude, unrelated to neural insults.

Effects of anesthetics on motor evoked potentials (MEPs)

Several different systems have been used to intraoperatively monitor motor pathways (Kalkman *et al.*, 1993). Stimulation can be applied at the motor cortex of the brain by transcranial electrical or magnetic stimulation, or at the spinal cord with the use of electrodes placed in the spinous process, the epidural space, or the intrathecal space. Evoked potentials can be recorded at the spinal cord, peripheral nerves, or muscles. Spinal responses after transcranial stimulation (TCS) are usually recorded from the epidural space and are relatively resistant to anesthetics, D-waves being especially so (Hicks *et al.*, 1992). Compound muscle action potentials (CMAPs) after electrical spinal stimulation are also well preserved in the presence of anesthetics and partial neuromuscular blockade (Adams *et al.*, 1993). Transcranial stimulation (TCS) combined with CMAP recording is least invasive but most sensitive to various anesthetics (Zentner *et al.*, 1989; Kalkman *et al.*, 1992). Therefore, we will focus on MEP monitoring using TCS and CMAP.

Intraoperative use of this technique is hampered by two reasons related to anesthetic management: (1) anesthetics can substantially influence outcome; and (2) when neuromuscular blocking agents are not used, surgery may be interrupted by movements of the patient in response to stimulation (Kalkman *et al.*, 1991; Kawaguchi *et al.*, 1993). Many commonly used anesthetics have been shown to strongly depress CMAPs after TCS (Zentner *et al.*, 1989; Kalkman *et al.*, 1992; Kalkman *et al.*, 1991). Low concentrations of isoflurane completely abolish responses when combined with nitrous oxide and opioids (Kalkman *et al.*, 1991). Nitrous oxide at 66% reportedly reduced the amplitude of thenal CMAP to 9% of the preoperative control response (Zentner *et al.*, 1989). Intravenous agents such as benzodiazepines, thiopental, and propofol also strongly depress the amplitude of CMAP (Zentner *et al.*, 1989; Kawaguchi *et al.*, 1993). Inhibition by propofol of the amplitude of the response is prominent 2 min after injection and persists for a prolonged period even after patients awaken (Kalkman *et al.*, 1992). Preservation of spinal responses after TCS suggests that the site of anesthetic action involves synapses with the spinal motor neurons, at least in part (Hicks *et al.*, 1992). As for differences between electrical and magnetic stimulation, CMAPs after magnetic stimulation appear to be slightly more susceptible to anesthetics than are those after electrical stimulation (Kalkman *et al.*, 1992). This may be because supramaximal stimulation is usually not possible with magnetic stimulation and because magnetic stimulation probably involves a synaptic event while electrical stimulation directly activates corticospinal tract neurons. It has been reported that ketamine, etomidate, and opioids have virtually no effects on MEP recording with TCS (Kalkman *et al.*, 1992;

Kalkman *et al.*, 1994). Many investigators seem to be able to record stable myogenic MEPs after TCS in patients given etomidate or ketamine as an induction agent in combination with nitrous oxide and opioids to maintain anesthesia (Kalkman *et al.*, 1995).

Although it was generally believed that neuromuscular blockade should be avoided during myogenic MEP recording, recent studies have shown that CMAP recording is possible under partial neuromuscular blockade (Kalkman *et al.*, 1995, 1992). The amplitude of twitch height is usually reduced to 10% to 25% of the preanesthetic value, and it is necessary to maintain stable levels of neuromuscular blockade, since the amplitude of CMAP depends on the degree of neuromuscular blockade. This technique provides two advantages: there is no need to interrupt surgical procedure during MEP recording because stimulation-induced movements are prevented; and there is a decreased risk of hazardous unexpected movements due to inadequate depth of anesthesia.

References

Adams DC, Emerson RG, Heyer EJ, McCormic PC, Carmel PW, Stein BM, Farcy JP, Gallo EJ (1993) Monitoring of intraoperative motor evoked potentials under conditions of controlled neuromuscular blockade. Anesth Analg 77: 913–916

Andoh T, Ohtsuka K, Okazaki K, Okutsu Y, Okumura F (1993) Effects of adenosine triphosphate (ATP) on somatosensory evoked potentials in humans anesthetized with isoflurane and nitrous oxide. Acta Anesthesiol Scand 37: 590–593

Cunningham JN, Laschinger JC, Spenser FC (1987) Monitoring of somatosensory evoked potentials during surgical procedures on the thoracoabdominal aorta. J Thorac Cardiovasc Surg 94: 275–285

Drummond JC, Todd MM, Hoi Sang U (1985) The effects of high dose sodium thiopental on brain stem auditory and median nerve somatosensory evoked responses in humans. Anesthesiology 63: 249–254

Drummond JC, Todd MM, Schubert A, Hoi Sang U (1987) Effect of the acute administration of high dose pentobarbital on human brain stem auditory and median nerve somatosensory evoked responses. Neurosurgery 20: 830–835

Fisher RS, Raudzens P, Nunemacher M (1995) Efficacy of intraoperative neurophysiological monitoring. J Clin Neurophysiol 12: 97–109

Freye E, Hartung E, Schenk GK (1989) Somatosensory-evoked potentials during block of surgical stimulation with propofol. Br J Anaesth 63: 357–359

Ganes T, Lundar T (1983) The effect of thiopentone on somatosensory evoked responses and EEGs in comatose patients. J Neurol Neurosurg Psychiatry 46: 509–514

Grundy BL, Brown RH, Greenberg PS (1979) Diazepam alters cortical evoked potentials. Anesthesiology 51, s38

Gugino V, Chabot RJ (1990) Somatosensory evoked potentials. Int Anesthesiol Clin 28: 154–164

Hicks RG, Woodforth I, Crawford MR, Stephen JP, Burke DJ (1992) Some effects of isoflurane on I waves of the motor evoked potential. Br J Anaesth 69: 130–134

Kalkman CJ, Been HD, Ongerboer de Visser (1993) Intraoperative monitoring of spinal cord function. Acta Orthop Scand 64: 114–123

Kalkman CJ, Drummond JC, Kennelly NA, Patel PM, Partridge BL (1992) Intraoperative monitoring of tibialis anterior muscle motor evoked responses to transcranial electrical stimulation during partial neuromuscular blockade. Anesth Analg 75: 584–588

Kalkman CJ, Drummond JC, Patel PM, Sano T, Chesnut RM (1994) Effects of droperidol, pentobarbital and ketamine on myogenic transcranial magnetic motor evoked responses in humans. Neurosurgery 35: 1066–1071
Kalkman CJ, Drummond JC, Ribberink A, Patel PM, Sano T, Reginald B (1992) Effects of propofol, etomidate, midazolam, and fentanyl on motor evoked responses to transcranial electrical or magnetic stimulation in humans. Anesthesiology 76: 502–509
Kalkman CJ, Drummond JC, Rierink AA (1991) Low concentrations of isoflurane abolish motor evoked responses to transcranial electrical stimulation during nitrous oxide/opioid anesthesia in humans. Anesth Analg 73: 410–414
Kalkman CJ, Traast H, Zuurmond WWA, Bovill JG (1991) Differential effects of propofol and nitrous oxide on posterior tibial nerve somatosensory cortical evoked potentials during alfentanyl anaesthesia. Br J Anaesth 66: 483–489
Kalkman CJ, Ubags LH, Been HD, Swaan A, Drummond JC (1995) Improved amplitude of myogenic motor evoked responses after paired transcranial electrical stimulation during sufentanyl/nitrous oxide anesthesia. Anesthesiology 83: 270–275
Kameyama Y (1994) Effect of isoflurane and sevoflurane on evoked potentials and EEG. Masui 43: 657–664 (in Japanese)
Kano T, Shimoji K (1974) The effects of ketamine and neuroleptanalgesia on evoked electrospinogram and electromyogram in man. Anesthesiology 40: 241–246
Kawaguchi M, Sakamoto T, Shimizu K, Ohnishi H, Karasawa J (1993) Effect of thiopentone on motor evoked potentials induced by transcranial magnet stimulation in humans. Br J Anaesth 71: 849–852
Koht A, Schutz W, Schmidt G, Schramm J, Watanabe (1988) Effects of etomidate, midazolam, and thiopental on median nerve somatosensory evoked potentials and the additive effects of fentanyl and nitrous oxide. Anesth Analg 67: 435–441
Lindsay KW, Carlin J, Kenedy I, Fry J, McInnes A, Teasdale GM (1981) Evoked potentials in severe head injury-analysis and relation to outcome. J Neurol Neurosurg Psychiatry 44: 796–802
Maruyama Y, Shimoji K, Shimizu H, Sato Y, Kuribayashi H, Kaieda R (1980) Effects of morphine on human spinal cord and peripheral nervous activity. Pain 8: 63–73
Maurette P, Simeon F, Castagnera L, Esposito J, Macouillard G, Heraut LA (1988) Propofol anaesthesia alters somatosensory evoked cortical potentials. Anaesthesia 43: 44–45
McPherson RW, Mahla M, Johnson R, Traystman RJ (1985) Effects of enflurane, isoflurane and nitrous oxide on somatosensory evoked potentials during fentanyl anesthesia. Anesthesiology 62: 626–633
McPherson RW, Sell B, Traystman RJ (1986) Effects of thiopental fentanyl, and etomidate on upper extremity somatosensory evoked potentials in humans. Anesthesiology 65: 584–589
Newlon PG, Greenberg RP, Enas GG, Becker DP (1983) Effects of therapeutic pentobarbital coma on multimodality evoked potentials recorded from severely head-injured patients. Neurosurgery 12: 613–619
American Electroencephalographic Society (1984) Guidelines for clinical evoked potential studies. J Clin Neurophysiol 1: 3–53
Nuwer MR, Dawson EG, Carlson LG, Kanim LEA, Sherman JE (1995) Somatosensory evoked potential spinal cord monitoring reduces neurological deficits after scoliosis surgery: results of a large multicenter survey. Electroencephalogr Clin Neurophysiol 96: 6–11
Pathak KS, Amaddio MD, Scoles PV, Shaffer JW, Mackay W (1989) Effects of halothane, enflurane, and isoflurane in nitrous oxide on multilevel somatosensory evoked potentials. Anesthesiology 70: 207–212
Pathak KS, Brown RH, Cascorbi HF, Nash CL (1984) Effects of fentanyl and morphine on intraoperative somatosensory cortical-evoked potentials. Anesth Analg 63: 833–837
Peterson DO, Drummond JC, Todd MM (1986) Effects of halothane, enflurane, isoflurane, and nitrous oxide on somatosensory evoked potentials in humans. Anesthesiology 65: 35–40

Samra SK, Vanderzant CW, Domer PA, Sackellares JC (1987) Differential effects of isoflurane on human median nerve somatosensory evoked potentials. Anesthesiology 66: 29–35

Scheepstra GL, De Lange JJ, Booij LHDJ, Ros HH (1989) Median nerve evoked potentials during propofol anaesthesia. Br J Anaesth 62: 92–94

Schubert A, Licina MG, Glaze GM, Paranandi L (1992) Systemic lidocaine and human somatosensory-evoked potentials during sufentanyl-isoflurane anesthesia. Can J Anaesth 39(6): 569–575

Schubert A, Licina MG, Lineberry PJ (1990) The effect of ketamine on human somatosensory evoked potentials and its modification by nitrous oxide. Anesthesiology 72: 33–39

Schubert A, Peterson DO, Drummond JC, Saidman LJ (1986) The effect of high-dose fentanyl on human median nerve somatosensory evoked responses. Anesth Analg 65: S1–S170

Sebel PS, Flynn PJ, Ingram DA (1984) Effect of nitrous oxide on visual, auditory and somatosensory evoked potentials. Br J Anaesth 56: 1403–1407

Seitz PA, Riet MT, Rush W, Merrell WJ (1990) Adenosine decreases the minimum alveolar concentration of halothane in dogs. Anesthesiology 72: 990–994

Shimoji K, Kano T, Nakashima H, Shimizu H (1974) The effects of thiamylal sodium on electrical activities of central and peripheral nervous systems in man. Anesthesiology 40: 234–240

Sloan TB, Ronai AK, Toleikis JR, Koht A (1988) Improvement of intraoperative somatosensory evoked potentials by etomidate. Anesth Analg 67: 582–585

Taniguchi M, Nadstawek J, Pechstein U, Schramm J (1992) Total intravenous anesthesia for improvement of intraoperative monitoring of somatosensory evoked potentials during aneurysm surgery. Neurosurgery 31: 891–897

Zentner J, Kiss I, Ebner A (1989) Influence of anesthetics-nitrous oxide in particular-on electromyographic response evoked by transcranial electrical stimulation of the cortex. Neurosurgery 24: 253–256

Correspondence: Dr. T. Andoh, Department of Anesthesiology, Yokohama City University School of Medicine, Kanazawa-ku, Yokohama, Japan.

Spinal Cord Monitoring
Eds: E. Stålberg, H. S. Sharma, Y. Olsson

Chapter 22

Intraoperative spinal cord monitoring – clinical overview

T. Tamaki

Department of Orthopedic Surgery, Wakayama Medical College, Wakayama, Japan

Introduction

Intraoperative spinal cord monitoring should provide reliable information to the surgeon at a critical scene during the operation and assist him or her to carry out aggressive surgery if necessary. The author, as an orthopedic surgeon who treats diseases of the spine and the spinal cord, describes the methodologies he is currently utilising and relying upon in the operating theatre. SEP (somatosensory cortical evoked potential) and "Wake up test" with neuroleptic analgesia will not be included in this chapter because of their imperfections to provide reliable information in the electrically hostile operating theatre and the shortcoming in providing continuous monitoring respectively.

Neurophysiological methods utilised in the operating theatre

To carry out continuous monitoring of the spinal cord function through the course of surgery, we have to utilise neurophysiological techniques. The value of these techniques lies in early recognition of spinal cord dysfunction before it may become permanent and while it still is possible to undertake immediate measures for prevention. Techniques for monitoring long tracts in the spinal cord are well established and regarded to be reliable to protect major spinal cord insult during the surgery of the spine and spinal cord (Brown and Nash, 1979; Burke *et al.*, 1995; Jones *et al.*, 1983; Jones *et al.*, 1996; Levy and York, 1983; Tamaki *et al.*, 1977; Tamaki *et al.*, 1984; Tsuyama *et al.*, 1978). However, there are reports of false negative cases (Lesser *et al.*, 1986; Nuwer *et al.*, 1995) and there is probably a great number of unreported false positive cases. To eliminate these false results, the author employs multimodality monitoring methods, i.e. applying multiple methods simultaneously during the surgery.

The electrical signals we are observing for the spinal cord monitoring are divided into two groups, namely:
those mainly reflecting the activity of sensory tracts
those used to monitor the motor related tracts
Characteristic features and methods of recording these potentials are described below.

Methods mainly reflecting the activity of sensory tracts

SCEP (spinal cord evoked potential) after stimulating the spinal cord. Tamaki *et al.* (1972) and Kurokawa *et al.* (1972) started to utilize this type of evoked potential for the purpose to examine the continuity of long tracts in the spinal cord in 1972 independently. The author and colleagues delivered stimulation to the spinal cord at the rostral site and recorded the signal from a caudal site (Tamaki *et al.*, 1972; Tamaki *et al.*, 1977; Tamaki *et al.*, 1984) while Kurokawa *et al.* recorded ascending volleys placing the recording electrode rostrally after stimulating at a caudal site of the spinal cord (Kurokawa, 1972; Tsuyama *et al.*, 1978). There was no essential difference between the recorded SCEPs. The initial spike shape potential (first component) has been considered to represent the electrical activities originating from the large diameter nerve fibers located in the postero-lateral region of the spinal cord by some authors (Harada, 1983; Imai, 1976; Tamaki *et al.*, 1977) or by nerve fibers in all quadrants of the spinal cord by others (Harada, 1983; Satomi and Nishimoto, 1985). The succeeding polyphasic component (second component) has been regarded to reflect the continuity of the dorsal column (Imai, 1988; Tamaki *et al.*, 1977). Several researchers (e.g. Cracco and Evans, 1978; Satomi and Nishimoto, 1985) have indicated contribution of post-synaptically transmitted potentials to this part of the signal. Since the contribution of post-synaptic electrical activities is small in size, the SCEP is regarded mainly to reflect the activity of sensory related tracts.

SSEP (spinal somatosensory evoked potential) after stimulating peripheral nerve trunks. The group of Royal National Orthopedic Hospital of England first described this type of evoked potential for the intra-operative spinal cord monitoring (Jones and Edger, 1983). They delivered electric stimulation to the tibial or peroneal nerve trunk on both sides simultaneously and recorded signals from the spinal cord placing the recording electrode in the epidural space rostrally to the site at risk. The SSEP is composed of polyphasic waves and mainly reflects the conductivity of the dorsal column. If the stimulation is performed on one side, laterality of the injury to the spinal cord may be expressed with this method.

Segmental SSEP is recorded with the recording electrode placed at the site of entry level of impulses to the spinal cord. This method can be employed to monitor the function of the conus medullaris.

Methods to monitor the motor related tracts

MEP (motor evoked potential) after stimulating the motor cortex. MEPs are recorded from the spinal cord after generating the descending volleys by electrical stimulation of the motor cortex of the brain. The techniques, which are most commonly employed in the operating theatre for electrical stimulation of the motor cortex trans-cranially, were developed by Merton *et al.* (1982), Matsuda *et al.* (1989) and Levy *et al.* (1983) independently. We are routinely using the method started by Matsuda and his colleagues (Matsuda, 1989) as subsequently described. The MEPs can alternatively be provoked by magnetic stimulation, but this technique has some drawbacks. Magnetic stimulation for continuous spinal cord monitoring intraoperatively is not a practical method because of the difficulty in maintaining the position of the coil throughout the procedure and the vulnerability of magnetically evoked MEPs to anesthesia. Observation of these volleys, which are transferred through motor related tracts mainly located in the anterior half of the spinal cord, is frequently used in conjunction with the recording of SCEP. Furthermore, this method is frequently used for the monitoring of the spinal cord function at the cranio-cervical junction (Kitagawa *et al.*, 1989).

CMAP (compound muscle action potential) evoked by the stimulation to the brain or the spinal cord. The CMAP is a potential generated by the contraction of striated muscle fibers. To elicit this response, motor neurones at the anterior horn of the gray matter of the spinal cord must fire. However, in the operation theatre it was difficult to excite the motor neurone by delivering electrical stimulation to the spinal cord or the brain. Recent advances in anesthesia have made it possible to some extent to diminish the suppressive effects of anesthesia on the motor neurone (Taylor *et al.*, 1993). Furthermore, the use of paired or train stimulation of the brain or the spinal cord facilitate the possibility to excite the motor neurone and record CMAP intraoperatively (Jones *et al.*, 1996; Mochida *et al.*, 1996; Taylor *et al.*, 1993; Zentner, 1989). If constant recording of this response after stimulating the brain is conceivable in every case, monitoring of all motor related pathways should be possible. However, in practical sense, it is not always possible to record CMAP, especially in lower extremities, after stimulating the brain or the spinal cord as the effect of anaesthesia cannot be sufficiently reduced. The variation of the size of the CMAP in amplitude and difficulty in maintaining stable level of neuromuscular blockage are causing problems in the interpretation of the obtained results. On the other hand the method is of great interest because of its sensitivity to detect motor dysfunction. There is a need to further improve this technique for its wider routine use as an indicator of functional changes in motor related pathways.

Techniques to obtain the SCEP, SSEP, MEP and CMAP

For the purpose of recording the above mentioned four kinds of potentials, we have to arrange stimulating and recording electrodes with regard to the spinal cord segment of interest, character of anticipated injury to the spinal cord and type of function modality to monitor. In other words, impulses provoked by the stimulation should be recorded after travelling through neural structures at risk so that their amplitudes and latencies reflect possible dysfunction. Furthermore, every effort to facilitate prompt recognition and response should be carried out.

Placement of the recording electrode

Placement of the recording electrode of spinal cord evoked potentials. Since 1972 the author and colleagues are utilising flexible tube type electrode introduced into subarachnoid space as the recording electrode (Fig.1). This electrode consists of two helical platinum wires of 1mm width attached to the end of a Teflon tube (Intermedics Co. Ltd. Japan). The distance between the two metal contacts is 15 mm. The electrode is adequately flexible not to injury neural tissues in the subarachnoid space, and stiff enough to be advanced in the intrathecal or extradural space. The

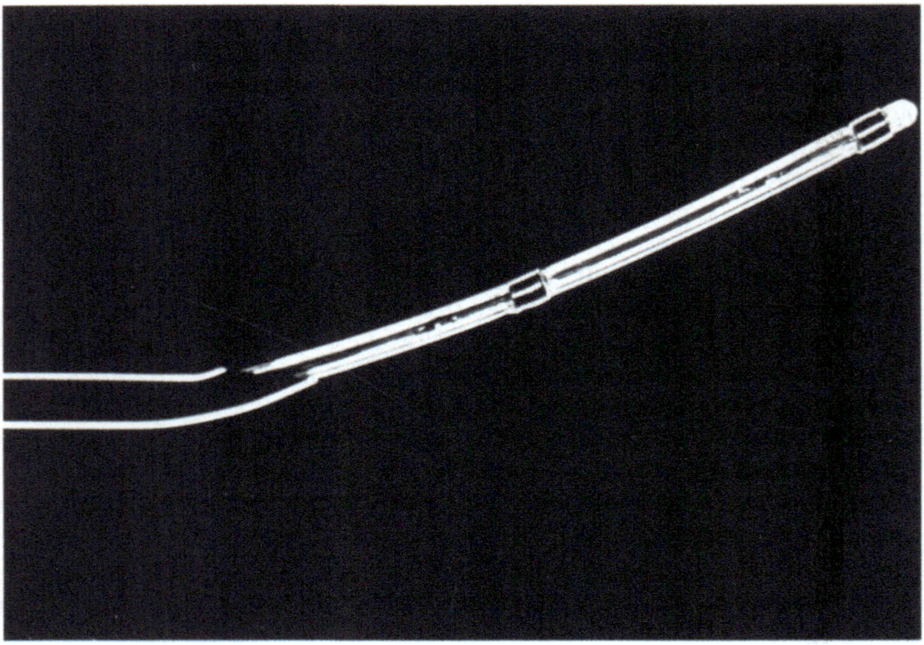

Fig. 1. Flexible tube type bipolar electrode. This electrode can be introduced into the intrathecal space as well as the extradural space

diameter of the electrode is 0.75 mm and it fits within the hole of a 17 gauge Tuohy needle. This electrode is also used for the purpose of stimulation. To record as large potential as possible, the recording electrode is introduced into the subarachnoid space utilising 17-gauge Tuohy needle after performing the lumbar puncture at lower lumbar level. Once the tip of the electrode is placed in the intrathecal space, it can be advanced rostrally by gently pushing. As the surface of the spinal cord is covered by pia mater the flexible tube type electrode does not penetrate into the spinal cord. For routine montage to record SCEP and MEP, the tip of the electrode is located at the thoracic level or close to the conus medullaris. This intrathecal placement of the recording electrode allows frequent monitoring runs with a rapid recognition of the response even in the hostile electrical environment of the operating theatre.

Epidural recording can be performed by inserting a similar type of electrode into the epidural space. If there is a necessity to place epidural electrode with percutaneous technique, we have made it a rule to perform percutaneous epidural puncture when the patient is alert. The amplitude of the electrical potential generated in the spinal cord is reduced to 1/3–1/2 of that obtained with the intrathecal recording.

We regard electrode introduction into the intrathecal space as safe and easy because lumbar puncture is always performed at the level of lower lumbar level and involves no risk to injure the spinal cord. The author and colleagues have accumulated more than 1800 cases that underwent monitoring with this electrode insertion. We have not encountered any serious complications, except for one case of curable meningitis in which the electrode insertion was performed in the X-ray room. Every effort should be made to introduce the electrode in an absolutely sterile fashion.

Placement of recording electrode for CMAP. Surface or needle electrodes are used to record CMAP from the muscles following the tendon–belly method placing one of the active electrodes on the muscle belly and the other on the tendon.

Placement of stimulation electrodes

Epidural stimulation electrode placement. A flexible tube type bipolar electrode, which is similar to the electrode to be placed in the intrathecal space, is used. The site to deliver stimulation from the epidural space to the spinal cord should be selected to document the delivery of neural impulses through the level where the spinal cord may be at risk during the surgery. For example, to monitor through the course of spinal instrumentation, a level rostrally to the proximal hook should be selected for the introduction of stimulation electrode when descending potentials are recorded with a caudal electrode. If the patient undergoes posterior approach, insertion into the epidural space shall be performed exposing the epidural space under direct visual control after removing ligamentum flavum.

Percutaneous introduction can be carried out after performing epidural puncture when the patient is alert. In a patient with scoliotic deformity, percutaneous introduction is not recommended because of the difficulty to perform epidural puncture.

Epidurally placed electrode for the purpose of stimulation can also be used to record SSEP after stimulating the peripheral nerve trunks.

Electrode placement for transcranial brain stimulation. Following Matsuda and his colleagues (Matsuda, 1989), two monopolar electrodes (DANTEC, Copenhagen) are positioned to deliver high current electric stimulation at a point 5 cm lateral from Cz and 2 cm anterior bilaterally.

Electrode placement for the peripheral nerve trunk stimulation. Disposable surface electrodes are routinely used. Needle electrode can be used if necessary. To send impulses in cepahalad direction, the rostral electrode should be cathode.

Recording of evoked potentials – practical aspects

Once the electrodes are placed in appropriate location, it is easy to switch stimulation site and recording site to record several kinds of evoked potentials.

SCEP (spinal cord evoked potential) elicited by spinal cord stimulation – stimulation and recording. Electrical stimulation is delivered to the spinal cord through the bipolar tube type electrode placed in the epidural space. The electrode, which is on the side of the recording electrode, is used as cathode. The electrical stimulation is 0.2–0.3 msec in duration, 3–13 mA in strength and 20–50 Hz in frequency. To improve signal to noise ratio, 30 to 50 signal sweeps are averaged. In cases with normal spinal cord function, averaging is not always necessary because the amplitude of SCEP is large enough. In these cases it is possible to perform real time monitoring observing potentials at every 1 second without averaging.

It is not possible to describe a standard waveform of the SCEP. The signal configuration differs from patient to patient and is influenced by delicate differences of relation between the stimulating and recording electrodes. Basically, the recorded SCEP consists of an initial spike wave followed by polyphasic waves (Fig. 2). We ordinarily use the amplitude decrement of the initial spike wave as an indicator of spinal cord insult (Tamaki *et al.*, 1977, 1984). As spinal cord volleys are highly reproducible in nature, Burke *et al.* suggested that an amplitude change of 20–30% is a sufficient indication to warn the surgeon (Burke *et al.*, 1995). On the other hand, many researchers have suggested that 50% attenuation of the amplitude is the sign of critical condition in patients with normal spinal cord function (Imai, 1988; Tamaki *et al.*, 1977, 1984; Tsuyama *et al.*, 1978). This standard is commonly accepted for MEP (Kitagawa *et al.*, 1989; Zentner, 1989), SSEP (Jones and Ransford, 1983) and SEP (Brown

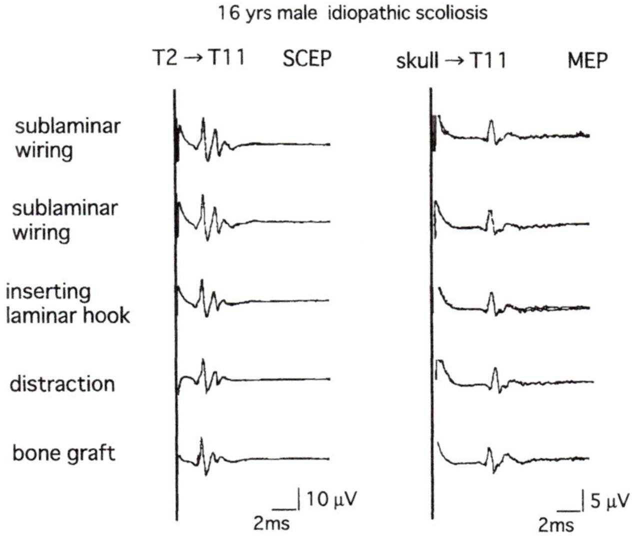

Fig. 2. SCEPs and MEPs recorded from the intrathecal space at the level of T11 after stimulating the spinal cord at T2 and the brain. The SCEPs are characterised by the initial spike wave and the following polyphasic wave component. The MEPs contain an initial D wave (spike shaped) and I waves. In this case SCEP degraded in amplitude after the distraction. This change was judged as the result of the movement of stimulating electrode. Notice the change of artefacts after the destructive manoeuvre, which indicate the change of the position of the stimulating electrode. Furthermore, there was no change in the amplitude of the D wave

and Nash, 1979). Our criterium to regard the spinal cord in critical condition is around 50% decrement of the initial spike wave potential. Attention should be paid to the polyphasic wave portion of the SCEP if intra-spinal cord manipulation is carried out. This part of the potential reflects the dorsal column pathway. A selective insult to the dorsal column, therefore, affects the wave configuration of polyphasic wave of SCEP.

SSEP stimulation and recording. Electrical stimuli delivered to the peripheral nerve trunk in the upper and lower extremities are typically 0.2 msec in duration and 2~3 Hz in frequency. The stimulation strength is 3–5 times of motor threshold. Signals recorded through the bipolar epidural electrode are small in amplitude and require about 100 times averaging to produce a reproducible pattern, which is typically a polyphasic potential.

MEP stimulation and recording. To excite motor cortex from the outside the cranium by single electrical stimulation, the stimulation strength should be high, around 200 mA with a pulse duration of 0.2 msec. Typically it is possible to record the MEP from the spinal cord through intrathecal or epidural electrode after averaging responses 20–50 times. A stimulation frequency of 2~3 Hz is used. Paired or train stimulation of 1~2 msec interstimulus interval and 99 mA in strength can elicit MEP as well (Jones

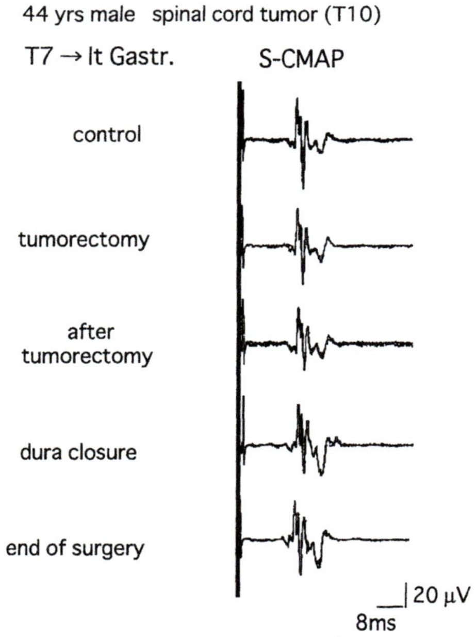

Fig. 3. CMAPs recorded from left gastrocnemius muscle after delivering paired stimulation to the spinal cord (S-CMAP) in a case of spinal cord tumor

et al., 1996; Kitagawa *et al.*, 1989). To reduce the risk of triggering a kindling effect, this method of stimulation should be avoided in patients with history of epileptic seizure and brain surgery.

Recorded potentials are composed of a spike shaped D wave and followed by polyphasic I waves (Fig. 3). The D wave is regarded as the action potential generated by direct stimulation of the cortical spinal motor neurone. The I wave is regarded as the potential generated by repetitive synaptic bombardment of the pyramidal neurones via chains of cortical inter-neurones (Patton and Amassian, 1954). In cases with motor tract dysfunction, MEPs frequently show a temporal dispersion and low amplitude.

The MEP is considered to be elicited under the anode of the stimulating electrode (Patton and Amassian, 1954). Therefore, by changing the polarity of stimulating electrode, it is possible to obtain information about the laterality of the insult to the motor related tracts.

CMAP – stimulation and recording. It is necessary to consider several facts of importance when recording this potential after stimulating the brain or spinal cord. First of all, it is deemed necessary to make appropriate selection of anesthetic substance to avoid disturbing suppression of the motor neurone activity. Presently, the combined use of propofol and fentanil is the method of choice (Jones *et al.*, 1996; Taylor *et al.*, 1993). It is known that the paired stimulation with an interstimulus interval of 1 or 2 msec to the brain or spinal cord can elicit effectively the CMAP in the extremities (Jones *et al.*, 1996; Taylor *et al.*, 1993). Using inhalation anesthetic substances train stimulation of 5 consecutive pulses has to be employed to produce CMAP (Mochida *et al.*, 1996). To achieve this we have a specially designed stimulator. At our institution, paired stimuli of 99 mA in strength, 1 msec ISI (interstimulus interval) and 0.2 msec duration are delivered to the brain with the anode placed on the contra-lateral side of the recording.

To generate CMAP by stimulating the spinal cord, we are using paired stimuli, 10~15 mA in strength, 0.2 msec duration and 1 msec ISI. Signals are averaged 10–30 times to obtain a reproducible wave pattern. (Fig. 3).

Despite utilizing these techniques our success rate of recording the CMAP in the lower extremities after stimulating the brain is around 50%. Therefore, further refinement of recording technique is necessary. Utilization of this potential during the course of spinal surgery can only be done with immobilization and under quantitatively controlled monitoring of the effect of the neuromuscular blocking agent leaving 10–20% of contraction measured with a Relaxograph (Datex Instrument Co. Helsinki, Finland). In practical sense the maintenance of a constant neuromuscular blockade is not easy to ascertain Therefore, the size of CMAP is very variable and judgement of critical change is difficult. As Machida *et al.* pointed out with animal experimental study, this potential is more sensitive than SCEP to detect a compressive insult to the spinal cord (Machida *et al.*, 1985). Still it may not be appropriate to use the recording of the CMAP as a single test in spinal cord monitoring. Animal experimental study at my institution has shown that complete disappearance of CMAP does not always mean a postoperative loss of motor function. Accordingly, we are utilising this method as an adjunct to the SCEP and MEP.

Practical comments

Preoperative composition of strategies concerning the type of surgery, neurological status of the patient and character of anticipated insult (for example vascular or mechanical) is fundamental for successful and effective monitoring. The best array of stimulation and recording sites should be selected based upon the strategy. Simultaneous recording of multiple potentials to accomplish reliable spinal cord monitoring should be performed if necessary. For example, to monitor the corrective surgery of spine in a case of congenital scoliosis, the combination of SCEP and MEP should be selected. If the patient had diastomatomyelia, SSEP may offer a good information because the spinal cord is divided into two parts being separated longitudinally. In a case with atlanto-axial dislocation, MEP after stimulating the brain is utilized simultaneously observing CMAP in upper extremities (Kitagawa *et al.*, 1989). SEP may be an appropriate adjunct, however not used by us.

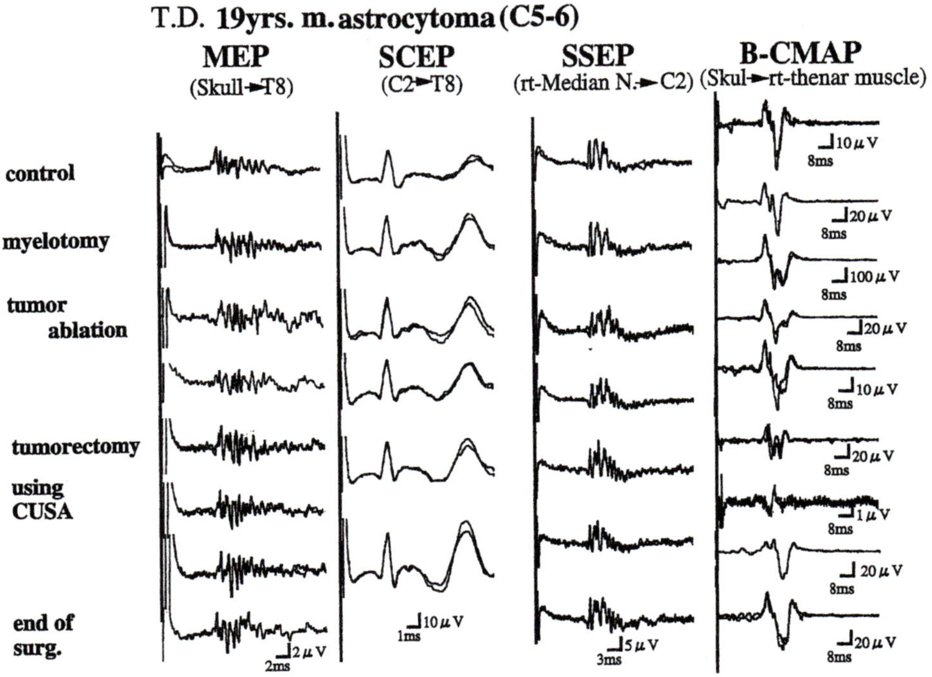

Fig. 4. Multimodality monitoring in a case with intramedullary spinal cord tumor. MEP, SCEP, SSEP and CMAP after stimulating the brain (B-CMAP). Notice the variability in amplitude and pattern of B-CMAP and the polyphasic pattern of MEP. When the *CUSA* (ultrasonic tissue removing instrument) was used, the amplitude of B-CMAP decreased, but there was no change in the other evoked potentials. This patient woke up without neurological sequele in upper extremities but slight degree of sensory disturbance at limited area in left lower extremity

Multimodality monitoring (Katayama *et al.*, 1986) using multiple types of potentials has been a routine measure for the past 5 years at our operating theatre. This is especially true in cases with possible parenchymal selective lesions, such as intramedullary spinal cord tumor. Here as many as possible of the neural structures should be monitored (Fig. 4).

In cases with normal spinal cord function, potentials recorded from the spinal cord with intrathecal electrode frequently have high amplitudes and can be observed without averaging. In cases with injured spinal cord, recording of the evoked potentials is not always possible even if the injury is not complete. For example, MEP is frequently difficult to record in patients with deteriorated motor function down to less than grade 2 evaluated with manual muscle testing. In these cases, we have to use atypical SCEP recordings by approximating the stimulation and recording sites. In these cases 10–20% deterioration in amplitude can be a sign of insult to the tracts (Tamaki *et al.*, 1985). Therefore, to monitor remnant functions in injured spinal cord, different criteria for the judgement have to be established.

Closing remarks

As described above, it has become possible to monitor not only the sensory and motor tracts but also the activity of motor neurones in the spinal cord. However, we can not observe every function of the spinal cord, nor can we detect and prevent small lesions such as sensory disturbance appearing in very limited area after the removal of an intramedullary tumor. In this meaning there might be a criticism that presently used technologies for the spinal cord monitoring are not yet perfect but need further development. Nevertheless, it is true that proper use of presently available methods can provide reliable and useful information to support aggressive spine and spinal cord surgery in the operating theatre.

References

Brown RH, Nash CL (1979) Current status of spinal cord monitoring. Spine 4: 466–470

Burke D, Hicks R, Stephen J, Woodforth I, Crawford M (1995) Trial-to-trial variability of corticospinal volleys in human subjects. Electroencephalogr Clin Neurophysiol 97: 231–237

Cracco RQ, Evans B (1978) Spinal evoked potentials in the cat: effects of asphyxia, strychinine, cord section and compression. Electroencephalogr Clin Neurophysiol 44: 187–201

Harada Y (1983) Study on experimental spinal cord injuries by distraction using spinal cord evoked potentials. J Jpn Orthop Ass 57: 685–701

Imai K (1988) A clinical study on Intra-operative spinal cord monitoring with spinal evoked potential for scoliosis. J Jpn Orthop Ass 62: 511–521

Imai T (1976) Human electrospinogram evoked by direct stimulation on the spinal cord through epidural space. J Jpn Orthop Ass 55: 1037–1056

Jones SJ, Edger MA, Ransford AO, Thomas NP (1983) A system for the electrophysiological monitoring of the spinal cord during operations for scoliosis. J Bone Joint Surg 65B: 134–139

Jones SJ, Harrison R, Koh KF, Mendoza N, Crockard HA (1996) Motor evoked potential monitoring during spinal surgery: responses of distal limb muscles to transcranial cortical stimulation with pulse trains. J Electroencephalogr Clin Neurophysiol 100: 375–383

Katayama Y, Tsubokawa T, Sugitani M, Maejima S, Hirayama T, Yamamoto T (1986) Assessment of spinal cord injury with multimodality evoked spinal cord potentials, Part I. localization of lesions in experimental spinal cord injury. Neuro-orthopedics 1: 130–141

Kitagawa H, Itoh T, Takano H, Takakuwa K, Yamamoto N, Yamada H, Tsuji H (1989) Motor evoked potential monitoring during upper cervical spine surgery. Spine 14(10): 1078–1083

Kurokawa T (1972) Spinal cord action potentials evoked by epidural stimulation of cord-a report of human and animal record. Jap J Electroenceph Electromyogra 1: 64–66

Lesser RP, Raudzens P, Lüders H, Nuwer MR, Goldie WD, Morris HH, Dinner DS, Klem G, Hahn JF, Shetter AG, Ginsburg HH, Gurd AR (1986) Postoperative neurological deficits may occur despite unchanged intraoperative somatosensory evoked potentials. Ann Neurol 19: 22–25

Levy W, York D (1983) Evoked potentials from the motor tracts in humans. Neurosurgery 12(4) 422–429

Machida M, Weinstein S, Yamada T, Kimura J (1985) Electrophysiological measures of sensory and motor function during spinal surgery. Spine 10(5): 407–413
Matsuda H (1989) Intraoperative spinal cord monitoring using electric responses to stimulation of caudal spinal cord or motor cortex. In: Desmedt JE (ed) Neuromonitoring in Surgery. Elsevier, Amsterdam, pp 175–190
Merton PA, Morton HB, Hill DK, Marsden CD (1982) Scope of a technique for electrical stimulation of the human brain, spinal cord and muscle. Lancet 2: 597–600
Mochida K, Shinomiya K, Komori H, Furuya K (1996) A new method of multisegment motor pathway monitoring using muscle potentials after train spinal stimulation. Spine 20: 2240–2246
Nuwer MR, Dawson EG, Carlson LG, Kanim LEA, Sherman JE (1995) Somatosensory evoked potential spinal cord monitoring reduces neurogenic deficits after scoliosis surgery: results of a large multicenter survay. Electroencephalogr Clin Neurophysiol 96: 6–11
Patton HD, Amassian VE (1954) Single multiple unit analysis of cortical stage of pyramidal tract activation. J Neurophysiol 17: 345–357
Satomi K, Nishimoto GI (1985) Comparison of evoked spinal potentials by stimulation of the sciatic nerve and the spinal cord. Spine 10: 884–890
Tamaki T, Yamashita T, Kobayashi H (1972) Spinal cord monitoring (abstract in Japanese) Jap J Electraenceph Electromyogra 1, p 196
Tamaki T, Kobayashi H, Yamane T, Egashira T, Tsuji H, Inoue S (1977) Clinical application of evoked spinal cord action potential. (in Japanese) Seikeigeka 28: 681–689
Tamaki T, Noguchi T, Takano H, Tsuji H, Nakagawa T, Imai K, Inoue S (1984) Spinal Cord monitoring as a clinical utilization of the spinal evoked potential. Clin Orthop 184: 58–64
Tamaki T, Takano H, Takakuwa H, Tsuji H, Nakagawa K, Imai K, Inoue S (1985) An assesment of the use of spinal cord evoked potentials in prognosis estimation of injured spinal cord. In: Schramm J, Jones SJ (eds) Spinal Cord Monitoring. Springer, Berlin Heidelberg New York Tokyo, pp 221–226
Taylor BA, Fennelly E, Taylor A, Farrell J (1993) Temporal summation – the key to motor evoked potential spinal cord monitoring in human. J Neurol Neurosurg Psychiatry 56: 104–106
Tsuyama N, Tsuzuki N, Kurokawa T, Imai, T (1978) Clinical application of spinal cord action potential measurement. Int Orthop 2: 951–965
Zentner J (1989) Noninvasive motor evoked potential monitoring during neurosurgical operations on the spinal cord. Neurosurgery 24: 709–712

Correspondence: Dr. T. Tamaki, Department of Orthopedic Surgery, Wakayama Medical College, Japan.

Subject index

Author index

Springer-Verlag and the Environment

WE AT SPRINGER-VERLAG FIRMLY BELIEVE THAT AN international science publisher has a special obligation to the environment, and our corporate policies consistently reflect this conviction.

WE ALSO EXPECT OUR BUSINESS PARTNERS – PRINTERS, paper mills, packaging manufacturers, etc. – to commit themselves to using environmentally friendly materials and production processes.

THE PAPER IN THIS BOOK IS MADE FROM NO-CHLORINE pulp and is acid free, in conformance with international standards for paper permanency.

GPSR Compliance

The European Union's (EU) General Product Safety Regulation (GPSR) is a set of rules that requires consumer products to be safe and our obligations to ensure this.

If you have any concerns about our products, you can contact us on ProductSafety@springernature.com

In case Publisher is established outside the EU, the EU authorized representative is:

Springer Nature Customer Service Center GmbH
Europaplatz 3
69115 Heidelberg, Germany

Batch number: 10726517

Printed by Printforce, the Netherlands